Mechanisms of Sensory Working Memory

MECHANISMS OF SENSORY WORKING MEMORY

ATTENTION AND PERFORMANCE XXV

PIERRE JOLICOEUR, PHD

Département de Psychologie, Université de Montréal,
Centre de recherche de l'Institut universitaire de gériatrie de Montréal (CRIUGM),
Montreal, QC, Canada

CHRISTINE LEFEBVRE, PHD

Centre de recherche de l'Institut universitaire de gériatrie de Montréal (CRIUGM),
Montreal, QC, Canada

JULIO MARTINEZ-TRUJILLO, MD, PHD

Department of Physiology and Pharmacology,
Robarts Research Institute,
Western University, Ontario, Canada

AMSTERDAM • BOSTON • HEIDELBERG • LONDON
NEW YORK • OXFORD • PARIS • SAN DIEGO
SAN FRANCISCO • SINGAPORE • SYDNEY • TOKYO
Academic Press is an imprint of Elsevier

Academic Press is an imprint of Elsevier
125 London Wall, London EC2Y 5AS, UK
525 B Street, Suite 1800, San Diego, CA 92101-4495, USA
50 Hampshire Street, 5th Floor, Cambridge, MA 02139, USA
The Boulevard, Langford Lane, Kidlington, Oxford OX5 1GB, UK

British Library Cataloguing-in-Publication Data
A catalogue record for this book is available from the British Library

Library of Congress Cataloging-in-Publication Data
A catalog record for this book is available from the Library of Congress

ISBN: 978-0-12-811042-3

For information on all Academic Press publications
visit our website at https://www.elsevier.com/

Publisher: Mica Haley
Acquisition Editor: Mica Haley
Editorial Project Manager: Kathy Padilla
Production Project Manager: Chris Wortley
Designer: Matthew Limbert

Typeset by TNQ Books and Journals
www.tnq.co.in

Contents

5. Neural Bases of the Short-term Retention of Visual Information

BRADLEY R. POSTLE

6. What are the Roles of Sensory and Parietal Activity in Visual Short-Term Memory?

STEPHEN M. EMRICH

7. Hemispheric Organization of Visual Memory: Analyzing Visual Working Memory With Brain Measures

GABRIELE GRATTON, EUNSAM SHIN AND MONICA FABIANI

8. Visual Working Memory and Attentional Object Selection

MARTIN EIMER

9. Individual Differences in Visual Working Memory Capacity: Contributions of Attentional Control to Storage

KEISUKE FUKUDA, GEOFFREY F. WOODMAN AND EDWARD K. VOGEL

10. Working Memory and Aging: A Review

MONICA FABIANI, BENJAMIN ZIMMERMAN AND GABRIELE GRATTON

11. Defining a Role for Lateral Prefrontal Cortex in Memory-Guided Decisions About Visual Motion

TATIANA PASTERNAK

12. Working Memory Representations of Visual Motion along the Primate Dorsal Visual Pathway

DIEGO MENDOZA-HALLIDAY, SANTIAGO TORRES AND JULIO MARTINEZ-TRUJILLO

13. Neurophysiological Mechanisms of Working Memory: Cortical Specialization and Plasticity

XUE-LIAN QI, XIN ZHOU AND CHRISTOS CONSTANTINIDIS

14. Neural and Behavioral Correlates of Auditory Short-Term and Recognition Memory

AMY POREMBA

15. Brain Activity Related to the Retention of Tones in Auditory Short-Term Memory

SOPHIE NOLDEN

Contributors

Claude Alain Rotman Research Institute, Baycrest Centre for Geriatric Care, Toronto, ON, Canada; Department of Psychology, University of Toronto, ON, Canada; Institute of Medical Sciences, University of Toronto, ON, Canada

Pedro B. Albuquerque School of Psychology, University of Minho, Braga, Portugal

Søren K. Andersen School of Psychology, University of Aberdeen, Aberdeen, United Kingdom

Stephen R. Arnott Rotman Research Institute, Baycrest Centre for Geriatric Care, Toronto, ON, Canada

Douglas Cheyne Neuromagnetic Imaging Laboratory, Hospital for Sick Children, Toronto, ON, Canada

Christos Constantinidis Department of Neurobiology & Anatomy, School of Medicine, Wake Forest University, Winston-Salem, NC, USA

Nelson Cowan Department of Psychological Sciences, University of Missouri, McAlester Hall, Columbia, MO, USA

Rhodri Cusack Brain and Mind Institute, Western University, London, ON, Canada

Martin Eimer Department of Psychological Sciences, Birkbeck College, University of London, London, UK

Stephen M. Emrich Department of Psychology, Brock University, St Catharines, ON, Canada

Monica Fabiani Department of Psychology, University of Illinois at Urbana-Champaign, Champaign, IL, USA

Alexandra M. Fernandes Institute for Energy Technology, Halden, Norway; School of Psychology, University of Minho, Braga, Portugal

Ulysse Fortier-Gauthier Centre de Recherche en Neuropsychologie et Cognition, Université de Montréal, Montreal, QC, Canada

Keisuke Fukuda Department of Psychological Sciences, Vanderbilt Vision Research Center, Center for Integrative and Cognitive Neuroscience, Vanderbilt University, Nashville, TN, USA

Susan Gillingham Rotman Research Institute, Baycrest Centre for Geriatric Care, Toronto, ON, Canada; Department of Psychology, University of Toronto, ON, Canada

Gabriele Gratton Department of Psychology, University of Illinois at Urbana-Champaign, Champaign, IL, USA

Su Keun Jeong Harvard University, Cambridge, MA, USA

Pierre Jolicoeur Centre de recherche de l'Institut universitaire de gériatrie de Montréal (CRIUGM), Montreal, QC, Canada; Experimental Cognitive Science Laboratory, Département de Psychologie, Université de Montréal, Montreal, QC, Canada

Tobias Katus Department of Psychology, Birkbeck College, London, United Kingdom

Simon Lacey Department of Neurology, Emory University, Atlanta, GA, USA

Rebecca Lawson Department of Experimental Psychology, University of Liverpool, UK

Christine Lefebvre Centre de recherche de l'Institut universitaire de gériatrie de Montréal (CRIUGM), Montreal, QC, Canada

Ada W.S. Leung Rotman Research Institute, Baycrest Centre for Geriatric Care, Toronto, ON, Canada; Department of Occupational Therapy, University of Alberta, Edmonton, AB, Canada; Centre for Neuroscience, University of Alberta, Edmonton, AB, Canada

Annika C. Linke Brain and Mind Institute, Western University, London, ON, Canada

René Marois Department of Psychology Vanderbilt University, Nashville, TN, USA

Julio Martinez-Trujillo Department of Physiology and Pharmacology, Robarts Research Institute, Western University, Ontario, Canada; Cognitive Neurophysiology Laboratory, Department of Physiology, McGill University, Montreal, QC, Canada

Diego Mendoza-Halliday Cognitive Neurophysiology Laboratory, Department of Physiology, McGill University, Montreal, QC, Canada

Sophie Nolden CERNEC, BRAMS, CRBLM, Département de Psychologie, Université de Montréal, Montreal, QC, Canada

Tatiana Pasternak Departments of Neurobiology & Anatomy, Brain and Cognitive Science, and Center for Visual Science, University of Rochester, Rochester, NY, USA

Amy Poremba Department of Psychology, Behavioral and Cognitive Neuroscience Division, University of Iowa, Iowa City, IA, USA

Bradley R. Postle Departments of Psychology and Psychiatry, University of Wisconsin–Madison, Madison, WI, USA

Xue-Lian Qi Department of Neurobiology & Anatomy, School of Medicine, Wake Forest University, Winston-Salem, NC, USA

Eunsam Shin Department of Psychology, Yonsei University, Yonsei, Korea

Santiago Torres Cognitive Neurophysiology Laboratory, Department of Physiology, McGill University, Montreal, QC, Canada

Edward K. Vogel Department of Psychology, University of Oregon, Eugene, OR, USA

Jeffrey Wong Rotman Research Institute, Baycrest Centre for Geriatric Care, Toronto, ON, Canada

Geoffrey F. Woodman Department of Psychological Sciences, Vanderbilt Vision Research Center, Center for Integrative and Cognitive Neuroscience, Vanderbilt University, Nashville, TN, USA

Yaoda Xu Harvard University, Cambridge, MA, USA

Xin Zhou Department of Neurobiology & Anatomy, School of Medicine, Wake Forest University, Winston-Salem, NC, USA

Benjamin Zimmerman Department of Psychology, University of Illinois at Urbana-Champaign, Champaign, IL, USA

Acknowledgments

This book originates from talks given at the 25th Attention & Performance meeting that took place in St. Hippolyte, Qc, in July 2013. We would like to thank all presenters and observers that made the meeting an unforgettable experience through their animated discussions, interventions, and overall cheerfulness.

We would also like to thank the following members of the Jolicœur lab for their help in preparing the meeting and welcoming the guests: Patrick Bermudez, Isabelle Corriveau, Brandi Lee Drisdelle, Shannon O'Malley, Manon Maheux, and Sandrine Mendizabal, as well as the employees of the Station de Biologie des Laurentides, who made everyone feel at home in these wonderful facilities.

We wish to acknowledge the generous support of Brain Vision and Brain Products, and particularly the understanding and support of Dr Patrick Britz (President of Brain Vision, and Marketing Director of Brain Products) for the Symposium. His timely and enthusiastic encouragement contributed significantly to the success of the meeting, for which we are very grateful.

Finally, we would like to thank the authors and the reviewers that carved chapters that are both relevant and interesting, helping us present a book of great quality.

The meeting was made possible by the generosity of our sponsors:

Brain vision

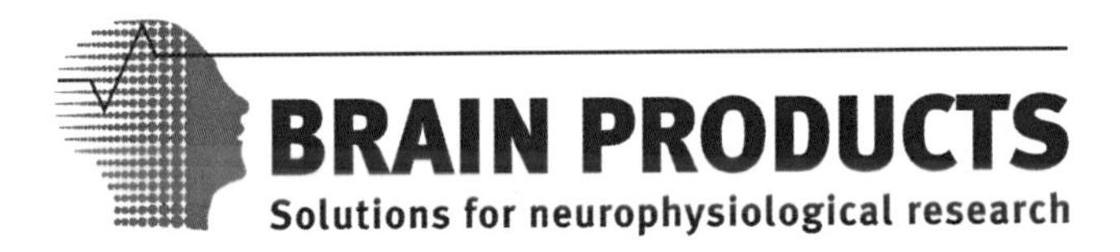

Brain products

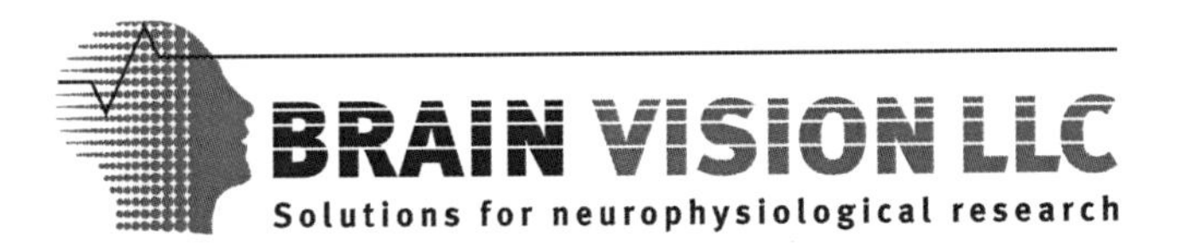

The Centre de recherche de l'Institut universitaire de gériatrie de Montréal (CRIUGM)

Vice Décanat à la recherche of l'Université de Montréal, The Département de psychologie, the Faculté des Arts et des Sciences

The Fonds de Recherche du Québec Nature et Technologie

Fonds de recherche
Nature et
technologies
Québec

Attention and Performance XXV

SAINT-HIPPOLYTE, QC, JULY 2013

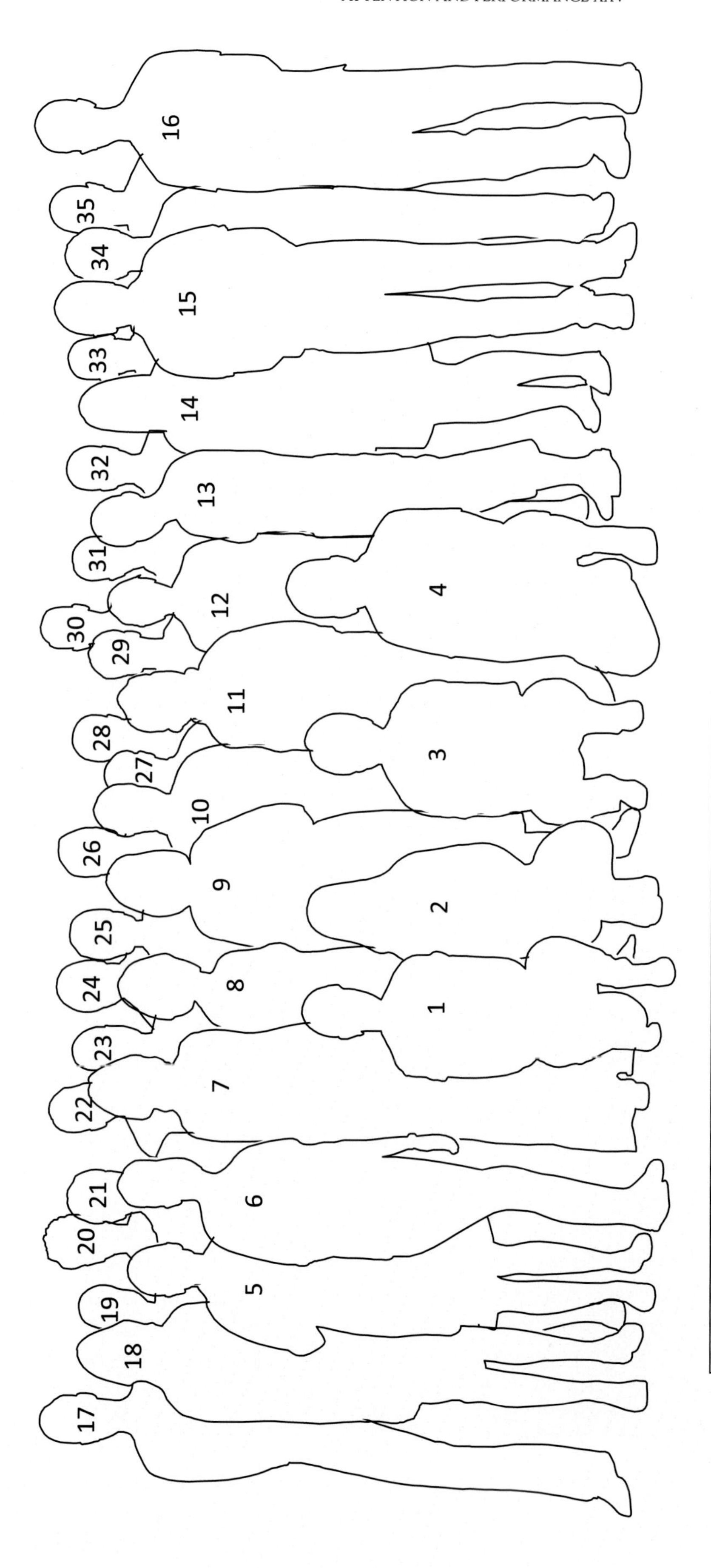
1 – Isabelle Corriveau
2 – Sandrine Mendizabal
3 – Keisuke Fukuda
4 – Julio Martinez-Trujillo
5 – Mina khoshnejad
6 – Véronique Taylor
7 – Sophie Nolden
8 – Manon Maheux
9 – Monica Fabiani
10 – Nelson Cowan
11 – Amy Poremba
12 – Tania Pasternak
13 – Yaoda Xu
14 – Shannon O'Malley
15 – Asher Cohen
16 – Ulysse Fortier-Gauthier
17 – Patrick Bermudez
18 – Christine Lefebvre
19 – Brandi Lee Drisdelle
20 – Martin Eimer
21 – Gabriele Gratton
22 – Pierre Rainville
23 – David Shore
24 – Rhodri Cusack
25 – Pierre Jolicœur
26 – René Marois
27 – Claude Alain
28 – Jöran Lepsien
29 – Tobias Katus
30 – Christos Constantinidis
31 – Stephen Emrich
32 – Ed Awh
33 – Brad Postle
34 – Jorge Armony
35 – Bernhard Hommel

CHAPTER

1

Introduction

Christine Lefebvre[1], *Julio Martinez-Trujillo*[3], *Pierre Jolicoeur*[1,2]

[1]Centre de recherche de l'Institut universitaire de gériatrie de Montréal (CRIUGM), Montreal, QC, Canada; [2]Experimental Cognitive Science Laboratory, Département de Psychologie, Université de Montréal, Montreal, QC, Canada; [3]Department of Physiology and Pharmacology, Robarts Research Institute, Western University, Ontario, Canada

Although the content of human conscious experience is fashioned from five basic sensory systems, there is growing evidence that conscious experience depends not only on mechanisms of perception, but also on attention, sensory short-term memory (STM), and working memory (WM). Attention weighs differentially some stimuli over others, and thus biases what is perceived and stored in memory. Passage into sensory STM may then be required for stimulation to become a conscious sensory experience, as well as to become the basis for controlled action. As such, sensory STM may act as the gateway between sensation, conscious experience, and control.[1]

For the 25th Attention and Performance symposium, we invited a group of experts in the study of short-term and working memory (WM) to share their ideas and evidence concerning sensory memory in a secluded biology research station operated by Université de Montréal. The goal was to foster exchanges and discussion of their respective research, coming from a broad range of testing and analysis methods as well as research topics, and to compare results. More specifically, we wanted to see what similarities and differences emerged when comparing the investigation of WM in different sensory modalities. From this symposium we fashioned the book you are now reading.

In this book we examine sensory WM in the auditory, tactile, and visual domains. We do so using numerous methods of investigation ranging from behavioral to extra-cellular recordings, as well as various imaging techniques such as electroencephalography (EEG), magnetoencephalography (MEG), and functional magnetic resonance (fMRI) imaging in human and nonhuman primates. Analyses of data also include localization methods, event-related averaging, and multivariate pattern analysis. Using these different approaches we aim to explore aspects of sensory WM that may be common across modalities, as well as those that may be distinct. In this way we strive for a more complete understanding of the behavioral and physiological basis of sensory WM, arguably one of the most important components of the human cognitive architecture.

The research presented in this book reflects the evolution of sensory short-term and WM investigations over several decades. Building on previous work, research questions became more precise and paradigms were refined, while new techniques and technological advancements made more comprehensive analyses possible. In early imaging studies, participants were exposed to the same stimulation but had to perform different tasks while their brain activity was recorded. Zatorre, Evans, and Meyer (1994), for example, asked participants to listen passively to a melody in one condition, or to remember the first two notes of the same melody in another condition. Activations measured in both conditions via a positron emission tomography scanner were then subtracted to isolate activations linked to the memory task from those linked to stimulation and passive listening. This subtraction revealed activations in the right superior temporal and right frontal cortex, which were believed to reflect mechanisms specific to the retention of pitch in auditory WM.

An important development in the field is illustrated by the work of Vogel and Machizawa (2004), as well as Todd and Marois (2004). Rather than compare conditions with and without memory, they added a parametric manipulation of

[1]Note that we use the term "short-term memory" (STM) when referring to the temporary storage of items and "working memory" (WM) for the additional processes that operate on information held in STM.

Mechanisms of Sensory Working Memory
http://dx.doi.org/10.1016/B978-0-12-811042-3.00001-3

memory load to the paradigm. By measuring activations during a retention interval devoid of other stimulation, and focusing on activations correlated with the increase in number of items held in memory, they were able to pinpoint activations exclusively linked to memory processes. More recently, Lefebvre et al. (2013), and Nolden et al. (2013, this book) extended this approach to the study of acoustic WM, whereas Fortier-Gauthier et al. (this book) focused on tactile WM processes. These approaches have been particularly useful in identifying brain regions participating in short-term sensory memory, but some recent work suggests that the networks revealed with these methods may be incomplete. Newly developed multivariate pattern analysis (MVPA) aims to decode activity patterns in measures of brain activity. Decoding means that the patterns of activity can be classified so as to distinguish which stimuli were perceived, or remembered, at the time of the brain recording. Successful decoding of stimulus identity from a brain region during retention provides evidence for a link to that region with mechanisms that mediate STM—either that region is the one (or one of several) that actually retains the information during the retention interval, or it receives input from a region that does. Several chapters in the book report recent results using MVPA and explore their implications for our understanding of sensory WM (see Chapters 4, 5, 6, and 17, by authors Xu, Postle, Emrich, and Cusack and Lincke, respectively).

A major contribution to the study of WM has been the work conducted by neurophysiologists using single-cell recordings in behaving monkeys. In 1971, Fuster and Alexander identified certain neurons in the prefrontal cortex of macaques that exhibited sustained activation selective for the contents of WM when the animals remembered an object. This sustained activation was hypothesized to be a neural correlate of WM. In the current book, Qi, Zhou, and Constantinidis (Chapter 13) explore brain regions where neural correlates of WM have been isolated, as well as the effect of cortical plasticity and specialization on the neural correlates of WM. In Chapter 11, Pasternak further explores how the activity of neurons in visual areas represents the contents of WM and how executive areas of the frontal lobe may encode different aspects of WM tasks. In Chapter 12, Mendoza-Halliday, Torres, and Martinez-Trujillo pinpoint the brain areas where the known correlates of visual WM emerge, and how single-unit data may relate to the findings reported in functional imaging studies. Finally, in Chapter 14, Poremba investigates the neural correlates of acoustic STM in macaques, highlighting differences between visual and acoustic STM.

The study of sensory STM and WM is not without its controversies. Indeed, two controversial elements were touched on in the preceding paragraph: namely, the existence of differences between sensory modalities and the role of the prefrontal cortex. A first matter of debate relates to the possibility that stimuli in different sensory modalities are maintained somehow differently in different structures of the brain, or, on the contrary, that all stimuli are maintained in the same way. For example, whereas in Chapter 15 Nolden reviews some interesting distinctions from electrophysiological and magnetoencephalographical correlates in the maintenance of acoustic stimuli versus correlates of visual or language-related stimuli, Cowan, in the second chapter, reviews evidence of similar processing for both types of stimuli. Many chapters of this book focus on how stimuli from different sensory modalities are processed in WM. This represents a departure from the literature, because WM is most often studied through the use of language-related material. In fact, in most studies, words are to be maintained and then compared or produced, or pictures or sounds have to be named. This is probably partly for convenience: it is easy and intuitive to devise tasks using verbal material. A second reason might reside in confusion in the definition of sensory: often, what is called "visual" is a word printed on a screen, whereas a spoken word is considered "auditory" stimulation. This labeling is not false in itself. Nevertheless, it is probable that in both cases the same representation is maintained in WM, because language-related material is highly practiced and used in the same way regardless of modality of presentation. One would not be surprised, then, to find that in both conditions, the structures identified as contributing to WM are the same. This is not proof, however, that visual and auditory STM representations are maintained and processed in the same way. To find out whether this is the case, one has to use stimuli that afford only one type of representation, namely a nonverbal sensory representation, in the task at hand. This is what was done in many of the chapters of this book: the authors used stimuli that cannot be labeled, or they controlled for the possible use of verbal labels. Data obtained using such methodology offer a different perspective.

A second point of contention discussed in this book is the role of the prefrontal cortex (PFC). Although many authors believe that WM recruits structures in PFC for the retention of representations of sensations in different modalities, this is not universally accepted. Although sustained activity was found through the retention interval of a delayed match-to-sample task for visual (motion) via extracellular recordings in Mendoza-Halliday's and Pasternak's chapters (12 and 11, respectively), Poremba (Chapter 14) found few cells showing such activity in the lPFC for an equivalent acoustic task, but instead found sustained activity mostly in primary auditory cortex. Gratton (Chapter 7) found evidence of lateralization of representations of visual stimuli in WM, which suggests that those WM representations are mostly processed at a lower level than the PFC. Although Qi et al. (Chapter 13) report activation in the PFC during a WM task, they suggest that the main role of PFC may be to filter out distraction. Finally, Postle, in his review, points out that the view that PFC is involved in short-term retention (STR) of stimuli that emerged with the first extracellular recordings does not have much support. Even the authors of the seminal studies (Fuster & Alexander, 1971; Kubota & Niki, 1971) were not convinced that the PFC was involved in retention, but rather in the focusing of attention on the maintained items. Postle argues instead that the concepts of the central executive and storage buffers from Baddeley's model were conflated, and that many studies have linked the PFC to central executive processing rather than to a storage buffer function. Moreover, studies using MVPA to analyze fMRI and EEG data show that PFC patterns of activation are not sufficient to recover which stimulus was retained in sensory STM. For example, Postle cites a study (Meyer, Qi, Stanford, & Constantinidis, 2011) that revealed that PFC activity tends to switch from representing the target stimulus to representing the trial's match or non-match status during a trial.

Many of Postle's arguments rely on a new way of analyzing fMRI and electrophysiological data: namely, MVPA. Many researchers use univariate analyses to study WM (see Chapters 15, 19, and 20, by Nolden; Fortier-Gauthier, Lefebvre, Cheyne, and Jolicœur; and Katus, respectively); others, such as Emrich (Chapters 6) and Xu (Chapter 4), are trying to find patterns of activity related to the maintenance of particular categories of stimuli. Although finding brain activation patterns with the univariate approach during a retention interval can demonstrate that a particular brain region is involved in the retention of memory representations, one cannot infer that regions that are not so activated are necessarily not part of the memory system under study. For example, cells within a given region could be more active during retention whereas other cells in the same region could be less active. If these two changes have equal weight, the net change could cancel out and make it appear as though the region has no role. Similarly, successful decoding of the identity of stimuli maintained during a retention interval suggests that the region for which decoding worked was involved in the memory network (or received input from such a region), but unsuccessful decoding does not mean that the region does not participate in active retention. Unsuccessful decoding in fMRI, for example, could occur if the cells representing different alternatives sought by the decoding algorithm were equally represented in all voxels. Instead of abandoning traditional univariate analysis methods in favor of MVPA, we propose that the two approaches complement each other. Our challenge is to find the truth in what we infer from each type of analysis by considering and comparing all available evidence. Readers of this book will find much to satisfy their curiosity concerning how sensory memories are encoded, stored, and retrieved in the brain.

References

Fuster, J. M., & Alexander, G. E. (1971). Neuron activity related to short-term memory. *Science, 173*, 652–654.

Kubota, K., & Niki, H. (1971). Prefrontal cortical unit activity and delayed alternation performance in monkeys. *Journal of Neurophysiology, 34*, 337–347.

Lefebvre, C., Vachon, F., Grimault, S., Thibault, J., Guimond, S., Peretz, I., et al. (2013). Distinct electrophysiological indices of maintenance in auditory and visual short-term memory. *Neuropsychologia, 51*, 2939–2952.

Meyer, T., Qi, X. L., Stanford, T. R., & Constantinidis, C. (2011). Stimulus selectivity in dorsal and ventral prefrontal cortex after training in working memory tasks. *The Journal of Neuroscience, 31*, 6266–6276.

Nolden, S., Bermudez, P., Alunni-Menichini, K., Lefebvre, C., Grimault, S., & Jolicœur, P. (2013). Electrophysiological correlates of the retention of tones differing in timbre in auditory short-term memory. *Neuropsychologia, 51*, 2740–2746.

Todd, J., & Marois, R. (2004). Capacity limit of visual short-term memory in human posterior parietal cortex. *Nature, 428*, 751–754.

Vogel, E. K., & Machizawa, M. G. (2004). Neural activity predicts individual differences in visual working memory capacity. *Nature, 428*, 748–751.

Zatorre, R. J., Evans, A. C., & Meyer, E. (1994). Neural mechanisms underlying melodic perception and memory for pitch. *The Journal of Neuroscience, 14*, 1908–1919.

CHAPTER

2

Sensational Memorability: Working Memory for Things We See, Hear, Feel, or Somehow Sense

Nelson Cowan

Department of Psychological Sciences, University of Missouri, McAlester Hall, Columbia, MO, USA

INTRODUCTION

I was pleased to be asked to visit a remote, beautiful, verdant, and earthy location to give the symposium lecture at a conference on something researchers rarely discuss as such: the sensory aspect of working memory. Is there such a thing? Why is it rarely referred to in this way? First I will discuss terms and what they mean, with a short modern history of the usage of these terms and a general theoretical framework to organize the data. Then I will discuss three categories of memory: (1) brief-duration sensory afterimages; (2) more persistent, processed sensory recollections; and (3) abstract units or chunks constructed across sensory modalities. In my theoretical framework, there is a great deal of commonality across sensory modalities despite the differences.

For someone like me with an interest in human experience, this sensory memory information is of obvious relevance. Sensory information ties together experience and memory. Imagine, for example, a child moving a flashlight in a circular motion in the dark. The trail of light that results appears to exist in the present, but it is actually a sensory record of the path that the light has taken in the previous second or so. Illustrating the general appeal of sensory processing to curious minds, it may have been the first aspect of psychology to capture my imagination, as shown in a selection of pages from a booklet that I assembled for a science assignment in grade school (Figure 1).

IS THERE A SENSORY WORKING MEMORY?

To address this new term, sensory working memory, one needs to consider in turn the definition of sensory memory and the definition of working memory.

Sensory Memory Definition

Sensory memory might be defined as any memory that preserves the characteristics of a particular sensory modality: the way an item looks, sounds, feels, and so on. Given that memories can be inaccurate, one might tend not to consider inaccurate information about sensory information to be sensory memory. For example, if one had a vivid memory of how an apple looked but incorrectly remembered the apple as red instead of green, one might consider that a constructed mental image and not a truly sensory memory. Yet, the constructed mental image is often useful; for example, reading a printed word may give rise to an acoustic image of how the word would be spoken. We know this because there are acoustic confusions among verbal items visually presented for recall (e.g., Conrad, 1964), but the acoustic imagery may be necessary for efficient covert verbal rehearsal.

There is evidence that items to be remembered that are presented in different modalities yield different patterns of neural response during a retention interval (e.g., Lefebvre et al., 2013), presumably with retention relying on at least

Mechanisms of Sensory Working Memory
http://dx.doi.org/10.1016/B978-0-12-811042-3.00002-5

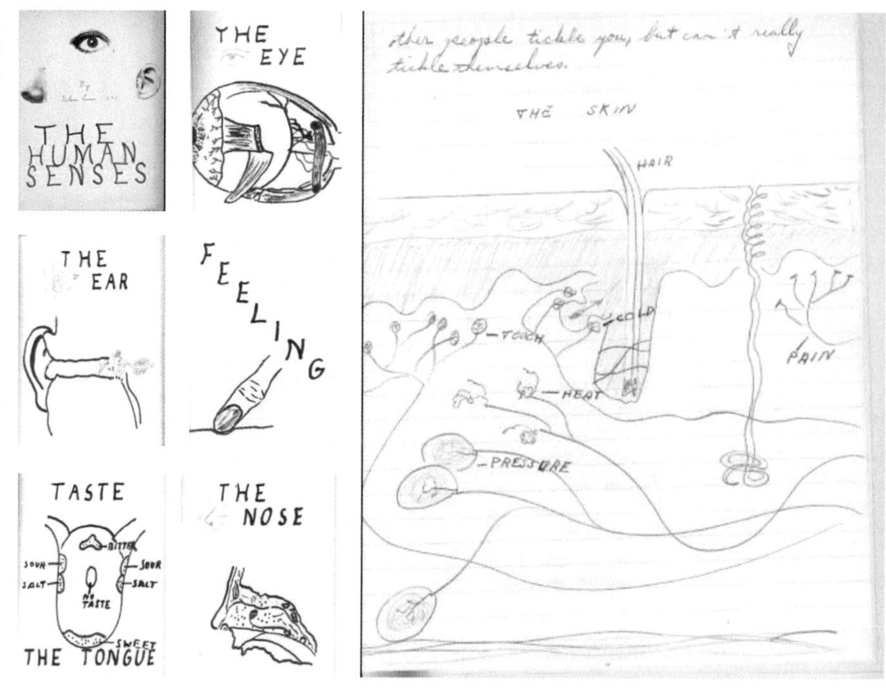

FIGURE 1 A selection of pages from a grade-school booklet by Nelson Cowan on the human senses. The page on human skin senses is enlarged to show the details.

some of the same regions in which the items were first processed, but there is also evidence that mental imagery yields patterns of neural activity similar to the imagined sensations if they had been actually perceived instead of imagined (Albers, Kok, Toni, Dijkerman, & de Lange, 2013; Farah, 1985). Mental imagery occurs not just in the case of visual stimuli, but also for other modalities (e.g., auditory stimuli such as tones). Crowder (1989) delivered a pure tone on each trial to be compared with a subsequent musical note in terms of pitch and found that the comparison was made more efficiently when participants were forewarned of the instrument in which the musical note would occur, which presumably allowed them to generate the appropriate timbre for the tone using auditory mental imagery.

It is not clear that it will always be possible to tell whether a memory is truly sensory in nature, given that mental imagery of sensations might sometimes get things right (e.g., correctly reconstructing through mental imagery that the apple was green). Given this theoretical difficulty, and given that I believe that memory of sensation over some seconds and mental imagery may play similar roles in processing, for the purpose of this chapter, I will consider both of them to be examples of sensory memory, which is thus broadly defined.

I should also add that I do not expect that there is an either/or situation as to whether a memory representation stemming from a stimulus is sensory or otherwise. Cowan (1988) introduced a rudimentary processing model in which there were several stages of memory. In the first, a brief afterimage of stimulation (in any modality) was said to last for several hundred milliseconds. Second, the afterimage impinging on the long-term memory system was said to give rise to activated features, which remain active for some seconds and include both sensory features such as color, angle, pitch, loudness, texture, smell, etc., depending on the nature of the stimulus, as well as more abstract or semantic features such as recognized phonemes, letters, shapes, or meanings (with many of these abstract or semantic features probably restricted to attended stimuli). The brain representation will depend on the constellation of activated features.

Sensory memory has traditionally been considered separate from working memory, in that sensory memory has a large capacity and a short duration, whereas working memory is more persistent but limited to a few items. For example, Sperling (1960) found that most characters in a briefly presented array were available for a short time when a partial report cue was presented (within a few hundred milliseconds), indicating the time limit of sensory memory, but he also found that the total number of characters that could be recalled on a trial was limited to about four regardless of the kind of report cue, indicating the item limit of working memory. According to Cowan (1988), Sperling (1960) was encountering the short phase of sensory memory, a literal sensory afterimage.

Capacity Limits for Sensory Memory?

It is likely that there are capacity limits for the second phase of sensory memory, defined as sets of sensory features that persist for some seconds. However, for several reasons, these capacity limits of sensory memory seem difficult to specify. It cannot be assumed that if a multifeatured item is presented, the observer's knowledge about the item is all or none. Luck and Vogel (1997) found that memory for one sensory feature implies memory for all perceived features, but this finding has been disconfirmed emphatically in subsequent work (Cowan, Blume, & Saults, 2013; Hardman & Cowan, in press; Oberauer & Eichenberger, 2013). These later articles show that if it is necessary to retain all features of each object in memory (e.g., color, orientation, length, and presence or absence of a gap), memory for any one feature is considerably impaired relative to trials in which only one feature must be retained.

Given this state of affairs, with the possibility of partial knowledge of an item, it is not clear how to measure the capacity limit in the longer phase of sensory information. Cowan (2001) suggested that capacity is limited to three to five items when each item is an integrated unit or chunk (Miller, 1956). When memory for one feature of an item is not accompanied by memory for other features of the same item, the item cannot be considered to be a single chunk; it may have to be remembered as multiple chunks (e.g., with orientation and color remembered separately), taking up more of the limited working memory capacity than would be the case if only one feature per object were needed. Also, there is the possibility of a sensory feature being remembered to some degree of specification, but not with enough precision to allow a correct response (e.g., memory for a spatial orientation of an item in an array to be remembered that is not specific enough to be distinguished from a probe item with which it is to be compared; see Bays & Husain, 2008; Zhang & Luck, 2008). So the characteristics of sensory memory, including its capacity, form a great area in need of further research.

Working Memory Definition

Working memory is a term that is used widely, but without much agreement between users. When I use the term, I try to do so in a general way, referring to any mechanisms of the mind that help to retain information temporarily. (For perhaps the first use of the term in this manner, see Newell & Simon, 1956, in their discussion of the use of computers for artificial intelligence.) Given that we cannot use the vast store of long-term memory information effectively all at once, but can only think about a small portion of it at any one time, there must be mechanisms to keep a small amount of information that is particularly task-relevant in a privileged state in which it can be accessed easily as needed for a short time. That information can come both from events recently experienced and from long-term memory of relevant past events and knowledge. Miller, Galanter, and Pribram (1960) used the term working memory in a way that does not seem too dissimilar from this, though they applied it specifically to memory for what one plans to do in the near future. Kane and Engle (2002) similarly have used the term working memory with specific reference to goal states to be maintained. In this sense of the term working memory, it is apparent that the activated sensory features of Cowan (1988) would count as part of working memory.

In contrast, in their seminal work, Baddeley and Hitch (1974) discussed working memory as a multicomponent system. By implication, the term included not only storage of information per se but also the manipulation of that information by central executive processes (an implication made more explicit by Baddeley, 1986). By definition, it appears that information coming directly from the senses cannot be manipulated or else it will no longer resemble what has come from the senses. Given that I have included sensory imagery as a kind of sensory memory, though, that kind of sensory memory could be manipulated. For example, Keller, Cowan, and Saults (1995) found that memory for the pitch of a tone was prolonged when the participant was free to rehearse the tone silently during the retention interval, rather than having to rehearse silently a distracting melody or a verbal sequence during that interval. The notion is that the participant was able to generate a mental image that matched the sensory memory and that the mental image could be maintained with attention, whereas the direct sensory memory of the tone otherwise would have decayed sooner.

Given my broad definition of "sensory," then, I agree that sensory working memory exists. Moreover, I would hasten to add that sensory working memory is a timely topic. It has been somewhat neglected by the standard body of work on working memory. That is probably because Baddeley and Hitch (1974) and others have been impressed with the finding that there are acoustic errors to printed verbal stimuli (e.g., Conrad, 1964), suggesting a nonmodality-specific form of memory coding. What has not received as much attention, however, is the well-researched finding that the recall of a verbal list is far superior at the end of the list (the recency effect) when that list is presented in a spoken as opposed to a printed form (auditory modality superiority effect: Cowan, Saults, & Brown, 2004; Murdock, 1968; Penney, 1989). Moreover, supporting the sensory memory basis of this effect, the auditory modality superiority is greatly reduced when the list is followed by a not-to-be-recalled final item, called a suffix, presented in the same voice as the list items (Crowder & Morton, 1969). The interference is less with a different-voice suffix and still less with a tone suffix. With a retention interval extended to 20 s, the voice-specific memory has faded, though the speech specificity of the memory remains (Balota & Duchek, 1986). Clearly, sensory memory plays an observable role.

I would see sensory memory, broadly defined here, as an integral part of working memory. Neurally, working memory may involve a limited number of attentional pointers from the parietal lobes that in themselves are abstract and not specific to a sensory modality, but that point to pools of neurons in sensory cortex and refer to them, integrating the sensory into the abstract (see, for example, Cowan, 2011; Harrison & Tong, 2009; Lewis-Peacock, Drysdale, Oberauer, & Postle, 2012; Serences, Ester, Vogel, & Awh, 2009).

Is Sensory Working Memory Veridical?

We have already discussed the fact that mental imagery of sensation is prone to being nonveridical, that is, different from the reality that it represents. Beyond that, it may be that all sensory memory is prone to error, just like other types of memory. Even the early stages of perception can be modified by attention (e.g., Woldorff et al., 1993), including modulation of the primary visual cortex (e.g., Pratte, Ling, Swisher, & Tong, 2013). Subjective properties as simple as the perceived duration of a stimulus can be affected by attention (Enns, Brehaut, & Shore, 1999). McCloskey and Watkins (1978) showed that when a line drawing was passed behind a narrow slit, what resulted was a perception of the entire object that was narrower than the actual object. Sensory or not, memory includes distortions that result from perceptual inferences; there is a little-explored topic of study called memory psychophysics, in which the distorting effects of memory on psychophysical functions are studied (Petrusic, Baranski, & Kennedy, 1998). Thus, if one distinguishes between the persistence of visible sensation and the persistence of information (Coltheart, 1980), one way in which they may differ is that the persisting sensation can incorporate some wrong information. For example, unlike sensory memory decay as a fading photograph, in Sperling's type of procedure, features from the array can survive but drift over time in memory relative to their initial positions (Mewhort, Campbell, Marchetti, & Campbell, 1981).

The terms *sense* and *sensation* are used in ways that are sometimes literal and sometimes metaphorical. The field of sensation is typically distinguished from perception on the grounds that sensations are not yet interpreted and categorized. One senses a red disk, perceives it to be a stoplight, and engages in further cognitive processing that leads to a decision to stop the car. This is the literal meaning of sense and sensation. More abstractly, instead of "in the literal meaning," I could have said, "in the literal sense," and that would reflect a usage of the word *sense* that is semantic rather than sensory. In *Sense and Sensibility*, a novel published "by a lady" in 1811 (in reality, Jane Austen), neither word is used in the sensory way; they mean, essentially, logic and emotion. The vocabulary usage demonstrates the special importance that people place on their sensory experiences, which they typically experience in a manner heavily infused with meaning and interpretation.

REVIEW OF SENSORY MEMORY RESEARCH

Having decided that sensory working memory broadly defined does exist, I will review the history of research on sensory memory because all of it seems important for the temporary maintenance of stimulation and is often neglected. Somewhere early in the history of cognitive psychology, sensory memory became divorced from other forms of information and excluded from what is typically called working memory. This can be traced to several influences. First, in a seminal investigation of temporary forms of memory, Sperling (1960) drew a clear distinction between a sensory memory that was time-limited but unlimited in capacity, and a short-term store that was limited to a small number of categorized items. When participants received an array of characters (3 rows × 4 columns in some critical experiments, for example) followed quickly by a tone cue indicating which row from the array was to be recalled, most of the row could be recalled. If the tone cue was delayed about a quarter of a second, performance declined to the point that only about four items from the entire array could be recalled (that is, just slightly more than one item per row in the 3 × 4 array mentioned). Without any cue, about four items could be recalled. This suggested that rich information about how the array looked was present, but that this information faded fast. Moreover, it could be used for recall only insofar as the information could be transferred into a short-term store limited to about four items. Subsequent experiments showed that for auditory stimuli, the sensory information lasted much longer yet still reached an asymptotic level of about four items when the sensory information faded (Darwin, Turvey, & Crowder, 1972; Rostron, 1974; Treisman & Rostron, 1972). Later I will present further analysis of this discrepancy between modalities, suggesting that there is actually a shorter and a longer store in all modalities.

In a second major influence on our understanding of sensory versus nonsensory forms of immediate memory, Conrad (1964) showed that our responses can be governed by something other than the sensory aspects of memory. He presented lists of letters to be recalled in the presented order and found that recall was impeded most by similarities, not in the way the letters looked (e.g., similarities in how *P* and *R* look), but in the way they would sound if pronounced (e.g., rhyming letter names for *B, C, D, E, G, P, T, V*, and *Z*). Based largely on this information, I think, Baddeley (1986) presented his model in a manner that tended to exclude sensory processes in favor of more processed phonological memory (regardless of the spoken or printed source of information) and visuo-spatial memory (which, despite its name, presumably could be formed even on the basis of the spatial qualities of nonvisual sensation).

This separation of sensory memory from the rest of working memory, however, leaves open the possibility that there are important commonalities between sensory and abstract types of working memory information. Cowan (1988) proposed that, regardless of sensory modality, there are two phases of sensory memory: a brief afterimage lasting several hundred milliseconds and a more processed phase lasting several seconds. Cowan further proposed that the latter phase consists of activated features from long-term memory, which can come from both past events and present sensory input. Activated features can include both sensory and categorical information.

Sensory memory of a very fleeting sort may be important in evolutionary terms because it allows the continued analysis of events that are very brief, such as flecks of moving color in the forest, brief growling sounds, or the first feeling of a spider crawling on your skin. A more persistent phase of sensory memory allows the careful comparisons of two stimuli, such as comparing your own trajectory when running with a rhinoceros coming in your direction, comparing the textures of two different stone surfaces to recognize your location in the dark, or learning a sound distinction in a new language. Sensory memory may be the glue that ties our experiences into a coherent stream of consciousness, allowing perception of the smooth motion of a deer as it runs between trees, the continuity of a talker's voice despite abrupt changes in the acoustics, or the feeling of your socks while you pull them up. This smoothness created out of discontinuity of the input is the basis of motion pictures.

Given these preliminary thoughts and definitions, I turn my attention to two questions that I will try to address. First, how much of working memory is sensory, that is, comprising the way things look, sound, feel, smell, etc.? The alternative is for much of working memory to be abstract, so that the modality of origin does not matter and sensory imagery does not dominate behavior. Second, how complicated is sensory memory? For instance, are there separate rules of operation for all modalities or are there common rules of operation across modalities?

What hinges on these questions is progress in scientific investigation of the mind. Finding regularities across modalities will allow us to establish more general principles of memory. Progress in understanding consciousness also depends on understanding sensory working memory. We somehow have to reconcile the vastness of the experienced world with the small limit of focal attention. The paradox can be seen, for example, in the phenomena of inattentional blindness (Bredemeiser & Simons, 2012; Simons & Rensink, 2005) or change blindness. Unattended aspects of the visual field, though clearly in plain view perceptually, can change without our noticing. If a flicker is introduced into the static showing of a photograph (which prevents transients from being used), one can make a dramatic change in the display without people perceiving it. For example, in a restaurant scene, a glass can alternate from being present to being absent, with most viewers needing quite some time to notice this alternation. The demonstrations and experiments of this sort highlight the difference between information that can be consciously perceived, as the entire scene in front of you can, in a holistic sense, and in contrast the selective entry of this information into working memory, which typically can occur for only a few separate objects in a scene at a time.

A Unifying Modeling Framework

The framework in which I will view sensory working memory is the one articulated by Cowan (1988), as shown in Figure 2. It is a framework rather than a well-worked out model, the intent being to summarize what we know and leave room for further details to be entered later, as we learn more.

Progression of Models

The model in Figure 2 can be put in context of the other models that preceded it. Stepping back, there are progressive changes in modeling conventions that are worth mentioning. (1) In the model of Sternberg (1966) and some other models of the era, phases of processing were depicted as if they occurred in a simple nonoverlapping sequence. Sternberg concluded that each item in a list is mentally scanned in turn. In many processing models of the era, first there was a large-capacity but brief-lived sensory store; then information was filtered by attention; then selected information entered a small-capacity short-term memory; and then that information was transferred to long-term memory for permanent storage. (2) Atkinson and Shiffrin (1968) kept that basic format but modified the modeling convention to focus on recurrent processes transferring information back from later stores to earlier ones, furthering a trend that was started in an offhand model sketched by Broadbent (1958). Presumably, information from long-term memory is used to enrich working memory, and information in short-term memory is recycled (refreshed) to

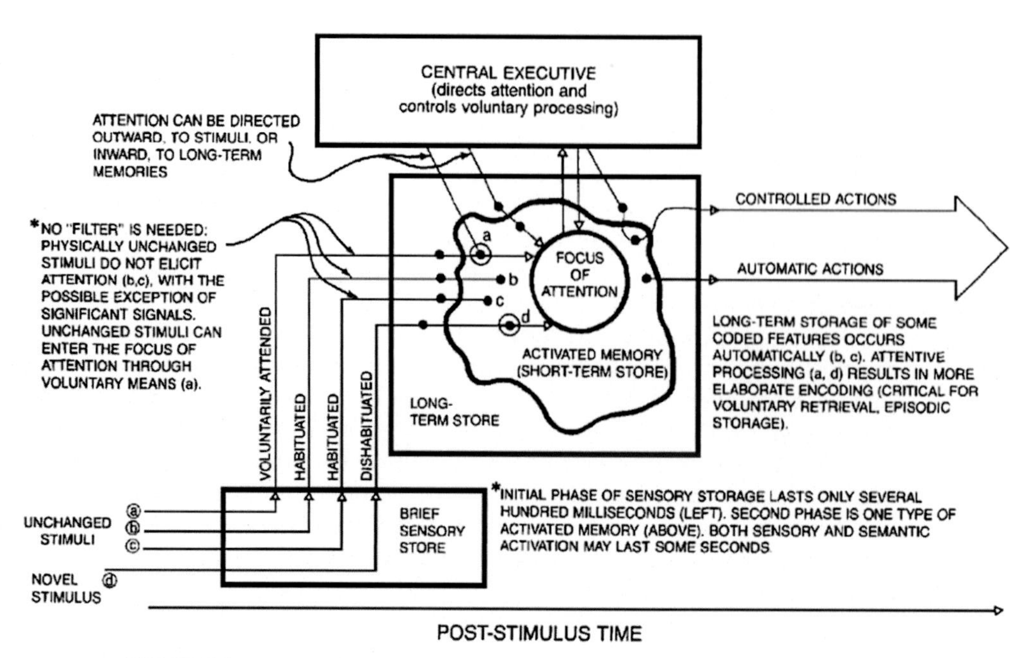

FIGURE 2 The modeling framework of Cowan (1988). *Reproduced from his Figure 1.*

counteract forgetting. (3) Baddeley and Hitch (1974), in developing the concept of working memory, changed the nature of transfer from one store to another. In their model, different aspects of information might be sent to the phonological store and the visuo-spatial store at the same time. The earlier notion of progression through the stores does not apply. Finally, (4) in the modeling framework of Cowan (1988), even the separation between components was abandoned, in ways that I will now explain.

The modeling framework of Cowan (1988), shown in Figure 2, leaves open the possibility that phonological and visuo-spatial memory operate separately but it leans toward the general principle that memory of any sort is susceptible to interference from subsequent items that are similar in their features. There are many forms of stimulation that clearly do not fit the phonological versus visuo-spatial distinction (e.g., spatial arrangements of tones, progressions of tactile or olfactory sensation). In the model, all forms are just represented as variants of activated sensory features from long-term memory. Moreover, until evidence can show otherwise, the assumption is that activated sensory features and semantic features of long-term memory may have similar properties, most notably decay over time and susceptibility to interference from subsequent input with similar features. Additionally, there is another important way in which components in this model are not separate. The limited-capacity storage system is conceived as the focus of attention, which is represented as a subset of the activated information from long-term memory. When information is no longer attended, the model states that it remains in an activated form for some seconds.

It is also the attentional filter of earlier models that is no longer separate in the model of Cowan (1988). In the modular way of thinking, it was unclear how information could sometimes get through the filter. Treisman (1964) suggested that there is an attenuating filter that only partly blocks unwanted stimulation. Instead, Cowan suggested that all information enters the memory system, whether it is attended or not. The information is used to form a neural model of the environment. That neural model includes sensory features of all sensory input streams but semantic features primarily of attended streams. When there is a change in the input that is perceived to differ from the neural model, either in perceived physical or perceived semantic characteristics, the result is a shift of attention accompanied by an orienting response (Sokolov, 1963). In the absence of an orienting response, the focus of attention remains on the task or sensory stream voluntarily selected. This model can account for why a change in the physical properties of an unattended aspect of the environment attracts attention. For example, the event can be a change from a low-pitched to a high-pitched voice in the background signifying that someone new just entered the room (see Cherry, 1953), or a sudden flash of light. There is no filter as such to exclude that information (at least, not when it results in the change in intensity of a feature; it may not apply when the abrupt change is in quality only, e.g., in color; see Folk, Remington, & Johnston, 1992). In short, the focus of attention is controlled partly by voluntary processes, but partly by orienting responses to changes in the environment.

Given evidence that there is not much semantic processing of truly unattended stimuli (Conway, Cowan, & Bunting, 2001; Cowan & Wood, 1997), Treisman's attenuation concept may not be needed. For example, Conway et al. showed

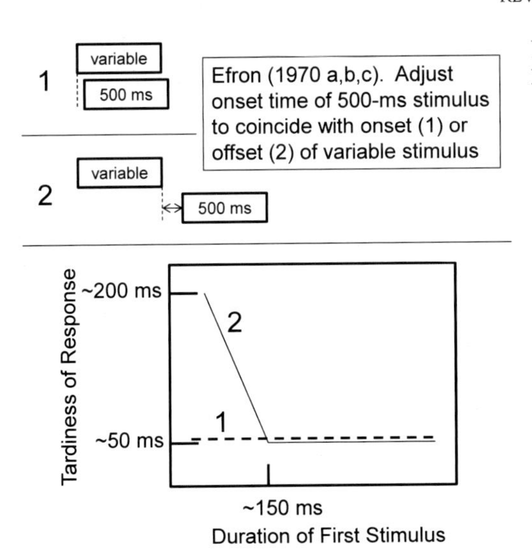

FIGURE 3 General method and schematic results of Efron (1970a, 1970b, 1970c). Number 1 reflects methods and results of onset estimation; number 2, of offset estimation.

that people noticing their names in an unattended channel in selective listening procedures (Moray, 1959) are primarily those with low working memory spans, whose attention may have wandered off of the assigned task to the channel with the name.

Within this framework, I will now provide more justification for the sensory and abstract aspects of working memory.

Phases of Working Memory Information, Both Sensory and Abstract

Cowan (1984) suggested that there are two forms of auditory sensory memory—a brief afterimage and a longer recollection of sensation—and Cowan (1988) extended that idea to all modalities (an idea presented also by Massaro, 1975). Why split something as elementary as sensory memory into two phases? The following research summary will explain why.

Brief Sensory Afterimages

There is abundant evidence of a kind of sensory afterimage. It is well known in the visual modality in which the chemical characteristics of the retina allow visual sensations to persist for some seconds, as in the case in which a candle is moved around in the dark. The kind of afterimage I mean, however, is not dependent on the retina, lasts only a fraction of a second, and seems to occur in every modality on which results have been reported.

One of the clearest indications of sensory afterimages was investigated in a remarkable series of experiments by Efron (1970a, 1970b, 1970c) on the minimal duration of a perception. His procedure and results are schematically illustrated in Figure 3. On every trial, two brief stimuli were presented. In some experiments, these were both visual or were both acoustic, whereas in other experiments there was one stimulus in each of these modalities. The stimulus being judged was of variable duration, whereas the other stimulus was of fixed duration. Sometimes, as shown in the top section of the figure, the task was to determine the temporal location of the fixed stimulus that would be perceived as synchronous with the onset of the variable stimulus. Other times, as shown in the second section of the figure, the task was to determine the temporal location of the fixed stimulus that would be perceived as synchronous with the offset of the variable stimulus. The difference between the two judgments was taken as the duration of the perception of the variable stimulus. The findings were similar regardless of the modalities of the stimuli: the point of offset of a very brief, variable stimulus was overestimated. The point of onset was only slightly overestimated and, putting together the two kinds of information (as shown schematically at the bottom of Figure 3), the minimum duration of a perception was about 200 ms. Therefore, the shorter the stimulus, the more its duration was overestimated. (A seasoned investigator may look at this schematic result and assume that the true result was much more variable, but it was actually surprisingly clean.)

The results of Efron (1970a, 1970b, 1970c) can be explained on the basis of a sensory afterimage of the variable stimulus. This sensory afterimage might occur as a side effect of the initial sensory processing of the stimulus. As shown at the top of Figure 4, in the critical condition, the indicator is supposed to be adjusted to the end of the variable-length stimulus but, unknown to the participant, it is actually adjusted to some point after the sensory memory of the variable stimulus has faded to a certain level at which it is no longer experienced as ongoing sensation.

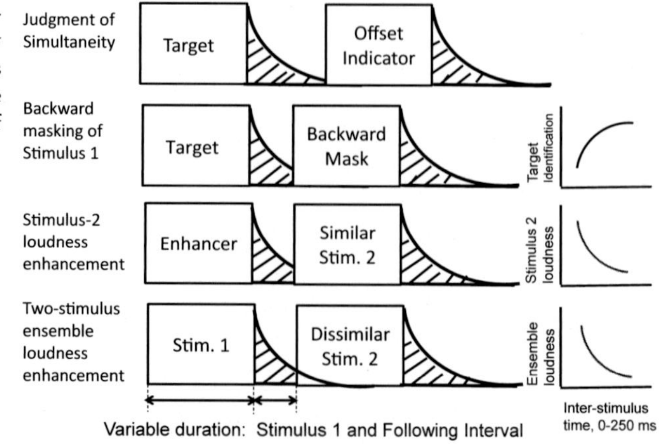

FIGURE 4 Four consequences of following a brief stimulus with a second stimulus, all of which I interpret as the consequence of a brief sensory afterimage. The box figures represent stimuli, the curved and striped figures represent the sensory afterimage, and the right side of the figure shows the typical result for the second through fourth procedures. For the results of the first procedure, see Figure 3.

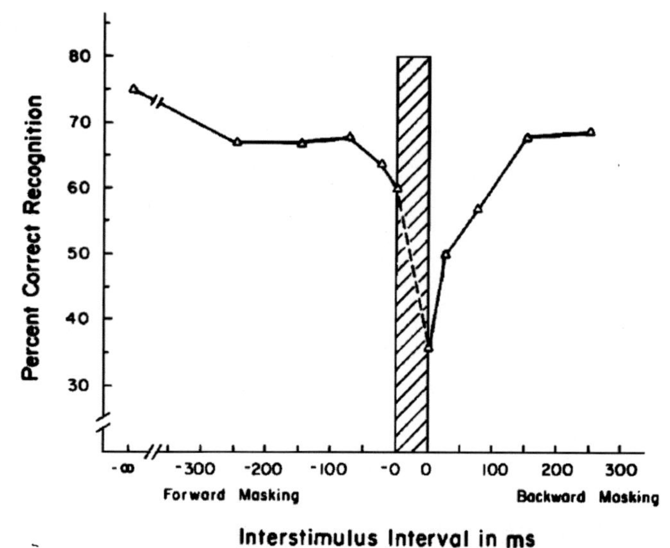

FIGURE 5 The results of Craig (1985, Figure 4) for forward and backward masking of vibrotactile sensation.

Figure 4 also shows other ways in which two-stimulus procedures are relevant to the notion that there is a brief sensory afterimage lasting several hundred milliseconds. In one type of procedure, recognition masking, two brief stimuli are presented in rapid succession and either the first stimulus is to be identified (backward recognition masking, shown in the figure) or the second stimulus is to be identified (forward recognition masking, not shown). The result is basically the same in vision (Turvey, 1973), audition (Massaro, 1975), or tactile sensation (Craig, 1985). To illustrate the finding, Figure 5 shows it for tactile sensation. Craig used pairs of very brief vibrotactile stimuli, each consisting of a pattern of pins that were thrust forward to stimulate a fingertip for 4.3 ms. The recognition of the second pattern was only minimally affected by presentation of the first pattern, though there was some effect when the patterns were within 50 ms of one another. In contrast, recognition of the first pattern was profoundly impaired by the second pattern (the backward masking effect) and this effect did not fully dissipate until the patterns were about 200 ms apart. The interpretation is that each pattern is followed by a sensory afterimage that contributes to the recognition process, and the second pattern interferes with recognition of the first if it interrupts the sensory afterimage of the first before that afterimage can be processed. Subsequent research has shown that the critical variable for the backward masking effect is the stimulus onset asynchrony, because for sufficiently long stimuli the recognition process can be completed while the stimulus is ongoing.

Cowan (1987) took the masking process a step further by demonstrating how the sensory afterimage might contribute to the process of loudness perception. On each trial in his Experiment 1, a pair of tones was presented, with a variable intertone interval. The loudness of each tone in the pair was to be judged on an eight-point scale. The tones came in three different brief durations (20, 30, and 40 ms), based on the fact that sensory integration results in louder judgments for longer tones. The expectation was that the sensory afterimage of each tone would contribute to that overall integration and that the second tone terminates the sensory afterimage of the first tone, limiting the neural

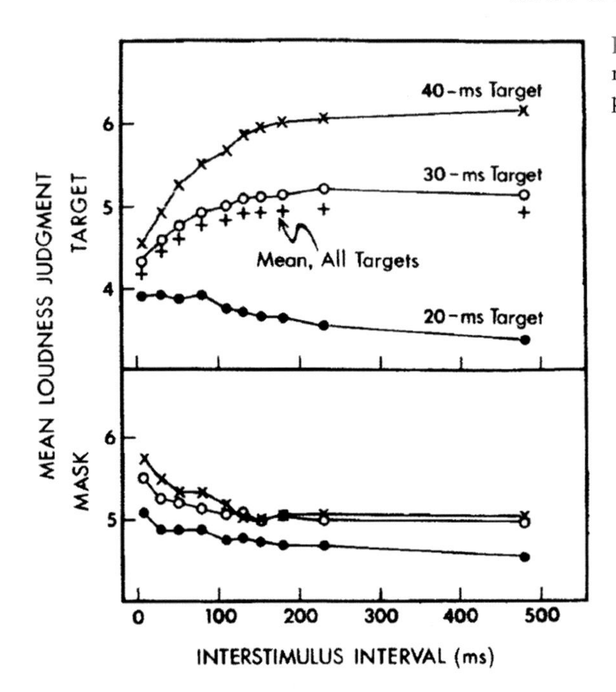

FIGURE 6 Masking of loudness, from Cowan (1987, Figure 2). The top panel represents the perception of the first tone in a pair; the bottom panel represents the perception of the second tone in a pair.

activity contributing to the perceived loudness of the first tone. If these assumptions are correct, the loudness of the first tone in a pair should increase as the interval between the two tones increases. Additionally, the scale should be used in such a way that the judgment of loudness for the three tone durations would diverge across intertone intervals, because each tone's loudness should be more clearly perceived at longer masking intervals. Those two principles together accounted well for the results. Figure 6 shows the results for the first tone in a pair (top panel) and for the second tone (bottom panel). For the first tone, the judgment of the 40-ms tone increased markedly as a function of the intertone interval, the judgment of the 30-ms tone increased more moderately, and the judgment of the 20-ms tone actually decreased slightly. The results were well-fit by a model that basically includes two factors as mentioned: (1) the integration of sensation across the stimulus and its sensory memory, up to the point at which another tone occurs to interrupt that sensory memory; and (2) a weighting function that produces more separation between the 20-, 30-, and 40-ms sounds as the interstimulus interval increases. For judgment of the second tone (or mask, as shown in the bottom panel of the figure), there was no backward masking, only forward masking, so responses were affected only at very short intervals, and not differentially for the three tone lengths. The results verify that the sensory afterimage adds to the sensation that is caused directly while the sound is ongoing.

Figure 4 also shows two other ways in which the sensory afterimage can be demonstrated using intensity perception (e.g., for tones, Arieh & Marks, 2003; for vibrotactile stimuli, Verrillo & Gescheider, 1975). When the two stimuli are very similar and close together, the sensory afterimage of the first stimulus can inappropriately contribute to the perceived intensity of the second stimulus, as shown in the third row of the figure. When the two stimuli are very different but close together, this effect on the judgment of the second stimulus does not occur. In that case, though, if participants are asked to judge the overall intensity of the two-stimulus complex, the judgment is higher when they are close together in time. Presumably, this phenomenon occurs because there are registers of stimulation (critical bands), and the overall perceived intensity is greater when more than one critical band is stimulated concurrently. I suggest that it is the concurrence of the sensory afterimage of the first stimulus with the reception of the second stimulus that causes this perceived intensity enhancement.

In sum, there is a brief phase of sensory memory lasting several hundred milliseconds from the onset of a stimulus. It appears to operate similarly across modalities, is tied to the process of stimulus recognition, and is experienced as the continuation of sensation in the modality of origin.

Processed Sensory Recollection

It seems to me that the field of sensory memory has always been the source of much confusion, largely because different procedures yield very different estimates of sensory memory. Whereas Sperling (1960) showed that there was a type of visual sensory memory lasting several hundred milliseconds, a much longer estimate of acoustic memory was obtained in a procedure similar to Sperling but using spatiotemporal arrays of spoken digits (Darwin et al., 1972).

Characters were spoken at left, center, and right locations concurrently in two successive temporal bursts for a total of six characters to be remembered per trial. A cue indicated which spatial location to recall. The finding was that sensory information took a lot longer to decline to a plateau than was found in vision: about 4 s. On the basis of this finding, it has often been argued that sensory information lasts a lot longer in the acoustic domain than in vision.

This interpretation of Darwin et al. (1972) as indicating a modality difference, however, would leave no explanation for the comparable results obtained for visual and acoustic stimuli by Efron (1970a, 1970b, 1970c), so there is a conundrum. The solution may be as follows. Results obtained by Massaro (1976) following up on Darwin et al. suggest that their sort of finding in the auditory domain is not comparable to what was obtained in the visual domain by Sperling. In vision, a postarray category cue (e.g., recall all letters but not numbers in the array) cannot be used to access a part of the sensory representation as efficiently as a postarray physical cue (e.g., recall the middle row). Yet, Massaro found that physical and category postarray cues worked equally well in audition. Darwin et al. carried out a comparable experiment with semantic cueing (their Experiment III) but it is difficult to compare with physical cueing because the whole-report results in these two experiments differ substantially for some reason. Overall, I conclude, with Massaro, that the ability to use the longer memory for acoustic stimuli may make this technique incapable of demonstrating the shorter memory of the type examined by Efron and others, and that there are two phases of sensory memory in each modality, with different time courses and characteristics (Cowan, 1984, 1988). It still remains to be discussed why the longer store does not similarly interfere with the findings in the visual modality.

Evidence of Two Phases of Sensory Memory

Although Sperling (1960) presented clean results on sensory memory, a further detail of the phenomenon leads to a different interpretation. Specifically, when one looks at the types of errors that occur with increasing cue delays, they turn out to be errors in the spatial location of items (Averbach & Coriell, 1961; Mewhort et al., 1981). This implies that information about the items continues to exist in memory, even in the visual modality, but not the same high-fidelity information that exists for a brief period of several hundred milliseconds and includes spatial location information.

Another well-known finding indicating two phases of visual storage, for a similar reason, was obtained by Phillips (1974). He presented on every trial a black-and-white grid with squares randomly assigned to black or white. After a variable retention interval, it was presented a second time, either intact or with one changed square color. The task was to indicate if the array was the same or changed. On some trials, however, the second array was also spatially shifted compared with the first array. The finding of interest here is the effect of the spatial shift. If the retention interval was about 300 ms or less, a spatial shift substantially impeded performance. If, however, the retention interval was longer (it could last up to about 600 ms), the spatial shift did not matter. This suggested that there is a literal representation that depends on an exact spatial match and lasts only about 300 ms, whereas there is a less literal representation that is more abstract but still contains information about many of the display elements. When the array contained as many as 64 squares, performance was still about 60% correct at the longest retention interval. The location-dependent and location-independent stores appear to be the same as the two phases of sensory memory that I have mentioned.

A different procedure has been used to suggest the existence of two stores in the visual modality. Kallman and Massaro (1979) presented three stimuli in the order *comparison-target-mask* or in an alternative order, *target-mask-comparison*. In either case, the target and comparison tones were to be compared. The time between the comparison and target tones was fixed but the time between the target and mask was varied. In the order *comparison-target-mask*, the masking interval could influence the recognition of the target by limiting access to its sensory afterimage. In the alternative order, *target-mask-comparison*, this was still the case but here there was an additional, second basis of interference. Specifically, regardless of the target-mask interval, the mask intervened between the target and comparison tones and could interfere with the longer phase of sensory memory of the target tone. Sure enough, the *comparison-target-mask* order showed an effect of the mask only at short masking intervals, whereas the *target-mask-comparison* order showed an effect of the mask regardless of the masking interval (because of the interference with both phases of sensory storage). The interference effects were seen as effects of similarity between the target and mask. This study provides clear evidence for the existence of two separate phases of sensory memory.

Although I cannot find any tactile study that similarly demonstrates the existence of two phases of sensory memory, there is evidence in the form of different procedures with different results. Recall that vibrotactile sensation yields evidence of recognition masking suggesting a store lasting several hundred milliseconds while the stimulus is identified. Other evidence, though, shows that the memory for a tactile stimulus is diminished across some seconds (e.g., across about 5 s for reproducing the location of the tactile event; Sullivan & Turvey, 1972).

Nature of the Longer Store

According to the model of Cowan (1988) shown in Figure 2, the second phase of sensory memory should be an activated portion of long-term memory. This point was demonstrated by Cowan, Winkler, Teder, and Näätänen (1993) using the mismatch negativity (MMN) response of event-related potentials. In the MMN, the participant is engaged in a visual task (such as reading a book) while ignoring acoustic stimuli. When a train of identical standard tones (e.g., 1 per second) is followed by a deviant tone, an MMN results. It is thought to result from the comparison of the standard representation to the deviant. However, this MMN is not produced under certain circumstances. If there is only one standard, there is no MMN to a change. If there are many standards but there is a long delay of over 10 s before the deviant is presented, it does not elicit an MMN. These results have been taken to suggest that the sensory memory representation of the standard must be built up across tones, and that this representation is lost by about 10 s. Cowan et al. reproduced this result but, in some conditions, after the long delay there was one reproduction of the standard tone, and then the deviant. Under these conditions, the MMN returned. The explanation was that the long phase of sensory memory is an activated set of elements from long-term memory, that the activation of the standard tone representation can be lost over the delay, and that a single reminder of the standard tone can reactivate this representation.

May and Tiitinen (2010) offered an alternative, non-memory basis of the MMN. In Cowan et al. (1993), they suggested that the neural populations that respond to the standard and deviant just following a long delay are in similar, unhabituated states, and therefore show no MMN. This interpretation appears to overlook the results of a roving standard control condition in which a standard train followed by a long delay had to be followed by not only one but several iterations of the standard before an MMN could occur. Perhaps the non-memory theory could be stretched to take that predelay basis of habituation into account. There are, however, subsequent data that would be even more challenging for a non-memory interpretation of the MMN. Winkler, Schröger, and Cowan (2001) found that the "long delay" that mattered was relative and not absolute. When standard tones presented in a fast train were followed by a 7-s delay, about half of the participants showed no MMN to a postdelay deviant. However, when standard tones separated by 7-s intervals were followed by a 7-s delay, those same participants did show an MMN to a postdelay deviant. Winkler et al. suggested that the processing system gauged the contextual relevance of the standards; a train of closely packed standards lost relevance after a long delay, but a train of separated standards did not because the context defined by the stimulus pace had not changed. It is difficult to see how the non-memory interpretation could account for those data.

The role of attention is probably different for the two phases of sensory memory (Winkler & Cowan, 2005). For the brief afterimage, Sperling (1960) showed that many items can be stored at once, though this information is short-lived except for attended items. The second phase of sensory memory lasts several seconds without attention, though it may be enhanced by attention. For example, Keller et al. (1995) found that same–different tone comparisons could be enhanced modestly by rehearsal across a retention interval of several seconds unless a distracting task was presented during that interval. This enhancement occurred even though the stimuli could not be easily categorized; tone differences still were less than a semitone.

In another investigation of the role of attention in the second phase of sensory memory, Cowan, Lichty, and Grove (1990) tested memory for syllables that were unattended during their presentation while the participant was busy reading. Memory for most syllables was not tested. After 1, 5, or 10 s of silent reading following the presentation of occasional target syllables, a light cue indicated that the participant should put down the reading and recall the most recent syllable. There was a marked decay rate of memory across this 10-s interval, with declining performance as a function of the retention interval. However, even a small amount of attention to the syllables abolished this decay function. Specifically, in one experiment, there was an added task to listen for instances of the syllable /dih/. This syllable was detected only 60% of the time, but this split attention was enough to abolish completely the decay of sensory information for the various syllables, which now held steady across 10 s. It seems likely that attention transfers information into a more abstract form, which we will discuss next.

It also might be, however, that the short phase of sensory persistence is influenced by attention as broadly defined. In Sperling's study, attention is distributed to the locations in space that contain all of the characters in the stimulus array, even though it may not be possible for the array items to be attended individually; there are too many of them. In other work, though, it has been shown that there is an effect of attention to versus away from part of a spatial field on the reported duration of a visual stimulus in that part of the field, a potential indication that attention extends the first phase of sensory memory (Enns et al., 1999). It is not clear, however, whether the measure of perceived duration that Enns et al. used provides insight into the same level of processing that Efron (1970a, 1970b, 1970c) provided by having participants compare the offsets of two stimuli. Enns et al. appear to have found an effect of attention from a 100-ms cue period on perception of a very short target, suggesting that the effect could result at an interpretive stage after the sensory afterimage has faded, rather than prolonging the sensory afterimage itself. This is fertile ground for further work.

In sum, the second phase of sensory memory is not literal; it is not an afterimage that continues the stimulus identification process, as the first phase is, but rather a vivid recollection of the sensation that can be used to compare stimuli. Examples are learning to distinguish between two slightly different words in a foreign language, and comparing two slightly different shades of paint (or two arrays, for that matter).

Constructed Abstract Units

It is abundantly clear that, in addition to sensory codes, there are abstract codes that do not directly depend on sensation after the stimuli have been encoded. For example, as mentioned previously, Conrad (1964) showed that memory for printed speech is still susceptible to acoustic confusions, suggesting that an acoustic-like code is internally generated. Even if this acoustic-like code is considered sensory, a more abstract phonemic category is also likely to be generated. Thus, words such as *metal* and *medal* can be acoustically identical and yet assigned to different categories. In other contexts, the categorical difference is not neutralized but heard (*metallic* vs. *medallion*).

One might suspect that all sensory codes are abstract, so that there would be no difference between the memory for acoustically presented speech and visually presented stimuli that are mentally converted to speech. That actually appears to be the expectation of Broadbent (1958), whose groundbreaking information-processing diagram showed a feedback arrow from working memory to what we now call sensory memory. An argument against this idea of equivalence, however, is that memory performance in the two cases is much different. In immediate free recall of lists of verbal items, memory for items near the end of the list is far better with acoustic presentation than with printed presentation (Murdock, 1968; Penney, 1989).

One commonality between the two phases of sensory memory is that they presumably each could include many elements at the same time (e.g., Phillips, 1974). In contrast to this, many believe that abstract information is limited to several distinct items at once. Cowan (2001) argued for this hypothesis. To identify working memory situations in which abstract units had to be relied upon, Cowan looked for procedures in which nothing else could be used: sensory memory could not be used (e.g., because of masking or a change in sensory qualities from stimulus to test), verbal rehearsal was prevented (e.g., by articulatory suppression), and items could not easily be combined to form larger chunks (e.g., because the presentation was too rapid). Under these circumstances, it was assumed that memory was based on abstract, nonsensory units. Normal adults could remember typically only three to five abstract units, with a somewhat smaller number in young children.

The term "abstract" is somewhat difficult to pin down, and the present section is an attempt to do so. The first part of the discussion covers a controversy regarding the capacity limit. One traditional distinction between sensory and abstract temporary memory has been that there is a capacity limit for abstract memory defined in terms of the number of chunks (after Miller, 1956) but no capacity limit for sensory memory; rather, a time limit (after Sperling, 1960). That distinction is challenged by a modern controversy in which some investigators assert that attention can be divided among any number of items in a stimulus field, with no mention of any abstract form of storage with a capacity limit.

After this controversy about capacity is discussed, the second part of the section asks how we might know that information is abstract. When verbal labeling of the information is possible it is easiest to know, but I argue that some information that cannot easily be labeled is still likely to be abstract as opposed to sensory. As mentioned previously, one reason why the distinction is difficult may be that we always have a sensory underpinning with any type of abstract processing. The two sections fit together inasmuch as it is possible that the belief in a capacity limited to several chunks is correct for abstract memory but that investigators who believe in a working memory based on attention spread thinly across the entire stimulus field are picking up on sensory aspects of memory.

Characteristics of Abstract Working Memory 1: Capacity Limits of Abstract Memory

Although the basic finding is clear, the interpretation is still an important issue. Much of the discussion is based on a procedure by Luck and Vogel (1997) in which a spatial array of objects is followed by a short retention interval and then a probe that is to be judged the same as in the array or different from it. Performance in adults is excellent with three items and declines markedly as the number of items increases. Simple formulas can be used to estimate the number of items held in working memory. Underlying them is the assumption that the participant answers correctly if the critical information is in working memory, and otherwise guesses with a certain bias. That bias depends on the test situation (see Cowan, 2001; Pashler, 1988; for an overall explanation of the formulas, see Rouder, Morey, Morey, & Cowan, 2011). The formulas show that on the average, between two and five items are held in working memory, and the average is typically close to three. Zhang and Luck (2008) modified the procedure to ask whether attention could actually be spread across an entire field. Participants received an array of items that could vary along a continuous dimension (color or orientation) and the answer was given by adjusting a wheel (color or orientation) to the correct value. The angle of disparity between the actual and recalled values was the imprecision of the answer. The findings along with a memory-plus-guessing model suggested that only a few items could be held in memory and that other responses were based on guessing.

Other recent evidence in favor of capacity limits comes from an investigation of reaction times and a sophisticated process model. Donkin, Nosofsky, Gold, and Shiffrin (2013) found that reaction times in change-detection tasks are as one would expect from a capacity-limited process: on some trials the participant knows the item being probed and responds quickly, whereas on other trials the participant must guess and responds more slowly.

In the other camp, Bays and Husain (2008) argued that working memory is a resource that actually can be spread across all items in a visual field. Anderson, Vogel, and Awh (2011) provided some strong evidence against that view, showing that imprecision increases when the number of stimuli increase from one to three, but then reaches a stable plateau, presumably because capacity has been filled. Zhang and Luck (2011) showed that participants could not be motivated to spread attention more thinly to more objects in the field. Nevertheless, recently, van den Berg, Shin, Chou, George, and Ma (2012) revived the hypothesis of working memory as a continuous resource, this time in a complex model in which the precision of the representation of the tested item can vary from trial to trial. van den Berg, Awh, and Ma (2014) took this line of argument further and tested a great number of models, finding that the winning model had variable capacity as well as variable precision. Nevertheless, there was an issue about some of the models being more flexible than others (i.e., more able to win the comparison even when fit to data generated by a process that does not match the model) so further work is needed. The jury is still out or, rather, is yet to be selected.

Although I favor the limited-capacity view, it is also the case that this view must be modified compared to where it was when defined by Luck and Vogel (1997). They carried out studies with objects that had up to four important features (color, orientation, length, and presence or absence of a gap) and concluded that if one had an object in working memory, it included all of the features. They found that people could remember three or four objects with all the features. In one sense, there could be nothing abstract about such a memory system because any item taking a slot in working memory would have to have attached to it all relevant sensory information. However, Hardman and Cowan (in press) investigated further with very similar materials and did not get the same results; the need to pay attention to all features in several objects consistently reduced the likelihood that any one feature of an object would be retained.

There are other findings concordant with Hardman and Cowan (in press), but note that none of them contradict the notion that there is a limit in the number of chunks that can cooccur in working memory. Cowan et al. (2013) found that attention to both color and shape resulted in fewer colors or shapes being available than if attention were directed only to color or only to shape. Oberauer and Eichenberger (2013) found that attention could be divided among even more than two features of an object, with tradeoffs between features. Both data sets, however, were fit to a model in which objects do play a role. In this model, the observer with capacity k apprehends k objects in one of the features (e.g., k colors) and then is able to encode other features, but only for these particular objects for which the first feature has already been encoded and, even then, often only for a subset of them. Therefore, there will be k objects with at least one feature encoded, though there also is a limit for the number of features that can be encoded concurrently. Hardman and Cowan (in press) also found that they could not rule out such a model in which there are two phases of working memory limits for complex objects: a phase in which features compete for attention, and a phase in which the incomplete objects in working memory are limited to about three objects.

Cowan (2001) summarized a great deal of other evidence suggesting that a capacity limit also applies in the case of the recognition or recall of verbal lists (further explored by Chen & Cowan, 2009; Cowan, Rouder, Blume, & Saults, 2012). There is additional evidence suggesting the possibility of similar capacity limits for tactile stimuli (e.g., Bliss, Crane, Mansfield, & Townsend, 1966) and even for odors (e.g., Laing & Francis, 1989).

Characteristics of Abstract Working Memory 2: The Distinction from Sensory Memory

For our understanding of human information processing it seems important to ask how much of what is considered abstract working memory is in fact sensory in nature. Abstract information would function in the same way regardless of the sensory modality of origin of the information. Abstract representations in working memory share neural areas even when the information comes from different modalities and codes (Cowan et al., 2011). Note though that there are sensory-specific areas of activation in these data too; it is just that a memory load area that cuts across sensory modalities can be found in the parietal lobes, the intraparietal sulcus (IPS).

One indication of abstractness comes in the nature of interference. Whereas memory for a tone's pitch sustains severe interference from additional presented tones, more than from other sounds (Deutsch, 1970), memory for a tone's pitch is indifferent to whether silent mental interference comes from an imagined digit sequence or an imagined tone sequence (Keller et al., 1995). Both interfere to the same degree relative to a no-interference condition in which imagery-based rehearsal is possible. This suggests that sensory input interferes with sensory memory but that abstract input from imagery may not interfere with sensory memory.

Under some situations, what Cowan (1984, 1988) called the second phase of sensory memory might be used to supplement abstract memory (e.g., see Sligte, Scholtel, & Lamme, 2008), but might not share some of the same capacity limits; the information could be spread over all array items. To investigate this issue, it is possible to impose a masking pattern between the array to be encoded into working memory and the test (cf. Saults & Cowan, 2007; Rouder et al., 2008). The masking pattern should be placed late enough to allow sufficient encoding of items into working memory in the first place; see Vogel, Woodman, and Luck (2006). Most of the studies in this field have not included a mask, so the data may in some ways suggest a continuous resource to some investigators because of the contribution of sensory memory, which makes the study inconclusive as to the properties of a more abstract working memory.

Last, one might wonder if the items that are abstract must be those that have verbal labels and therefore might be retained through verbal coding. This is a difficult issue because in the absence of an available verbal label, it is difficult to know whether the participant truly has a category for a stimulus or whether the stimulus might be perceived as an ensemble of parts, not integrated chunks or members of categories. Olsson and Poom (2005) argued that categories were needed, on the grounds that when their visual stimuli were difficult to categorize, participants could retain only about one such item.

Li, Cowan, and Saults (2013) addressed this question of categorical, yet nonverbal, working memory by developing a set of 12 tones that had musical timbres, which were generally too close together to be labeled as members of particular instruments; participants generally reported not labeling them. The frequencies of the tones did not fall on a musical scale, either. The presence of timbres presumably allowed categorization of the 12 sounds following repeated exposure to the sounds. In experiments using tones with and without the timbres, a sequence of tones to be remembered was followed by a mask to limit sensory memory. The presence of timbres increased the number of tones that could be remembered, with the more capable participants able to remember about three of them. Thus, there is some evidence suggesting that the use of abstract memory requires categorization even when the stimuli cannot be labeled, though further work is needed.

Brain studies could help to clarify what information is abstract and what information is sensory. On one hand, function magnetic resonance imaging studies with auditory verbal and visual spatial objects show that there are brain areas (most specifically the IPS) that respond to a working memory load regardless of its modality (Chein, Moore, & Conway, 2011; Cowan et al., 2011; Majerus et al., 2010). This leads to the notion that the stimuli of both types give rise to a very general neural representation of abstract information about the stimuli. On the other hand, there are certainly sense-specific areas of memory activation, even in these same studies. Lefebvre et al. (2013) emphasized very different representations for visual versus nonverbal auditory short-term memory (nonmusical tones) in electrophysiological studies. Looking across all of the research, it seems possible that the brain's parietal representation requires abstract categories, whereas the representation of sensory information that is not categorized requires more sustained attention, activating frontal areas. More work is needed crossing stimulus methods with imaging methods before we will have a full picture.

Another way to show nonverbal, yet categorical memory is to require categorical, verbal responses even for nonverbal items and to show that the representations remain distinctly nonverbal despite the verbal responses. Rowe and colleagues took this approach. Rowe, Philipchalk, and Cake (1974) presented identifiable sound effects or their labels and interfered with the memory with either poetry or music. The poetry interference hurt label memory more whereas the music interference hurt sound effect memory more, indicating two different sources of memory underlying the verbal response. (See also Rowe & Rowe, 1976.)

One could imagine a similar study of working memory for complex visual stimuli that become familiar over trials, yet cannot be named; I cannot think of such work. However, given that short-term memory for visual objects has similar properties in people and pigeons (Gibson, Wasserman, & Luck, 2011), it must not be completely dependent on verbal labeling. Further showing that verbal labeling cannot always account for capacity-limited performance, human studies of memory for simple objects in arrays show little or no effect of a rehearsal suppression task (e.g., Morey & Cowan, 2004).

Tradeoffs between Sensory Modalities

A key difference between sensory and abstract types of working memory is that abstract memory is supposedly based on a limited resource that is shared in common among modalities. Note that this is not necessarily true of sensory and abstract *long-term* memories. There are long-term memories that seem sensory in nature, such as one's memory for a person's voice that sometimes allows recognition over a telephone. One can remember the voice, however, presumably without impeding long-term memory for the way the person looks. In working memory, though, there is likely to be such a tradeoff across modalities. Remembering four exact colors should interfere with remembering four exact tones. There is some evidence in favor of that hypothesis (e.g., Cowan & Morey, 2007; Morey & Bieler, 2013; Morey, Cowan, Morey, & Rouder, 2011; Saults & Cowan, 2007). Still, it is a hypothesis that is in dispute; in his talk at the conference, Marois suggested that the tradeoff between modalities would disappear if all types of structural similarity between modalities were removed (e.g., a tone series versus a spatial layout of visual locations). The issue remains under debate.

Functional magnetic resonance imaging may help. Todd and Marois (2004) showed that there is a brain area that responds more strongly to a larger visual memory load, but only up to the capacity limit measured behaviorally: specifically, the IPS, bilaterally. Cowan et al. (2011) looked for areas across the brain that responded to memory loads consisting of arrays of colored squares, series of spoken letters, or combinations of both. Only one area was found that robustly responded regardless of modality: the left IPS (see Chein et al., 2011; Majerus et al., 2010). This area also responds heavily to attention demands even when working memory is not required (for a review, see Cowan, 2011). These data taken together may suggest that attention is used to form and hold abstract representations of items from different modalities, and that there is a limit in how many abstract representations can be held in this way. More work is needed to verify the prediction that this resource has a fixed limit of several abstract items regardless of the modality of origin, provided that sensory memory is disallowed.

Last, there are some complementary types of evidence suggesting abstract memory for information that may trade off when other modalities are combined (e.g., visual and tactile information: Gallace, Tan, & Spence, 2007).

CONCLUDING OBSERVATIONS

In sum, I have made the argument that there are striking similarities in the operation of a short-lived sensory memory across modalities. There is a brief-lived afterimage that lasts several hundred milliseconds from the beginning of the stimulus and is tied to the minimum duration of a perception (Efron, 1970a, 1970b, 1970c) and of the recognition process (Craig, 1985; Turvey, 1973; Massaro, 1975). There is also a longer recollection of sensation that consists of activated sensory features that seem to lose activation in a matter of seconds. The more common belief that sensory memory simply lasts far longer in audition than in vision (e.g., Darwin et al., 1972) appears problematic, inasmuch as it overlooks many procedures in which the very brief-lived afterimage exists in both of these modalities with comparable time constants.

Abstract representations are formed that have to do with the underlying type of information; for example, phonological representations are formed on the basis of either spoken or written items (Conrad, 1964). I have argued that there is a limit in this type of abstract working memory to three to five items (Anderson et al., 2011; Cowan, 2001; Luck & Vogel, 1997; Zhang & Luck, 2011).

More uncharted territory is in the memory of information that cannot be verbalized or easily categorized. We do not know, for example, if there is a fixed number of unknown voices or visual textures that can be held in mind at some level of precision, if memory for them depends on a fluid attentional resource, or if the limits are defined in some other way. The information of this sort can come either directly from a presented stimulus or perhaps, as I have suggested in this chapter, from sensory imagery.

There are also undeniable differences between the ways the modalities work. For example, information appears to be encoded with better spatial resolution in vision versus better temporal resolution in hearing (see Penney, 1989). The apparent duration of a sensory representation may depend on the type of information required, for example, on tests of temporal versus spatial distinctions (superior in the auditory vs. visual modalities, respectively). Sensory modalities also differ on whether the information basically pertains to what is going on inside the body (in the cases of proprioception, internal feelings, and pain), at the boundary of the organism and the environment (in the cases of touch, taste, and smell), and outside of the organism in the environment (in the cases of hearing and vision). A successful outcome of the symposium would be a reawakening of the field to the importance of sensory information for a better understanding of consciousness and behavior.

Acknowledgments

This work was completed with support from NIH Grant R01-HD21338. Thanks to Pierre Jolicœur, Christine Lefebvre, and an anonymous reviewer for important feedback on an earlier draft of this chapter.

This chapter based on the Symposium Lecture of 25th International Symposium on Attention and Performance, Station de biologie des Laurentides, Quebec, entitled "Mechanisms of Sensory Working Memory." Proceedings editors: Christine LeFebvre, Pierre Jolicœur, and Julio Martinez-Trujillo.

References

Albers, A. M., Kok, P., Toni, I., Dijkerman, H. C., & de Lange, F. P. (2013). Shared representations for working memory and mental imagery in early visual cortex. *Current Biology, 23*, 1427–1431.

Anderson, D. E., Vogel, E. K., & Awh, E. (2011). Precision in visual working memory reaches a stable plateau when individual item limits are exceeded. *The Journal of Neuroscience, 31*, 1128–1138.

Arieh, Y., & Marks, L. E. (2003). Time course of loudness recalibration: Implications for loudness enhancement. *Journal of the Acoustical Society of America, 114*, 1550–1556.

Atkinson, R. C., & Shiffrin, R. M. (1968). Human memory: a proposed system and its control processes. In K. W. Spence, & J. T. Spence (Eds.), *The psychology of learning and motivation: Advances in research and theory* (Vol. 2) (pp. 89–195). New York: Academic Press.

Averbach, E., & Coriell, A. S. (1961). Short-term memory in vision. *Bell System Technical Journal, 40*, 309–328.

Baddeley, A. D. (1986). Working memory. Oxford, England: Clarendon Press.

Baddeley, A. D., & Hitch, G. (1974). Working memory. In G. H. Bower (Ed.), *The psychology of learning and motivation* (Vol. 8) (pp. 47–89). New York: Academic Press.

Balota, D. A., & Duchek, J. M. (1986). Voice-specific information and the 20-second delayed suffix effect. *Journal of Experimental Psychology: Learning, Memory, and Cognition, 12*, 509–516.

Bays, P. M., & Husain, M. (2008). Dynamic shifts of limited working memory resources in human vision. *Science, 321*, 851–854.

Bliss, J. C., Crane, H. D., Mansfield, P. K., & Townsend, J. T. (1966). Information available in brief tactile presentations. *Perception & Psychophysics, 66*, 273–283.

Bredemeiser, K., & Simons, D. J. (2012). Working memory and inattentional blindness. *Psychonomic Bulletin & Review, 19*, 239–244.

Broadbent, D. E. (1958). *Perception and communication*. New York: Pergamon Press.

Chein, J. M., Moore, A. B., & Conway, A. R. A. (2011). Domain-general mechanisms of complex working memory span. *NeuroImage, 54*, 550–559.

Chen, Z., & Cowan, N. (2009). Core verbal working memory capacity: the limit in words retained without covert articulation. *Quarterly Journal of Experimental Psychology, 62*, 1420–1429.

Cherry, E. C. (1953). Some experiments on the recognition of speech, with one and with two ears. *The Journal of the Acoustical Society of America, 25*, 975–979.

Coltheart, M. (1980). Iconic memory and visible persistence. *Perception & Psychophysics, 27*, 183–228.

Conrad, R. (1964). Acoustic confusion in immediate memory. *British Journal of Psychology, 55*, 75–84.

Conway, A. R. A., Cowan, N., & Bunting, M. F. (2001). The cocktail party phenomenon revisited: the importance of working memory capacity. *Psychonomic Bulletin & Review, 8*, 331–335.

Cowan, N. (1984). On short and long auditory stores. *Psychological Bulletin, 96*, 341–370.

Cowan, N. (1987). Auditory sensory storage in relation to the growth of sensation and acoustic information extraction. *Journal of Experimental Psychology: Human Perception and Performance, 13*, 204–215.

Cowan, N. (1988). Evolving conceptions of memory storage, selective attention, and their mutual constraints within the human information processing system. *Psychological Bulletin, 104*, 163–191.

Cowan, N. (2001). The magical number 4 in short-term memory: a reconsideration of mental storage capacity. *Behavioral and Brain Sciences, 24*, 87–185.

Cowan, N. (2011). The focus of attention as observed in visual working memory tasks: Making sense of competing claims. *Neuropsychologia, 49*, 1401–1406.

Cowan, N., Blume, C. L., & Saults, J. S. (2013). Attention to attributes and objects in working memory. *Journal of Experimental Psychology: Learning, Memory, and Cognition, 39*, 731–747.

Cowan, N., Lichty, W., & Grove, T. R. (1990). Properties of memory for unattended spoken syllables. *Journal of Experimental Psychology: Learning, Memory, & Cognition, 16*, 258–269.

Cowan, N., Li, D., Moffitt, A., Becker, T. M., Martin, E. A., Saults, J. S., et al. (2011). A neural region of abstract working memory. *Journal of Cognitive Neuroscience, 23*, 2852–2863.

Cowan, N., & Morey, C. C. (2007). How can dual-task working memory retention limits be investigated? *Psychological Science, 18*, 686–688.

Cowan, N., Rouder, J. N., Blume, C. L., & Saults, J. S. (2012). Models of verbal working memory capacity: what does it take to make them work? *Psychological Review, 119*, 480–499.

Cowan, N., Saults, J. S., & Brown, G. D. A. (2004). On the auditory modality superiority effect in serial recall: separating input and output factors. *Journal of Experimental Psychology: Learning, Memory, and Cognition, 30*, 639–644.

Cowan, N., Winkler, I., Teder, W., & Näätänen, R. (1993). Memory prerequisites of the mismatch negativity in the auditory event-related potential (ERP). *Journal of Experimental Psychology: Learning, Memory, & Cognition, 19*, 909–921.

Cowan, N., & Wood, N. L. (1997). Constraints on awareness, attention, and memory: some recent investigations with ignored speech. *Consciousness and Cognition, 6*, 182–203.

Craig, J. C. (1985). Tactile pattern perception and its perturbations. *Journal of the Acoustical Society of America, 77*, 238–246.

Crowder, R. G. (1989). Imagery for musical timbre. *Journal of Experimental Psychology: Human Perception & Performance, 15*, 472–478.

Crowder, R. G., & Morton, J. (1969). Precategorical acoustic storage. *Perception & Psychophysics, 5*, 365–373.

Darwin, C. J., Turvey, M. T., & Crowder, R. G. (1972). An auditory analogue of the Sperling partial report procedure: evidence for brief auditory storage. *Cognitive Psychology, 3*, 255–267.

Deutsch, D. (1970). Tones and numbers: specificity of interference in immediate memory. *Science, 168*, 1604–1605.

Donkin, C., Nosofsky, R. M., Gold, J. M., & Shiffrin, R. M. (2013). Discrete slot models of visual working-memory response times. *Psychological Review, 4*, 873–902.

Efron, R. (1970a). The relationship between the duration of a stimulus and the duration of a perception. *Neuropsychologia, 8*, 37–55.

Efron, R. (1970b). The minimum duration of a perception. *Neuropsychologia, 8*, 57–63.

Efron, R. (1970c). Effects of stimulus duration on perceptual onset and offset latencies. *Perception and Psychophysics, 8*, 231–234.

Enns, J. T., Brehaut, J. C., & Shore, D. I. (1999). The duration of a brief event in the mind's eye. *Journal of General Psychology, 126*, 355–372.

Farah, M. J. (1985). Psychophysical evidence for a shared representational medium for mental images and percepts. *Journal of Experimental Psychology: General, 114*, 91–103.
Folk, C. L., Remington, R. W., & Johnston, J. C. (1992). Involuntary covert orienting is contingent on attentional control settings. *Journal of Experimental Psychology: Human Perception and Performance, 18*, 1030–1044.
Gallace, A., Tan, H. Z., & Spence, C. (2007). Multisensory numerosity judgments for visual and tactile stimuli. *Perception & Psychophysics, 69*, 487–501.
Gibson, B., Wasserman, E., & Luck, S. J. (2011). Qualitative similarities in the visual short-term memory of pigeons and people. *Psychonomic Bulletin & Review, 18*, 979–984.
Hardman, K., Cowan, N. Remembering complex objects in visual working memory: do capacity limits restrict objects or features? *Journal of Experimental Psychology: Learning, Memory, and Cognition* http://dx.doi.org/10.1037/xlm0000031.
Harrison, S. A., & Tong, F. (2009). Decoding reveals the contents of visual working memory in early visual areas. *Nature, 458*, 632–635.
Kallman, H. J., & Massaro, D. W. (1979). Similarity effects in backward recognition masking. *Journal of Experimental Psychology: Human Perception and Performance, 5*, 110–128.
Kane, M. J., & Engle, R. W. (2002). The role of prefrontal cortex in working-memory capacity, executive attention, and general fluid intelligence: an individual-differences perspective. *Psychonomic Bulletin & Review, 9*, 637–671.
Keller, T. A., Cowan, N., & Saults, J. S. (1995). Can auditory memory for tone pitch be rehearsed? *Journal of Experimental Psychology: Learning, Memory, & Cognition, 21*, 635–645.
Laing, D. G., & Francis, G. W. (1989). The capacity of humans to identify odors in mixtures. *Physiology & Behavior, 46*, 809–814.
Lefebvre, C., Vachon, F., Grimault, S., Thibault, J., Guimond, S., Peretz, I., et al. (2013). Electrophysiological indices of maintenance in auditory and visual short-term memory. *Neuropsychologia, 51*, 2939–2952.
Lewis-Peacock, J. A., Drysdale, A. T., Oberauer, K., & Postle, B. R. (2012). Neural evidence for a distinction between short-term memory and the focus of attention. *Journal of Cognitive Neuroscience, 24*, 61–79.
Li, D., Cowan, N., & Saults, J. S. (2013). Estimating working memory capacity for lists of nonverbal sounds. *Attention, Perception, & Psychophysics, 75*, 145–160.
Luck, S. J., & Vogel, E. K. (1997). The capacity of visual working memory for features and conjunctions. *Nature, 390*, 279–281.
Majerus, S., D'Argembeau, A., Martinez Perez, T., Belayachi, S., Van der Linden, M., Collette, F., et al. (2010). The commonality of neural networks for verbal and visual short-term memory. *Journal of Cognitive Neuroscience, 22*, 2570–2593.
Massaro, D. W. (1975). *Experimental psychology and information processing*. Chicago: Rand McNally.
Massaro, D. W. (1976). Perceptual processing in dichotic listening. *Journal of Experimental Psychology: Human Learning & Memory, 2*, 331–339.
May, J. C., & Tiitinen, H. (2010). Mismatch negativity (MMN), the deviance-elicited auditory deflection, explained. *Psychophysiology, 47*, 66–122.
McCloskey, M., & Watkins, M. J. (1978). The seeing more than is there phenomenon: Implications for the locus of iconic storage. *Journal of Experimental Psychology: Human Perception & Performance, 4*, 553–564.
Mewhort, D. J. K., Campbell, A. J., Marchetti, F. M., & Campbell, J. I. D. (1981). Identification, localization, and "iconic memory": an evaluation of the bar-probe task. *Memory & Cognition, 9*, 50–67.
Miller, G. A. (1956). The magical number seven, plus or minus two: some limits on our capacity for processing information. *Psychological Review, 63*, 81–97.
Miller, G. A., Galanter, E., & Pribram, K. H. (1960). *Plans and the structure of behavior*. New York: Holt, Rinehart and Winston, Inc.
Moray, N. (1959). Attention in dichotic listening: affective cues and the influence of instructions. *Quarterly Journal of Experimental Psychology, 11*, 56–60.
Morey, C. C., & Bieler, M. (2013). Visual short-term memory always requires attention. *Psychonomic Bulletin & Review, 20*, 163–170.
Morey, C. C., & Cowan, N. (2004). When visual and verbal memories compete: evidence of cross-domain limits in working memory. *Psychonomic Bulletin & Review, 11*, 296–301.
Morey, C. C., Cowan, N., Morey, R. D., & Rouder, J. N. (2011). Flexible attention allocation to visual and auditory working memory tasks: manipulating reward induces a tradeoff. *Attention, Perception, & Psychophysics, 73*, 458–472.
Murdock, B. B., Jr. (1968). Modality effects in short-term memory: storage or retrieval? *Journal of Experimental Psychology, 77*, 79–86.
Newell, A., & Simon, H. A. (1956). *The logic theory machine: A complex information processing system*. Santa Monica, CA: Rand Corporation.
Oberauer, K., & Eichenberger, S. (2013). *Visual working memory declines when more features must be remembered for each object. Memory & Cognition*. http://dx.doi.org/10.3758/s13421-013-0333-6. E-publication ahead of print.
Olsson, H., & Poom, L. (2005). Visual memory needs categories. *PNAS, 102*, 8776–8780.
Pashler, H. (1988). Familiarity and visual change detection. *Perception & Psychophysics, 44*, 369–378.
Penney, C. G. (1989). Modality effects and the structure of short-term verbal memory. *Memory & Cognition, 17*, 398–422.
Petrusic, W. M., Baranski, J. V., & Kennedy, R. (1998). Similarity comparisons with remembered and perceived magnitudes: memory psychophysics and fundamental measurement. *Memory & Cognition, 26*, 1041–1055.
Phillips, W. A. (1974). On the distinction between sensory storage and short-term visual memory. *Perception & Psychophysics, 16*, 283–290.
Pratte, M. S., Ling, S., Swisher, J. D., & Tong, F. (2013). How attention extracts objects from noise. *Journal of Neurophysiology, 119*, 1346–1356.
Rostron, A. B. (1974). Brief auditory storage: some further observations. *Acta Psychologica, 38*, 471–482.
Rouder, J. N., Morey, R. D., Cowan, N., Zwilling, C. E., Morey, C. C., & Pratte, M. S. (2008). An assessment of fixed-capacity models of visual working memory. *Proceedings of the National Academy of Sciences (PNAS), 105*, 5975–5979.
Rouder, J. N., Morey, R. D., Morey, C. C., & Cowan, N. (2011). How to measure working-memory capacity in the change-detection paradigm. *Psychonomic Bulletin & Review, 18*, 324–330.
Rowe, E. J., Philipchalk, R. P., & Cake, L. J. (1974). Short-term memory for sounds and words. *Journal of Experimental Psychology, 102*, 1140–1142.
Rowe, E. J., & Rowe, W. G. (1976). Stimulus suffix effects with speech and nonspeech sounds. *Memory & Cognition, 4*, 128–131.
Saults, J. S., & Cowan, N. (2007). A central capacity limit to the simultaneous storage of visual and auditory arrays in working memory. *Journal of Experimental Psychology: General, 136*, 663–684.
Serences, J. T., Ester, E., Vogel, E., & Awh, E. (2009). Stimulus-specific delay activity in human primary visual cortex. *Psychological Science, 20*, 207–214.

Simons, D. J., & Rensink, R. A. (2005). Change blindness: past, present, and future. *Trends in Cognitive Sciences*, *9*, 16–20.
Sligte, I. G., Scholtel, H. S., & Lamme, V. A. F. (2008). Are there multiple visual short-term memory stores? *PLOS One*, *3*(2), 1–9.
Sokolov, E. N. (1963). *Perception and the conditioned reflex*. NY: Pergamon Press.
Sperling, G. (1960). The information available in brief visual presentations. *Psychological Monographs*, *74* (Whole No. 498.).
Sternberg, S. (1966). High-speed scanning in human memory. *Science*, *153*, 652–654.
Sullivan, E., & Turvey, M. (1972). Short-term retention of tactile information. *Quarterly Journal of Experimental Psychology*, *24*, 253–261.
Todd, J. J., & Marois, R. (2004). Capacity limit of visual short-term memory in human posterior parietal cortex. *Nature*, *428*, 751–754.
Treisman, A. M. (1964). Selective attention in man. *British Medical Bulletin*, *20*, 12–16.
Treisman, M., & Rostron, A. B. (1972). Brief auditory storage: a modification of Sperling's paradigm. *Acta Psychologica*, *36*, 161–170.
Turvey, M. T. (1973). On peripheral and central processes in vision: inferences from an information processing analysis of masking with patterned stimuli. *Psychological Review*, *80*, 1–52.
van den Berg, R., Awh, E., & Ma, W. J. (2014). Factorial comparison of working memory models. *Psychological Review*, *121*, 124–149.
van den Berg, R., Shin, H., Chou, W.-C., George, R., & Ma, W. J. (2012). Variability in encoding precision accounts for visual short-term memory limitations. *Proceedings of the National Academy of Sciences*, *109*, 8780–8785.
Verrillo, R. T., & Gescheider, G. A. (1975). Enhancement and summation in the perception of two successive vibrotactile stimuli. *Perception & Psychophysics*, *18*, 128–136.
Vogel, E. K., Woodman, G. F., & Luck, S. J. (2006). The time course of consolidation in visual working memory. *Journal of Experimental Psychology: Human Perception and Performance*, *32*, 1436–1451.
Winkler, I., & Cowan, N. (2005). From sensory to long term memory: evidence from auditory memory reactivation studies. *Experimental Psychology*, *52*, 3–20.
Winkler, I., Schröger, E., & Cowan, N. (2001). The role of large-scale memory organization in the mismatch negativity event-related brain potential. *Journal of Cognitive Neuroscience*, *13*, 59–71.
Woldorff, M. G., Gallen, C. C., Hampson, S. A., Hillyard, S. A., Pantev, C., Sobel, D., et al. (1993). Modulation of early sensory processing in human auditory cortex. *Proceedings of the National Academy of Science*, *90*, 8722–8726.
Zhang, W., & Luck, S. J. (2008). Discrete fixed-resolution representations in visual working memory. *Nature*, *453*, 23–35.
Zhang, W., & Luck, S. J. (2011). The number and quality of representations in working memory. *Psychological Science*, *22*, 1434–1441.

CHAPTER

3

The Brain Mechanisms of Working Memory: An Evolving Story

René Marois

Department of Psychology Vanderbilt University, Nashville, TN, USA

WORKING MEMORY: FUNDAMENTAL CHARACTERISTICS

Working memory (WM) typically refers to the online maintenance and manipulation of information that is no longer present in the environment. It is thought to be central to cognition and has been dubbed "the workspace of consciousness." It is not just a lab curiosity, because its ecological validity in the dynamic perception of natural scenes is well-established (Hollingworth & Hollingworth, 2004; Hollingworth, Williams, & Henderson, 2001; Irwin, 1996). It is distinguished from sensory memory and from long-term memory by three fundamental and interdependent characteristics: time course, limited capacity, and attention-dependence. Although sensory memory (previously named "iconic" memory for visual memory, now usually called visible persistence; see Coltheart, 1980) has a high capacity, it is extremely labile and easy to disrupt and lasts only a few hundred milliseconds (Sperling, 1960). Long-term memory, on the other hand, can be long-lived and its capacity is vast (Von Neumann, 1958). WM is distinct from those two other forms of memory in its duration and limited capacity: it can hold up to only about three or four items, at least in the visual domain (Cowan, 2001; Luck & Vogel, 1997), and it does so for seconds. As such, it is often seen as having a distinct time scale from those of sensory and long-term memory. What is responsible for its limited capacity is not yet clear, but it seems to be related to its third characteristic: its active, attention-dependence. WM performance deteriorates significantly when attention is diverted away from it (Awh & Jonides, 2001; Awh, Jonides, & Reuter-Lorenz, 1998).

The other "working" aspect of working memory—which could be considered its fourth characteristic—is that unlike the other forms of memory, WM does not correspond simply to the maintenance of information, but also to its manipulation (sometimes referred to as "executive" working memory), such as the updating or reordering of information (Baddeley, 1986; Engle, 2002; Fougnie & Marois, 2007). As such, working memory is dubbed the "workbench of consciousness" because it is the cognitive space in which information can be manipulated in accordance to task goals.

As the dissociation between maintenance and manipulation of information in WM implies, it is important to consider that working memory is not a unitary process. At a minimum, there is good evidence to suggest that the stage of encoding information in WM is at least partly distinct from the stage of actively maintaining this information (Fougnie & Marois, 2009; Woodman & Vogel, 2005), and perhaps also from retrieving information from visual WM (VWM) (Fougnie & Marois, 2009).

WORKING MEMORY: THEORETICAL MODELS

It is hard to think about WM in the absence of theoretical constructs (Miyake & Shaw, 1999) since Baddeley and colleagues proposed their influential models of WM, in which WM is compartmentalized into a central executive and slave systems each involved in maintaining specific types (e.g., verbal and visual) information in separate rehearsal

Mechanisms of Sensory Working Memory
http://dx.doi.org/10.1016/B978-0-12-811042-3.00003-7

loops (Baddeley, 1986; Baddeley & Hitch, 1974). More recent versions of that model also incorporate a buffer system that allows WM to bind information across modalities and time (Baddeley, 2000; Baddeley & Logie, 1999). Although Baddeley's model may have been the first and perhaps still most influential WM model, it is not the only one. Another prominent model is that of Cowan (1995, 2001, 2006), in which WM consists of a central focus of attention that gives this memory its limited capacity, and of the activated portion of WM. A similar models is that of Oberauer (2002, 2003), who purported an additional focus of attention can only handle one item at a time.

WORKING MEMORY: NEURAL BASIS

Although some neurobiological studies have been inspired by, or were designed to specifically test, theoretical models of WM (e.g. Buchsbaum & D'Esposito, 2008; Cowan et al., 2011; Nee & Jonides, 2013; Smith & Jonides, 1997, 1999), most of the work has been done independent of, or at least impartial to, the main theoretical constructs. Indeed, the pioneering studies on the brain basis of WM predate any influential theoretical constructs. As early as 1935, pioneering bilateral prefrontal lesion studies in monkeys by Jacobsen (1935, 1936) demonstrated deficits in a simple delayed response task that required the short-term retention of a sensory cue for the performance of a choice response. With these neuropsychological results in mind, the first neurophysiological studies of WM carried out by Fuster and colleagues focused on the prefrontal cortex (Fuster, 1992; Fuster & Alexander, 1971; Goldman-Rakic, 1987). These single-cell studies also used simple delayed response tasks in which the monkey was trained to make a delayed eye movement to the remembered location of a previously presented peripheral cue after the disappearance of a central fixation point (Fuster & Alexander, 1971; Goldman-Rakic, 1995; Kubota & Niki, 1971). The results showed a number of cells in lateral prefrontal cortex with sustained activity during the delay interval, and these were interpreted as single-cell activity correlates of WM. Moreover, trials in which monkeys failed to make a saccade to the correct location were often associated with cessation of delay activity. However, neurophysiological studies also indicated that the prefrontal cortex was not the sole site associated with WM; inferotemporal neurons also showed similar delay activity (Chelazzi, Duncan, Miller, & Desimone, 1988; Fuster & Jervey, 1981).

The emergence of functional neuroimaging confirmed the neurophysiological findings: using positron emission tomography and functional magnetic resonance imaging (fMRI), such studies have revealed robust activity in prefrontal regions during the performance of WM tasks (see, for example, Cohen et al., 1994; D'Esposito et al., 1995; Fiez et al., 1996; Jonides et al., 1993; McCarthy et al., 1994; Petrides, Alivisatos, Meyer, & Evans, 1993; Smith, Jonides, Koeppe, & Awh, 1995). At the same time, however, these studies showed that activity was also present in other cortical brain regions, particularly in the parietal and visual cortex, at least for VWM tasks, leading to the idea that it is a complex network that underlies WM. Importantly, these tasks—such as the n-back task in which the subject must judge whether a currently observed stimulus is the same as the nth-presented stimulus before the current one—were typically much more complex than tasks that simply require the maintenance of memoranda, because they called on several executive processes (Nee et al., 2013). Hence, follow-up studies aimed to investigate WM maintenance independently of WM manipulations (e.g., Courtney, Ungerleider, Keil, & Haxby, 1997) and to dissociate manipulation and maintenance. Such studies consistently show prefrontal activation with executive processes, often accompanied with parietal (intra- or superior) cortex activity. By contrast, maintenance-related activity is much reduced, if at all present, in prefrontal cortex, whereas it is strongly observed in parietal and sensory cortices.

The above findings do not imply that there is no regional processing of different modalities of WM information in the prefrontal cortex (Gruber & von Cramon, 2003). For example, verbal WM processing tends to be left-lateralized, whereas visuospatial WM engages the right prefrontal cortex more (Wager & Smith, 1993), although such activity differences are relative and perhaps even subtle, rather than dramatic. Much debate in the 1990s focused on whether a separate prefrontal dichotomy could also be observed for visuospatial and object WM, with a dorsoventral pattern of representation that reflected the one observed posteriorly in superior parietal and ventral temporal cortices, respectively (e.g., Ungerleider et al., 1998; Smith & Jonides, 1999). However, alternative models of dorsoventral organizations have been proposed, such as WM maintenance/manipulation versus retrieval accounts (D'Esposito, Postle, Ballard, & Lease, 1999; Petrides, 1996). No clear answer arose from this debate (if anything, the extant data favor more complex patterns than those postulated by these simple dichotomies; e.g. Rao, Rainer, & Miller, 1997) and the field has moved on to focusing lately on understanding the rostrocaudal pattern of prefrontal organization (Badre & D'Esposito, 2009; Koechlin, Ody, & Kouneiher, 2003). Moreover, the notion that the prefrontal cortex is critical for WM, at least for WM maintenance, has since been challenged. A review of patients with prefrontal lesions demonstrated that they do not exhibit impairment in the maintenance of information in WM (Müller & Knight, 2006).

SIMPLE EXPERIMENTAL MODELS OF VISUAL WM

Much of the progress in our understanding of visual WM, both from perceptual and neurobiological standpoints, has come from the adoption of simpler paradigms that allowed not only the isolation of WM storage mechanisms from executive processes, but also the manipulation of stimulus parameters to characterize the nature of WM. Thus, such issues as the elementary units of WM storage, the capacity limit of WM storage, the mechanisms underlying the WM maintenance, and the number of WM stores could now be addressed.

The paradigms are variations of a delayed change detection task in which subjects are briefly presented with a number of simple visual objects, followed by a delay of a second or more during which subjects must hold the information in mind before they are to compare that memorandum to a probe display (e.g., Luck & Vogel, 1997). Several variants of this basic delayed matching design have been employed in which the information can be presented sequentially instead of simultaneously, for example, but the principle is the same (e.g., Courtney et al., 1997; Harrison, Jolicœur, & Marois, 2010; Lefebvre et al., 2013).

Using a change detection paradigm with oriented bars as objects, Pessoa, Gutierrez, Bandettini, and Ungerleider (2002) isolated a complex network of frontal, parietal, and occipital brain regions involved in the simple maintenance of information (see also Courtney et al., 1997). Thus, even simply maintaining information may recruit a large neural network. A comprehensive understanding of the brain basis of WM will require, however, that we elucidate the computational contributions associated with each of the components of this network. For example, the neural correlates of the capacity limit of VWM have been investigated in an fMRI experiment by Todd and Marois (2004) using a change detection paradigm by varying the number of colored circled subjects had to maintain in WM. The results revealed a strong correlation between VWM capacity and parietal cortex activity, but not with prefrontal activity. Moreover, the capacity of individual subjects co-varied with posterior parietal activity (Todd & Marois, 2005). Vogel and Machizawa (2004), using an ingenious electrophysiological approach, found similar results, both at the group and individual subject level, at electrode sites in parieto-occipital cortex. Although these studies have pinpointed a neural substrate for VWM capacity, they do not reveal the mechanisms responsible for these limits. Several accounts have been postulated, but there is as yet no consensus on this topic. One possibility is that the capacity of WM is associated with that of attention (e.g., Cowan, 1995, 2001, 2006). In that scheme, the brain basis of VWM capacity should essentially be the same as that associated with attention. Correspondingly, the parietofrontal network isolated with the capacity limits of attending to multiple items (Culham, Cavanagh, & Kanwisher, 2001) overlaps at least partly, with those of VWM capacity (Todd & Marois, 2004). It is possible that attention is a limited resource, and it gets progressively depleted as the number of items or coverage of attention is increased (Alvarez & Cavanagh, 2004; Culham et al., 2001; Müller, Bartelt, Donner, Villringer, & Brandt, 2003). Another possibility is that WM capacity is a result of neural noise (Bays, 2014). Yet another theory of WM capacity is that it is caused by mutual inhibitory interactions between items (Edin et al., 2009), as it may occur in perception (Reynolds, Chelazzi, & Desimone, 1999; Scalf & Beck, 2010; Scalf, Torralbo, Tapia, & Beck, 2013). Although attention may lead to a winner-take-all when only one of the items is attended to (Desimone & Duncan, 1995), it may have little effect on these competitive interactions if spread out across all items (Scalf & Beck, 2010).

Alternatively, and analogously to the mechanisms by which periodic rehearsal of verbal material is thought to be necessary to prevent the verbal items from fading away (Baddeley, 1992; Baddeley, Lewis, & Vallar, 1984), it is possible that attention-based refreshing of visual items in WM is limited by a combination of the time it takes to switch between items, to refresh each item, and the rapidity at which each item may fade away between periods of rehearsal (Baddeley & Logie, 1999; Logie, 1995). A final hypothesis is that WM capacity is dictated by the number of objects that can be maintained in distinct phases of oscillatory neural frequencies (Siegel, Warden, & Miller, 2009), consistent with evidence that there may be a finite number of discrete items that can be maintained in visual WM (Zhang and Luck, 2008). Some of these models are not mutually exclusive, although clearly more work remains before adjudicating between any or some of them. Whatever the mechanism(s) turns out to be, it is clear that it will have to account for the differences in WM capacity within, relative to between, hemispheres (Buschman, Siegel, Roy, & Miller, 2011; Delvenne, 2005).

ATTENTION AND WM

The role of attention in WM has long been established. Diverting attention away from the location or object that is kept in WM during a delay leads to performance impairment (Awh & Jonides, 2001). It is therefore not surprising that the neural correlates of WM overlap extensively with those of attention (Corbetta, Kincade, & Shulman, 2002; Ikkai & Curtis, 2011; LaBar, Gitelman, Parrish, & Mesulam, 1999; Mayer et al., 2007). However, although some theories purport that attention may fully explain WM capacity (Cowan, 2006), other work suggests that this may not be the case (Fougnie & Marois, 2006). Indeed taking attention away from a WM task does not bring performance down to chance (Awh & Jonides, 2001; Awh et al., 1998), and WM maintenance may not strongly depend on sustained attention (Hollingworth & Maxcey-Richard, 2013).

If attention does not fully explain WM capacity, what else may contribute to WM performance? It is possible that passive short-term representations contribute to subjects' WM performance (Sligte, Scholte, & Lamme, 2008), although other studies have cast doubt on the existence of such a memory subsystem (Matsukura & Hollingworth, 2011). Rather, even though many WM studies have used a small pool of items (e.g., six possible colors) in their task to prevent the contribution of long-term memory to WM performance, nevertheless it cannot be ruled out that recent presentations of visual stimuli, even if not in the focus of attention, can persist in a non-working form of memory (Maxcey & Woodman, 2014), perhaps owing to a lingering of activity associated with their presentations through some reverberating cell assembly mechanisms (Hebb, 1949). Hence, a full account of VWM capacity may have to address both the attention-based and passive short-term/long-term memory contributions to VWM.

A SINGLE OR SEVERAL CAPACITY-LIMITED STORES?

If central attention underlies WM capacity, one would expect this capacity to be supramodal or amodal (Cowan, 1995, 2006) and that such a central capacity store may rely on a common brain region or neural network across WM modalities. Consistent with this hypothesis, several studies have shown significant activation overlap across modalities not only for attention tasks (Ivanoff, Branning, & Marois, 2009; Krumbholz, Nobis, Weatheritt, & Fink, 2009; Salmi, Rinne, Degerman, Salonen, & Alho, 2007; Shomstein & Yantis, 2006), but for (auditory and visual) WM tasks as well, at least in the left parietal cortex (Cowan et al., 2011). However, because of the limited spatial resolution of fMRI, activation overlap in neuroimaging studies should not be interpreted as strong evidence for the engagement of common neural ensembles for both modalities. Consistent with this cautionary note, an fMRI study using multi-voxel pattern analysis demonstrated significant sensorimotor modality decoding in a response-selection task in several brain regions including the parietal cortex, which showed activation overlap with univariate analysis (Tamber-Rosenau, Dux, Tombu, Asplund, & Marois, 2013). This suggests that modalities may be coded by at least partly distinct neural ensembles in these brain regions (only two regions of prefrontal cortex consistently failed to decode modalities; see Tamber-Rosenau et al., 2013). Moreover, even with univariate analysis, modality-specific regional activations in the intra-parietal sulcus (IPS) has been observed (Anderson, Ferguson, Lopez-Larson, & Yurgelun-Todd, 2010).

These neurobiological findings of distinct neural coding of sensory modalities are mostly corroborated by behavioral studies. From a behavioral standpoint, if WM capacity is centrally regulated, there should be similar dual-task costs when two concurrently performed WM tasks call on the same or distinct modalities (Saults & Cowan, 2007). Contrary to this expectation, however, most dual-task WM studies have shown lesser interference costs between two WM tasks when they originated from different modalities rather than from the same ones (Cocchini, Logie, Della Sala, MacPherson, & Baddeley, 2002; Fougnie & Marois, 2006; Morey & Cowan, 2004, 2005; Saults & Cowan, 2007; Scarborough, 1972); this suggests that a single amodal store cannot fully account for dual WM task performance (Fougnie & Marois, 2011). Furthermore, a recent study found no dual-task cost at all during the concurrent performance of a visuo-spatial WM task and an auditory object WM task when potential sources of dual-task interference other than competition for a central, capacity-limited store—such as task preparation/coordination, overlap in representational content (e.g., object vs. space-based), or cognitive strategies (e.g., verbalization, stimulus chunking)—were eliminated (Fougnie, Zughni, Godwin, & Marois, 2014). These results are inconsistent with the use of a domain-independent storage system, and suggest instead that there is nothing intrinsic about the functional architecture of the human mind that prevents it from storing two distinct representations in WM. With that viewpoint, interference between two tasks arises only when there is functional overlap between their task sets (similar to the idea of functional distance in dual-task interference, Kinsbourne & Hicks, 1978). Moreover, these behavioral results are highly convergent with event-related potential results, suggesting that the maintenance of auditory and visual information in WM is associated with distinct electrophysiological signals (Lefebvre et al., 2013).

WHERE ARE THE SENSORY WM REPRESENTATIONS STORED?

The notion that WM storage is modality-specific suggests that these representations may be stored in sensory cortex. Consistent with this idea, both Blood Oxygen-level Dependent (BOLD) amplitude measurements (Postle, Stern, Rosen, & Corkin, 2000) and multi-voxel pattern analysis (Christophel, Hebart, & Haynes, 2012; Harrison & Tong, 2009;

Linden, Oosterhof, Klein, & Downing, 2012; Riggall & Postle, 2012; Serences, Ester, Vogel, & Awh, 2009) suggest that VWM representations are held in visual cortex, even at the earliest stages of cortical processing. There is also neurophysiological evidence in support of this assertion (Pasternak & Greenlee, 2005). This accords well with the sensory recruitment hypothesis that VWM representations are perceptual representations maintained after stimulus offset (Ester, Serences, & Awh, 2009; Serences et al., 2009; also see Tsubomi, Fukuda, Watanabe, & Vogel, 2013). In another study, however, Mendoza-Halliday, Torres, and Martinez-Trujillo (2014) argued that WM for motion is not observed in middle temporal (MT) but in an area immediately downstream of MT, namely medial superior temporal cortex. This finding—the dissociation between early sensory visual areas and downstream association areas that also encode WM representations—is inconsistent with a strong form of the sensory recruitment hypothesis. This result nevertheless remains to be reconciled with neuroimaging studies, indicating the recruitment of the earliest sensory cortical area with mnemonic representations of static stimuli (e.g. Harrison & Tong, 2009; Serences et al., 2009).

WHAT ARE THE PARIETAL AND PREFRONTAL CONTRIBUTIONS TO VWM?

Although there is convergent evidence for VWM representations maintained in visual cortex, there is much debate about whether such representations are also held in parietal and prefrontal cortex. Using multi-voxel pattern analysis (MVPA) some authors have failed to detect object representation-specific activity in parietal cortex (Linden et al., 2012; Riggall & Postle, 2012) whereas others have observed such activity (Christophel et al., 2012). It could be argued that such MVPA decoding in parietal cortex is not due to WM representations per se, but rather to the maintenance of task rules, such as for object categories. However, that argument cannot be applied to the results of Christophel et al. (2012) because they used circles with complex spatial patterns of colors as objects. It is conceivable that decoding of object identity in parietal cortex may result from the use of spatial information to distinguish between objects. This hypothesis is consistent with the idea that the parietal cortex is primarily involved in spatial rather than object-based information processing, including WM tasks (Harrison & Tong, 2009). However, neuroimaging studies have shown object-based representations in the parietal cortex (Konen & Kastner, 2008; see also Xu & Jeong, in press), even when at fixation (Xu & Chun, 2006). Moreover, modality-specific decoding has been observed in parietal cortex during the performance of response selection tasks (Tamber-Rosenau et al., 2013), which hints at the prospect that sensory coding of VWM information may also be held in posterior parietal cortex. Alternatively, the parietal cortex activity may reflect the attentional basis of VWM (Emrich, Riggall, LaRocque, & Postle, 2013). In that context, the parietal activation observed with VWM capacity may reflect the capacity limits of attention in VWM. However, it is also possible that the identity of brain regions involved in holding VWM representations depends on the specific representational format to be held in VWM. Thus, the visual cortex and inferior temporal cortex may hold object-specific representations, whereas the parietal cortex may be important for the maintenance of location-specific information or the binding of location- and object identity-specific information. Consistent with this idea, a study has demonstrated that the locus of the storage of task-relevant information in WM is flexibly determined by set goals (Lee, Kravitz, & Baker, 2013).

While much debate rages on the role of the parietal cortex in VWM, there is a greater consensus of findings in the prefrontal cortex. So far, all MVPA studies have failed to see significant object-based decoding in prefrontal cortex (Christophel et al., 2012; Linden et al., 2012; Riggall & Postle, 2012). These results suggest that VWM representations may not be observed in prefrontal cortex. Consistent with these neuroimaging studies, neurophysiological work (Lara & Wallis, 2014) has also failed to show object (color)-specific encoding. Instead, the predominant encoding was spatial attention, which suggests that the lateral prefrontal cortex (LPFC) is involved in the allocation of attention to support WM rather than storing object-relevant information. Moreover, a study has claimed that the LPFC may not even show sustained WM-related activity (Sneve, Magnussen, Alnæs, Endestad, & D'Esposito, 2013), although other work suggests that there is small but significant activation in lateral prefrontal cortex during both the maintenance of attention and working memory (Todd, Han, Harrison, & Marois, 2011; Tamber-Rosenau, Asplund, & Marois, submitted for publication).

When taking the ensemble of results in visual, parietal, and prefrontal cortex, an emerging theory is that the visual cortex stores VWM representations, and frontoparietal cortex activity corresponds to attention that holds the VWM representations held in visual cortex (Emrich et al., 2013; Lara & Wallis, 2014). This model is consistent with the idea that top-down control of attention originates from the frontal, prefrontal, and dorsal parietal cortex (Corbetta & Shulman, 2002; Pessoa, Kastner, Ungerleider, 2003), and with findings that transcranial magnetic stimulation of the posterior parietal cortex affects attentional selection in visual cortex (Blankenburg et al., 2010; Moos, Vossel, Weidner, Sparing, & Fink, 2012; Ruff et al., 2008). There is also evidence, however, for an alternative hypothesis in which the visual and parietal cortex are both involved in storing WM representations but for different stimulus formats, with the visual cortex storing object shape and the parietal cortex storing object location (Harrison et al., 2010; Wager & Smith, 2003; Todd & Marois, 2004). In that context, studies that have associated the parietal cortex with VWM capacity (e.g., Todd & Marois, 2004) may actually reflect the spatial components of the tasks (but see Xu & Chun, 2006) or feature-binding requirements (Shafritz, Gore, & Marois, 2002). This schema is naturally consistent with the sensory system hypothesis.

Importantly these theories are not mutually exclusive. Indeed, it could well be that different regions of the parietal and frontal cortex exhibit different roles in VWM (e.g., Xu & Chun, 2006). After all, the prefrontal cortex has been implicated not only in top-down attention, but also in several executive processes used for manipulating information in WM (Wager & Smith, 2003). In particular, the LPFC has been involved in the encoding and retrieval of information from WM (Todd, Han, Harrison, & Marois, 2011; Sneve et al., 2013). The prefrontal cortex may also come into action in WM tasks when those tasks require maintenance of information in the face of distractors that scramble the representations in visual cortex (Feredoes, Heinen, Weiskopf, Ruff, & Driver, 2011; Katsuki & Constantinidis, 2012; Miller, Erickson, & Desimone, 1996; Qi, Katsuki, Meyer, Rawley, & Zhou, 2010).

CONCLUSIONS

Much progress in our understanding of the neural basis of VWM has taken place in the past two decades. In particular, our view of WM is much richer than it was in the pioneering times of Fuster and Goldman-Rakic. Thus, although the prefrontal cortex is still implicated in WM to this day, recent findings have at once redefined its role (perhaps less seen as important to WM storage) and greatly expanded it (e.g., in executive WM, resistance to distractor interference, attention-based support of WM). At the same time, the prominences of the parietal cortex and visual cortex have surged in the past decade. Working memory reflects brain activity distributed over a widespread network (Postle, 2006), but the computations brought about by each of the core components of this network, and how these components interact to produce WM and explain its fundamental characteristics of active dependence, low capacity, and time-scale remain to be fully understood. In that respect, I surmise that this field will benefit from recent advances in connectivity methods (e.g., Stevens, Tappon, Garg, & Fair, 2012) and frequency analyses (Hsu, Tseng, Liang, Cheng, & Juan, 2014) to develop a global view of how WM emerges from these network interactions. Evidently, whereas the past two decades have led to great strides, there is much to be optimistic about the next decade of research on the neural basis of WM.

References

Alvarez, G. A., & Cavanagh, P. (2004). The capacity of visual short-term memory is set both by visual information load and by number of objects. *Psychological Science, 15*, 106–111.

Anderson, J. S., Ferguson, M. A., Lopez-Larson, M., & Yurgelun-Todd, D. (2010). Topographic maps of multisensory attention. *Proceedings of the National Academy of Sciences of the United States of America, 107*, 20110–20114.

Awh, E., & Jonides, J. (2001). Overlapping mechanisms of attention and spatial working memory. *Trends in Cognitive Sciences, 5*, 119–126.

Awh, E., Jonides, J., & Reuter-Lorenz, P. A. (1998). Rehearsal in spatial working memory. *Journal of Experimental Psychology: Human Perception and Performance, 24*, 780–790.

Baddeley, A. D. (1986). *Working memory*. New York: Oxford University Press.

Baddeley, A. (1992). Working memory. *Science, 255*, 556–559.

Baddeley, A. (2000). The episodic buffer: a new component of working memory? *Trends in Cognitive Sciences, 4*(11), 417–423.

Baddeley, A., & Hitch, D. J. (1974). Working memory. In G. H. Bower (Ed.), *The psychology of learning and motivation: Advances in research and theory* (Vol. 8) (pp. 47–89). New York: Academic Press.

Baddeley, A., Lewis, V., & Vallar, G. (1984). Exploring the articulatory loop. *Quarterly Journal of Experimental Psychology, 36*, 233–252.

Baddeley, A. D., & Logie, R. (1999). Working memory: the multiple component model. In A. Miyake & P. Shah (Eds.), *Models of working memory: Mechanisms of active maintenance and executive control* (pp. 28–61). New York: Cambridge University Press.

Badre, D., & D'Esposito, M. (2009). Is the rostro-caudal axis of the frontal lobe hierarchical? *Nature Reviews Neuroscience, 10*, 659–669.

Bays, P. M. (2014). Noise in neural populations accounts for errors in working memory. *Journal of Neuroscience, 34*, 3632–3645.

Blankenburg, F., Ruff, C. C., Bestmann, S., Bjoertomt, O., Josephs, O., Deichmann, R., et al. (2010). Studying the role of human parietal cortex in visuospa- tial attention with concurrent TMS-fMRI. *Cerebral Cortex, 20*, 2702–2711.

Buchsbaum, B. R., & D'Esposito, M. (2008). The search for the phonological store: from loop to convolution. *Journal of Cognitive Neuroscience, 20*, 762–778.

Buschman, T. J., Siegel, M., Roy, J. E., & Miller, E. K. (2011). Neural substrates of cognitive capacity limitations. *Proceedings of the National Academy of Sciences of the United States of America, 108*, 11252–11255.

Chelazzi, L., Duncan, J., Miller, E. K., & Desimone, R. (1988). Responses of neurons in inferior temporal cortex during memory-guided visual search. *Journal of Neurophysiology, 80*, 2918–2940.

Christophel, T. B., Hebart, M. N., & Haynes, J. D. (2012). Decoding the contents of visual short-term memory from human visual and parietal cortex. *Journal of Neuroscience, 32*, 12983–12989.

Cocchini, G., Logie, R. H., Sala, S. D., MacPherson, S. E., & Baddeley, A. D. (2002). Concurrent performance of two memory tasks: evidence for domain-specific working memory systems. *Memory and Cognition*, *30*, 1086–1095.

Cohen, J. D., Forman, S. D., Braver, T. S., Casey, B. J., Servan-Schreiver, D., & Noll, D. C. (1994). Activation of the prefrontal cortex in a nonspatial working memory task with functional MRI. *Human Brain Mapping*, *1*, 293–304.

Coltheart, M. (1980). Iconic memory and visible persistence. *Perception Psychophys*, *27*, 183–228.

Corbetta, M., Kincade, J. M., & Shulman, G. L. (2002). Neural systems for visual orienting and their relationships to spatial working memory. *Journal of Cognitive Neuroscience*, *14*, 508–523.

Corbetta, M., & Shulman, G. L. (2002). Control of goal-directed and stimulus-driven attention in the brain. *Nature Reviews Neuroscience*, *3*, 201–215.

Courtney, S. M., Ungerleider, L. G., Keil, K., & Haxby, J. V. (1997). Transient and sustained activity in a distributed neural system for human working memory. *Nature*, *386*, 608–611.

Cowan, N. (1995). *Attention and memory*. New York: Oxford University Press.

Cowan, N. (2001). The magical number 4 in short-term memory: a reconsideration of mental storage capacity. *Behavioral and Brain Sciences*, *24*(1), 87–185.

Cowan, N. (2006). *Working memory capacity*. New York: Psychology Press.

Cowan, N., Li, D., Moffitt, A., Becker, T. M., Martin, E. A., Saults, J. S., et al. (2011). A neural region of abstract working memory. *Journal of Cognitive Neuroscience*, *23*, 2852–2863.

Culham, J. C., Cavanagh, P., & Kanwisher, N. G. (2001). Attention response functions: characterizing brain areas using fMRI activation during parametric variations of attentional load. *Neuron*, *32*, 737–745.

D'Esposito, M., Detre, J. A., Alsop, D. C., Shin, R. K., Atlas, S., & Grossman, M. (1995). The neural basis of the central executive system of working memory. *Nature*, *378*, 279–281.

D'Esposito, M., Postle, B. R., Ballard, D., & Lease, J. (1999). Maintenance versus manipulation of information held in working memory: an event-related fMRI study. *Brain Cognition*, *41*, 66–86.

Delvenne, J. F. (2005). The capacity of visual short-term memory within and between hemifields. *Cognition*, *96*, B79–B88.

Desimone, R., & Duncan, J. (1995). Neural mechanisms of selective visual-attention. *Annual Review of Neuroscience*, *18*, 193–222.

Edin, F., Klingberg, T., Johansson, P., McNab, F., Tegnér, J., & Compte, A. (2009). Mechanism for top-down control of working memory capacity. *Proceedings of the National Academy of Sciences of the United States of America*, *106*, 6802–6807.

Emrich, S. M., Riggall, A. C., LaRocque, J. J., & Postle, B. R. (2013). Distributed patterns of activity in sensory cortex reflect the precision of multiple items maintained in visual short-term memory. *Journal of Neuroscience*, *33*, 6516–6523.

Engle, R. W. (2002). Working memory capacity as executive attention. *Current Directions in Psychological Science*, *11*, 19–23.

Ester, E. F., Serences, J. T., & Awh, E. (2009). Spatially global representations in human primary visual cortex during working memory maintenance. *Journal of Neuroscience*, *29*, 15258–15265.

Feredoes, E., Heinen, K., Weiskopf, N., Ruff, C., & Driver, J. (2011). Causal evidence for frontal involvement in memory target maintenance by posterior brain areas during distracter interference of visual working memory. *Proceedings of the National Academy of Sciences of the United States of America*, *108*, 17510–17515.

Fiez, J. A., Raife, E. A., Balota, D. A., Schwarz, J. P., Raichle, M. E., & Petersen, S. E. (1996). A positron emission tomography study of the short-term maintenance of verbal information. *Journal of Neuroscience*, *16*, 808–822.

Fougnie, D., Zughni, S., Godwin, D., & Marois, R. (2014, November 10). Working memory storage is intrinsically domain specific. *Journal of Experimental Psychology: General*. Advance online publication. http://dx.doi.org/10.1037/a0038211.

Fougnie, D., & Marois, R. (2006). Distinct capacity limits for attention and working memory: evidence from attentive tracking and visual working memory paradigms. *Psychological Science*, *17*(6), 526–534.

Fougnie, D., & Marois, R. (2007). Executive working memory load induces inattentional blindness. *Psychonomic Bulletin & Review*, *14*, 142–147.

Fougnie, D., & Marois, R. (2009). Dual-task interference in visual working memory: a limitation in storage capacity but not in encoding or retrieval. *Attention, Perception, & Psychophysics*, *71*, 1831–1841.

Fougnie, D., & Marois, R. (2011). What limits working memory capacity? Evidence for modality-specific sources to the simultaneous storage of visual and auditory arrays. *Journal of Experimental Psychology–Learning Memory and Cognition*, *37*, 1329.

Fuster, J. M. (1992). Prefrontal cortex and memory in primates. In D. Eckroth (Ed.), *Encyclopedia of learning and memory* (pp. 532–536). New York: Macmillan.

Fuster, J. M., & Alexander, G. E. (1971). Neuron activity related to short-term memory. *Science*, *173*, 652–654.

Fuster, J. M., & Jervey, J. P. (1981). Inferotemporal neurons distinguish and retain behaviorally relevant features of visual stimuli. *Science*, *212*, 952–955.

Goldman-Rakic, P. S. (1987). Circuitry of the prefrontal cortex and the regulation of behavior by representational memory. In F. Plum & V. Mountcastle (Eds.), *Handbook of physiology Section 1. The nervous system: Vol. 5.* (pp. 373–417). Bethesda, MD: American Physiological Society.

Goldman-Rakic, P. S. (1995). Cellular basis of working memory. *Neuron*, *14*, 477–485.

Gruber, O., & von Cramon, D. Y. (2003). The functional neuroanatomy of human working memory revisited. Evidence from 3-t fmri studies using classical domain-specific interference tasks. *Neuroimage*, *19*, 797–809.

Harrison, A., Jolicœur, P., & Marois, R. (2010). 'What' and 'Where' in the intraparietal sulcus: an fMRI study of object identity and location in visual short-term memory. *Cerebral Cortex*, *20*, 2478–2485.

Harrison, S. A., & Tong, F. (2009). Decoding reveals the contents of visual working memory in early visual areas. *Nature*, *458*, 632–635.

Hebb, D. (1949). *The organization of behavior; a neuropsychological theory*. New York: Wiley.

Hollingworth, A., & Hollingworth, A. (2004). Constructing visual representations of natural scenes: the roles of short- and long-term visual memory. *Journal of Experimental Psychology: Human Perception and Performance*, *30*, 519–537.

Hollingworth, A., & Maxcey-Richard, A. M. (2013). Selective maintenance in visual working memory does not require sustained visual attention. *Journal of Experimental Psychology: Human Perception and Performance*, *39*, 1047–1058.

Hollingworth, A., Williams, C. C., & Henderson, J. M. (2001). To see and remember: visually specific information is retained in memory from previously attended objects in natural scenes. *Psychonomic Bulletin & Review*, *8*, 761–768.

Hsu, T. Y., Tseng, P., Liang, W. K., Cheng, S. K., & Juan, C. H. (2014). Transcranial direct current stimulation over right posterior parietal cortex changes prestimulus alpha oscillation in visual short-term memory task. *Neuroimage, 98*, 306–313.

Ikkai, A., & Curtis, C. E. (2011). Common neural mechanisms supporting spatial working memory, attention and motor intention. *Neuropsychologia, 49*, 1428–1434.

Irwin, D. E. (1996). Integrating information across saccadic eye movements. *Current Directions in Psychological Science, 5*, 94–100.

Ivanoff, J. I., Branning, P., & Marois, R. (2009). Mapping the pathways of information processing from sensation to action in four distinct sensorimotor tasks. *Human Brain Mapping, 30*, 4167–4186.

Jacobsen, C. F. (1935). Functions of frontal association areas in primates. *Archives of Neurology and Psychiatry, 33*, 558–560.

Jacobsen, C. F. (1936). The functions of the frontal association areas in monkeys. *Comparative Psychology Monographs, 13*, 1–60.

Jonides, J., Smith, E. E., Koeppe, R. A., Awh, E., Minoshima, S., & Mintun, M. A. (1993). Spatial working memory in humans as revealed by PET. *Nature, 363*, 623–625.

Katsuki, F., & Constantinidis, C. (2012). Early involvement of prefrontal cortex in visual bottom-up attention. *Nat Neurosci, 15*, 1160–1166.

Kinsbourne, M., & Hicks, R. E. (1978). Functional cerebral space: a model for overflow, transfer and interference effects in human performance. *Attention and Performance, VII*, 345–362.

Koechlin, E., Ody, C., & Kouneiher, F. (2003). The architecture of cognitive con-trol in the human prefrontal cortex. *Science, 302*, 1181–1185.

Konen, C. S., & Kastner, S. (2008). Two hierarchically organized neural systems for object information in human visual cortex. *Nature Neuroscience, 11*, 224–231.

Krumbholz, K., Nobis, E. A., Weatheritt, R. J., & Fink, G. R. (2009). Executive control of spatial attention shifts in the auditory compared to the visual modality. *Human Brain Mapping, 30*, 1457–1469.

Kubota, K., & Niki, H. (1971). Prefrontal cortical unit activity and delayed alternation performance in monkeys. *Journal of Neurophysiology, 34*, 337–347.

LaBar, K. S., Gitelman, D. R., Parrish, T. B., & Mesulam, M. (1999). Neuroanatomic overlap of working memory and spatial attention networks: a functional MRI comparison within subjects. *NeuroImage, 10*, 695–704.

Lara, A. H., & Wallis, J. D. (2014). Executive control processes underlying multi-item working memory. *Nature Neuroscience, 17*, 876–883.

Lee, S. H., Kravitz, D. J., & Baker, C. I. (2013). Goal-dependent dissociation of visual and prefrontal cortices during working memory. *Nature Neuroscience, 16*, 997–999.

Lefebvre, C., Vachon, F., Grimault, S., Thibault, J., Guimond, S., Peretz, I., et al. (2013). Distinct electrophysiological indices of maintenance in auditory and visual short-term memory. *Neuropsychologia, 51*, 2939–2952.

Linden, D. E., Oosterhof, N. N., Klein, C., & Downing, P. E. (2012). Mapping brain activation and information during category-specific visual working memory. *Journal of Neurophysiology, 107*, 628–639.

Logie, R. H. (1995). *Visuo-spatial working memory*. Hove, UK: Erlbaum.

Luck, S. J., & Vogel, E. K. (1997). The capacity of visual working memory for features and conjunctions. *Nature, 390*, 279–281.

Matsukura, M., & Hollingworth, A. (2011). Does visual short-term memory have a high-capacity stage? *Psychonomic Bulletin & Review, 18*, 1098–1104.

Maxcey, A. M., & Woodman, G. F. (2014). Can we throw information out of visual working memory and does this leave informational residue in long-term memory? *Frontiers in Psychology, 5*, 294.

Mayer, J. S., Bittner, R. A., Nikolić, D., Bledowski, C., Goebel, R., & Linden, D. E. (2007). Common neural substrates for visual working memory and attention. *Neuroimage, 36*, 441–453.

McCarthy, G., Blamire, A. M., Puce, A., Nobre, A. C., Bloch, G., Hyder, F., et al. (1994). Functional magnetic resonance imaging of human prefrontal cortex activation during a spatial working memory task. *Proceedings of the National Academy of Sciences of the United States of America, 91*, 8690–8694.

Mendoza-Halliday, D., Torres, S., & Martinez-Trujillo, J. C. (2014). Sharp emergence of feature-selective sustained activity along the dorsal visual pathway. *Nature Neuroscience, 17*, 1255–1262.

Miller, E. K., Erickson, C. A., & Desimone, R. (1996). Neural mechanisms of visual working memory in prefrontal cortex of the macaque. *Journal of Neuroscience, 16*, 5154–5167.

Miyake, A., & Shah, P. (Eds.). (1999). *Models of working memory*. Cambridge, UK: Cambridge Univesity Press.

Moos, K., Vossel, S., Weidner, R., Sparing, R., & Fink, G. R. (2012). Modulation of top-down control of visual attention by cathodal tDCS over right IPS. *Journal of Neuroscience, 32*, 16360–16368.

Morey, C. C., & Cowan, N. (2004). When visual and verbal memories compete: evidence of cross-domain limits in working memory. *Psychonomic Bulletin & Review, 11*, 296–301.

Morey, C. C., & Cowan, N. (2005). When do visual and verbal memories conflict? The importance of working-memory load and retrieval. *Journal of Experimental Psychology: Learning, Memory, and Cognition, 31*, 703–713.

Müller, N. G., Bartelt, O. A., Donner, T. H., Villringer, A., & Brandt, S. A. (2003). A physiological correlate of the "zoom lens" of visual attention. *Journal of Neuroscience, 23*, 3561–3565.

Müller, N. G., & Knight, R. T. (2006). The functional neuroanatomy of working memory: contributions of human brain lesion studies. *Neuroscience, 139*, 51–58.

Nee, D. E., Brown, J. W., Askren, M. K., Berman, M. G., Demiralp, E., Krawitz, A., et al. (2013). A meta-analysis of executive components of working memory. *Cerebral Cortex, 23*, 264–282.

Nee, D. E., & Jonides, J. (2013). Neural evidence for a 3-state model of visual short-term memory. *NeuroImage, 74*, 1–11.

Oberauer, K. (2002). Access to information in working memory: exploring the focus of attention. *Journal of Experimental Psychology Learning Memory and Cognition, 28*, 411–421.

Oberauer, K. (2003). Selective attention to elements in working memory. *Experimental Psychology, 50*, 257–269.

Pasternak, T., & Greenlee, M. W. (2005). Working memory in primate sensory systems. *Nature Reviews Neuroscience, 6*, 97–107.

Pessoa, K., Gutierrez, E., Bandettini, P. A., & Ungerleider, L. G. (2002). Neural correlates of visual working memory: fMRI amplitude predicts task performance. *Neuron, 35*, 975–987.

Pessoa, L., Kastner, S., & Ungerleider, L. G. (2003). Neuroimaging studies of attention: from modulation of sensory processing to top-down control. *Journal of Neuroscience, 23*, 3990–3998.

Petrides, M. (1996). Specialized systems for the processing of mnemonic information within the primate frontal cortex. *Philosophical Transactions of the Royal Society of London B Biological Sciences, 351*, 1455–1461.

Petrides, M., Alivisatos, B., Meyer, E., & Evans, A. C. (1993). Functional activation of the human frontal cortex during the performance of verbal working memory tasks. *Proceedings of the National Academy of Sciences of the United States of America, 90*, 878–882.

Postle, B. R. (2006). Working memory as an emergent property of the mind and brain. *Neuroscience, 139*, 23–38.

Postle, B. R., Stern, C. E., Rosen, B. R., & Corkin, S. (2000). An fMRI investigation of cortical contributions to spatial and nonspatial visual working memory. *NeuroImage, 11*, 409–423.

Qi, X. L., Katsuki, F., Meyer, T., Rawley, J. B., Zhou, X., Douglas, K. L., et al. (2010). Comparison of neural activity related to working memory in primate dorsolateral prefrontal and posterior parietal cortex. *Frontiers in Systems Neuroscience, 4*, 12.

Rao, S. C., Rainer, G., & Miller, E. K. (1997). Integration of what and where in the primate prefrontal cortex. *Science, 276*, 821–824.

Reynolds, J. H., Chelazzi, L., & Desimone, R. (1999). Competitive mechanisms subserve attention in macaque areas V2 and V4. *Journal of Neuroscience, 19*, 1736–1753.

Riggall, A. C., & Postle, B. R. (2012). The relation between working memory storage and elevated activity, as measured with functional magnetic resonance imaging. *Journal of Neuroscience, 32*, 12990–12998.

Ruff, C. C., Bestmann, S., Blankenburg, F., Bjoertomt, O., Josephs, O., Weiskopf, N., et al. (2008). Distinct causal influences of parietal versus frontal areas on human visual cortex: evidence from concurrent TMS-fMRI. *Cerebral Cortex, 18*, 817–827.

Salmi, J., Rinne, T., Degerman, A., Salonen, O., & Alho, K. (2007). Orienting and maintenance of spatial attention in audition and vision: multimodal and modality-specific brain activations. *Brain Structure and Function, 212*, 181–194.

Saults, J. S., & Cowan, N. (2007). A central capacity limit to the simultaneous storage of visual and auditory arrays in working memory. *Journal of Experimental Psychology: General, 136*, 663–684.

Scalf, P. E., & Beck, D. M. (2010). Competition in visual cortex impedes attention to multiple items. *Journal of Neuroscience, 30*, 161–169.

Scalf, P. E., Torralbo, A., Tapia, E., & Beck, D. M. (2013). Competition explains limited attention and perceptual resources: implications for perceptual load and dilution theories. *Frontiers in Psychology, 10*, 243.

Scarborough, D. L. (1972). Memory for brief visual displays of symbols. *Cognitive Psychology, 3*, 408–429.

Serences, J. T., Ester, E. F., Vogel, E. K., & Awh, E. (2009). Stimulus-specific delay activity in human primary visual cortex. *Psychological Science, 20*, 207–214.

Shafritz, K. M., Gore, J. C., & Marois, R. (2002). The role of the parietal cortex in visual feature binding. *Proceedings of the National Academy of Sciences of the United States of America, 99*, 10917–10922.

Shomstein, S., & Yantis, S. (2006). Parietal cortex mediates voluntary control of spatial and nonspatial auditory attention. *Journal of Neuroscience, 26*, 435–439.

Siegel, M., Warden, M. R., & Miller, E. K. (2009). Phase-dependent neuronal coding of objects in short-term memory. *Proceedings of the National Academy of Sciences of the United States of America, 106*, 21017–21018.

Sligte, I. G., Scholte, H. S., & Lamme, V. A. (2008). Are there multiple visual short term memory stores? *PLoS One, 3*(2), e1699.

Smith, E. E., & Jonides, J. (1997). Working memory: a view from neuroimaging. *Cognitive Psychology, 33*, 5–42.

Smith, E. E., & Jonides, J. (1999). Neuroscience – storage and executive processes in the frontal lobes. *Science, 283*, 1657–1661.

Smith, E. E., Jonides, J., Koeppe, R. A., & Awh, E. (1995). Spatial versus object working memory: PET investigations. *Journal of Cognitive Neuroscience, 7*, 337–356.

Sneve, M. H., Magnussen, S., Alnæs, D., Endestad, T., & D'Esposito, M. (2013). Top-down modulation from inferior frontal junction to FEFs and intraparietal sulcus during short-term memory for visual features. *Journal of Cognitive Neuroscience, 25*, 1944–1956.

Sperling, G. (1960). The information available in brief visual presentations. *Psychological Monographs: General and Applied, 74*(11), 1–29.

Stevens, A. A., Tappon, S. C., Garg, A., & Fair, D. A. (2012). Functional brain network modularity captures inter- and intra-individual variation in working memory capacity. *PLoS One, 7*(1), e30468.

Tamber-Rosenau, B. J., Asplund, C. L., & Marois R. *Functional dissociation of the lateral prefrontal cortex and the dorsal attention network in top-down attentional control*. Department of Psychology, Vanderbilt University, submitted for publication. Available from the authors upon request.

Tamber-Rosenau, B. J., Dux, P. E., Tombu, M. N., Asplund, C. L., & Marois, R. (2013). Amodal processing in human prefrontal cortex. *Journal of Neuroscience, 33*, 11573–11587.

Todd, J. J., & Marois, R. (2005). Posterior parietal cortex activity predicts individual differences in visual short-term memory capacity. *Cognitive, Affective & Behavioral Neuroscience, 5*, 144–155.

Todd, J. J., Han, S. W., Harrison, S., & Marois, R. (2011). The neural correlates of visual working memory encoding: a time-resolved fMRI study. *Neuropsychologia, 49*, 1527–1536.

Todd, J. J., & Marois, R. (2004). Capacity limit of visual short-term memory in human posterior parietal cortex. *Nature, 428*(6984), 751–754.

Tsubomi, H., Fukuda, K., Watanabe, K., & Vogel, E. K. (2013). Neural limits to representing objects still within view. *Journal of Neuroscience, 33*, 8257–8263.

Ungerleider, L. G., Courtney, S. M., & Haxby, J. V. (1998). A neural system for human visual working memory. *Proceedings of the National Academy of Sciences of the United States of America, 95*, 883–890.

Vogel, E. K., & Machizawa, M. G. (2004). Neural activity predicts individual differences in visual working memory capacity. *Nature, 428*, 748–751.

Von Neumann, J. (1958). *The computer and the brain*. Yale University Press.

Wager, T. D., & Smith, E. E. (2003). Neuroimaging studies of working memory: a meta-analysis. *Cognitive, Affective & Behavioral Neuroscience, 3*, 255–274.

Woodman, G. F., & Vogel, E. K. (2005). Fractionating working memory: consolidation and maintenance are independent processes. *Psychological Science, 16*, 106–113.

Xu, Y. D., & Chun, M. M. (2006). Dissociable neural mechanisms supporting visual short-term memory for objects. *Nature, 440*, 91–95.

Xu, Y., & Jeong, S. The contribution of human superior intra-parietal sulcus to visual short-term memory and perception. In P. Jolicœur & J. Martinez-Trujillo (Eds.), *Mechanisms of sensory working memory: Attention and performance XXV*, in press.

Zhang, W., & Luck, S. J. (2008). Discrete fixed-resolution representations in visual working memory. *Nature, 453*, 23–35.

CHAPTER

4

The Contribution of Human Superior Intraparietal Sulcus to Visual Short-Term Memory and Perception

Yaoda Xu, Su Keun Jeong

Harvard University, Cambridge, MA, USA

INTRODUCTION

Visual short-term memory (VSTM) is a short-term memory buffer that temporarily stores visual information (Phillips, 1974). It has a durable but very limited capacity (Luck & Vogel, 1997; Pashler, 1988; Phillips, 1974). Although VSTM has often been studied in isolation, it is an integral part of visual perception (Xu, 2002). This is because, theoretically, it would not be possible to separate perception from VSTM, because for perception to occur, visual information has to be encoded in some kind of short-term memory buffer before further processing can take place. In practice, the paradigms employed to study perception often ask observers to report a briefly presented visual stimulus, necessarily engaging VSTM, making any results obtained a reflection of both sensory processing and VSTM characteristics. In cognitive neuroscience research, the brain areas engaged in sensory processing are found to also participate in VSTM information maintenance (e.g., Harrison & Tong, 2009; Xu & Chun, 2006). As such, VSTM is tightly integrated with visual perception.

In everyday visual perception, we are often faced with a huge number of visual inputs, some being essential to task performance and some being just pure distractions. To ensure proper task performance, it is critical that our visual system selectively retains and processes what is most relevant to the current goals and thoughts of the observer. As such, by examining what is stored in VSTM and the control process that determines what is stored there, we can gain an in-depth understanding of how goal-directed visual perception is accomplished. VSTM is therefore not an isolated cognitive operation. But rather, understanding the characteristics of VSTM will provide us with better knowledge of how visual perception works in general.

Over the years, researchers have attempted to understand what limits VSTM capacity. Although some have proposed a slot-like representation in which a fixed number of about three or four visual objects can be represented in VSTM (Cowan, 2001; Luck & Vogel, 1997; Vogel, Woodman, & Luck, 2001; Zhang & Luck, 2008), others have argued instead that its resources can be flexibly divided among objects and, depending on the encoding demands, information from more than three or four objects may be represented in VSTM (Alvarez & Cavanagh, 2004; Bays & Husain, 2008; van den Berg, Shin, Chou, George, & Ma, 2012; Wilken & Ma, 2004). As supporting experimental evidence exists for both accounts, it seems that neither account alone can fully accommodate all the experimental findings, how should we then understand what determines VSTM capacity limit?

The human brain has evolved to accommodate the demands of information processing within. As such, examining brain activation associated with a particular task can provide us with vital clues about the processing algorithm employed in accomplishing the task. Prior neuroimaging research has identified a brain region in human intraparietal sulcus (IPS) that tracks the number of items stored in VSTM (Todd & Marois, 2004, 2005; see also a related event-related brain potential finding from Vogel & Machizawa, 2004). The discovery of this neural substrate provided a unique opportunity for us to

Mechanisms of Sensory Working Memory
http://dx.doi.org/10.1016/B978-0-12-811042-3.00004-9

understand whether slots or flexible resources would best characterize information representation in VSTM. By manipulating the complexity of the visual objects encoded and thereby the encoding demands of each object, Xu and Chun (2006) found that although a region in the inferior IPS tracks a fixed number of about four objects regardless of object complexity, a region in superior IPS (corresponding to the IPS area identified by Todd & Marois, 2004) tracked the number of objects successfully retained in VSTM, which was variable depending on the encoding demands of each object. Consistent with this functional magnetic resonance imaging (fMRI) finding, a recent event-related brain potential study reported that electrodes at different sites showed dissociable effects corresponding to both the slot-like and the flexible resource-based representations in VSTM (Wilson, Adamo, Barense, & Ferber, 2012). Thus, both types of representations can play a role in VSTM information representation, with slot-like representation involved in object selection and flexible resource-based representation involved in the encoding and retention of information in VSTM.

Based on this initial neuroimaging result and subsequent findings (Xu, 2007, 2008, 2009; Xu & Chun, 2006, 2007) and existing ideas and results from behavioral studies, Xu and Chun (2009) proposed the neural object file theory and argued that there exists two distinctive stages of visual information processing whenever multiple visual objects need to be selected and encoded. The first stage is object individuation and involves the inferior IPS. Here a fixed number of about four objects from a crowded scene are selected based on their spatial information. The second stage is object identification and involves superior IPS and higher visual areas. Here details of the selected objects are encoded and retained. This theory not only resolves the debate regarding what determines VSTM capacity limit, but also accounts for a number of other (sometimes puzzling) behavioral results, such as those obtained in multiple visual object tracking in which observers failed to detect obvious feature changes on successfully attended and tracked objects (Bahrami, 2003). The neural object file theory also bridges studies on object perception in humans after parietal brain lesion (Coslett & Saffran, 1991) and the development of object concepts in infants (Leslie, Xu, Tremoulet, & Scholl, 1998) (for details of how the neural object file theory accounts for all of these, see Xu & Chun, 2009).

The involvement of superior IPS in VSTM information encoding and retention suggests that this brain region can represent a variety of visual information dynamically based on the task demands. This echoes findings from monkey neurophysiological studies in which neurons in lateral intraparietal (LIP) sulcus have been shown to dynamically encode behaviorally relevant visual stimuli (Gottlieb, Kusunoki, & Goldberg, 1998; Toth & Assad, 2002). Meanwhile, existing neuropsychological and neuroimaging studies have associated human parietal cortex primarily with attention-related processing (Corbetta & Shulman, 2002; Szczepanski, Konen, & Kastner, 2010; Wojciulik & Kanwisher, 1999; Yantis et al., 2002). As such, although Xu and Chun (2009) imply that information can be directly represented in IPS during visual processing, it is also possible that parietal activation simply tracks the deployment of attentional resources without actually carrying detailed visual representations.

Recent development in fMRI multivoxel pattern analysis (MVPA) has enabled researchers to decode fMRI response patterns and gain a better understanding of the nature of information representation in specific brain regions (Cox & Savoy, 2003; Haxby et al., 2001; Haynes & Rees, 2006; Norman, Polyn, Detre, & Haxby, 2006; Peelen & Downing, 2007). Using this approach, here we aim to understand whether the content of VSTM could be successfully decoded from fMRI response patterns in the superior IPS or whether superior IPS response patterns are oblivious to what is stored in VSTM. We were able to successfully decode object information in the superior IPS, showing that visual information can be directly represented in the superior IPS. Importantly, this representation is task dependent and reflects the encoding of visual information that is needed for the successful performance of the current task. These results are thus consistent with those obtained from fMRI response amplitude measures (e.g., Todd & Marois, 2004; Xu & Chun, 2006) and support the notion that the superior IPS likely plays a key role in mediating the moment-to-moment, goal-directed visual information representation in the human brain.

MATERIALS AND METHODS

Participants

Five observers (3 females, mean age = 31.8 years, standard deviation = 1.92) participated in Experiments 1 and 2, with at least a two-month separation. One additional observer (1 male) participated in Experiment 2 in an effort to see whether adding more observers would change the results of Experiment 2 in any substantial way (which it did not). Because this observer then left the area and was unable to come back to participate in Experiment 1, his data from Experiment 2 were excluded from the final analyses. One more observer was scanned in Experiment 2, but excluded from data analysis because of excessive head motion (greater than 5 mm). In Experiment 2, two observers' behavioral data for the main experiment were not recorded because of equipment failure.

All observers had normal, or corrected-to-normal, visual acuity and reported no history of neurological impairment. They were recruited from the Harvard University community and received payments for their participation in the experiments. All observers gave their informed consent before their participation in the experiment. The study was approved by the institutional review board of Harvard University.

Experimental Design

Main Experiments

During both Experiments 1 and 2, observers viewed a sequential presentation of 10 unique object images either all above or all below the central fixation dot (see Figure 1). The 10 images were drawn from the same object category and shared the same general shape contour (e.g., 10 side-view shoe images, see Figure 1). Different shape categories were viewed in different trial blocks. In Experiment 1, observers viewed the images and detected an immediate repetition of the same image (one-back task), requiring them to store each image in VSTM. In Experiment 2, instead of the one-back task, observers viewed the images and detected the direction of object motion that occurred randomly twice in each block of trials, making neither object shape nor location task relevant.

Two square-shaped white placeholders were present above and below the central fixation during each block to mark the two object locations. The white placeholders subtended 7.0° × 4.7°, with the distance between the fixation and the center of each placeholder being 3.2°. Each object exemplar subtended approximately 5.5° × 2.8°. Each stimulus block lasted 8 s and contained 10 unique images, with each image appearing for 300 ms followed by a 500 ms blank display. Fixation blocks, lasting 8 s, were inserted at the beginning and end of the run, as well as between adjacent stimulus blocks. The presentation order of the different stimulus blocks and that of the images within each block were chosen randomly for each run.

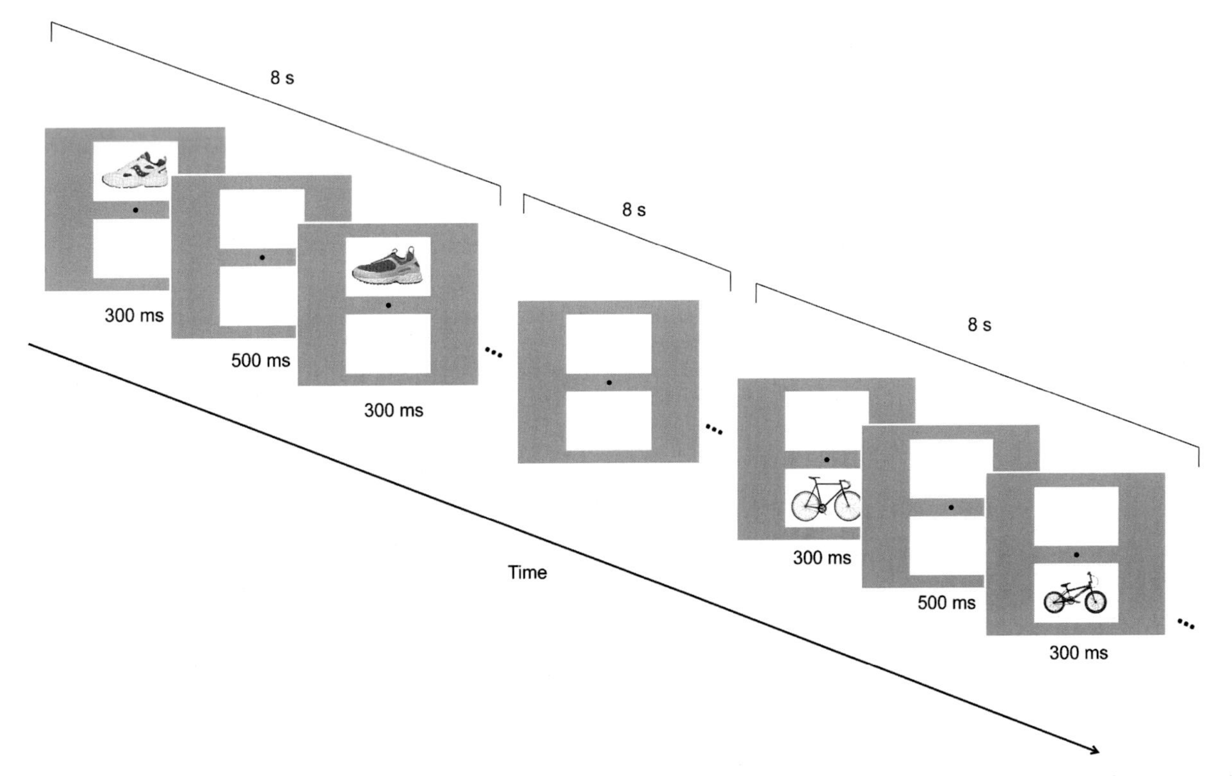

FIGURE 1 Example stimuli and trial structure. In both Experiments 1 and 2, observers viewed a sequential presentation of 10 unique object images either all above or all below the central fixation dot. The 10 images were drawn from the same object category and shared the same general shape contour. Different shape categories were viewed in different trial blocks. In Experiment 1, observers viewed the images and detected an immediate repetition of the same image, making object shape, but not location, task-relevant. In Experiment 2, observers passively viewed the images and detected the direction of an occasional image jitter (either horizontal or vertical), making neither object shape nor location task relevant.

In Experiment 1, two object categories were used, namely shoe and bike. Each run contained three blocks of trials for each category at each location. Each observer was tested with six runs, each lasting 3 min 28 s. Thus, for each observer, a total of 72 blocks were collected with 18 blocks for each object category at each of the two locations.

In Experiment 2, four object categories were used, namely shoe, bike, guitar, and couch. Each stimulus block appeared once in each run. There were also stimulus conditions in which two objects appeared simultaneously with one above and one below the fixation. These conditions were intended for a different study and were excluded from the present analysis. Each observer had 8 to 10 runs, each lasting 5 min 36 s. Thus, for each observer, a total of 64–80 blocks were collected with 8–10 blocks for each object category at each of the two locations.

Superior IPS Localizer

To localize superior IPS, following previous studies (Todd & Marois, 2004; Xu & Chun, 2006), we conducted an object VSTM experiment. Observers were asked to remember which object category was shown at which location and to judge whether the probe object (a new object) shown in the test display matched the category of the object shown at the same location in the sample display. A given sample display contained one to four unique objects appearing in four possible locations around the central fixation, with each object being an exemplar from a different category. The four possible object locations were located above, below, or to the right or left of the central fixation. These locations were marked with white placeholders visible throughout the trial. For no-change trials, the probe object in the test display would match the object category shown at the same location in the sample display but would be a different exemplar from the same category; and for change trials, the probe object would be an exemplar from a different category. Gray-scale photographs of everyday objects from four categories (shoes, bikes, guitars, and couches) were used as stimuli.

Each trial lasted 6 s and included a fixation period (1000 ms), a sample display (200 ms), a delay period (1000 ms), a test display/response period (2500 ms), and a feedback (1300 ms). The display subtended approximately 12° × 12°, with the distance between the fixation and the center of each object being 4° and the size of the placeholder being 4.5° × 3.6°. With a counterbalanced trial history design, each run contained 15 stimulus trials for each set size and 15 fixation trials in which only the central fixation dot appeared for 6 s. Three filler trials were added for practice and trial history balancing purposes (two at the beginning and one at the end of the run), but were excluded from data analysis. Each observer was tested with two runs, each lasting 8 min.

Inferior IPS/Lateral Occipital (LO) Localizer

Following previous studies (Xu & Chun, 2006, 2007; Xu, 2007, 2008, 2009), observers viewed blocks of objects and noise images. The object images were the set size four displays from the superior IPS localizer. For the noise images, we took the object images but phase-scrambled each component object. Each block lasted 16 s and contained 20 images, each appearing for 500 ms followed by a 300-ms blank display. Observers were asked to detect the direction of a slight spatial jitter (either horizontal or vertical), which occurred randomly once in every 10 images. Each run contained eight object blocks and eight noise blocks. Each observer was tested with two or three runs, each lasting 4 min 40 s.

To ensure proper fixation, we monitored observers' eye movement in the main experiments by using an EyeLink 1000 eye tracker.

fMRI Methods

fMRI data were acquired from a Siemens Tim Trio 3T scanner. Observers viewed images back projected onto a screen at the rear of the scanner bore through an angled mirror mounted on the head coil. All experiments were controlled by an Apple MacBook Pro running Matlab with Psychtoolbox extensions (Brainard, 1997). For anatomical images, high-resolution 144 T1-weighted images were acquired (echo time, 1.54 ms; flip angle, 7°; matrix size, 256 × 256; repetition time, 2200 ms; voxel size, 1 × 1 × 1 mm). For the functional images, T2*-weighted echo-planar gradient echo sequence was used (Huettel, Song, & McCarthy, 2009). For the two main experiments and the inferior IPS/lateral occipital (LO) localizer runs, 31 near axial slices (3 × 3 mm in plane, 3 mm thick, 0 skip, interleaved acquisition) were acquired (echo time, 30 ms; flip angle, 90°; matrix size, 72 × 72; repetition time, 2000 ms). The number of volumes collected for Experiment 1, Experiment 2, and the inferior IPS/LO localizer runs were 168, 104, and 140, respectively. For superior IPS localizer runs, 24 slices (3 × 3 mm in plane, 5 mm thick, 0 skip) parallel to the AC-PC line were acquired (volumes, 320; echo time, 29 ms; flip angle, 90°; matrix size, 72 × 72; repetition time, 1500 ms).

Data Analysis

fMRI data were analyzed in native space with BrainVoyager QX (http://www.brainvoyager.com). Data preprocessing included three-dimensional motion correction, slice acquisition time correction, linear trend removal, and the removal of the first two volumes of each functional run. No spatial smoothing or other data preprocessing was applied.

fMRI data from the localizer runs were analyzed using general linear models. All regions of interest (ROIs) were defined separately in each individual observer. Following previously established procedures (Xu & Chun, 2006), LO and inferior IPS ROIs were defined as voxels showing higher activations to the object than to the noise displays (false discovery rate (FDR) $q < 0.05$, corrected for serial correlation) in lateral occipital cortex and the inferior part of the IPS bordering the traverse occipital sulcus, respectively. As was done before (Todd & Marois, 2004; Xu & Chun, 2006), superior IPS was defined as voxels tracking each observer's behavioral VSTM capacity. This was achieved by first calculating each observer's behavioral VSTM capacity based on Cowan's K formula (Cowan, 2001), and then in a multiple regression analysis weighing the regression coefficient for each set size with that observer's behavioral VSTM capacity for that set size. Superior IPS was defined as voxels showing significant activations in the regression analysis (FDR $q < 0.05$, corrected for serial correlation).

During MVPA analysis, for each observer, we overlaid the ROIs onto the raw fMRI data from the main experiments, applied a general linear model to the data, and extracted the resulting beta-weights for each stimulus condition in each voxel of each ROI. These beta-weights then served as the input to our MVPA analysis. Following Haxby et al. (2001), we first divided the input data into odd and even runs and then normalized the data within each ROI by removing the mean across all stimulus conditions. This normalization procedure was carried out based on the assumption that the neural response pattern for each stimulus condition is a linear combination of the shared pattern across all conditions and a distinctive pattern unique to that condition. As such, subtracting the mean response would eliminate the contribution of the shared pattern and amplify neural pattern differences among the different stimulus conditions. This normalization method could result in negative correlation between conditions. But because the main dependent variable of interest is the relative differences between correlations rather than the absolute values of the correlations themselves, this normalization procedure provides a good way to illustrate the results and has been used successfully in previous studies (Haxby et al., 2001; Schwarzlose, Swisher, Dang, & Kanwisher, 2008). Because this normalization procedure may distort results when the number of stimulus conditions is small (Garrido, Vaziri-Pashkam, Nakayama, & Wilmer, 2013), we also analyzed data without normalization and with z-scoring to remove amplitude differences among the different conditions. Both analyses revealed similar results as those obtained with the normalization procedure.

To decode the nature of representation contained in each ROI, we correlated voxel response patterns from odd runs with those from even runs across all stimulus condition pairs. This resulted in four correlation conditions: both same (same location and same shape pairs), location change (same shape and different location pairs), shape change (different shape and same location pairs), and both change (different shape and different location pairs). The resulting correlation coefficients were Fisher-transformed to ensure normal distribution of the values and then subjected to a repeated-measures analysis of variance with shape (same, different) and location (same, different) as factors. To directly show shape and location representation in each ROI, we also obtained the main effect of shape by calculating the difference between conditions that shared the same shape and those that did not, averaging over both the location same and change conditions. Similarly, we obtained the main effect of location by calculating the difference between conditions that shared the same location and those that did not, averaging over both the shape same and change conditions.

RESULTS

In Experiment 1, while undergoing an fMRI scan, observers viewed a sequential presentation of 10 unique object images either all above or all below the central fixation dot (see Figure 1). The 10 images were drawn from the same object category and shared the same general shape contour (e.g., 10 side-view shoe images, see Figure 1). Different shape categories (shoes and bikes) were viewed in different trial blocks. Observers viewed the images and detected an immediate repetition of the same image, requiring them to store each image in VSTM. Object shape, but not location, was thus task-relevant. Observers were fairly accurate in detecting shape repetitions and had an average behavioral accuracy of 88.40 ± 6.10%.

To examine information representations in a brain region, following Haxby and colleagues (Haxby et al., 2001), after normalizing fMRI response amplitudes to the mean of all the stimulus conditions, we calculated correlation coefficient between fMRI response patterns from odd and even runs (see Methods). If a brain region contains distinct neural representations resulting in distinct fMRI response patterns for different object shapes, then higher correlation coefficients of

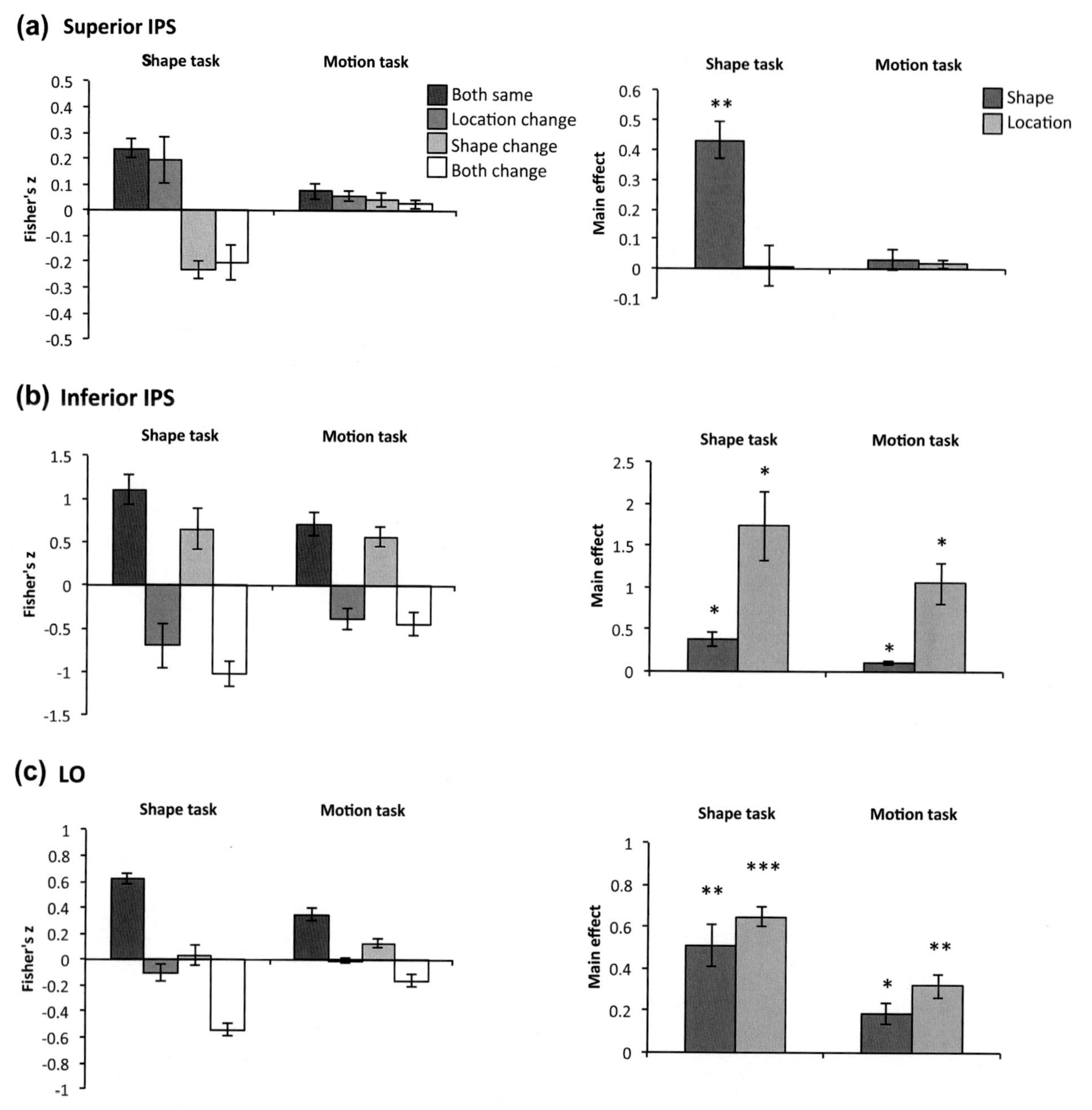

FIGURE 2 (Left column) Results from Experiments 1 and 2 showing Fisher-transformed correlation coefficients of functional magnetic resonance imaging response pattern correlations (i.e., decoding) between pairs of stimulus conditions from odd and even runs. Each pair could share both shape and location (both), shape but not location (location change), location but not shape (shape change), or neither features (both change). (Right column) Main effects of shape and location decoding in Experiments 1 and 2(a)–(c) Results for the superior intraparietal sulcus (IPS), inferior IPS, and lateral occipital (LO), respectively. In the superior IPS, shape decoding was observed in Experiment 1 but not in Experiment 2. No location decoding was observed in either experiments. In the LO and inferior IPS, shape and location decoding was observed in both experiments with decoding being more robust in Experiment 1 than 2. In the left column, black, dark gray, light gray, and white bars indicate correlation coefficients for "Both Same," "Location Change," "Shape Change," and "Both Change" pairs, respectively. In the right column, dark gray and light gray bars indicate main effect of shape and location, respectively. Error bars indicate within-subject standard error of the mean. *$p < .05$; **$p < .01$; ***$p < .001$.

fMRI response patterns would be expected when the same than different object shapes were shown in odd and even runs. Likewise, if spatial locations (i.e., above and below fixation) are uniquely represented in a brain region, then higher correlation coefficients should be observed when objects appeared in the same than in different locations in odd and even runs.

In superior IPS, successful decoding emerged for object shapes, with the correlation coefficient being higher between the same than different shapes shown in odd and even runs ($F_{1,4} = 50.80$, $p = .002$; see Figure 2(a)). This indicates that superior IPS carries the content of VSTM. Such decoding success, however, was absent for object locations ($F < 1$,

$p > .89$). The interaction between shape and location decoding was not significant ($F < 0.19, p > .68$), showing that shape decoding was not modulated by whether shapes appeared in the same or different locations.

To understand whether shape representation in superior IPS is task-dependent, in Experiment 2, we repeated the design of Experiment 1, but asked the same group of observers from Experiment 1 to passively view the images and detected the direction of an occasional image jitter (either horizontal or vertical). This task requirement made neither object shape nor location task-relevant. To ensure that our result from Experiment 1 was not due to the two unique object categories (shoes and bikes) selected, in addition to shoes and bikes, we also included guitars and couches in Experiment 2 (see Methods). There were no main decoding difference between shoes and bikes and between guitars and couches ($F < 1, p > .90$) and no interaction between the two object groups and shape or location decoding ($F < 1$, $p > .52$). As such, results from all four categories were combined in Experiment 2.

Overall, observers were again fairly accurate in their behavioral performance (93.68 ± 5.75%). In superior IPS, no decoding of either shape or location was obtained (no main effects of shape and location, $F < 1.57, p > .28$; see Figure 2(a)). Direct comparison between the two experiments revealed shape decoding by experiment interaction ($F_{1,4} = 24.25, p = .008$), but not location decoding by experiment interaction ($F < 1, p > .90$). These results were not influenced by minor design differences between the two experiments. When we only included the same number of shoes/bikes trials from the two experiments, despite a reduction of power, shape decoding was still present in Experiment 1 ($F_{1,4} = 8.86, p = .041$), but absent in Experiment 2 ($F_{1,4} < 1, p > .55$), with an almost significant interaction between shape decoding and experiment ($F_{1,4} = 4.52, p = .06$). As before, location decoding never reached significance in either experiments in this analysis ($Fs < 2.48, ps > .19$). Thus, superior IPS carries shape representation, but only when such information was required by the task.

To determine whether task-dependent object shape representation seen in superior IPS is present in other brain regions involved in visual object representation, we also examined responses in the LO and another parietal region, the inferior IPS. The LO plays an important role in visual object shape representation and damage to it can lead to severe deficits in object shape perception (Milner et al., 1991). Consistent with this neuropsychological finding, both fMRI response amplitudes and patterns from the LO have been correlated with success in object shape identification (Grill-Spector, Kushnir, Hendler, & Malach, 2000; Williams, Dang, & Kanwisher, 2007). Interestingly, recent studies have also shown that both object shape and location can be decoded from this brain region (Kravitz, Kriegeskorte, & Baker, 2010; Schwarzlose, et al., 2008). As mentioned previously, the inferior IPS has previously been shown to participate in object selection and individuation via location and may contain coarse object shape information necessary for carrying out these operations (Jeong & Xu, 2013; Xu & Chun, 2006, 2009).

To examine shape and location encoding in the LO and inferior IPS, to increase power, we first combined data from all categories from Experiment 2 in our analyses. Consistent with prior findings, in both experiments, object shape and location information could be reliably decoded in both the inferior IPS and LO (Figure 2(b) and 2(c), $Fs > 14.42, ps < .019$). In Experiment 1, shape and location decoding did not interact in either the LO or inferior IPS ($Fs < 2.42, ps > .19$). In Experiment 2, shape and location did not interact in the LO ($F < 1.23, p > .32$) but had a marginally significant interaction in the inferior IPS ($F_{1,4} = 7.4, p = .053$). These results show that, regardless of the task demands, both shape and location could be decoded in LO and inferior IPS, with shape and location encoding being largely independent of each other. Direct comparison between experiments revealed a shape by experiment interaction ($Fs > 9.49, ps < .037$) in both inferior IPS and LO. Location by experiment interaction was also significant in LO ($F_{1,4} = 14.90, p = .018$) and marginally significant in inferior IPS ($F_{1,4} = 6.12, p = .069$). When the number of object categories and trials were matched between the two experiments, weaker but similar results were obtained in both the LO and inferior IPS (shape by experiment interaction, $Fs > 4.91, ps < .09$; and location by experiment interaction, $Fs > 7.13, ps < .056$). Overall, these results show that, in the LO and inferior IPS, shape and location features are represented more strongly when the task engages shape processing and hence more attention to the object. This was true for both the task-relevant feature (shape) and the task-irrelevant feature (location). When attention was engaged in neither shape nor location processing, these features were still encoded robustly despite the fact that their encodings were completely task irrelevant and unnecessary.

In Experiment 2, in LO, decoding difference between shoes and bikes and between guitars and couches was not significant ($Fs < 1, ps > .67$). In inferior IPS, decoding between guitars and couches was better than decoding between shoes and bikes ($F_{1,4} = 11.31, p = .028$), with no other main effect or interactions ($Fs < 1.5, ps > .28$). It thus seems that differences between guitars and couches were better represented than those between shoes and bikes in inferior IPS. However, because this difference was not found in superior IPS (see previous results), the lack of shape encoding in superior IPS in the motion task could not therefore be attributed to the specific shape categories used, but must indicate a general lack of shape encoding in this brain area when shape information was task irrelevant.

DISCUSSION

Using fMRI MVPA and a VSTM task, we found that the human superior IPS represents object shape information in a task-dependent manner. Specifically, shape representation was present when it was required by a task and absent when it was task irrelevant. In other words, the nature of the task has a strong influence on the nature of object representation in superior IPS. This is consistent with results obtained in fMRI response amplitude measures in which superior IPS was found to track the encoding of task-relevant items in VSTM (Todd & Marois, 2004; Xu & Chun, 2006) and ignore task irrelevant object features or distractors (Jeong & Xu, 2013; Xu, 2010). Thus, shape representation in superior IPS is dynamic and task driven, similar to responses observed in LIP neurons in monkey neurophysiology studies (Fitzgerald, Swaminathan, & Freedman, 2012; Freedman & Assad, 2006, 2009; Swaminathan & Freedman, 2012; Toth & Assad, 2002). Although the exact human homologue of LIP is still under debate, our results suggest possible functional correspondence between the human superior IPS and monkey LIP.

Human parietal cortex has been implicated in attention-related processing (Corbetta & Shulman, 2002; Szczepanski et al., 2010; Wojciulik & Kanwisher, 1999; Yantis et al., 2002) and has the capability of directing encoding resources in posterior sensory regions. By showing the encoding of visual information in a task-dependent manner in superior IPS, the present finding puts forward an alternative, perhaps equally important, mechanism by which parietal attention control mechanism can exert its influence on visual information processing in the brain. By directly encoding task-relevant visual information, superior IPS functions similarly as the random access memory in a computer in which information important for the current task and goal is gathered and possibly integrated.

Although task also enhanced shape representation in other regions such as inferior IPS and LO, this enhancement was not specific to the task-relevant feature as it also boosted the encoding of a task-irrelevant feature. Moreover, shape representation persisted even when it was task irrelevant. This is qualitatively different from what was observed in superior IPS in which without a shape-task, shape representation was absent. Thus, task and attention may play two primary roles during visual information transmission: Modulating but without altering what is intrinsically represented in a brain region, such as those in LO and inferior IPS; or determining the nature of representation in a brain region, such as that in superior IPS.

Previous fMRI studies have reported attentional modulation of visual information representation in human parietal cortex (Liu, Hospadaruk, Zhu, & Gardner, 2011; Thompson & Duncan, 2009; Woolgar, Hampshire, Thompson, & Duncan, 2011). By either treating parietal cortex as a single functionally uniform region or failing to find distinctions among topographically defined subregions within parietal cortex, these studies either assumed or argued for functional homogeneity of the human parietal cortex. However, using a functionally defined ROI-based approach, here we show distinctive response profiles from inferior and superior IPS and illustrate different ways in which attention could influence visual information processing. Thus, with appropriate localizers, functionally heterogeneous brain regions can be unveiled in human parietal cortex, providing us with a richer understanding of the precise and unique roles of parietal cortex in visual cognition.

Functional heterogeneity in IPS regions may also explain inconsistencies across published studies. Although some recent studies have found successful decoding of VSTM contents in IPS (Christophel & Haynes, 2014; Christophel, Hebart, & Haynes, 2012), some failed to do so (Emrich, Riggall, Larocque, & Postle, 2013; Linden, Oosterhof, Klein, & Downing, 2012; Riggall & Postle, 2012). It is possible that these studies examined different IPS regions, and depending on whether or not the superior IPS was included, the decoding results varied. In the current study, we functionally localized voxels within the IPS that tracked VSTM capacity in individual observers and found decoding of task relevant visual information in the superior IPS. Studies using a searchlight approach to locate informative voxels have also reported decoding of VSTM representation in the parietal cortex (Christophel et al., 2012; Christophel & Haynes, 2014). It thus seems that methods using an information-based approach have all showed the involvement of parietal cortex in VSTM representation. Beyond functional heterogeneity in parietal cortex, factors such as the stimuli used may also contribute to whether or not VSTM contents may be decoded from IPS. It seems that studies using motion stimuli have not found VSTM decoding in parietal cortex (Emrich et al., 2013; Riggall & Postle, 2012), whereas studies using nonmotion stimuli have (Christophel et al., 2012; Christophel & Haynes, 2014). This suggests that distinct neural mechanisms may exist for VSTM representations of motion and nonmotion stimuli.

In sum, our findings show that the human parietal cortex plays a greater role than simply directing attentional resources during visual perception. A subregion in human parietal cortex, the superior IPS, is capable of directly representing incoming visual information in a task-dependent manner. This is consistent with a growing body of evidence suggesting that the human parietal cortex is part of a brain network involved in the flexible and dynamic representation and processing of incoming information (Cole et al., 2013; Cusack & Owen, 2008; Emrich, Burianová,

& Ferber, 2011; Fedorenko, Duncan, & Kanwisher, 2013; Vincent, Kahn, Snyder, Raichle, & Buckner, 2008). We would like to argue that the superior IPS may be a key node in this network that mediates the moment-to-moment, goal-directed visual information representation in the human brain.

Acknowledgments

This research was supported by NSF grant 0855112 and NIH grant 1R01EY022355 to Y.X.

References

Alvarez, G. A., & Cavanagh, P. (2004). The capacity of visual short-term memory is set both by visual information load and by number of objects. *Psychological Science*, *15*(2), 106–111.

Bahrami, B. (2003). Object property encoding and change blindness in multiple object tracking. *Visual Cognition*, *10*(8), 949–963. http://dx.doi.org/10.1080/13506280344000158.

Bays, P. M., & Husain, M. (2008). Dynamic shifts of limited working memory resources in human vision. *Science*, *321*(5890), 851–854. http://dx.doi.org/10.1126/science.1158023 (New York, N.Y.).

Brainard, D. H. (1997). The psychophysics toolbox. *Spatial Vision*, *10*(4), 433–436.

Christophel, T. B., & Haynes, J.-D. (2014). Decoding complex flow-field patterns in visual working memory. *NeuroImage*, *91*, 43–51. http://dx.doi.org/10.1016/j.neuroimage.2014.01.025.

Christophel, T. B., Hebart, M. N., & Haynes, J.-D. (2012). Decoding the contents of visual short-term memory from human visual and parietal cortex. *The Journal of Neuroscience: The Official Journal of the Society for Neuroscience*, *32*(38), 12983–12989. http://dx.doi.org/10.1523/JNEUROSCI.0184-12.2012.

Cole, M. W., Reynolds, J. R., Power, J. D., Repovs, G., Anticevic, A., & Braver, T. S. (2013). Multi-task connectivity reveals flexible hubs for adaptive task control. *Nature Neuroscience*, *16*(9), 1348–1355. http://dx.doi.org/10.1038/nn.3470.

Corbetta, M., & Shulman, G. L. (2002). Control of goal-directed and stimulus-driven attention in the brain. *Nature Reviews. Neuroscience*, *3*(3), 201–215. http://dx.doi.org/10.1038/nrn755.

Coslett, H. B., & Saffran, E. (1991). Simultanagnosia. To see but not two see. *Brain: A Journal of Neurology*, *114*(Pt 4), 1523–1545.

Cowan, N. (2001). The magical number 4 in short-term memory: a reconsideration of mental storage capacity. *The Behavioral and Brain Sciences*, *24*(1), 87–114; discussion 114–185.

Cox, D. D., & Savoy, R. L. (2003). Functional magnetic resonance imaging (fMRI) "brain reading": detecting and classifying distributed patterns of fMRI activity in human visual cortex. *NeuroImage*, *19*(2 Pt 1), 261–270.

Cusack, R., & Owen, A. M. (2008). Distinct networks of connectivity for parietal but not frontal regions identified with a novel alternative to the "resting state" method. *Proceedings of Cognitive Neuroscience Society*, 1.

Emrich, S. M., Burianová, H., & Ferber, S. (2011). Transient perceptual neglect: visual working memory load affects conscious object processing. *Journal of Cognitive Neuroscience*, *23*(10), 2968–2982. http://dx.doi.org/10.1162/jocn_a_00028.

Emrich, S. M., Riggall, A. C., Larocque, J. J., & Postle, B. R. (2013). Distributed patterns of activity in sensory cortex reflect the precision of multiple items maintained in visual short-term memory. *Journal of Neuroscience*, *33*(15), 6516–6523. http://dx.doi.org/10.1523/JNEUROSCI.5732-12.2013.

Fedorenko, E., Duncan, J., & Kanwisher, N. G. (2013). Broad domain generality in focal regions of frontal and parietal cortex. *Proceedings of the National Academy of Sciences of the United States of America*, *110*(41), 16616–16621. http://dx.doi.org/10.1073/pnas.1315235110.

Fitzgerald, J. K., Swaminathan, S. K., & Freedman, D. J. (2012). Visual categorization and the parietal cortex. *Frontiers in Integrative Neuroscience*, *6*, 18. http://dx.doi.org/10.3389/fnint.2012.00018.

Freedman, D. J., & Assad, J. A. (2006). Experience-dependent representation of visual categories in parietal cortex. *Nature*, *443*(7107), 85–88. http://dx.doi.org/10.1038/nature05078.

Freedman, D. J., & Assad, J. A. (2009). Distinct encoding of spatial and nonspatial visual information in parietal cortex. *The Journal of Neuroscience: the Official Journal of the Society for Neuroscience*, *29*(17), 5671–5680. http://dx.doi.org/10.1523/JNEUROSCI.2878-08.2009.

Garrido, L., Vaziri-Pashkam, M., Nakayama, K., & Wilmer, J. (2013). The consequences of subtracting the mean pattern in fMRI multivariate correlation analyses. *Frontiers in Neuroscience*, *7*, 174. http://dx.doi.org/10.3389/fnins.2013.00174.

Gottlieb, J. P., Kusunoki, M., & Goldberg, M. E. (1998). The representation of visual salience in monkey parietal cortex. *Nature*, *391*(6666), 481–484. http://dx.doi.org/10.1038/35135.

Grill-Spector, K., Kushnir, T., Hendler, T., & Malach, R. (2000). The dynamics of object-selective activation correlate with recognition performance in humans. *Nature Neuroscience*, *3*(8), 837–843. http://dx.doi.org/10.1038/77754.

Harrison, S. A., & Tong, F. (2009). Decoding reveals the contents of visual working memory in early visual areas. *Nature*, *458*(7238), 632–635. http://dx.doi.org/10.1038/nature07832.

Haxby, J. V., Gobbini, M. I., Furey, M. L., Ishai, A., Schouten, J. L., & Pietrini, P. (2001). Distributed and overlapping representations of faces and objects in ventral temporal cortex. *Science*, *293*(5539), 2425–2430. http://dx.doi.org/10.1126/science.1063736 (New York, N.Y.).

Haynes, J.-D., & Rees, G. (2006). Decoding mental states from brain activity in humans. *Nature Reviews. Neuroscience*, *7*(7), 523–534. http://dx.doi.org/10.1038/nrn1931.

Huettel, S. A., Song, A. W., & McCarthy, G. (2009). *Functional magnetic resonance imaging* (2nd ed.). Sunderland, MA: Sinauer Associates.

Jeong, S. K., & Xu, Y. (2013). Neural representation of Targets and distractors during object individuation and identification. *Journal of Cognitive Neuroscience*, *25*(1), 117–126. http://dx.doi.org/10.1162/jocn_a_00298.

Kravitz, D. J., Kriegeskorte, N., & Baker, C. I. (2010). High-level visual object representations are constrained by position. *Cerebral Cortex*, *20*(12), 2916–2925. http://dx.doi.org/10.1093/cercor/bhq042 (New York, N.Y.: 1991).

Leslie, A. M., Xu, F., Tremoulet, P. D., & Scholl, B. J. (1998). Indexing and the object concept: developing 'what' and 'where' systems. *Trends in Cognitive Sciences*, *2*(1), 10–18.

Linden, D. E. J., Oosterhof, N. N., Klein, C., & Downing, P. E. (2012). Mapping brain activation and information during category-specific visual working memory. *Journal of Neurophysiology*, *107*(2), 628–639. http://dx.doi.org/10.1152/jn.00105.2011.

Liu, T., Hospadaruk, L., Zhu, D. C., & Gardner, J. L. (2011). Feature-specific attentional priority signals in human cortex. *The Journal of Neuroscience: The Official Journal of the Society for Neuroscience*, *31*(12), 4484–4495. http://dx.doi.org/10.1523/JNEUROSCI.5745-10.2011.

Luck, S. J., & Vogel, E. K. (1997). The capacity of visual working memory for features and conjunctions. *Nature*, *390*(6657), 279–281. http://dx.doi.org/10.1038/36846.

Milner, A. D., Perrett, D. I., Johnston, R. S., Benson, P. J., Jordan, T. R., Heeley, D. W., et al. (1991). Perception and action in 'visual form agnosia'. *Brain: A Journal of Neurology*, *114*(Pt 1B), 405–428.

Norman, K. A., Polyn, S. M., Detre, G. J., & Haxby, J. V. (2006). Beyond mind-reading: multi-voxel pattern analysis of fMRI data. *Trends in Cognitive Sciences*, *10*(9), 424–430. http://dx.doi.org/10.1016/j.tics.2006.07.005.

Pashler, H. (1988). Familiarity and visual change detection. *Perception & Psychophysics*, *44*(4), 369–378.

Peelen, M. V., & Downing, P. E. (2007). Using multi-voxel pattern analysis of fMRI data to interpret overlapping functional activations. *Trends in Cognitive Sciences*, *11*(1), 4–5. http://dx.doi.org/10.1016/j.tics.2006.10.009.

Phillips, W. A. (1974). On the distinction between sensory storage and short-term visual memory. *Perception & Psychophysics*, *16*, 283–290.

Riggall, A. C., & Postle, B. R. (2012). The relationship between working memory storage and elevated activity as measured with functional magnetic resonance imaging. *The Journal of Neuroscience: The Official Journal of the Society for Neuroscience*, *32*(38), 12990–12998. http://dx.doi.org/10.1523/JNEUROSCI.1892-12.2012.

Schwarzlose, R. F., Swisher, J. D., Dang, S., & Kanwisher, N. G. (2008). The distribution of category and location information across object-selective regions in human visual cortex. *Proceedings of the National Academy of Sciences of the United States of America*, *105*(11), 4447–4452. http://dx.doi.org/10.1073/pnas.0800431105.

Swaminathan, S. K., & Freedman, D. J. (2012). Preferential encoding of visual categories in parietal cortex compared with prefrontal cortex. *Nature Neuroscience*, *15*(2), 315–320. http://dx.doi.org/10.1038/nn.3016.

Szczepanski, S. M., Konen, C. S., & Kastner, S. (2010). Mechanisms of spatial attention control in frontal and parietal cortex. *The Journal of Neuroscience: The Official Journal of the Society for Neuroscience*, *30*(1), 148–160. http://dx.doi.org/10.1523/JNEUROSCI.3862-09.2010.

Thompson, R., & Duncan, J. (2009). Attentional modulation of stimulus representation in human fronto-parietal cortex. *NeuroImage*, *48*(2), 436–448. http://dx.doi.org/10.1016/j.neuroimage.2009.06.066.

Todd, J. J., & Marois, R. (2004). Capacity limit of visual short-term memory in human posterior parietal cortex. *Nature*, *428*(6984), 751–754. http://dx.doi.org/10.1038/nature02466.

Todd, J. J., & Marois, R. (2005). Posterior parietal cortex activity predicts individual differences in visual short-term memory capacity. *Cognitive, Affective & Behavioral Neuroscience*, *5*(2), 144–155.

Toth, L. J., & Assad, J. A. (2002). Dynamic coding of behaviourally relevant stimuli in parietal cortex. *Nature*, *415*(6868), 165–168. http://dx.doi.org/10.1038/415165a.

van den Berg, R., Shin, H., Chou, W.-C., George, R., & Ma, W. J. (2012). Variability in encoding precision accounts for visual short-term memory limitations. *Proceedings of the National Academy of Sciences of the United States of America*, *109*(22), 8780–8785. http://dx.doi.org/10.1073/pnas.1117465109.

Vincent, J. L., Kahn, I., Snyder, A. Z., Raichle, M. E., & Buckner, R. L. (2008). Evidence for a frontoparietal control system revealed by Intrinsic functional connectivity. *Journal of Neurophysiology*, *100*(6), 3328–3342. http://dx.doi.org/10.1152/jn.90355.2008.

Vogel, E. K., & Machizawa, M. G. (2004). Neural activity predicts individual differences in visual working memory capacity. *Nature*, *428*(6984), 748–751. http://dx.doi.org/10.1038/nature02447.

Vogel, E. K., Woodman, G. F., & Luck, S. J. (2001). Storage of features, conjunctions and objects in visual working memory. *Journal of Experimental Psychology. Human Perception and Performance*, *27*(1), 92–114.

Wilken, P., & Ma, W. J. (2004). A detection theory account of change detection. *Journal of Vision*, *4*(12), 1120–1135. http://dx.doi.org/10.1167/4.12.11.

Williams, M. A., Dang, S., & Kanwisher, N. G. (2007). Only some spatial patterns of fMRI response are read out in task performance. *Nature Neuroscience*, *10*(6), 685–686. http://dx.doi.org/10.1038/nn1900.

Wilson, K. E., Adamo, M., Barense, M. D., & Ferber, S. (2012). To bind or not to bind: addressing the question of object representation in visual short-term memory. *Journal of Vision*, *12*(8), 14. http://dx.doi.org/10.1167/12.8.14.

Wojciulik, E., & Kanwisher, N. G. (1999). The generality of parietal involvement in visual attention. *Neuron*, *23*(4), 747–764.

Woolgar, A., Hampshire, A., Thompson, R., & Duncan, J. (2011). Adaptive coding of task-relevant information in human frontoparietal cortex. *The Journal of Neuroscience: The Official Journal of the Society for Neuroscience*, *31*(41), 14592–14599. http://dx.doi.org/10.1523/JNEUROSCI.2616-11.2011.

Xu, Y. (2002). Encoding color and shape from different parts of an object in visual short-term memory. *Perception & Psychophysics*, *64*(8), 1260–1280.

Xu, Y. (2007). The role of the superior intraparietal sulcus in supporting visual short-term memory for multifeature objects. *The Journal of Neuroscience: The Official Journal of the Society for Neuroscience*, *27*(43), 11676–11686. http://dx.doi.org/10.1523/JNEUROSCI.3545-07.2007.

Xu, Y. (2008). Representing connected and disconnected shapes in human inferior intraparietal sulcus. *NeuroImage*, *40*(4), 1849–1856. http://dx.doi.org/10.1016/j.neuroimage.2008.02.014.

Xu, Y. (2009). Distinctive neural mechanisms supporting visual object individuation and identification. *Journal of Cognitive Neuroscience*, *21*(3), 511–518. http://dx.doi.org/10.1162/jocn.2008.21024.

Xu, Y. (2010). The neural fate of task-irrelevant features in object-based processing. *The Journal of Neuroscience: The Official Journal of the Society for Neuroscience*, *30*(42), 14020–14028. http://dx.doi.org/10.1523/JNEUROSCI.3011-10.2010.

Xu, Y., & Chun, M. M. (2006). Dissociable neural mechanisms supporting visual short-term memory for objects. *Nature*, *440*(7080), 91–95. http://dx.doi.org/10.1038/nature04262.

Xu, Y., & Chun, M. M. (2007). Visual grouping in human parietal cortex. *Proceedings of the National Academy of Sciences of the United States of America*, *104*(47), 18766–18771. http://dx.doi.org/10.1073/pnas.0705618104.

Xu, Y., & Chun, M. M. (2009). Selecting and perceiving multiple visual objects. *Trends in Cognitive Sciences*, *13*(4), 167–174. http://dx.doi.org/10.1016/j.tics.2009.01.008.

Yantis, S., Schwarzbach, J., Serences, J. T., Carlson, R. L., Steinmetz, M. A., Pekar, J. J., et al. (2002). Transient neural activity in human parietal cortex during spatial attention shifts. *Nature Neuroscience*, *5*(10), 995–1002. http://dx.doi.org/10.1038/nn921.

Zhang, W., & Luck, S. J. (2008). Discrete fixed-resolution representations in visual working memory. *Nature*, *453*(7192), 233–235. http://dx.doi.org/10.1038/nature06860.

CHAPTER

5

Neural Bases of the Short-term Retention of Visual Information

Bradley R. Postle

Departments of Psychology and Psychiatry, University of Wisconsin–Madison, Madison, WI, USA

PREAMBLE: DEFINING CONCEPTS AND TERMINOLOGY

To make progress in any scientific endeavor, it is necessary for there to be conceptual clarity about the parameters of the phenomenon being studied and for there to be terminological clarity such that there is an unambiguous mapping between the concepts and the words that are used to describe them. Although considerable progress has been made in our understanding of the cognitive and neural factors underlying of the behaviors that fall under the rubric of "working memory," this progress has sometimes been hindered by a lack of systematicity at the terminological level that has, in turn, sometimes promoted a lack of conceptual clarity. Within the theme "Mechanisms of Sensory Working Memory," this chapter will focus on the neural processes involved in the short-term retention (STR) of visual information—the neural representation of visual information when that information is no longer present in the environment. Discussions at the 25th International Symposium on Attention and Performance made it clear that this will be done most effectively if we first take the time to disambiguate the STR of information from two related concepts and their labels—short-term memory (STM) and working memory (WM).

Short-term Retention (STR) versus Short-term Memory (STM) versus Working Memory (WM)

In this chapter, the terms "STM" and "WM" will be used to refer to classes of behavior and to tasks that measure performance and ability within these classes of behavior. Importantly, they will not be used to refer to hypothetical cognitive or neural mechanisms—further along in this chapter, it will be illustrated how their use in this latter way can lead to a lack of conceptual clarity and, sometimes, to erroneous inferences about neural functioning. Tests of STM entail the presentation of a small amount of to-be-remembered information (referred to alternately as the "target(s)" or as the "sample(s)"), followed by a delay that can last from hundreds of milliseconds to tens of seconds or longer, followed by a second stimulus that requires a response that is related to the first. On tests of delayed response, the second stimulus is a cue to execute the action that had been signaled by the target. On tests of delayed recall, the second stimulus cues the subject to reproduce the target(s). On tests of delayed recognition, the second stimulus (the "probe") requires yes/no recognition, n-alternative forced choice (if more than one probe is presented), or some other recognition decision about its match to the first stimulus. Thus, STM results from the STR of information that was recently either presented or cued, but is not currently present in the environment, and it is tested by one's ability to guide behavior with that information.

Theoretical Bases for the STM-WM Distinction

The concept of WM as distinct from STM was proposed and developed by Baddeley and Hitch (1974), in part from their findings that performance on each of two STM tasks under dual-task conditions could approach levels of performance that one would see when the tasks were administered individually, so long as the two engaged different

Mechanisms of Sensory Working Memory
http://dx.doi.org/10.1016/B978-0-12-811042-3.00005-0

domains of information.[1] Because, in their early research, the two domains were verbal and visuospatial, early versions of their model called for two STM buffers (dubbed the "phonological loop" and the "visuospatial sketchpad," respectively) that could operate independently of each other (Baddeley, 1986). The "work" of coordinating the simultaneous engagement of the two buffers was ascribed to a cognitive control mechanism that they dubbed the "Central Executive." Baddeley (1986) has written subsequently that he construed the Central Executive as being akin to the Norman and Shallice's (1980) Supervisory Attentional System and, as such, its function was not assumed to be restricted to only tasks with an overt memory component.[2] In the context of this "multiple component model" (Baddeley, 1986), tasks that only require the STR of information are assumed to only engage one of the buffers, and are thus considered tests of STM (e.g., Logie & Niven, 2012). Tasks that require coordination between the buffers, or manipulation of information being held in the buffers—such as mental navigation in a spatial array, or mental arithmetic—in contrast, are considered tests of WM. Beyond the context of the multiple component model, many cognitive models that do not relate explicitly to this framework also use a similar convention for categorizing tasks as tests of STM or WM: the former require only the STR of information, the latter require additional cognitive operations that entail the manipulation of the information that is being retained, and/or the flexible use of that information to guide behavior in tasks that are more complicated than simple recognition or recall (e.g., Daneman & Hannon, 2007; Engle, Tuholski, Laughlin, & Conway, 1999; Konstantinou & Lavie, 2013).

There is a perspective from which this distinction of STM corresponding to the "simple" STR of information versus WM corresponding to STR plus additional executive control can be called into question, and this will be taken up in the final subsection of this preamble. First, however, a review of the most compelling basis for maintaining this distinction.

Neurobiological Bases for the STM-WM Distinction

At the level of neural systems, there is strong evidence that tests of WM require contributions from the prefrontal cortex (PFC) that are not necessary for STM. In some cases, this suggests an organization in which WM can be construed as, to oversimplify, "STM+PFC." In others, full double dissociations have been demonstrated. One example of the latter is that in the monkey, the STR of a single object for up to 120s (a test of STM) depends on the integrity of the anterior inferotemporal cortex, but not of the mid-dorsolateral (dl) PFC, whereas the serial selection of two different items, separated in time by 10s, from a set of three, four, or five items ("self-ordered choosing," a test of WM) shows the opposite pattern (Petrides, 2000). In humans, delayed recognition of the ordinal position of five randomly ordered letters is not affected by delay-period repetitive transcranial magnetic stimulation (rTMS) of the dl PFC, but it is impaired by rTMS to parietal cortex. rTMS of dlPFC does impair performance, however, when subjects are required reorder these letters into alphabetical order during the delay period (Postle et al., 2006). Similarly, although large lesions of PFC generally leave verbal and spatial span performance intact, the same patients are impaired on tests of immediate serial recall of items in the reverse of the presentation order (i.e., "backward span") (D'Esposito & Postle, 1999).

As indicated in the previous paragraph, WM differs from STM in that the former entails operations that transform the remembered information, or that require control operations beyond its simple STR. A real-world example of the former is performing mental arithmetic on the remembered total on a restaurant bill, so as to determine the amount of money to leave as a tip. A real-world example of the latter might be keeping track, during a conversation about the grown children of someone who one is meeting for the first time, of which of several children live in which of several cities. There are three broad classes of WM task. Delay tasks are operationally similar to tests of STM, but require some mental transformation of the remembered information, such as reordering it from order of presentation to some cardinal order (e.g., alphabetical or numerical). Continuous (or "running") tasks entail the serial presentation of multiple items, each of which updates the mental representation of the memory set (as in tests of running span), and, for some tasks, requires a decision that is contingent upon what came earlier in the series of stimuli (e.g., "n-back" and "AX-CPT"[3]). Finally, dual tasks require switching between two tasks being performed in parallel, with one task often being retention of a span of items being generated by the other task (e.g., sentence span and operation span).

[1] At a conference in 2014, Alan Baddeley attributed his adoption of the term "working memory" to a label in a diagrammatic figure of the Atkinson and Shiffrin model (although not appearing in their highly cited (1968) chapter), and explicitly not to Miller, Galanter, and Pribram (1960), which he professed to have not read prior to 1974 (A. Baddeley, personal communication).

[2] Indeed, an early neuroimaging study designed to isolate activity attributable to the Central Executive (and identifying it in the prefrontal cortex) employed a dual-task procedure in which neither of the individual tasks was a memory task (D'Esposito et al., 1995).

[3] This is a "continuous performance task" (CPT) featuring the serial presentation of individual letters, and the instructions that the appearance of the letter X requires one type of response if it was immediately preceded by the letter *A*, but a different response if it was immediately preceded by any other letter.

STM and WM both depend on the STR of information. They differ from long-term memory (LTM) in that they do not depend on the integrity of the medial-temporal lobe diencephalic memory system (hereafter abbreviated MTL)—patients with pure anterograde amnesia will typically perform normally on tests of STM and on some tests of WM. This means that the STR of information is independent of the processes that encode information into LTM. Importantly, however, there are two things that this does not mean. First, it does not mean that information being held for a test of STM or WM cannot also be incidentally processed by the MTL (in neurologically healthy individuals), and thus encoded into LTM (Ranganath & Blumenfeld, 2005; Ranganath, Cohen, & Brozinsky, 2005). Second, it does not rule out the possibility that the STR of information may be supported by the temporary activation, via attention, of LTM representations that existed prior to the performance of the STM or WM test (Lewis-Peacock & Postle, 2008; Postle, 2007).

The Role of PFC in the Control of Interference on Tests of STM and WM

One perspective from which the distinction between STM and WM becomes murky is that of the control of proactive interference (PI) in STM. It has been known at least since the 1960s that every trial after the very first on a test of STM is more prone to error because of interference from material that was processed on previous trials (Keppel & Underwood, 1962). The control of PI in STM has been localized to left inferior PFC via brain imaging (D'Esposito, Postle, Jonides, & Smith, 1999; Jonides, Smith, Marshuetz, Koeppe, & Reuter-Lorenz, 1998), neuropsychology (Thompson-Schill et al., 2002), and rTMS (Feredoes & Postle, 2010; Feredoes, Tononi, & Postle, 2006). Relatedly, findings in the monkey suggest that the delayed-response impairment that follows PFC lesions is better understood as increased susceptibility to interference, and/or poor control of behavior with remembered information, rather than as an increased impairment in the STR, per se, of information (Malmo, 1942; Tsujimoto & Postle, 2012). Indeed, there is ongoing debate about the extent to which the high correlations that are found between visual STM and general fluid intelligence (reviewed, e.g., in Luck & Vogel, 2013) reflect the STR, per se, of information versus such operations as the selection of information into STM (e.g., Linke, Vicente-Grabovetsky, Mitchell, & Cusack, 2011), the control of the effects of interference in STM (sometimes called "filtering," e.g., Vogel, McCollough, & Machizawa, 2005), and/or the retrieval of information in the face of PI (Shipstead & Engle, 2013). Independent of this scientific debate, it is undoubtedly the case that in the "real world" outside of the laboratory there are likely very few occasions in which STM is not supported by some level of PFC-based control. Does this mean that it is not useful to distinguish between STM and WM? The answer to this question is an unequivocal "no." After all, "some level of PFC-based control" is involved in virtually all classes of behavior—for example, motor control, language, social behavior, perceptual decision-making—yet, we nonetheless find it useful to distinguish between these classes of behavior because of important differences between them. Similarly, the arguments laid out earlier in this section provide a conceptual basis, at several levels of analysis, for distinguishing between STM and WM. Indeed, later, we shall see that failing to do so can lead to erroneous interpretation of neural data.

THE NEUROANATOMICAL BASES OF THE STR OF VISUAL INFORMATION

Evidence from Lesions and Stimulation Experiments

There is broad consensus that the STR of information depends on the same networks that are responsible for the processing of that information in contexts that do not require memory. For language-based information, this implicates regions that are involved in speech production (Acheson, Hamidi, Binder, & Postle, 2011; Koenigs et al., 2011; Leff et al., 2009; Richardson et al., 2011); for sensory information, this implicates regions involved in the perception of that sensory modality (e.g., Pasternak & Greenlee, 2005). The strongest evidence for this view comes from the demonstration that perturbations of the critical tissue influences STM in systematic, and selective, ways. For example, the studies of verbal STM cited previously have shown that both speech production and verbal STM are sensitive to the integrity of gray matter in the left posterior superior temporal gyrus (Koenigs et al., 2011; Leff et al., 2009; Richardson et al., 2011) or to delay-period rTMS of this region (Acheson et al., 2011). Delay-period rTMS of left middle temporal gyrus affects other language-based processes, but neither speech production-specific processes nor verbal STM (Acheson et al., 2011). Within the visual modality, similar evidence has been marshaled for the STR of spatial location with the demonstration that delay-period rTMS of intraparietal sulcus (IPS), superior parietal lobule, and the frontal eye field selectively influences performance on tests of STM for object location but not for object identity. Further, delay-period rTMS of neither dorsolateral prefrontal cortex (PFC) nor postcentral gyrus has this effect (Hamidi, Tononi, & Postle, 2008).

A more nuanced approach, taken for the STR of visual motion, leverages the fact that TMS of the anatomically adjacent visual motion processing areas MT and MST (hereafter "MT+") can produce the percept of a "moving phosphene"—a flash of light that contains coherent motion within the area of the flash. The perceived direction of motion is reproducibly toward the periphery, away from the fovea, in the visual field contralateral to the side of stimulation. Silvanto and Cattaneo (2010) demonstrated that this percept is systematically influenced when TMS is delivered while the subject is engaged in STM for the direction of motion of a target stimulus. When the target motion is in the same direction as the motion of the phosphene, its perception is enhanced. However, when the target motion is in the opposite direction, perception of the delay-period phosphene is reduced. These results indicate that the physiological state of MT varies systematically as a function of the direction of motion being remembered, just as it does, when a stimulus is present, as a function of the direction of motion being perceived.

Evidence from Multivariate Analyses of fMRI Data

Compelling evidence for what has been dubbed the "sensory recruitment" model (e.g., Ester, Anderson, Serences, & Awh, 2013) has also been generated by recent functional magnetic resonance imaging (fMRI) studies of visual STM that have employed multivariate analysis techniques. These represent an important advance over traditional univariate approaches, because their superior specificity and sensitivity support tests of whether neural activity in a brain region may support the STR of the stimulus information being retained on a trial-by-trial basis. Multivariate pattern analysis (MVPA) has been reviewed in many places (e.g., Haynes & Rees, 2006; Kriegeskorte, Goebel, & Bandettini, 2006; Norman, Polyn, Detre, & Haxby, 2006; Pereira, Mitchell, & Botvinick, 2009); reviews of its applications in STM research can be found in Postle (2015) and Sreenivasan, Curtis, and D'Esposito et al. (2014).) For example, two studies have demonstrated that primary visual cortex (V1) supports the delay period-spanning representation of the color or orientation of target stimuli (Harrison & Tong, 2009; Serences, Ester, Vogel, & Awh, 2009). Further, and of relevance for the later section on *The Neurophysiological Bases of the STR of Visual Information*, both of these findings were made despite the absence of sustained, elevated levels of signal intensity during the delay period. This pattern of results has been replicated with other classes of stimulus: the STR of motion can be decoded from lateral extrastriate cortex, including area MT+, as well as from medial calcarine and extracalcarine cortex (Emrich, Riggall, Larocque, & Postle, 2013; Riggall & Postle, 2012); the STR of complex visuospatial patterns can be decoded from occipital and parietal cortex (Christophel, Hebart, & Haynes, 2012); and the STR of familiar objects, faces, houses, scenes, and body stimuli can be decoded from the delay-period activity of ventral occipitotemporal cortex (Han, Berg, Oh, Samaras, & Leung, 2013; Lee, Kravitz, & Baker, 2013; Nelissen, Stokes, Nobre, & Rushworth, 2013; Sreenivasan, Vytlacil, & D'Esposito, 2014).

Many of these studies have also established the specificity of these functions to these regions by failing to decode target identity from other regions. The studies of the STR of visual motion did this, for example, by also applying a standard univariate analysis to identify brain regions that showed elevated activity that was sustained across the delay period. This identified portions of superior and lateral frontal cortex and of IPS that are invariably identified with such analyses. When MVPA was applied to these regions, however, it was unable to recover trial-specific stimulus information from these delay-active regions (Emrich et al., 2013; Riggall & Postle, 2012) (Figure 1). It is unlikely that the frontal and parietal regions were somehow "less classifiable," because in the Riggall and Postle (2012) data set, pattern classifiers were able to recover trial-specific task instruction-related information from these regions. That is, MVPA showed that the frontal and parietal regions encoded whether the instructions on a particular trial were to remember the speed or the direction of the moving dots that had been presented as the sample stimulus. Analogous patterns of specificity have been reported for STM for other classes of stimuli. With familiar objects, when the task required STM for stimulus features, the STR of stimulus identity could be decoded from ventral occipitotemporal cortex, but not from dorsolateral PFC; when it required STM for stimulus category, however, the reverse was true (Lee et al., 2013). Further, as was the case in the STM-of-motion studies, this pattern was independent of delay-period signal intensity, which did not distinguish among tasks in either region. A study of delayed recognition of faces, scenes, or faces and scenes that decoded at the level of category found evidence for delay-period representation of stimulus category in both dorsolateral PFC and ventral occipitotemporal cortex. However, a clever analysis of the classifier's "confusion matrix"—that is, the pattern of guesses that it made on trials that were classified erroneously—indicated that that classifier's performance with signal from ventral occipitotemporal cortex, but not from PFC, was consistent with the STR of a sensory representation (Sreenivasan, Vytlacil, & D'Esposito, 2014).

Finally, three studies have linked the precision of the delay-period neural representation of target stimuli in sensory cortex with behavioral estimates of mnemonic precision, showing that "the relative 'quality' of ... patterns [of activity in sensory cortex] determine the clarity of an individual's memory" (Ester et al., 2013, p. 754). Ester et al. (2013) did so for the STR of orientation by using multivariate encoding models to relate individual differences in delay-period neural representation of target stimuli in visual areas V1 and V2v to individual differences in

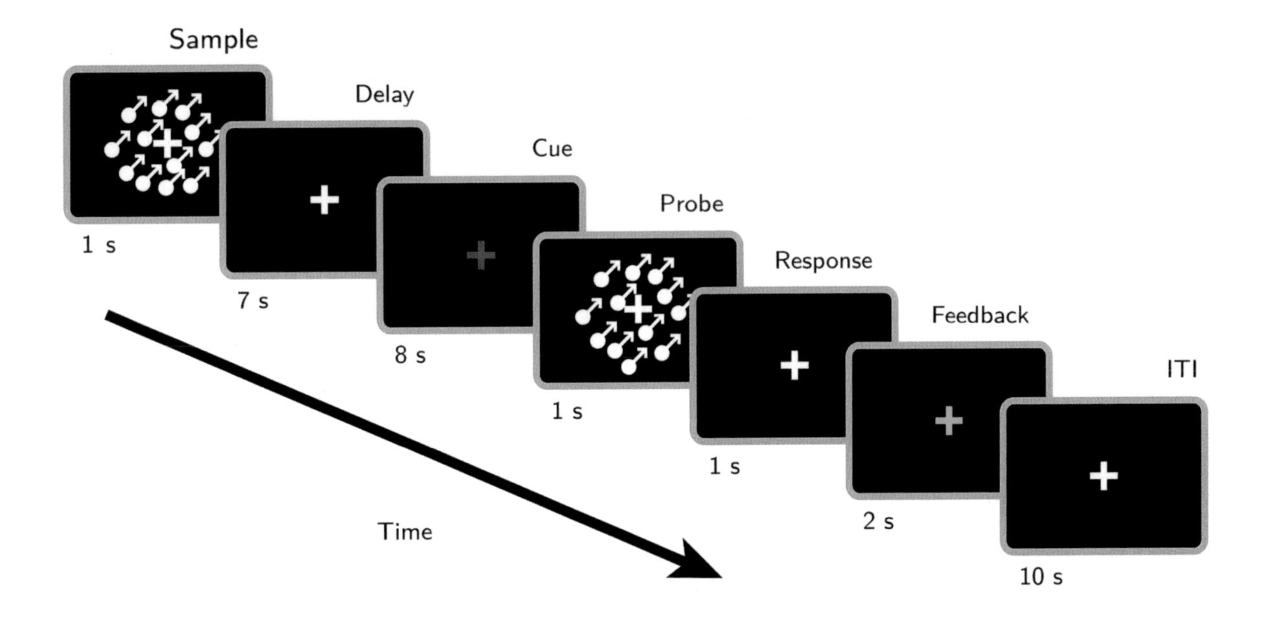

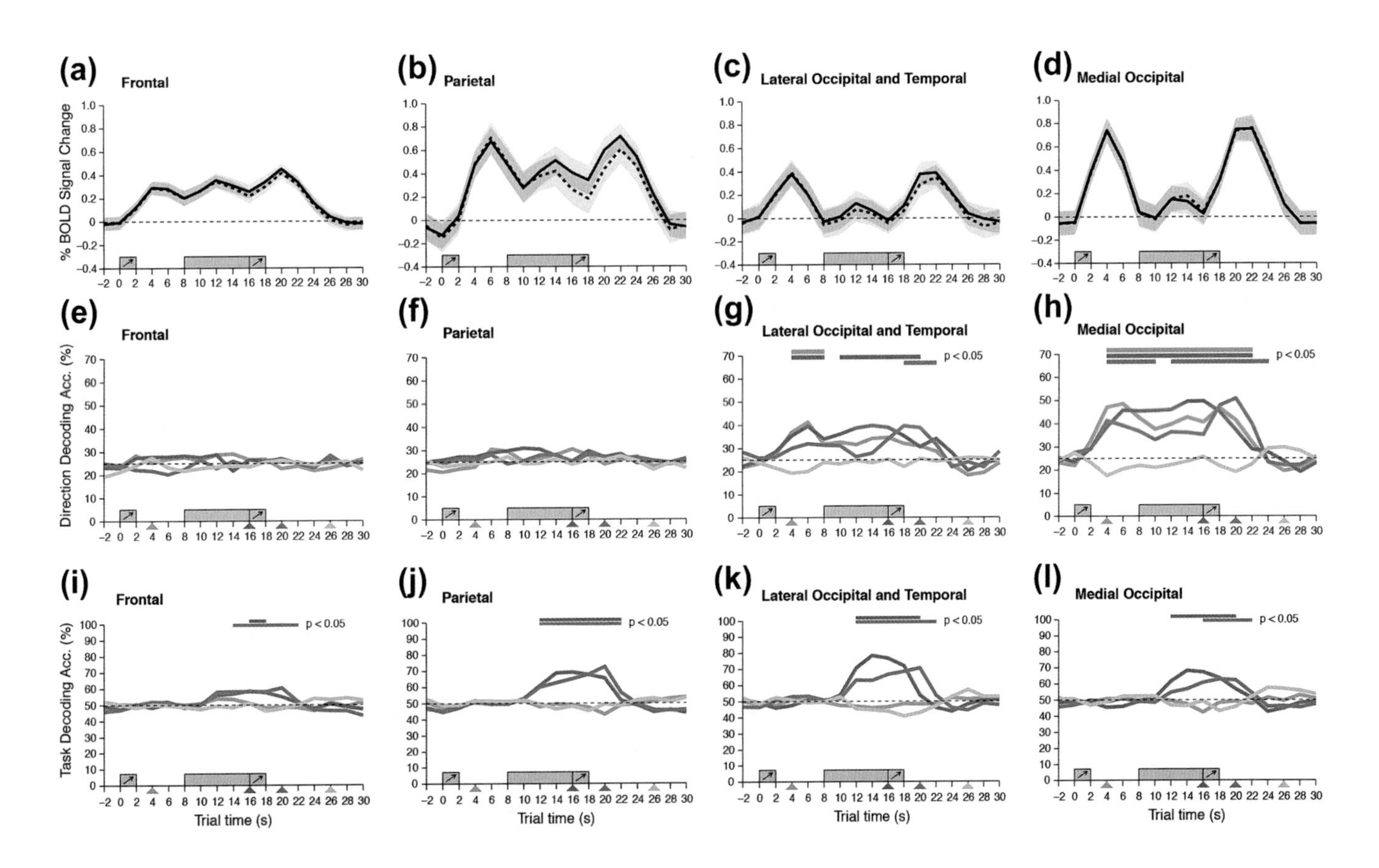

FIGURE 1 Behavioral task from Riggall and Postle (2012). Subjects maintained the direction and speed of a sample motion stimulus over a 15-s-long delay period. Midway through the delay period, they were cued as to the dimension on which they would be making an upcoming comparison against the remembered sample, either direction or speed. At the end of the delay period, they were presented with a probe motion stimulus and had to indicate with a button press whether it did or did not match the sample stimulus on the cued dimension. BOLD and MVPA time courses from four ROIs. Sample presentation occurred at 0 s, and at 8 s subjects were cued that either the direction or speed of sample motion would be tested on that trial. (a–d) Average ROI BOLD activity. Data from direction-cued trials use solid lines and speed-cued trials use dashed lines, bands cover average standard error across subjects. (e–h) ROI stimulus-direction decoding results and (i–l) ROI trial-dimension decoding results. Each waveform represents the mean direction-decoding accuracy across subjects (n = 7) for a classifier trained with data limited to a single time point in the trial and then tested on all time points in the hold out trials (e.g., the green line illustrates the decoding time course from a classifier trained on only data from time point 4, indicated by the small green triangle along the x axis.) Horizontal bars along the top indicate points at which the decoding accuracy for the corresponding classifier was significantly above chance ($p < .05$, permutation test). Schematic icons of trial events are shown at the appropriate times along the x-axis. Data are unshifted in time. *The Journal of Neuroscience: the Official Journal of the Society for Neuroscience by Society for Neuroscience. Copyright 2012. Reproduced with permission of Society for Neuroscience in the format Other Book via Copyright Clearance Center.*

behavioral performance. More specifically, they showed that the precision of population tuning curves estimated from delay-period signal from these regions predicted the accuracy with which a subject was able to reconstruct the target orientation at the end of the delay period. (See Serences and Saproo (2012) for a nice tutorial review of multivariate encoding models.) Emrich et al. (2013), in work that is detailed in Emrich's contribution to this volume, did so by varying from trial to trial the number of directions of motion that had to be remembered, and demonstrating a reliable within-subject correlation between the load-related decline in delayed-recall precision and a load-related decline in MVPA decoding performance from calcarine, pericalcarine, and extrastriate regions. Anderson, Serences, Vogel, and Awh (2014) have demonstrated that the precision of the STR of visual information is reflected in patterns of stimulus-induced activity in the alpha band, measured primarily from posterior electrodes.

Because function follows from structure, the previously summarized evidence for *where* in the brain stimulus information is retained during tests of STM constitutes a necessary and important first step for addressing the broader question of *how* the STR of visual information may be accomplished. The next section reviews the evolution in thinking about the mechanisms that may underlie the STR of visual information.

THE NEUROPHYSIOLOGICAL BASES OF WORKING MEMORY

The thesis of this section is that conflation of the concepts of the STR of information and of WM has led to mistaken assumptions about the neural bases of the former. Thus, disentangling this confused state of affairs, and specifying some of what we know about the neural bases of WM performance, is a necessary step for making progress in understanding the STR of visual information.

Rate Coding Models

There is a seductively powerful, intuitive tendency to ascribe a first-order rate-code interpretation to the information-processing function of a neuron. To experience this first hand, do an internet search for "Hubel and Wiesel simple cell" and watch a movie of one of their early, pioneering experiments. What one sees is the projection screen from the perspective of the anesthetized cat, and, like the cat's fanciful homunculus, one also listens in on the report from one V1 neuron while also observing the full range of movements and shape changes of the bar of light while it sweeps about the screen, both inside and outside of the neuron's receptive field. The selectivity of this neuron's responses seems clear: It only responds to a luminance contrast-defined edge within a narrow range of orientations, and only at a specific location on the screen. The inference of the function being carried out by this neuron follows automatically: *The neuron is acting as a detector of this narrowly defined feature, and only when the feature appears in this small region of the visual field.* When this process is repeated across several neurons, and each is observed to have different tuning properties, one can then formulate a model about the function of the region from which these neurons were sampled: *The primary visual cortex functions as banks of feature detectors that report the presence of a finite number of elemental features to higher levels of cortex; integration across receptive fields at these higher levels leads to object recognition.* This would be a reasonable, if grossly oversimplified, one-sentence summary of "how primary visual cortex works." Note, however, that the initial intuitive interpretation of the function of a V1 simple cell is necessarily univariate, in that one only has access to the activity of one unit in the brain and, further, to only one dimension along which that unit's activity is varying (i.e., firing rate). Thus, it was not only "lucky" for Hubel and Wiesel that the edge of a glass slide became stuck, at just the right angle, in their projector. It was also lucky for them that the system that they were studying turned out to be amenable to this intuitive analysis. The same would not have been true, for example, if they had been seeking to "crack the code" of olfactory perception (Wilson & Mainen, 2006). And, of primary relevance for this chapter, the preponderance of evidence indicates that same is also not true for the functions of neurons of the PFC. Nonetheless, for historical as well as expository reasons, a review of how we have come to know that the STR of visual information is not supported by a rate code in highly specialized PFC neurons will be an instructive way to begin this section.

As reviewed elsewhere (e.g., Postle, 2006), in the decades before the advent of extracellular recording from neurons in the PFC, the first of which were reported in 1971, the consensus view from lesion experiments was that the dorsolateral PFC is necessary for performance on tasks of WM, as defined in the *Preamble* of this chapter, but not for the STR, per se, of information. For example, the seminal volume from Warren and Akert (1964) on *The Frontal Granular Cortex and Behavior*, which followed from a 1962 symposium featuring the who's-who of the day, emphasizes the role for PFC in learning novel tasks, controlling perseveration on tasks that periodically reverse reward contingencies (including delayed alternation in monkeys (Mishkin, 1964) and the Wisconsin Card Sorting Task in humans (Milner, 1964)), and guiding behavior with information held in STM. With regard to the delayed-response deficit that is sometimes observed after lesions of the PFC, Pribram, Ahumada, Hartog, and Roos (1964) summarized studies from the 1930s and 1940s that "found experimentally that aspects other than the trace of memory [in our parlance, the STR of information] were involved by the frontal procedure [i.e., were compromised by PFC lesions]: action at the time of stimulus presentation and distractability were found to be important. These results were interpreted to indicate that the delayed-response task tested for one trial learning and retroactive inhibition rather than for memory-trace decay" (p. 29). Consistent with this review of the (then) two decades-old data from monkeys, in that same volume, Milner also reported on the performance of PFC-lesioned patients in tests of 60-second delayed recognition. When tested with novel nonsense shapes, with no reuse of stimuli across trials (i.e., an "open" stimulus set), performance of these patients was intact. With other types of stimuli, however, the patients were tested on "closed sets," in which items were reused across trials, and with these their performance was impaired. Thus, the impairment of the patients was interpreted as one of impaired control of PI, rather than of impaired STR of information. These decades of accumulated knowledge were largely neglected, however, with the advent of single-unit electrophysiology.

In 1971, the groups of Joaquin Fuster and Garrett Alexander at the University of California Los Angeles, and of Kisou Kubota and Hiroaki Niki at Kyoto University, published the results of recordings from neurons in the PFC while monkeys performed delayed response and delayed alternation, respectively. During delayed-response performance, Fuster and Alexander (1971) found that many neurons in the PFC displayed elevated firing rates that spanned the duration of the delay period, and which varied in length, unpredictably, within a range of 15–65 s. During delayed-alternation performance, Kubota and Niki (1971) observed two classes of task-related activity: neurons with elevated activity during the delay ("D"); and neurons that became active just before and during the response period ("E"; presumably because their activity predicted the onset of activity in the electromyogram that was recording from muscles in the animal's arm). Interestingly, the response profile of many E cells changed quantitatively when the animals performed a simple alternation task with no delay period interposed between responses—preresponse bursts were of lower intensity (i.e., slower firing rate) and they preceded the motor response by a smaller period of time. Of particular relevance for our present purposes, both sets of authors provided interpretations of their data that was compatible with the preexisting understanding of the neuropsychology of the PFC.

Fuster and Alexander (1971):

> "The temporal pattern of firing frequency observed in prefrontal and thalamic units during cue and delay periods suggest the participation of these units in the acquisition and temporary storage of sensory information which are implicated in delay response performance. Their function, however, does not seem to be the neural coding of information contained in the test cues, at least according to a frequency code, for we have not found any unit showing differential reactions to the two positions of the reward.
>
> It is during the transition from cue to delay that apparently the greatest number of prefrontal units discharge at firing levels higher than the intertrial baseline... We believe that the excitatory reactions of neurons in MD and granular frontal cortex during delayed response trials are specifically related to the focusing of attention by the animal on information that is being or has been placed in temporary memory storage for prospective utilization." (p. 654)

Kubota and Niki (1971):

> ... in the delay task, periprincipal E unit may be causally coupled with the initiation of the voluntary lever pressing. During delay task the E unit activity may be coupled with the initiation and sustaining of [...] lever pressing rather than [a] memory retrieval process.
>
> ... D units are hardly correlated with the memory storage (1) or retrieval of the memory for the lever pressing (2). Frequency of D unit during delay phases is not apparently different between right and left lever pressings. However, this interpretation does not exclude the possibility that the activities of neurons in the prefrontal cortex represent a memory function related to the choice of the correct performance on the basis of immediate past experience, i.e., remembering the spatially directed response on the preceding trial" (p. 346)

Despite these interpretations from the authors themselves, what seems to have captured the attention of many who read these reports was that they suggested a physiological correlate of the first of Hebb's (1949) dual traces, the "reverberatory" mechanism for "a transient 'memory' of [a] stimulus" (p. 61) that would sustain a representation until it could be encoded into a more permanent state via synaptic strengthening.

The attribution of explicitly mnemonic interpretations to the findings of Fuster and Alexander (1971) and of Kubota and Niki (1971) became more prevalent with the subsequent explosion interest in the construct of WM that arose from the introduction of the multiple-component model of Baddeley and Hitch (1974). Particularly influential was the view of Goldman-Rakic (1987, 1992), that the sustained delay-period activity in the PFC of the monkey and the storage buffers of the multiple-component model of WM from cognitive psychology were cross-species homologues of the same fundamental mental phenomenon: the STR of information in the service of STM and WM. The basic research protocol employed to test this idea, taken from visual neurophysiology research, proceeded in two stages: first, determine the tuning properties of a neuron; second, study its delay-period activity during trials when the animal is remembering that neuron's preferred stimulus versus when it is remembering a nonpreferred stimulus. This was exemplified by the studies of oculomotor delayed response by Funahashi, Bruce, and Goldman-Rakic (1989, 1990), in which neurons in the PFC showed elevated firing rates that were delay-spanning and retinotopically specific. The model that emerged from these data was of a population of PFC neurons, each tuned for a narrow region of retinotopically defined space, supporting the STR of a location in space via a rate code.

The flaw with this reasoning, however, was that it conflated two constructs from the multiple component model: the storage buffers and the Central Executive. In the parlance of this chapter, it conflated concepts of the STR of information and of WM. It is the Central Executive that would be expected to support performance on tests of WM, as defined at the beginning of this chapter, and that might reasonably be localized, to a first order of approximation, to the PFC. However, even though the delay-spanning activity of PFC neurons was often referred to as "spatial WM activity," conceptually it was being interpreted as a neural correlate of a storage buffer, rather than as a neural correlate of the Central Executive. This, therefore, propagated the idea that the STR of information is accomplished via the rate code-based activity of memory-specialized neurons in the PFC.

Beginning in the late 1990s, many studies have been performed that have shown that sustained delay-period activity in PFC neurons is better understood as supporting one or more aspects of the control of memory and/or behavior (what might be portrayed as the functions of the Central Executive), rather than the STR, per se, of information. For example, PFC neurons have been shown to not be specially tuned for any particular kind of information, but, instead, to modify their response properties to reflect changing environmental exigencies (Duncan, 2010; Duncan & Miller, 2002; Fuster, 2002; Rao, Rainer, & Miller, 1997). The activity of these neurons is not restricted to delay periods or even to tasks that require the STR of information, in that dl PFC neurons with delay-period activity in the oculomotor delayed-response task also exhibit similar sustained activity during the "delay" period of a visually guided saccade task (i.e., when the target remains visible throughout the trial, and so no memory is needed, Tsujimoto & Sawaguchi, 2004). These neurons can dynamically change during a single delay period from retrospectively representing the location of the sample stimulus to prospectively representing the target of the impending saccade, on a task in which the saccade must be made to a location that is a rotated transformation of the sample location (Takeda & Funahashi, 2002, 2004, 2007). In a task that dissociates the focus of spatial attention from the focus of spatial memory, the majority of dlPFC neurons are seen to track the former (Lebedev et al., 2004). Additional evidence against the view that the PFC acts as an STM buffer—from extracellular electrophysiology, neuroimaging, and perturbation studies using lesions, rTMS, and neuropharmacological interventions—has been reviewed elsewhere, for example, by Postle (2006) and Tsujimoto and Postle (2012).

Given these more recent findings, how can one then understand the activity of PFC neurons that had seemed to be supporting the STR of visual information with a rate code? One answer comes from the recent emphasis on the multivariate nature of neural coding to which we now turn.

Dynamic and Distributed Coding in the PFC Supports WM Performance

A simple statement of the rationale behind multivariate analyses of neural signals, which are similar in principle to those introduced in the earlier section of this chapter on *Evidence from Multivariate Analyses of Functional Magnetic Resonance Imaging Data*, is that the brain supports the simultaneous activity of many, many millions of processing elements all at the same time, and that the overall influence of this multiplicity of activity is very likely to be different from what one could deduce by studying any single processing element in isolation. This has been demonstrated very convincingly at the level of neuronal activity in the PFC via multivariate reanalyses of extant datasets that retrospectively assemble all of the records of single neurons recorded while the animal performed identical trials, and analyzes them all together as though all had been recorded in parallel during a single "virtual trial." In one such study, Meyers, Freedman, Kreiman, Miller, and Poggio (2008) reanalyzed a data set in which monkeys had viewed images of three-dimensional computer-generated animals and judged whether each was more cat-like or dog-like. The original study had compared neuronal activity in PFC versus inferior temporal (IT) cortex, finding that IT seemed "more involved in the analysis of currently viewed shapes, whereas the PFC showed stronger category signals, memory effects, and a greater tendency to encode information in terms of its behavioral meaning" (Freedman, Riesenhuber, Poggio, & Miller, 2003, p. 5235). The multivariate reanalysis, although generally confirming these broader patterns, uncovered novel, surprising information about the interpretation of sustained, elevated firing rates: the representation of stimulus category information in a delayed-match-to-category task "is coded by a nonstationary pattern of activity that changes over the course of a trial with individual neurons … containing information on much shorter time scales than the population as a whole" (Meyers et al., 2008, p. 1407). That is, this information was not carried for extended periods by any individual neuron, regardless of whether or not its activity level was elevated at a sustained level across the delay period. Thus, a first-order, intuitive rate-coding interpretation of the activity of a neuron will often lead to a faulty understanding of how its activity is contributing to the representation of behaviorally relevant information.

A second example of a multivariate reanalysis of a single-unit data set, also led by Meyers (Meyers, Qi, & Constantinidis, 2012), illustrates this point very clearly. Whereas the original univariate ("rate code–based") analysis of these data indicated that there was elevated sustained activity in feature-tuned neurons, which could be interpreted within the framework of a rate-coding model (Meyer, Qi, Stanford, & Constantinidis, 2011), the multivariate analyses of the same neurons revealed new information that was incompatible with such a model. First, it revealed that, at the population level, the PFC transitioned during the trial from representing the target stimulus to representing the trial's status as a "match" or "nonmatch" trial. Second, the first-order activity profile of any particular neuron could not be interpreted at face value, because "task-relevant information in several neurons was present for only short periods of time relative to the duration of the ... delay period" (p. 4652). This led to the conclusion that "task-relevant information is incorporated into PFC by interleaving/overlapping new information into ongoing dynamic activity that is carrying information about other variables and consequently the absolute firing rate level of a single neuron at a particular time point is often highly ambiguous if the context of the larger population is not taken into account" (Meyers et al., 2012, p. 4652). Other aspects of these studies are considered in Constantinidis's contribution to this volume.

With this understanding of how rate-coding interpretations of the delay-period activity of PFC neurons are being reinterpreted in the context of dynamic, distributed coding models, we can now turn our attention to mechanisms that may underlie the STR of visual information.

THE NEUROPHYSIOLOGICAL BASES OF THE STR OF VISUAL INFORMATION

Distributed Patterns of Activity in Sensory Cortex

The literature reviewed in the previous section (*The Neuroanatomical Bases of the STR of Visual Information*) helps to reconcile the MVPA-based fMRI findings with the results from earlier studies of single-unit activity in posterior regions. For example, recordings from MT in monkeys performing delayed recognition of the direction of motion have failed to find delay-spanning elevated activity in directionally tuned neurons (i.e., no evidence for a sustained rate code, Zaksas & Pasternak, 2006), yet MVPA analyses of fMRI data are able to detect a delay-spanning representation of the target stimulus from MT+ (Emrich et al., 2013; Riggall & Postle, 2012). It may well be that, as with the representation of stimulus information in the PFC (e.g., Meyers et al., 2012; Stokes et al., 2013), inferior temporal cortex (Meyers et al., 2008), and posterior parietal cortex (Crowe, Averbeck, & Chafee, 2010), the STR of motion in MT+ is supported by a dynamic, distributed code that is not evident from the inspection of the activity of single units. A related possibility is considered in some detail in the chapter in this volume by Martinez-Trujillo, which includes evidence that the STR of motion information may be supported by distributed patterns in local field potentials (LFPs) in MT. That is, the delay-period neural code in MT may be subthreshold, in the sense that it is maintained in patterns of oscillating membrane potentials, but not in spiking activity that could be read out by other brain regions. Consistent with this oscillatory account of the STR of information are the electroencephalograph (EEG) findings of Anderson et al. (2014), that the STR of visual orientation information, including its precision, is reflected in patterns of stimulus-induced activity in the alpha band. Because the blood oxygenation level-dependent (BOLD) signal to which fMRI is sensitive corresponds more closely to dynamics of the LFP than to firing rates (e.g., Logothetis, Pauls, Augath, Trinath, & Oeltermann, 2001), it is plausible that successful delay-period decoding of target motion from MT+ with fMRI (Emrich et al., 2013; Riggall & Postle, 2012) may be due to distributed patterns in the LFP in this region.

For the remainder of this chapter, however, we will consider evidence that the MVPA decoding of delay-period activity may not reflect the STR of information, per se, but rather the allocation of attention to this information.

Does the STR of Sensory Information Fundamentally Depend on a Weight-Based Scheme?

It is unequivocally the case that MVPA techniques feature superior sensitivity and specificity than univariate analysis methods (e.g., Lewis-Peacock & Postle, 2012). These facts embolden the articulation of the question posed in the title of this subsection, which derives from the last set of data to be reviewed in this chapter.

The first of three studies to be reviewed here, Lewis-Peacock, Drysdale, Oberauer, and Postle (2012), was an fMRI study of a multistep delayed-recognition task (adapted from Oberauer, 2005) presenting two sample stimuli, then retrocues informing the subject which sample was relevant for each of the two successively presented memory probes. More specifically, each trial began with the presentation of two sample stimuli, always selected from two of three categories (oriented lines, words, and pronounceable pseudowords), one in the top half of the screen and one in the bottom half. After offset of the stimulus display and an initial delay period, a retrocue indicated which sample was relevant for the first recognition probe, followed by a second delay, followed by an initial yes/no recognition probe (and response). Critically, up until this point in the trial, both items needed to be kept in STM, because the first probe was followed by a second retrocue that, with equal probability, would indicate that the same item (a "repeat" trial) or the previously uncued item (a "switch" trial) would be tested by the trial-ending second recognition probe. Thus, the first delay was assumed to require the active retention of two items, whereas the second delay would feature the retention of an "attended memory item" (AMI) and an "unattended memory item" (UMI). (Behavioral data indicate that subjects remove items from the focus of attention following the first retrocue, even though they know that $p=.5$ that the uncued memory item will be cued by the second retrocue (LaRocque, Lewis-Peacock et al., 2013; Oberauer, 2005).) The third delay would only require the retention of an AMI, because it was certain that memory for the item not cued by the second retrocue would not be tested. This design therefore allowed us to unconfound the STR of information from the effects of attending to information in STM. (Note that the vast majority of tasks assessing STM and WM do confound these two factors. Exceptions include the dual-task paradigms summarized in this chapter's *preamble*, the study of Lebedev et al. (2004) that was cited in the section on *The Neurophysiological Bases of the STR of Visual Information*, and the studies reviewed in this subsection.)

Before performing the multistep delayed-recognition task, subjects were first scanned while performing a simple one-delay delayed-recognition task, and the data from this phase 1 scan were used to train the classifier that was then applied to the data from the multistep task. For phase 1, subjects were trained to indicate whether the probe stimulus matched the sample according to a category-specific criterion—synonym judgment for words, rhyme judgment for pseudowords, and an orientation judgment for line segments. Our rationale was that by training the classifiers (separately for each subject) on data from the delay period of this task, we would be training it on patterns of brain activity related to the STR of just a single representational code: phonological (pseudoword trials), semantic (word trials), or visual (line trials). This, in turn, would provide the most unambiguous decoding of delay periods entailing the STR of two AMIs (delay 1) versus of one AMI and one UMI (delay 2) versus of one AMI (delay 3).

As illustrated in Figure 2, for all trial types, classifier evidence for both trial-relevant categories rose steeply at trial onset and remained at the same elevated level until the onset of the first retrocue. This indicated that both items were encoded and sustained in the focus of attention across the initial memory delay, while it was equiprobable that either could be relevant for the first memory response. Following onset of the first retrocue, however, classifier evidence for the two memory items diverged. Postcue brain activity patterns were classified as highly consistent with the category of the cued item, whereas evidence for the uncued item dropped precipitously, becoming indistinguishable from the classifier's evidence for the stimulus category not presented on that trial (i.e., not different from baseline). If the second cue was a repeat cue, classifier evidence for the already-selected memory item remained elevated and that of the uncued item remained indistinguishable from baseline (Figure 2, repeat). If, in contrast, the second cue was a switch cue, classifier evidence for the previously uncued item was reinstated, and evidence for the previously cued item dropped to baseline (Figure 2, switch).

These results raise the possibility that only AMIs are held in a neurally active state. Importantly, despite the apparent loss of sustained activity, UMIs were nonetheless easily remembered on "cue-switch" trials, meaning that they remained "in" STM throughout the trial. Thus, it may be that the STR of information does not depend on the active neural representation of that item. Possible alternative mechanisms, such as the STR of information in a synaptic weight-based format, will be considered toward the end of this section. First, however, we have to consider possible alternative reasons for what amounts to a null finding of multivariate evidence for an active neural representation of UMIs.

One important caveat about the Lewis-Peacock et al. (2012) findings is that, because they were derived from fMRI data, the failure to detect the active representation of UMIs may have been because the STR of the UMI is accomplished via a mechanism to which the BOLD signal is not sensitive. One candidate for such a mechanism is neuronal oscillations, as considered in an earlier subsection (*Distributed Patterns of Activity in Sensory Cortex*). Although the point there was that an oscillatory code may be used as an alternative to a spiking code, and that the BOLD signal is sensitive to dynamics in the LFP, it is nonetheless the case that the BOLD signal is most reliably sensitive to LFP dynamics within certain ranges of frequencies, the gamma band (≥30 Hz) being the most important. Thus, if the STR of information were supported by neuronal oscillations at a lower frequency (e.g., Fuentemilla, Penny, Cashdollar,

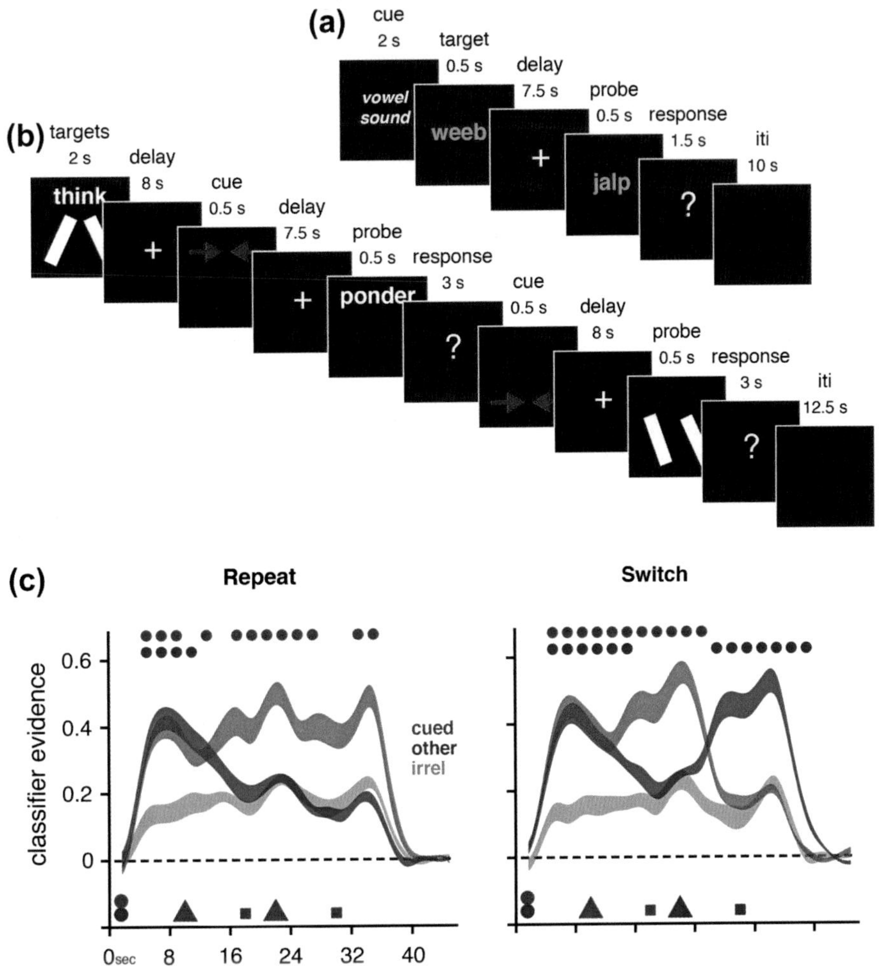

FIGURE 2 Images from experiment 2 of Lewis-Peacock et al. (2012). (a) In the first phase, participants performed short-term recognition of a pseudoword (phonological STM), a word (semantic STM), or two lines (visual STM). (b) In the second phase, during the same scanning session, participants performed short-term recognition with two stimuli (between-category combinations of pseudowords, words, and lines). In half of the trials, the same memory item was selected as behaviorally relevant by the first and second cues (repeat trials), and on the other half of trials the second cue selected the previously uncued item (switch trials). (c) Classifier decoding from experiment 2 of Lewis-Peacock et al. (2012). Results are shown separately for repeat (left) and switch (right) trials. Classifier evidence values for phonological, semantic, and visual were relabeled and collapsed across all trials into three new categories: *cued* (red, the category of the memory item selected by the first cue), *other* (blue, the category of the other memory item), and *irrel* (gray, the trial-irrelevant category). The colored shapes along the horizontal axis indicate the onset of the targets (red and blue circles, 0 s), the first cue (red triangle, 10 s), the first recognition probe (red square, 18 s), the second cue (red or blue triangle, 22 s), and the final recognition probe (red or blue square, 30 s). Data for each category are shown as ribbons whose thicknesses indicate ±1 SEM across participants, interpolated across the 23 discrete data points in the trial-averaged data. Statistical comparisons of evidence values focused on within-subject differences. For every 2-s interval throughout the trial, color-coded circles along the top of each graph indicate that the classifier's evidence for the *cued* or *other* categories, respectively, was reliably stronger ($p < .002$, based on repeated measures t tests, corrected for multiple comparisons) than the evidence for the *irrel* category. (Classification was performed at the whole-brain level, not in restricted ROIs. "Importance maps" showing which brain regions contributed most importantly to the discrimination of each stimulus category are presented in Figure 2 of Lewis-Peacock and Postle (2012).). *Reprinted with permission from Lewis-Peacock, J. A., Drysdale, A. T., Oberauer, K., & Postle, B. R. (January, 2012). Neural Evidence for a Distinction between Short-term Memory and the Focus of Attention.* Journal of Cognitive Neuroscience, *24(1), 61–79. © 2012 by the Massachusetts Institute of Technology.*

Bunzeck, & Düzel, 2010; Jensen, Gelfand, Kounios, & Lisman, 2002; Palva & Palva, 2011; Sauseng et al., 2009; Uhlhaas et al., 2009), fMRI might not be expected to pick this up reliably. Therefore, LaRocque, Lewis-Peacock et al. (2013) designed a follow-up study to replicate the critical features of Lewis-Peacock et al. (2012), with the exception that neural activity was concurrently measured with EEG, rather than with fMRI. EEG is sensitive to neuronal oscillations across a broad, physiologically relevant range of frequencies, and its efficacy for studying STM and WM is highlighted, for example, in Anderson et al. (2014).

The results from the EEG study of LaRocque, Lewis-Peacock et al. (2013) replicated the principal finding from fMRI (Lewis-Peacock et al., 2012): MVPA of the EEG signal failed to find evidence that information that was outside the focus of attention, but nonetheless in STM (i.e., UMIs), was retained in an active state. An additional analysis also ruled out the possibility that a neural representation is represented differently when being retained as a UMI versus when being retained as an AMI. (If this were the case, MVPA of data trained on AMIs from phase 1 might be expected to fail to detect UMIs during phase 2.) This was achieved by implementing MVPA by training and testing on data from delay 2 (i.e., following the first retrocue, when there was one AMI and one UMI) via the leave-one-out cross-validation procedure. These results, illustrated in Figure 3, qualitatively replicated those from the train-on-phase-1-test-on-phase-2 analysis.

Although the findings from the Lewis-Peacock et al. (2012) and the LaRocque, Lewis-Peacock et al. (2013) studies both failed to find evidence that UMIs are maintained in an active state, there is a caveat that applies to both of them because they employed decoding at the level of category (i.e., semantics versus phonology versus visual features), rather than at the level of remembered item. Because of this, they cannot rule out the possibility that the sustained activation of the neural representation of an individual item may not be detectable by a classifier trained to discriminate among categories of stimuli. The most likely reason for this would boil down to sensitivity: absent the boost of selective attention, the activity presumed to underlie the STR of an individual item may simply not be detectable via MVPA of fMRI and EEG data. Although we can never rule out this possibility by conducting additional studies with fMRI or EEG, LaRocque, Riggall et al. (2013) are in the process of carrying out the strongest possible test of this idea with fMRI, by replicating the procedure of Lewis-Peacock et al. (2012), but doing so at the level of individual items all drawn from the same stimulus category. Instead of drawing, for each trial, on two stimuli from each of three possible categories, each trial in the LaRocque, Riggall et al. (2013) study draws from two of three possible directions of motion. The results from this study, look qualitatively identical to the results from Lewis-Peacock et al. (2012) which are illustrated in Figure 1. That is, MVPA evidence for an active neural representation for an item drops to baseline as soon as a retrocue indicates that that item will not be relevant for the impending memory probe. Thus, across two+ studies, we have consistently failed to find evidence for the active neural representation of UMIs in STM. Although this record can be construed as a string of null results, each was, nonetheless, produced with a method boasting superior sensitivity and specificity than any study of STM that preceded it. This invites us to consider the plausibility of the proposition that mechanisms other than sustained activity that may support the STR of visual information.

There is near-universal acceptance among neuroscientists that LTM for past episodes is encoded in the brain via distributed patterns of synaptic weights that, when activated, produce the retrieval of information about that episode. Less often considered, but just as computationally plausible, is that information that must be held for seconds, in STM, might also be encoded in a distributed pattern of synaptic weights. Such a mechanism has also been inferred from multivariate studies of extracellular activity in PFC (Barak, Sussillo, Romo, Tsodyks, & Abbott, 2013; Barak,

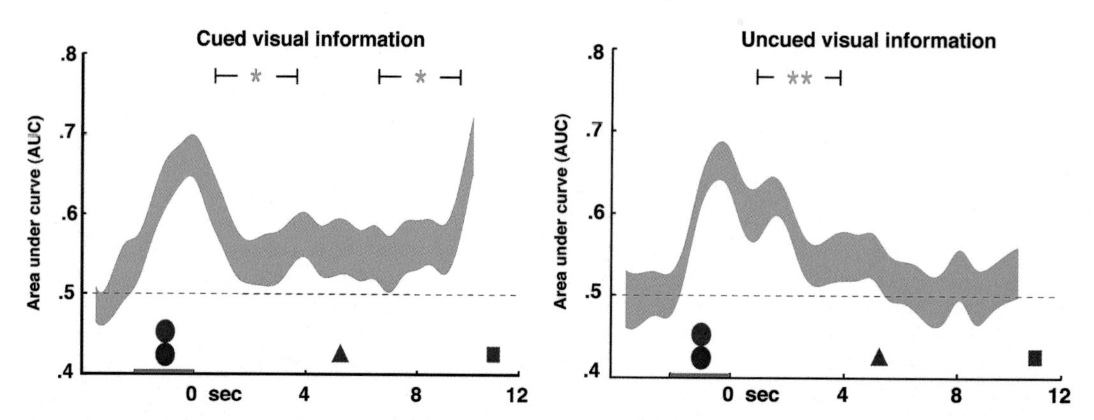

FIGURE 3 Performance of MVPA cross-validation decoding of trials, from LaRocque, Lewis-Peacock et al. (2013), when a visual stimulus was the AMI during delay 2 ("Cued visual information") and when it was the UMI during delay 2 ("Uncued visual information"). Classifier sensitivity to the visual category, determined by calculating the area under the receiver operating characteristic curve (AUC), is plotted for each k-fold cross-validation analysis time window, averaged across subjects. The width of the ribbon corresponds to the standard error of the AUC, calculated across subjects. Width of brackets surrounding significance markers indicates extent of delay period used for statistical analysis. Time is represented on the horizontal axis, with stimulus presentation (circles) from −2 to 0 s, the first cue (triangle) at 5 s, and the first probe (square) at 10.5 s $*p<.05$, $**p<.005$. *Reprinted with permission from LaRocque, J. J., Lewis-Peacock, J. A., Drysdale, A. T., Oberauer, K., & Postle, B. R. (January, 2013). Decoding Attended Information in Short- Term Memory: An EEG Study* Journal of Cognitive Neuroscience, *25(1), 127–142. © 2013 by the Massachusetts Institute of Technology..*

Tsodyks, & Romo, 2010; Stokes et al., 2013). For example, Stokes et al. (2013) offered the following account of activity recorded during performance of a delayed paired-association task, in which the target could be followed by one or more neutral or foil stimuli before its associate appeared:

> "If patterned activity leaves behind a patterned change in the synaptic weights of the network (i.e., hidden state), then subsequent stimulation will be patterned according to the recent stimulation history of a network. Thus, any driving input to the system will trigger a systematic population response that could be used to decode the recent stimulation history of the network. Exactly this phenomenon is seen in our data during the presentation of the neutral stimulus ... Although this stimulus was fixed across trials, the population response was patterned according to the identity of the previous cue, providing a more reliable readout of the memory content than the population response observed during the relatively quiescent delay period." (p. 372)

Relatedly, Sugase-Miyamoto, Liu, Wiener, Optican, and Richmond (2008) have proposed that regions of inferior temporal cortex support visual STM for objects by acting as a "matched filter" that holds a "static copy" of the target in a pattern of synaptic weights. The physiological mechanism(s) that would implement such a transient pattern of weights would be different from those that implement MTL-mediated LTM consolidation. Plausible mechanisms include GluR1-dependent short-term potentiation, which effectively lasts in a tissue slice for a few tens of seconds (Erickson, Maramara, & Lisman, 2009), and transient presynaptic increases in calcium ion concentration, which is the basis for a computational simulation of STM without sustained activity (Mongillo, Barak, & Tsodyks, 2008).

CONCLUSION

In summary, a review of decades-old lesion data, combined with the results of recent neurostimulation studies and multivariate analyses of fMRI and electrophysiological studies in human and nonhuman primates, supports the following propositions about of the STR of visual information:

- It depends on the same neural circuits that are necessary for the real-time perception of visual features, objects, and locations (i.e., "sensory recruitment").
- It may not depend on the sustained activity of neural representations; instead, it may depend on transiently configured patterns of synaptic weights.
- Rather than reflecting the STR of information, per se, the active delay-period representation of target stimuli may be attributable to selective attention, a mechanism that is often experimentally confounded with the STR of information.
- Therefore, although there may, indeed, be multiple states in which information can exist in STM (e.g., Cowan, 1988; Oberauer, 2002; Olivers, Peters, Houtkamp, & Roelfsema, 2011), psychological states of "activation," such as "activated LTM," need not always correspond to a neurophysiological state of activation.

Future Research

There are many important topics for future research. One is how systems that act as the source of attentional control select and interact with neural representations of sensory information (e.g., Nelissen et al., 2013). A second is, assuming that some variant of the structural-trace model of the STR of visual information can be confirmed, an understanding of the circumstances under which such a trace will or will not be "overwritten" by trial-irrelevant information that is presented during the delay period would be invaluable (Stokes et al., 2013; Sugase-Miyamoto et al., 2008). A third is understanding the function of the (metabolically expensive) increases in STM and WM task-related activity that are commonly observed in PFC and IPS, even when MVPA suggests that these regions aren't directly engaged in stimulus representation (e.g., Woolgar, Hampshire, Thompson, & Duncan, 2011).

References

Acheson, D. J., Hamidi, M., Binder, J., & Postle, B. R. (2011). A common neural substrate for language production and verbal working memory. *Journal of Cognitive Neuroscience*, *23*, 1358–1367.

Anderson, D. E., Serences, J. T., Vogel, E. K., & Awh, E. (2014). Induced alpha rhythms track the content and quality of visual working memory representations with high temporal precision. *The Journal of Neuroscience*, *34*, 7587–7599.

Atkinson, R. C., & Shiffrin, R. M. (1968). Human memory: a proposed system and its control processes. In *The psychology of learning and motivation*. New York: Academic Press.

Baddeley, A. D. (1986). *Working memory*. London: Oxford University Press.

Baddeley, A. D., & Hitch, G. J. (1974). Working memory. In G. H. Bower (Ed.), *The psychology of learning and motivation: 8* (pp. 47–89). New York: Academic Press.

Barak, O., Sussillo, D., Romo, R., Tsodyks, M., & Abbott, L. F. (2013). From fixed points to chaos: three models of delayed discrimination. *Progress in Neurobiology, 103*, 214–222.

Barak, O., Tsodyks, M., & Romo, R. (2010). Neuronal population coding of parametric working memory. *The Journal of Neuroscience, 30*(28), 9424–9430.

Christophel, T. B., Hebart, M. N., & Haynes, J. D. (2012). Decoding the contents of visual short-term memory from human visual and parietal cortex. *The Journal of Neuroscience, 32*(38), 12983–12989.

Cowan, N. (1988). Evolving conceptions of memory storage, selective attention, and their mutual constraints within the human information processing system. *Psychological Bulletin, 104*, 163–171.

Crowe, D. A., Averbeck, B. B., & Chafee, M. V. (2010). Rapid sequences of population activity patterns dynamically encode task-critical spatial information in parietal cortex. *The Journal of Neuroscience, 30*, 11640–11653.

D'Esposito, M., Detre, J. A., Alsop, D. C., Shin, R. K., Atlas, S., & Grossman, M. (1995). The neural basis of the central executive system of working memory. *Nature, 378*, 279–281.

D'Esposito, M., & Postle, B. R. (1999). The dependence of span and delayed-response performance on prefrontal cortex. *Neuropsychologia, 37*, 1303–1315.

D'Esposito, M., Postle, B. R., Jonides, J., & Smith, E. E. (1999). The neural substrate and temporal dynamics of interference effects in working memory as revealed by event-related functional MRI. *Proceedings of the National Academy of Sciences United States of America, 96*, 7514–7519.

Daneman, M., & Hannon, B. (2007). What do working memory span tasks like reading span really measure? In N. Osaka, R. H. Logie, & M. D'Esposito (Eds.), *The cognitive neuroscience of working memory* (pp. 21–42). Oxford: Oxford University Press.

Duncan, J. (2010). The multiple-demand (MD) system of the primate brain: mental programs for intelligent behaviour. *Trends in Cognitive Sciences, 14*(4), 172–179.

Duncan, J., & Miller, E. K. (2002). Cognitive focus through adaptive neural coding in the primate prefrontal cortex. In D. Stuss & R. Knight (Eds.), *Principles of frontal lobe function* (pp. 278–291). Oxford: Oxford University Press.

Emrich, S. M., Riggall, A. C., Larocque, J. J., & Postle, B. R. (2013). Distributed patterns of activity in sensory cortex reflect the precision of multiple items maintained in visual short-term memory. *The Journal of Neuroscience, 33*(15), 6516–6523.

Engle, R. W., Tuholski, S. W., Laughlin, J. E., & Conway, A. R. A. (1999). Working memory, short-term memory, and general fluid intelligence: a latent-variable approach. *Journal of Experimental Psychology: General, 128*, 309–331.

Erickson, M. A., Maramara, L. A., and Lisman, J. (2009). A single brief burst induces GluR1-dependent associative short-term potentiation: a potential mechanism for short-term memory. *Journal of Cognitive Neuroscience 22*, 2530–2540.

Ester, E. F., Anderson, D. E., Serences, J. T., & Awh, E. (2013). A neural measure of precision in visual working memory. *Journal of Cognitive Neuroscience, 25*(5), 754–761.

Feredoes, E., & Postle, B. R. (2010). Prefrontal control of familiarity and recollection in working memory. *Journal of Cognitive Neuroscience, 22*, 323–330.

Feredoes, E., Tononi, G., & Postle, B. R. (2006). Direct evidence for a prefrontal contribution to the control of proactive interference in verbal working memory. *Proceedings of the National Academy of Science United States of America, 103*, 19530–19534.

Freedman, D. J., Riesenhuber, M., Poggio, T., & Miller, E. K. (2003). A comparison of primate prefrontal and inferior temporal cortices during visual categorization. *The Journal of Neuroscience, 23*(12), 5235–5246.

Fuentemilla, L., Penny, W. D., Cashdollar, N., Bunzeck, N., & Düzel, E. (2010). Theta-coupled periodic replay in working memory. *Current Biology, 20*, 606–612.

Funahashi, S., Bruce, C. J., & Goldman-Rakic, P. S. (1989). Mnemonic coding of visual space in the monkey's dorsolateral prefrontal cortex. *Journal of Neurophysiology, 61*, 331–349.

Funahashi, S., Bruce, C. J., & Goldman-Rakic, P. S. (1990). Visuospatial coding in primate prefrontal neurons revealed by oculomotor paradigms. *Journal of Neurophysiology, 63*, 814–831.

Fuster, J. M. (2002). Physiology of executive functions: the perception-action cycle. In D. T. Stuss & R. T. Knight (Eds.), *Principles of frontal lobe function* (pp. 96–108). Oxford: Oxford University Press.

Fuster, J. M., & Alexander, G. E. (1971). Neuron activity related to short-term memory. *Science, 173*, 652–654.

Goldman-Rakic, P. S. (1987). Circuitry of the prefrontal cortex and the regulation of behavior by representational memory. In V. B. Mountcastle, F. Plum, & S. R. Geiger (Eds.), *Handbook of neurobiology* (pp. 373–417). Bethesda: American Physiological Society.

Goldman-Rakic, P. S. (1992). Working memory and the mind. *Scientific American, 267*, 110–117.

Hamidi, M., Tononi, G., & Postle, B. R. (2008). Evaluating frontal and parietal contributions to spatial working memory with repetitive transcranial magnetic stimulation. *Brain Research, 1230*, 202–210.

Han, X., Berg, A. C., Oh, H., Samaras, D., & Leung, H. C. (2013). Multi-voxel pattern analysis of selective representation of visual working memory in ventral temporal and occipital regions. *Neuroimage, 73*, 8–15.

Harrison, S. A., & Tong, F. (2009). Decoding reveals the contents of visual working memory in early visual areas. *Nature, 458*, 632–635.

Haynes, J.-D., & Rees, G. (2006). Decoding mental states from brain activity in humans. *Nature Reviews Neuroscience, 7*, 523–534.

Hebb, D. O. (1949). *The organization of behavior: A neuropsychological theory*. New York, NY: John Wiley & Sons, Inc.

Jensen, O., Gelfand, J., Kounios, J., & Lisman, J. E. (2002). Oscillations in the alpha band (9-12 Hz) increase with memory load during retention in a short-term memory task. *Cerebral Cortex, 12*, 877–882.

Jonides, J., Smith, E. E., Marshuetz, C., Koeppe, R. A., & Reuter-Lorenz, P. A. (1998). Inhibition of verbal working memory revealed by brain activation. *Proceedings of the National Academy of Sciences, 95*, 8410–8413.

Keppel, G., & Underwood, B. J. (1962). Proactive inhibition in short-term retention of single items. *Journal of Verbal Learning and Verbal Behavior, 1*, 153–161.

Koenigs, M., Acheson, D. J., Barbey, A. K., Solomon, J., Postle, B. R., & Grafman, J. (2011). Areas of left perisylvian cortex mediate auditory-verbal short-term memory. *Neuropsychologia, 49*(13), 3612–3619.

Konstantinou, N., & Lavie, N. (2013). Dissociable roles of different types of working memory load in visual detection. *Journal of Experimental Psychology: Human Perception and Performance, 39*, 919–924.

Kriegeskorte, N., Goebel, R., & Bandettini, P. A. (2006). Information-based functional brain mapping. *Proceedings of the National Academy of Science United States of America, 103*, 3863–3868.

Kubota, K., & Niki, H. (1971). Prefrontal cortical unit activity and delayed alternation performance in monkeys. *Journal of Neurophysiology, 34*(3), 337–347.

LaRocque, J. J., Lewis-Peacock, J. A., Drysdale, A., Oberauer, K., & Postle, B. R. (2013). Decoding attended information in short-term memory: an EEG study. *Journal of Cognitive Neuroscience, 25*, 127–142.

LaRocque, J. J., Riggall, A. C., Emrich, S. M., & Postle, B. R. (2013). Active representations of individual items in short-term memory: a matter of attention, not retention. *Presentation at the Annual Meeting of the Society for Neuroscience.*

Lebedev, M. A., Messinger, A., Kralik, J. D., & Wise, S. P. (2004). Representation of attended versus remembered locations in prefrontal cortex. *PloS Biology, 2*, 1919–1935.

Lee, S. H., Kravitz, D. J., & Baker, C. I. (2013). Goal-dependent dissociation of visual and prefrontal cortices during working memory. *Nature Neuroscience, 16*(8), 997–999.

Leff, A. P., Schofield, T. M., Crinion, J. T., Seghier, M. L., Grogan, A., Green, D. W., et al. (2009). The left superior temporal gyrus is a shared substrate for auditory short-term memory and speech comprehension: evidence from 210 patients with stroke. *Brain, 132*(Pt 12), 3401–3410.

Lewis-Peacock, J. A., Drysdale, A., Oberauer, K., & Postle, B. R. (2012). Neural evidence for a distinction between short-term memory and the focus of attention. *Journal of Cognitive Neuroscience, 23*, 61–79.

Lewis-Peacock, J. A., & Postle, B. R. (2008). Temporary activation of long-term memory supports working memory. *The Journal of Neuroscience, 28*, 8765–8771.

Lewis-Peacock, J. A., & Postle, B. R. (2012). Decoding the internal focus of attention. *Neuropsychologia, 50*, 470–478.

Linke, A. C., Vicente-Grabovetsky, A., Mitchell, D. J., & Cusack, R. (2011). Encoding strategy accounts for individual differences in change detection measures of VSTM. *Neuropsychologia, 49*, 1476–1486.

Logie, R. H., & Niven, E. (2012). Working memory: an ensemble of functions in on-line cognition. In V. Gyselinck & F. Pazzaglia (Eds.), *From mental imagery to spatial cognition and language: Essays in honour of Michel Denis* (pp. 77–105). Hove: Psychology Press.

Logothetis, N. K., Pauls, J., Augath, M., Trinath, T., & Oeltermann, A. (2001). Neurophysiological investigation of the basis of the fMRI signal. *Nature, 412*, 150–157.

Luck, S. J., & Vogel, E. K. (2013). Visual working memory capacity: from psychophysics and neurobiology to individual differences. *Trends in Cognitive Sciences, 17*, 391–400.

Malmo, R. B. (1942). Interference factors in delayed response in monkey after removal of the frontal lobes. *Journal of Neurophysiology, 5*, 295–308.

Meyer, T., Qi, X. L., Stanford, T. R., & Constantinidis, C. (2011). Stimulus selectivity in dorsal and ventral prefrontal cortex after training in working memory tasks. *The Journal of Neuroscience, 31*(17), 6266–6276.

Meyers, E. M., Freedman, D. J., Kreiman, G., Miller, E. K., & Poggio, T. (2008). Dynamic population coding of category information in inferior temporal and prefrontal cortex. *Journal of Neurophysiology, 100*(3), 1407–1419.

Meyers, E. M., Qi, X. L., & Constantinidis, C. (2012). Incorporation of new information into prefrontal cortical activity after learning working memory tasks. *Proceedings of the National Academy of Science United States of America, 109*(12), 4651–4656.

Miller G. A., Galanter E., Pribham K. H. (1960). Plans and the Structure of Behavior. New York: Holt, Rinehart andWinston.

Milner, B. (1964). Some effects of frontal lobectomy in man. In J. M. Warren & K. Akert (Eds.), *The frontal granular cortex and behavior* (pp. 313–334). New York: McGraw-Hill.

Mishkin, M. (1964). Perseveration of central sets after frontal lesions in monkeys. In J. M. Warren & K. Akert (Eds.), *The frontal granular cortex and behavior* (pp. 219–237). New York: McGraw-Hill.

Mongillo, G., Barak, O., & Tsodyks, M. (2008). Synaptic theory of working memory. *Science 319*, 1543–1546.

Nelissen, N., Stokes, M., Nobre, A. C., & Rushworth, M. F. (2013). Frontal and parietal cortical interactions with distributed visual representations during selective attention and action selection. *Journal of Neuroscience, 33*(42), 16443–16458.

Norman, K. A., Polyn, S. M., Detre, G. J., & Haxby, J. V. (2006). Beyond mind-reading: multi-voxel pattern analysis of fMRI data. *Trends in Cognitive Sciences, 10*, 424–430.

Norman, D. A., & Shallice, T. (1980). *Attention to action: Willed and automatic control of behavior*. San Diego: University of California.

Oberauer, K. (2002). Access to information in working memory: exploring the focus of attention. *Journal of Experimental Psychology: Learning, Memory, and Cognition, 28*, 411–421.

Oberauer, K. (2005). Control of the contents of working memory—A comparison of two paradigms and two age groups. *Journal of Experimental Psychology: Learning, Memory, and Cognition, 31*, 714–728.

Olivers, C. N. L., Peters, J., Houtkamp, R., & Roelfsema, P. R. (2011). Different states in visual working memory: when it guides attention and when it does not. *Trends in Cognitive Sciences, 15*, 327–334.

Palva, S., & Palva, J. M. (2011). Functional roles of alpha-band phase synchronization in local and large-scale cortical networks. *Frontiers in Psychology, 2*. http://dx.doi.org/10.3389/fpsyg.2011.00204.

Pasternak, T., & Greenlee, M. W. (2005). Working memory in primate sensory systems. *Nature Reviews Neuroscience, 6*, 97–107.

Pereira, F., Mitchell, T., & Botvinick, M. M. (2009). Machine learning classifiers and fMRI: a tutorial overview. *NeuroImage, 45*, S199–S209.

Petrides, M. (2000). Dissociable roles of mid-dorsolateral prefrontal and anterior inferotemporal cortex in visual working memory. *Journal of Neuroscience, 20*, 7496–7503.

Postle, B. R. (2006). Working memory as an emergent property of the mind and brain. *Neuroscience, 139*, 23–38.

Postle, B. R. (2007). Activated long-term memory? the bases of representation in working memory. In N. Osaka, R. H. Logie, & M. D. Esposito (Eds.), *The cognitive neuroscience of working memory* (pp. 333–350). Oxford, U.K: Oxford University Press.

Postle, B. R. (2015). The cognitive neuroscience of visual short-term memory. *Current Opinion in Behavioral Sciences, 1*, 40–46.

Postle, B. R., Ferrarelli, F., Hamidi, M., Feredoes, E., Massimini, M., Peterson, M., et al. (2006). Repetitive transcranial magnetic stimulation dissociates working memory manipulation from retention functions in prefrontal, but not posterior parietal, cortex. *Journal of Cognitive Neuroscience, 18*, 1712–1722.

Pribram, K. H., Ahumada, A., Hartog, J., & Roos, L. (1964). A progress report on the neurological processes disturbed by frontal lesions in primates. In J. M. Warren & K. Akert (Eds.), *The frontal granular cortex and behavior* (pp. 28–55). New York: McGraw-Hill Book Company.

Ranganath, C., & Blumenfeld, R. S. (2005). Doubts about double dissociation between short-and long-term memory. *Trends in Cognitive Sciences, 9*, 374–380.

Ranganath, C., Cohen, M. X., & Brozinsky, C. J. (2005). Working memory maintenance contributes to long-term memory formation: neural and behavioral evidence. *Journal of Cognitive Neuroscience, 17*, 994–1010.

Rao, S. C., Rainer, G., & Miller, E. K. (1997). Integration of what and where in the primate prefrontal cortex. *Science, 276*, 821–824.

Richardson, F. M., Ramsden, S., Ellis, C., Burnett, S., Megnin, O., Catmur, C., et al. (2011). Auditory short-term memory capacity correlates with gray matter density in the left posterior STS in cognitively normal and dyslexic adults. *Journal of Cognitive Neuroscience, 23*(12), 3746–3756.

Riggall, A. C., & Postle, B. R. (2012). The relationship between working memory storage and elevated activity as measured with funtional magnetic resonance imaging. *The Journal of Neuroscience, 32*, 12990–12998.

Sauseng, P., Klimesch, W., Heise, K. F., Gruber, W. P., Holz, E., Karim, A. A., et al. (2009). Brain oscillatory substrates of visual short-term memory capacity. *Current Biology, 19*, 1846–1852.

Serences, J. T., Ester, E. F., Vogel, E. K., & Awh, E. (2009). Stimulus-specific delay activity in human primary visual cortex. *Psychological Science, 20*, 207–214.

Serences, J. T., & Saproo, S. (2012). Computational advances towards linking BOLD and behavior. *Neuropsychologia, 50*(4), 435–446.

Shipstead, Z., & Engle, R. W. (2013). Interference within the focus of attention: working memory tasks reflect more than temporary maintenance. *Journal of Experimental Psychology: Learning, Memory, and Cognition, 39*, 277–289.

Silvanto, J., & Cattaneo, Z. (2010). Transcranial magnetic stimulation reveals the content of visual short-term memory in the visual cortex. *Neuroimage, 50*(4), 1683–1689.

Sreenivasan, K. K., Curtis, C. E., & D'Esposito, M. (2014). Revisiting the role of persistent neural activity in working memory. *Trends in Cognitive Sciences, 18*, 82–89.

Sreenivasan, K., Vytlacil, J., & D'Esposito, M. (2014). Distributed and dynamic storage of working memory stimulus information in extrastriate cortex. *Journal of Cognitive Neuroscience, 26*, 1141–1153.

Stokes, M. G., Kusunoki, M., Sigala, N., Nili, H., Gaffan, D., & Duncan, J. (2013). Dynamic coding for cognitive control in prefrontal cortex. *Neuron, 78*(2), 364–375.

Sugase-Miyamoto, Y., Liu, Z., Wiener, M. C., Optican, L. M., & Richmond, B. J. (2008). Short-term memory trace in rapidly adapting synapses of inferior temporal cortex. *PLoS Computational Biology, 4*(5), e1000073.

Takeda, K., & Funahashi, S. (2002). Prefrontal task-related activity representing visual cue location or saccade direction in spatial working memory tasks. *Journal of Neurophysiology, 87*(1), 567–588.

Takeda, K., & Funahashi, S. (2004). Population vector analysis of primate prefrontal activity during spatial working memory. *Cerebral Cortex, 14*(12), 1328–1339.

Takeda, K., & Funahashi, S. (2007). Relationship between prefrontal task-related activity and information flow during spatial working memory performance. *Cortex, 43*(1), 38–52.

Thompson-Schill, S., Jonides, J., Marshuetz, C., Smith, E. E., D'Esposito, M., Kan, I. P., et al. (2002). Effects of frontal lobe damage on interference effects in working memory. *Cognitive, Affective, and Behavioral Neuroscience, 2*, 109–120.

Tsujimoto, S., & Postle, B. R. (2012). The prefrontal cortex and delay tasks: a reconsideration of the "mnemonic scotoma". *Journal of Cognitive Neuroscience, 24*, 627–635.

Tsujimoto, S., & Sawaguchi, T. (2004). Properties of delay-period neuronal activity in the primate prefrontal cortex during memory- and sensory-guided saccade tasks. *European Journal of Neuroscience, 19*(2), 447–457.

Uhlhaas, P. J., Pipa, G., Lima, B., Melloni, L., Neuenschwander, S., Nikolic, D., et al. (2009). Neural synchrony in cortical networks: history, concept and current status. *Frontiers in Integrative Neuroscience, 3*. http://dx.doi.org/10.3389/neuro.3307.3017.2009.

Vogel, E. K., McCollough, A. W., & Machizawa, M. G. (2005). Neural measures reveal individual differences in controlling access to working memory. *Nature, 438*, 368–387.

Warren, J. M., & Akert, K. (Eds.). (1964). *The frontal granular cortex and behavior*. New York: McGraw-Hill Book Company.

Wilson, R. I., & Mainen, Z. F. (2006). Early events in olfactory processing. *Annual Review of Neuroscience, 29*, 163–201.

Woolgar, A., Hampshire, A., Thompson, R., & Duncan, J. (2011). Adaptive coding of task-relevant information in human frontoparietal cortex. *Journal of Neuroscience, 31*(41), 14592–14599.

Zaksas, D., & Pasternak, T. (2006). Directional signals in the prefrontal cortex and in area MT during a working memory for visual motion task. *The Journal of Neuroscience, 26*, 11726–11742.

CHAPTER

6

What are the Roles of Sensory and Parietal Activity in Visual Short-Term Memory?

Stephen M. Emrich

Department of Psychology, Brock University, St Catharines, ON, Canada

INTRODUCTION

Although its duration is perhaps the most obvious feature that delineates short-term memory from other memory systems (i.e., iconic memory, long-term memory), the feature that has received the most attention is the limited capacity of short-term memory. Relative to the near infinite capacities of iconic and long-term memory (Brady, Konkle, Alvarez, & Oliva, 2008; Sperling, 1960), the representation of information in short-term memory appears to be subject to severe limitations on information processing (Cowan, 2001). The interest in capacity limits has been perhaps most notable in the study of visual short-term memory (VSTM), which mediates the representation of visual information after it has been removed from view. This approach to VSTM capacity limits has been incredibly fruitful, with substantial research dedicated to investigating the nature and quality of VSTM capacity. These capacity estimates have been very useful in furthering our understanding the role of VSTM in cognition because performance on a range of cognitive tasks, from measures of fluid intelligence (Engle, Kane, & Tuholski, 1999; Engle, Tuholski, Laughlin, & Conway, 1999) to visual search (Emrich, Al-Aidroos, Pratt, & Ferber, 2010; Kundu, Sutterer, Emrich, & Postle, 2013), has been linked to individual variation in VSTM performance, as measured by estimates of VSTM capacity.

Over the past decade, numerous studies using functional magnetic resonance imaging (fMRI) as well as electro- and magnetoencephalography (EEG/MEG) have provided strong converging evidence in support of specific capacity limits in VSTM, both at a group level and across individual differences in VSTM capacities. Specifically, much of the neural data appeared to parallel the behavioral models, in that neural activity associated with VSTM maintenance appeared to similarly be limited to a small number of items (Todd & Marois, 2004; Vogel & Machizawa, 2004). Thus, the study of VSTM over the past decade has been arguably one of the major successes in the field of cognitive neuroscience, as the behavioral and neural evidence largely converged in support of a unified, integrated model—namely, that VSTM is a limited-capacity system in which a fixed number of objects can be represented that is mediated by a specific neural substrate.

Although the evidence in support of this model was never equivocal, there have been two recent and noteworthy developments which have provided significant challenges to the notion of a fixed capacity limit to VSTM and/or to its underlying neural mechanisms. First, developments in the behavioral and psychophysical models that explain VSTM performance challenge the notion that VSTM stores a fixed number of discrete objects, and instead conceptualize VSTM as a limited pool of resources that are allocated across memoranda (Bays, Catalao, & Husain, 2009; Bays & Husain, 2008; Wilken & Ma, 2004). According to these models, VSTM representations will vary in their precision or fidelity as a function of the proportion of resources they receive. Second, the use of multivoxel pattern analysis (MVPA) in fMRI has revealed that the representation of information in VSTM may occur entirely independently of those areas previously thought to support the maintenance of a fixed number of items in VSTM (Harrison & Tong, 2009; Serences, Ester, Vogel, & Awh, 2009).

Mechanisms of Sensory Working Memory
http://dx.doi.org/10.1016/B978-0-12-811042-3.00006-2

Although a new picture of VSTM is beginning to emerge, it is unclear how these novel findings fit in with the extant models of VSTM capacity limitations or how neural activity associated with VSTM may relate to variable precision models. Consequently, the goal of this chapter is twofold: first, to examine how recent findings in both behavioral models and neuroimaging may be inconsistent with extant models of VSTM capacity limitations and to examine ways in which these data may provide a novel framework. Specifically, I will present recent evidence from fMRI that provides a plausible neural mechanism for variable-precision/resource-based models. Second, given these novel findings, I will examine data that may provide alternative interpretations to the traditional neural signatures of VSTM maintenance.

EVIDENCE FOR A FIXED CAPACITY OF VSTM

Behavioral Evidence

Before examining the current state of the VSTM field, it is worth outlining the behavioral and neuroimaging evidence that has shaped the predominant models of VSTM capacity over the past few decades. There have been a number of recent reviews that have covered this literature much more extensively (Brady, Konkle, & Alvarez, 2011; Luck & Vogel, 2013).

The study of VSTM capacity largely began with the seminal study by Luck and Vogel (Luck & Vogel, 1997), although earlier studies also examined the nature of VSTM capacity limits (Palmer, 1990; Phillips, 1974). To determine the capacity of VSTM, Luck and Vogel (1997) employed a change-detection paradigm developed by Phillips (1974). In the change-detection task, an initial sample array of a number of simple visual objects (e.g., colored squares) is briefly presented. Following a short delay period, a test array is presented, and participants must indicate whether or not there has been a change between the sample and test displays.

The study of Luck and Vogel (Luck & Vogel, 1997; see also, Vogel, Woodman, & Luck, 2001) provided two lines of evidence in support of a specific capacity of a limited number storage "slots," each of which could maintain a fixed-resolution item. First, when the number of sample items was between one and three items, change-detection performance was close to ceiling, indicating that participants had nearly perfect memory for up to around three objects, suggesting a capacity limit of roughly three to four objects. Beyond this limit, items are assumed to no longer be stored in memory, resulting in an increase in the number of change-detection errors. In addition, change-detection performance remained constant regardless of whether participants had to detect changes to a single feature (i.e., color, orientation) or multiple features (i.e., color + orientation, two colors). This finding suggested that the number of items that could be maintained in VSTM was not only stable, but that multiple features of an "object" were maintained as bound representations, each of which was stored with a fixed-resolution "slot."

fMRI Evidence

Although other studies found evidence inconsistent with a pure fixed-resolution model of VSTM capacity, instead suggesting that the number of items that can be stored is also limited by the amount of information that must be maintained, or the resolution with which items must be represented (Alvarez & Cavanagh, 2004; Olson & Jiang, 2002), the evidence from both fMRI and electrophysiology largely supported the fixed-capacity model.

Using fMRI, a number of studies examined the effect of VSTM load on delay-period activity, and found load-dependent changes of blood-oxygen level dependent (BOLD) signal across a range of regions (Linden et al., 2003; Munk et al., 2002). Todd and Marois (2004) were the first to vary VSTM load systematically in a change-detection paradigm similar to that of Luck and Vogel (1997). To identify those areas associated with the number of items maintained, rather than the number of items presented, a *K*-weighted multiple regression was performed. That is, this analysis isolated those areas in which activity increased as a function of individual VSTM capacity (*K*), as defined by the formula established by Pashler and modified by Cowan (Cowan, 2001; Pashler, 1987). The results of this analysis revealed that activation in bilateral intraparietal sulcus (IPS) and intraoccipital sulcus (IOS) increased with increasing VSTM load and reached an asymptote around three to four items. Importantly, load-dependent activity in this region was also sustained throughout the delay period, providing strong evidence that this activity was involved in the online representation of information in VSTM.

The results of the study by Todd and Marois (2004) were largely consistent with the limited-capacity "slot" based model of VSTM: activity increased monotonically with increasing VSTM load, and reached an apparent plateau at an individual's given VSTM capacity. Moreover, individual differences in VSTM capacity were found to correlate with individual differences in IPS/IOS activity (Todd, Fougnie, & Marois, 2005), providing further evidence that this activity is linked to the capacity limit of VSTM. Later studies by Xu and Chun (Xu & Chun, 2006; Xu, 2007) provided evidence that activity in the IPS dissociates different aspects of VSTM performance: whereas activity in the inferior IPS appears to index a limited number of objects in different spatial locations, the superior IPS (as well as the lateral occipital complex) appears to reflect the complexity of the objects stored. Thus, although these data do not support a fixed-precision model (because both behavioral performance and fMRI delay-period BOLD signal was affected by the complexity of objects stored in VSTM), it is consistent with a model in which a limited number of items are stored in VSTM. Subsequent studies, however, have called into question this dissociation of the IPS (Harrison, Jolicœur, & Marois, 2010), suggesting instead that IPS activity represents the number of locations selected.

Although the precise nature of IPS activity remains unclear (e.g., Mitchell & Cusack, 2008), the results of these studies, together with the strong behavioral evidence in support of a fixed capacity of VSTM, provided strong evidence in support of an integrated model of VSTM in which VSTM is supported by those regions where activity is sustained throughout the delay period. Moreover, studies using electrophysiological techniques have isolated activity similar to that observed in fMRI, providing converging evidence in support of this model.

Evidence from Electrophysiology

Early EEG studies of working memory identified a slow-wave event-related potential (ERP) over posterior electrodes that varied with working memory load (Ruchkin, Canoune, Johnson, & Ritter, 1995; Ruchkin, Johnson, Canoune, & Ritter, 1990). Moreover, a similar slow-wave ERP was found to be greater over contralateral channels (Klaver, Talsma, Wijers, Heinze, & Mulder, 1999) and varied with load, but not with the number of features (Klaver, Smid, & Heinze, 1999), consistent with a fixed-resolution VSTM mechanism. Vogel and Machizawa (2004) were the first to examine these slow-wave ERPs relative to VSTM capacity using a change-detection task similar to the ones used by Luck and Vogel (1997) as well as the one used in the fMRI study by Todd and Marois (2004). The results of this study revealed a slow-wave ERP, referred to as the sustained posterior contralateral negativity (SPCN)[1], which was present over posterior electrodes and sustained throughout the delay period. Moreover, the amplitude of the SPCN increased with increasing VSTM load, suggesting that it played a role in the online maintenance of information in VSTM.

Importantly, there were two features of the SPCN that paralleled the results of Todd and Marois (2004) and that provided evidence in support of the three- to four-item capacity limit of VSTM: first, although the amplitude of the SPCN increases with VSTM load, it reaches an asymptote at around three to four items; second, individual differences in VSTM capacity (as measured by K) were predicted by individual differences in SPCN amplitude (specifically, the amplitude difference between the four- and two-item conditions). In other words, "high-capacity" individuals showed greater SPCN changes between low- and high-load conditions, suggesting that the amplitude of the SPCN reflects the number of items that are *stored* in VSTM. Thus, the extent of activity observed in both IPS BOLD activity and the SPCN is predicted by individual differences in VSTM capacity.

Though the SPCN seems to provide strong evidence in favor of a limited capacity mechanism of VSTM storage, the evidence in support of a fixed-precision mechanism is less clear. One study by Luria and Vogel (2011a) presented objects with a varying number of features and found that although the amplitude of the SPCN tracked largely with the number of objects, there were differences between, for example, two objects with a single color and two multicolored objects. Wilson and colleagues (Wilson, Adamo, Barense, & Ferber, 2012) found evidence that the sensitivity of the SPCN to features or number of objects may depend on the specific electrode pair examined. Specifically, the amplitude of the SPCN at electrode sites P1/P2 scaled with the number of discrete objects, whereas at the more lateral electrode pair P7/P8, the amplitude of the SPCN was greater for three bound features than for three separate features. Thus, this finding provides an important parallel to the fMRI findings of Xu and Chun (2006), providing converging evidence that although VSTM may be able to store a limited number of objects, the information load present in those objects is also a limiting factor of both behavioral performance and neural activity. Importantly, several studies using MEG, as well as EEG and fMRI, have isolated the source of the SPCN to the IPS, although other generators may be present in regions of parietal occipital and ventral occipital areas, and activity observed across these different methodologies may not be identical in nature (Mitchell & Cusack, 2011; Robitaille et al., 2010; Robitaille, Grimault, & Jolicœur, 2009).

[1] The SPCN is also referred to as the contralateral delay activity or the contralateral negative slow wave.

CHALLENGES TO THE INTEGRATED FIXED-CAPACITY MODEL

Flexible-Resource Models

Although a decade of behavioral, electrophysiological, and neuroimaging studies converged on a common model—namely, that sustained activity in the IPS could support the maintenance of a limited number of objects in VSTM—recent findings challenge some of the core aspects of this model. In particular, several recent studies reexamining the nature of VSTM capacity limitations have made use of a recall task developed by Wilken and Ma (2004), in which participants have to report an item from VSTM along a continuous dimension (e.g., color, orientation, spatial frequency, motion direction). Instead of measuring accuracy by determining whether participants could detect a change to a sample stimulus, VSTM performance is measured by the amount of error in the participant's response around the target value (i.e., the item that was to be reported). That is, the amount of error in the response is assumed to reflect the noise in the underlying VSTM representation. Using this technique, Wilken and Ma (2004) demonstrated that, across a range of memory stimuli, the amount of error increased as a function of set size. Importantly, the amount of error continued to increase at sizes greater than four, suggesting there was no apparent capacity limit on the number of items that could be maintained in VSTM. Bays and Husain (Bays & Husain, 2008) also measured the error for locations and orientations and found that the amount of error in responses increased according to a power law, even at set sizes greater than four items. Thus, these studies suggest that the number of items that can be stored in VSTM is not fixed to an upper limit of three to four items, but rather that VSTM precision decreases as a function of load. Specifically, these models conceptualize VSTM as a limited pool of (neural) resources that are allocated across the to-be-remembered information; as the number of items increases, fewer resources are allocated to each item, resulting in increased encoding noise and greater response error.

Subsequent studies have added additional factors into the models that describe the data in tasks similar to the ones used by Wilken and Ma (2004). Zhang and Luck (2008) added a uniform distribution of responses to capture "guesses:" those trials on which the error is not systematically related to the target, but rather randomly distributed. The results of this study suggested that although the probability that a response would fall onto a random uniform distribution (i.e., a "guess") continued to increase beyond four items, the error term for the correctly recalled items (i.e., precision) reached a plateau around three items. Thus, Zhang and Luck (2008) argued that although the precision of representations could vary when the number of items was low, VSTM performance was still best explained by three to four discrete, fixed-resolution storage slots. Bays and colleagues (Bays et al., 2009) later added a third component into the mixture model, which attributed some of the apparent guessing to nontarget errors (responses in which participants reported one of the items present in the memory sample that was not the target), and found no evidence for a fixed limit on the number of items that could be recalled. Other authors have added even more factors into the models, including aspects such as variability in encoding precision (Van den Berg, Shin, Chou, George, & Ma, 2012) or random fluctuations in attentional allocation (Fougnie, Suchow, & Alvarez, 2012) to explain behavioral performance most accurately. The question of how best to characterize the limited capacity of VSTM is still being addressed (Van den Berg, Awh, & Ma, 2014), and as such it remains unclear whether there exists a fixed capacity on the number of items that can be remembered. At a minimum, it is clear that a comprehensive neural model of VSTM must be able to account for changes in VSTM precision (Ma, Husain, & Bays, 2014).

Sensory Representation of VSTM Stimuli

Further challenges to the classic model of VSTM capacity limits have come from advances in neuroimaging analysis techniques. Most notably, several recent studies have used applied multivariate analysis techniques to fMRI studies of VSTM to attempt to decode the specific items being held in VSTM from BOLD signal. In one study, Harrison and Tong (2009) presented participants with images of two oriented gratings, and participants were subsequently instructed to remember either the first or the second grating over a delay period. By training a multivariate pattern classifier on the different stimuli, the authors could then attempt to "decode" which of the two gratings was being maintained in VSTM on a given trial. Using this MVPA technique, the authors revealed that it is possible to identify from the patterns of an fMRI BOLD signal which of the two stimuli was being remembered. Similar findings have been found for color (Serences et al., 2009) and visual motion stimuli (Riggall & Postle, 2012).

Several other authors have outlined the advantages of these multivariate approaches relative to univariate analysis of fMRI BOLD signal (Coutanche, 2013; Davis & Poldrack, 2013; Serences & Saproo, 2012), but for the purposes of investigating VSTM, there are two key advantages worth mentioning: first, by training classifiers for particular stimuli, these studies are dealing at the level of the representation. That is, unlike studies investigating gross level activity across a range of stimuli, these studies address the extent to which evidence for a *particular* item held in VSTM can be found in the fMRI BOLD signal. Second, MVPA makes no assumptions about the level of BOLD signal, but rather identifies patterns of activity across a number of voxels that are consistent to a condition. Thus, finding evidence for a maintained representation does not depend on an assumption about the presence of elevated, sustained, BOLD signal during the delay period. Maintenance-related activity can still be identified, for example, by identifying the presence of consistent patterns of activity across the delay period (Harrison & Tong, 2009; Riggall & Postle, 2012; Serences et al., 2009).

Given these advantages, the studies mentioned above present a significant challenge to the "classic" model of VSTM maintenance. Namely, the specific contents of VSTM have primarily been decoded from areas of sensory visual cortex (e.g., V1–V4, MT). In contrast, attempts to decode the specific contents of VSTM from areas of the IPS (i.e., those regions purportedly involved in the maintenance of a limited number of items in VSTM) have largely failed (Emrich, Riggall, Larocque, & Postle, 2013; Linden, Oosterhof, Klein, & Downing, 2012; Riggall & Postle, 2012; but see, Christophel, Hebart, & Haynes, 2012; Christophel & Haynes, 2014). Thus, these findings challenge the classical view of VSTM maintenance by revealing that VSTM maintenance may largely occur in sensory areas, rather than being supported by memory-specific activity in the IPS. This interpretation is in line with an argument made by Postle (2006), who reviewed a number of findings from the field of working memory and argued that working (short-term) memory is an emergent property of sensory and motor systems, rather than a unique cognitive process with independent neural resources.

UPDATING THE CLASSIC MODEL OF VSTM

Together, the recent developments in both neuroimaging techniques and behavioral modeling present two strong challenges to the classic model of VSTM. First, they indicate that for VSTM there is significant variability in the precision of maintained representations, perhaps reflecting a shared pool of resources that support VSTM maintenance. Second, they reveal that VSTM may not be mediated by elevated, sustained activity in the IPS, but instead may be supported by patterns of activity in sensory visual cortex.

These findings suggest that updates to the classic view of VSTM are required to account for changes in VSTM precision as a function of VSTM load as well as to identify the role of sensory cortex in the capacity-limited maintenance of information in VSTM. That is, a comprehensive model of VSTM maintenance must also be able to address at least two outstanding problems: first, although recent fMRI findings suggest that VSTM maintenance is mediated by regions completely distinct from those that have traditionally been associated with VSTM maintenance, it is not yet clear how activity in these regions relates to behavioral models of VSTM performance (including their role in change-detection). Second, if activity in sensory areas is found to be the locus of VSTM maintenance, it is necessary to identify what role, if any, activity previously associated with VSTM maintenance plays in the maintenance and resolution of VSTM representations. In other words, if the classic model of a fixed-capacity model is no longer parsimonious with the extant neural and behavioral data, what aspects of behavior does the activity previously associated the maintenance of three to four items actually account for? In the following section, I will outline evidence suggesting that activity in sensory visual is linked to changes in precision across set sizes, and outline some alternative interpretations to the observed IPS delay-period activity.

Evidence for a Sensory Account of VSTM Precision

As mentioned previously, several recent studies have used MVPA to decode the contents of VSTM from sensory visual cortex (Linden et al., 2012; Riggall & Postle, 2012). These early studies, however, did not speak to the issue of VSTM load or capacity/precision, instead examining classification performance for only a single item. The first study to explicitly examine the effect of VSTM load using a decoding approach was performed recently by Emrich et al. (2013). In this study, stimuli were moving dot patterns, and the directions of motion were to be maintained over an 8-s delay period. Participants were always presented with three patterns on every trial and the number of moving dot patterns was varied between one, two, or three of the presented stimuli. Thus, memory load was varied on across conditions while keeping the number of presented stimuli constant. The majority of trials contained one of three critical (orthogonal) motion directions that were used to train and test classification performance.

Critically, the task allowed us to assess response error using the detection method approach pioneered by Wilken and Ma (2004). At the end of each trial, participants were cued to report the direction of a single sample stimulus by rotating a line toward the remembered direction of motion. Responses were then modeled using the three-factor model of Bays et al. (2009), allowing us to assess across all three loads the probability that an item was recalled

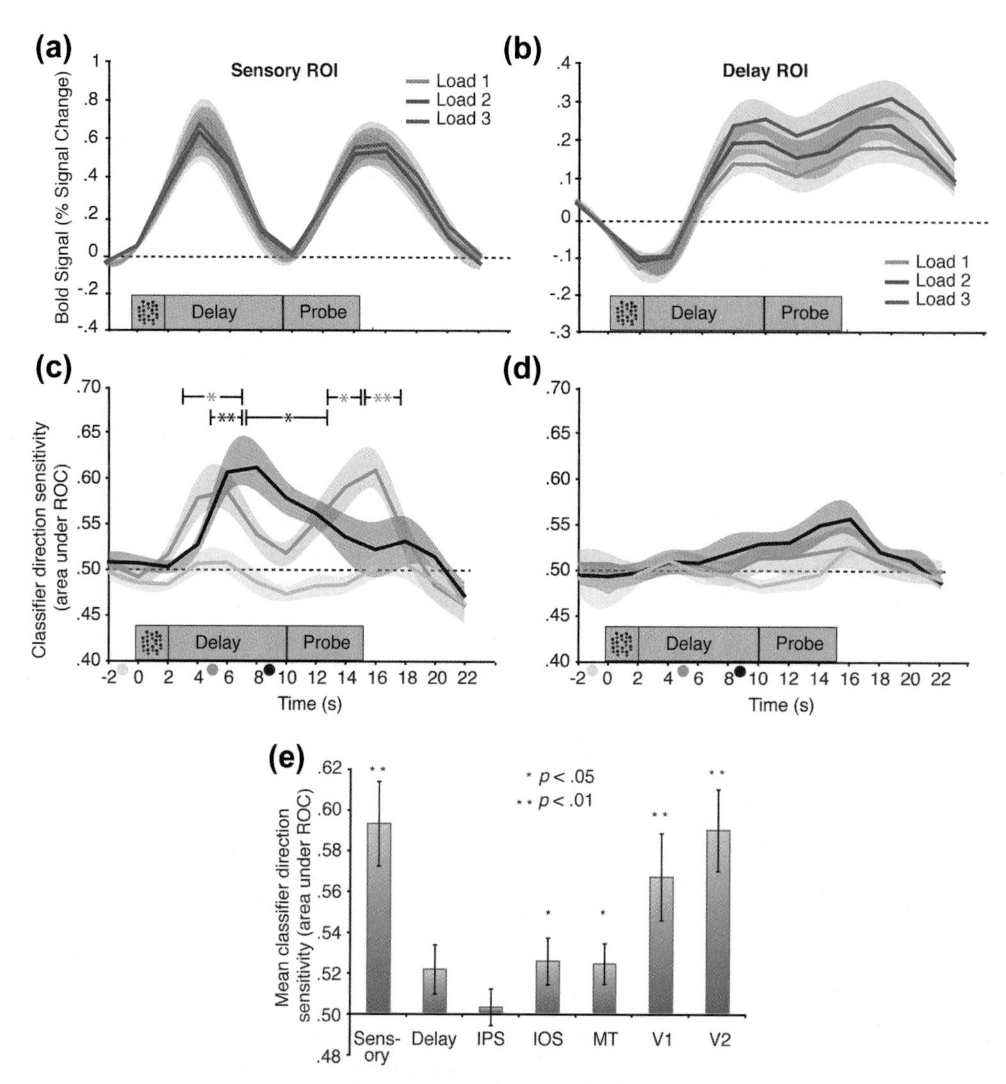

FIGURE 1 **fMRI BOLD signal (a-b) and classification performance (c-e) during a VSTM task.** Each trial began with a 2-s sample of one to three patterns of moving dots, followed by an 8-s delay. (a) and (b) depict the BOLD signal increase at each set size in sensory areas (those defined by increased activity in response to the sample) and delay areas (those defined by increased activity during the delay), respectively. (c) and (d) indicate the sensitivity of decoding the remembered direction of motion, with the classifier trained patterns observed in the sample (blue) and delay (red), or a control time point (gray), and tested across the trials. Despite the absence of sustained increases in BOLD activity in the sample ROI (a) the direction of motion could be successfully decoded from this activity (c). In contrast, despite the presence of sustained, load-sensitive BOLD activity in the delay ROI (b), the direction of motion could not be decoded form this region (d). (e) Classification performance from anatomical ROIs, averaged over the delay period (2–12s). *Used with Permission from Emrich et al. (2013).*

correctly, and the amount of error (noise) in the representation (as measured by the concentration parameter of the normal circular von Mises distribution) as well as the proportion of nontarget errors.

There were three critical results of this study which indicated that patterns of activity in the sensory visual cortex not only support the maintenance of information in VSTM, but may also have reflected the amount of noise (or the precision) of the maintained representation. First, classification performance was only above chance in areas that showed greater BOLD signal intensity in response to the sample than during the delay period (e.g., V1, V2, MT; see Figure 1). This was true regardless of whether the classifier was trained to patterns observed during encoding (4s after the onset of the stimuli) or those observed during maintenance (8s after stimulus onset). Moreover, analysis of anatomically defined regions of interest (ROIs) revealed that direction of motion could not be decoded from the IPS at better than chance. Thus, as with previous studies, this finding suggests that the actual representation of the contents of VSTM does not occur in the IPS, despite the presence of elevated, sustained, delay-period activity. Instead, this finding suggests that the representation of information occurs primarily in sensory visual cortex. Consequently, if the fidelity of VSTM representations varies as a function of load, it is these representations that should be degraded at high VSTM load because the amount of noise in the encoded stimulus increases.

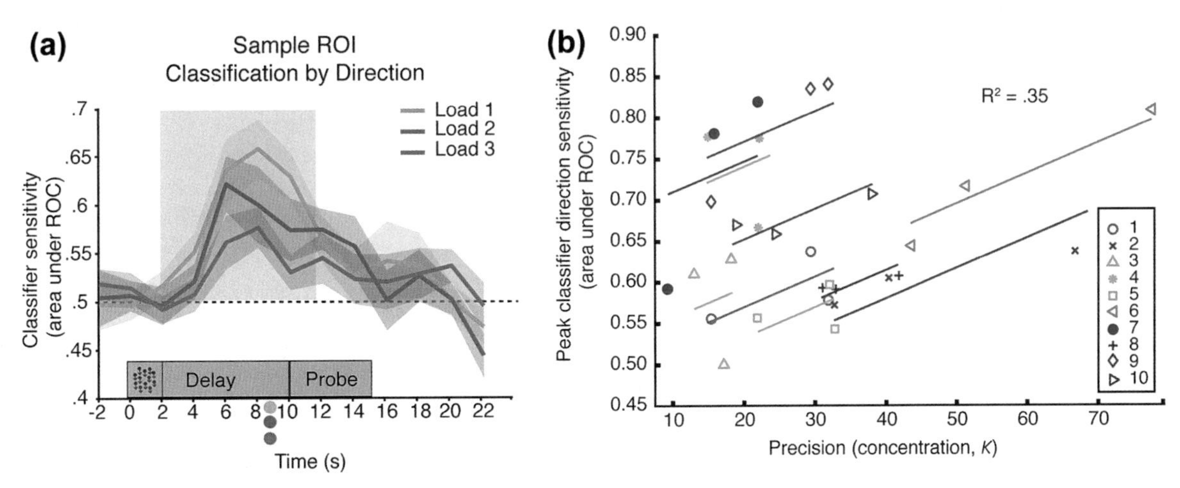

FIGURE 2 **Effect of load on classification performance.** (a) Classification performance in sensory areas decreased with increasing VSTM load (as tested using the mean of those points in the shaded region). (b) Within-subject correlations between classification performance and VSTM precision. For each subject, the decrease in classification sensitivity as a function of load was predicted by the decline in the precision parameter (i.e., the fidelity of the reported representation). *Adapted from Emrich et al. (2013).*

Consistent with this prediction, we also observed that classification accuracy in the sensory visual cortex decreased as a function of VSTM load. That is, the amount of information present in the patterns of fMRI BOLD signal was greater when only a single motion direction was held in VSTM compared with when two or three stimuli were being maintained (Figure 2(a)). This finding is consistent with a resource account of VSTM: when only a single item is being maintained in VSTM, the entire pool of neural resources are dedicated to representing the specific identity of that item. The resulting representation should be encoded with relatively little noise, resulting in high levels of signal from which to decode the maintained representation. In contrast, when multiple items are being maintained by the same (finite) pool of neurons, representations should be encoded with more noise, resulting in a decrease in decoding accuracy from the lower signal-to-noise ratio present in the pattern of neural activity.

Critically, we also examined within-subject changes in precision across loads one to three relative to the within-subject changes in classification sensitivity in the sensory visual cortex, with the prediction that if classification performance does in fact reflect the fidelity or noise of the encoded stimulus, then changes in classification performance across load should be accompanied by a concomitant change in response precision. Consistent with this prediction, we found a significant within-subject correlation between classification sensitivity and response precision. Thus, the patterns of activity in the sensory visual cortex appear to reflect the fidelity of the maintained representation.

One other important result of this study that is consistent with resources models comes from the finding that the amount of delay-period BOLD signal in sensory ROIs did not change as a function of VSTM load. That is, unlike the IPS, there was no greater activity observed in V1 or other sensory areas at load three compared with load one trials. This is consistent with a resource account of VSTM in that it suggests that a similar pool of resources was being allocated to maintain information in VSTM, regardless of the number of items present. According to the resource account, precision should decrease precisely because the same pool of resources is being allocated across a larger number of stimuli. In contrast, the traditional model of increasing IPS activity with increasing load (up to three to four items) is inconsistent with a resource account, in that it suggests more resources are being allocated to maintain two or three items, compared to when only a single item is to be represented.

Similar findings to those of Emrich et al. (2013) have been found in other recent studies using a related forward encoding approach. This approach involves modeling the response patterns of each voxel according to its selectivity for a particular feature dimension (e.g., orientation). Ester and colleagues (Ester, Anderson, Serences, & Awh, 2013) used this approach in a VSTM task by having participants maintain the orientations of a grating stimulus and reporting the orientation by rotating a test stimulus to the initial sample orientation. They found that the amount of dispersion (i.e., the width) of an individual's orientation response profile was significantly correlated with the amount of error in the recall response. In other words, the more narrow (precise) the response profile of neurons tuned to particular orientations, the more precise the VSTM response. In a follow-up study, they reported that attending to two oriented gratings resulted in a significant loss in encoding precision (i.e., less precisely tuned response profiles), although it is unclear whether there was a similar loss in behavioral precision (Anderson, Ester, Serences, & Awh, 2013). Importantly, both of these studies examined the response profiles of neurons in early sensory areas (V1–V2). Thus, these findings corroborate the three key claims of Emrich et al. (2013): (1) The representation of the contents of VSTM occurs in the sensory visual cortex; (2) the amount of noise encoded in the representation increases with set size; and (3) the amount of noise present in the patterns of activity that encode information in VSTM predicts the behavioral precision of VSTM responses.

Consequently, the results of these studies are consistent with a model in which the sensory visual cortex mediates the maintenance of representations in VSTM (Postle, 2006), and in which the amount of noise present in the encoded representations is directly reflected in the precision of VSTM. There are several issues, however, that these studies do not or cannot address. For example, although the data of Emrich et al. (2013) indicate that both precision and classification accuracy decrease with increasing load, they cannot speak to whether there are upper limits on the number of items that can be stored, as a maximum of only three items were tested. The observed decreases could be consistent with either a pure resource model, or hybrid model that incorporates both changes in precision and a fixed upper capacity on the number of items that can be maintained because both of these models could predict a change in precision at set sizes below four items. Moreover, although classification sensitivity and precision were correlated within subjects, it is not entire clear what is driving between-subject differences in precision and classification performance in the study of Emrich et al. (2013). One possibility is that between-subject differences relate to differences in encoding strategy. For example, a recent study found that encoding strategy in a VSTM task predicted the strength of the activity in the sensory visual cortex (Linke, Vicente-Grabovetsky, Mitchell, & Cusack, 2011).

Another issue these studies do not address is the role of IPS activity in VSTM maintenance and precision. That is, although the study by Emrich et al. (2013) and other similar studies have failed to find evidence for stimulus-specific coding in the IPS (Linden et al., 2012; Riggall & Postle, 2012), it is unclear from these findings what role, if any, the elevated IPS activity plays in creating or establishing the precision of maintained representations (i.e., those in the sensory visual cortex). Thus, a critical step in understanding the precise neural mechanisms that support VSTM maintenance will be to understand the relationship between IPS activity and the representations in early sensory cortex that appear to reflect the precision of VSTM representations. In the following section, I will review some findings that hint toward potential ways in which integration could be achieved.

REEVALUATING IPS ACTIVITY AND THE SPCN

Integrating Precision Models with Sustained Parietal Activity

To date, few studies have attempted to integrate the current model of IPS-mediated VSTM maintenance with flexible-resource or precision-based models of VSTM performance. One study was performed by Anderson and colleagues (Anderson, Vogel, & Awh, 2011) in which they examined SPCN amplitude as a function of VSTM precision. Importantly, the results of this study were more consistent with discrete capacity model (Zhang & Luck, 2008) than with a pure resource model: that is, they observed an asymptote in the precision of responses around three to four items, and no strong evidence for an increase in the number of responses toward unprobed (nontarget) items. Critically, the set size at which the SPCN reached an asymptote was correlated with the point at which the precision of VSTM responses reached an apparent asymptote. Thus, the authors concluded that VSTM is in fact a discrete resource and that the precision of maintained representations is related to the amplitude of the activity traditionally associated with VSTM maintenance (namely, the SPCN).

Although this study established a relationship between SPCN amplitude and VSTM precision, it is not clear from these data what role the SPCN plays in establishing or maintaining the precision of representations. Given that the amplitude of the SPCN was related to both the precision of representations and the number of representations maintained, it is unclear to what extent the SPCN reflects one or both of these measures. It is possible, for example, that these measures are nonindependent, or that they are both related to other aspects of task performance (e.g., attention; see the following section). In contrast, classification performance of early sensory activity was correlated only with precision and not the number of items maintained (Emrich et al., 2013). In addition, recent studies have called into question the validity of establishing a relationship between the SPCN and precision by measuring the inflection point of the SPCN and the standard deviation of the von Mises (normal) distribution (i.e., the precision measure; Van den Berg & Ma, 2014). Thus, although this study provides early evidence that parietal activity associated with VSTM maintenance is related to precision, more research is required to understand the exact nature of this relationship.

As mentioned previously, there are also a small number of studies that have found evidence that the amount of activation in areas of the IPS (in particular, the superior region of the IPS) may relate to the amount of information stored in VSTM. When more features or more complex items must be stored, this area exhibits greater activity and reaches an asymptote sooner compared with when fewer features or less complex stimuli are to be remembered (Xu & Chun, 2006, 2007; Xu, 2007). However, given that attempts to decode the contents of VSTM from the IPS have largely been unsuccessful, it is unclear whether activity in this region directly reflects the precision of maintained representations, or whether it instead reflects activity that is related to other aspects of task performance (see the following section).

At this point, the relationship between IPS activity and the precision of maintained representations is still hazy. Importantly, none of the studies mentioned in this section can speak to the relationship between sustained IPS activity and the early sensory activity from which VSTM representations can be decoded. Thus, one of the major aims of future research should be to attempt understand this relationship.

An Attentional Account of Delay-Period Activity

Although sustained activity in the IPS and the SPCN has been associated with VSTM maintenance, there have been several studies that have suggested that this activity is related to attentional processes, rather than to VSTM maintenance per se. One such study (Mitchell & Cusack, 2008) compared IPS activity in a change-detection task with a task in which items remained present on the screen. The participants' objective was to monitor the items for changes, and thus attention was allocated across both space (i.e., the sample items) and time. Importantly, because the stimuli remained present on the display, there was no explicit requirement for the items to be maintained in VSTM. The results of this study showed that IPS activity reached a similar asymptote in the change-detection and the attentional conditions. Thus, the authors concluded from these findings that IPS activity reflects attentional processes rather than VSTM maintenance. Specifically, the IPS may subserve the attentional selection of a limited number of objects and locations in a visual scene, regardless of whether those items are sustained in memory (internally directed attention) or present in the visual display (externally directed attention).

Interestingly, a recent ERP study (Tsubomi, Fukuda, Watanabe, & Vogel, 2013) appears to present converging evidence that VSTM is not a requirement for sustained activity (i.e., activity typically associated with the delay period of VSTM tasks). In this study, participants were presented with bilateral displays of colored squares. In one condition, the stimuli were removed from the display, and participants had to report the color of a test item at the end of a delay period. In a second condition, stimuli remained present on the display. The results of this study revealed that the amplitude of the SPCN was nearly identical across the two tasks: regardless of whether the objects remained on the display or not, the amplitude of the SPCN increased with set size and reached an asymptote at around four items. Interestingly, the behavioral performance was nearly identical across the two tasks, suggesting that participants were only able to attend to a limited number of items, even as they remained in the display. Thus, the authors concluded that participants could only represent a limited amount of task-relevant information, regardless of the demands on VSTM. This study provides converging evidence with studies that have demonstrated the presence of SPCN-like activity in a number of tasks in which stimuli remain present in the display, namely multiple object tracking (MOT; Drew & Vogel, 2008) and visual search (Emrich, Al-Aidroos, Pratt, & Ferber, 2009; Luria & Vogel, 2011b). Although these tasks may depend to some extent on VSTM processing, these findings all indicate that the SPCN and sustained IPS activity may not be specific to VSTM maintenance.

Thus, these findings suggest that sustained parietal activity could reflect selective attentional mechanisms directed toward a limited number of items. This would be consistent with the idea that the visual system contains an indexing mechanism with pointers directed toward a limited number of objects (Pylyshyn, 1989) as well as with evidence that IPS activity only increases in response to an increasing number of locations to remember (Harrison et al., 2010). This mechanism, therefore, would not only be present during tasks of VSTM, but also during perceptual, MOT, visual search, or any number of tasks in which discrete objects need to be individuated. Thus, one possible way in which IPS activity could be integrated with that observed in the sensory visual cortex is to view sustained, delay-period activity as part of a selection or indexing mechanism that could be used to bias or prioritize some of the information encoded by sensory areas of visual cortex. This could also potentially account for both an apparent capacity limit in VSTM, while still allowing for a resource-based mechanism: high-threshold tests of VSTM and perception may be sensitive to a limited number of items that can be indexed or prioritized, whereas detection-based approaches may still demonstrate evidence for representations outside of this limit.

An Awareness Account of Delay-Period Activity

On the other hand, it is possible that the activity observed during both VSTM and tasks in which items remain present in the display is related to perceptual processing or perceptual load, rather than to attentional indexing or selection. Specifically, one possibility that has not been explored great detail is the possibility that load-dependent activity observed in IPS reflects the amount of information contained within conscious awareness. One aspect that is common to perceptual tasks, VSTM, MOT, and visual search, is that visual information is held in conscious awareness. Moreover, the SPCN is also observed in tasks such as mental rotation (Prime & Jolicœur, 2010) and curve tracing

(Lefebvre, Jolicœur, & Dell'acqua, 2010). In each of these cases, objects must be encoded, identified/monitored, and reported on. Critically, although the mechanisms that support conscious awareness remain elusive, it is known that there is much more information in the environment than gains access to awareness (Marois & Ivanoff, 2005). Thus, conscious awareness is subject to similar capacity limitations as VSTM and attention. It is therefore possible that sustained activity observed during these tasks reflects the contents of awareness for a limited number of stimuli. This account may be similar to the attentional account in that the contents of awareness may often reflect those objects that are currently selected by attention; however, attention and awareness are likely not one and the same (Koch & Tsuchiya, 2007).

Although this hypothesis has not been thoroughly explored, there are a few indications that activity associated with VSTM is closely linked with conscious awareness. For example, some studies have demonstrated that in conditions of masking (Robitaille & Jolicœur, 2006) or during the attentional blink (Dell'Acqua, Sessa, Jolicœur, & Robitaille, 2006), the amplitude of the SPCN is related to the probability that a particular item is reported. For example, during an attentional blink task, the amplitude of the SPCN is much lower (or nonexistent) during the "blink" (i.e., when targets are missed) (Dell'Acqua et al., 2006). Thus, under these conditions, the presence of the contralateral delay activity appears to be linked with the ability to consciously perceive and report a target in a stream of letters.

Although these studies are indicative that parietal activity is closely tied to the ability to consciously report a stimulus, it is difficult to dissociate the process of conscious perception from those of VSTM (or, in some cases, attention). This question was tested more explicitly in recent study by Pun and colleagues (Pun, Emrich, Wilson, Stergiopoulos, & Ferber, 2012), in which we examined the amplitude of the SPCN relative to the perceived awareness of recognizable objects (Figure 3(a)). In this shape-from-motion task, images of recognizable objects are created out of line segments and are embedded in a background of randomly oriented lines. In order to perceive the object, the object must initially move in counterphase with the background. That is, coherent direction acts as the segregation cue, allowing subjects to become aware of the presence of an object. An interesting phenomenon occurs when the motion is stopped: participants typically continue to perceive the object for a brief period of a few seconds, before the percept of the object fades back into the background.

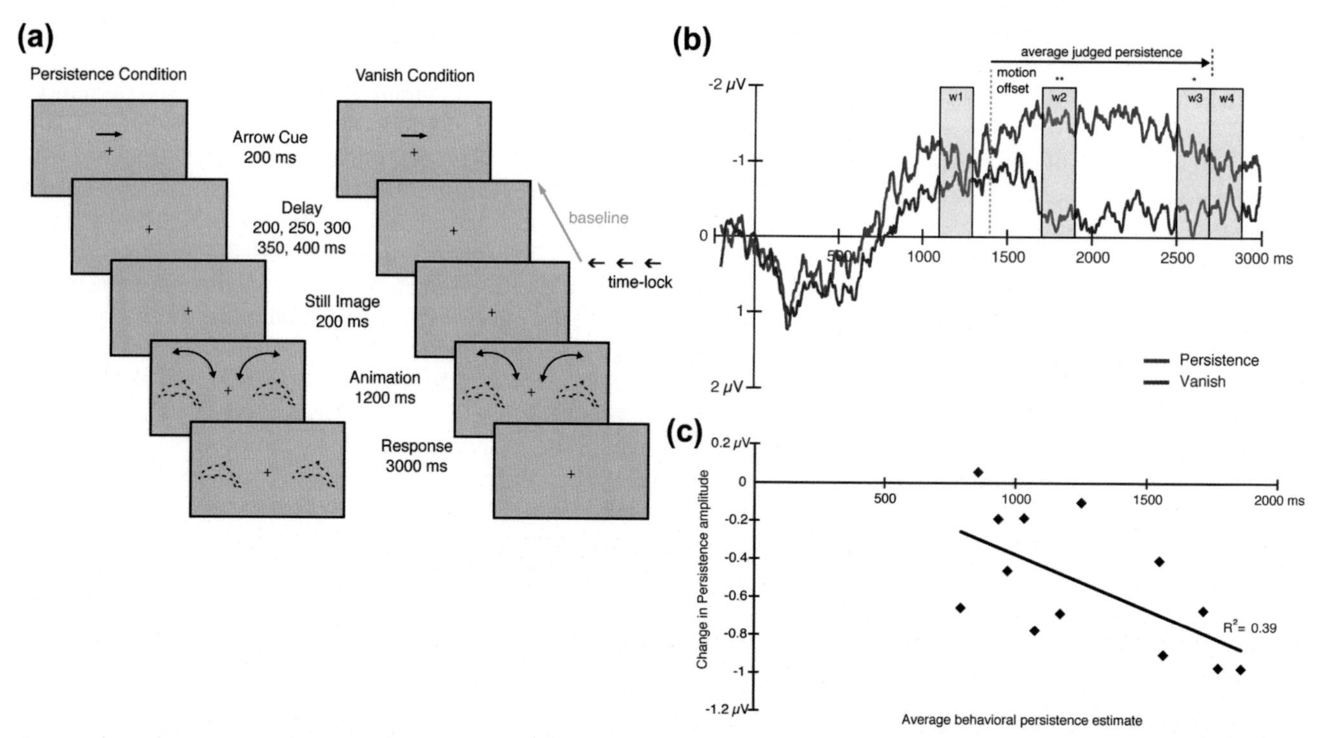

FIGURE 3 **Evidence for the role of the SPCN in conscious awareness.** (a) Shape from motion paradigm. Subjects were cued to attend a moving object against a noise background (not depicted). When the object remains on screen after the motion stops, subjects report continuing to see the object for roughly 1–2 s before it fades into the background. When the object is removed from the display when the motion stops, subjects report the object vanishing immediately. (b) The SPCN was observed when the object persisted (blue), but not after it vanished (red). That is the presence of the SPCN seemed to reflect the perceptual awareness of the observer. (c) The SPCN amplitude was correlated with the perceived duration of awareness, with longer persisters exhibiting larger SPCN amplitudes later in the trial (Figure 3(b), "w3"). *Adapted from Pun et al. (2012).*

In the study by Pun et al. (2012), we presented these shape-from-motion stimuli using a lateralized display, allowing us to examine the amplitude of the SPCN in response to the attended stimulus. After the motion was initially presented, a slow-wave ERP similar to the SPCN immediately became apparent and was present throughout the period when the motion cue was available. Thus, once the object gained access to conscious awareness, the SPCN emerged. The interesting question was what would happen after the motion stopped. If the SPCN was related only to the presence of the motion cue, then it should decrease immediately after the motion stopped. Although the amplitude of the SPCN did eventually return to baseline levels, it remained elevated for a number of seconds after the offset of the motion cue (Figure 3(b)). In contrast, when the object was removed from the background simultaneous with the offset of motion, there was an immediate and significant decrease in the amplitude of the SPCN. Thus, the SPCN persisted only in the condition in which the percept of the object remained in conscious awareness, suggesting that the SPCN was tied to this experience. Moreover, the amplitude of the SPCN was significantly correlated with individuals' subjective reports of when the object disappeared. That is, those subjects who experienced the percept of the object for longer exhibited a greater SPCN amplitude during the period of persistence (Figure 3(c)). Thus, this finding suggests a strong relationship between the slow-wave ERP associated with VSTM and visual awareness: the SPCN was present only when participants were aware of the object (either with motion or without), and the strength of this activity predicted the strength of the perceptual experience.

The results of this study largely decouple the effects of VSTM and perceptual awareness, because the SPCN was observed while the objects remained in the display and was correlated with duration of perceptual awareness. Moreover, given that the objects were made up of dozens of line segments that were perceived as a coherent gestalt, this suggests that the SPCN is not only related to the indexing of a fixed number of objects in the scene. Instead, this finding suggests a novel way of interpreting previous findings: that instead of reflecting the maintenance, tracking, or other perceptual processing of a limited number of items, the SPCN reflects processes related to conscious awareness. These processes may often be coupled: the items that are within conscious awareness are likely to be the ones that are within the focus of attention or VSTM; however, the SPCN may be a by-product of this relationship, rather than reflecting VSTM maintenance.

The proposal that the SPCN is related to perceptual awareness (at least in the visual domain) is also consistent with a number of findings demonstrating the role of parietal activity in conscious perception. For example, in a recent study, Emrich, Burianová, and Ferber (2011) demonstrated that under low and high VSTM loads, the likelihood of accurately reporting an object was directly linked with activity in areas of the superior IPS (as well as lateral occipital and inferior temporal regions). That is, the more objects that could be accessed for conscious report, the higher the activity in these areas, suggesting that IPS activity reflects the amount of information available for report in awareness (Figure 4). Other studies have similarly demonstrated that the perception of stimuli in the attentional blink (Kihara et al., 2007; Marois, Chun, & Gore, 2000) and binocular rivalry (Zaretskaya, Thielscher, Logothetis, & Bartels, 2010) depends on IPS activity, thus lending further support to the hypothesis that sustained IPS activity observed during VSTM tasks may reflect the awareness of those stimuli held in VSTM, rather than maintenance related processes per se. Similar arguments have been made by other authors; for example, Jolicœur, Brisson, and Robitaille (2008) argued that passage into VSTM is required to have conscious control over responses made on the basis of visual percepts.

Critically, an awareness account of IPS activity could be parsimonious with the finding that VSTM maintenance is supported by activity in the sensory visual cortex. That is, the initial encoding of stimulus specific information (color, orientation, direction) likely occurs in highly specialized regions of the sensory visual cortex. Similarly, the ability to sustain that information will likely depend on processes that originate in the same regions. Whether and which of that information gets access to high-level cognitive processes may depend on feedback and feed-forward interactions with the IPS (Lamme, 2006). Thus, parietal activity may also serve to stabilize information encoded in sensory cortex, perhaps relying on synchronized activity: in the absence of synchronization between these regions, representations encoded in sensory cortex may become too noisy or unstable to be used during tasks of working memory or perception. Thus, this proposal could lend itself to a model of VSTM in which visual representations themselves are maintained in the sensory visual cortex, and capacity-limited activity in IPS reflects the limited selection, enhancement, or awareness of a small number of objects and/or features in the visual display.

A Binding Account of Delay-Period Activity

Another suggestion that has been raised by a number of authors is that the activity observed in the IPS reflects the binding of perceptual information to specific objects and locations (Mitchell & Cusack, 2008; Sander, Lindenberger, & Werkle-Bergner, 2012). That is, to successfully perform tasks of both VSTM and perception, featural

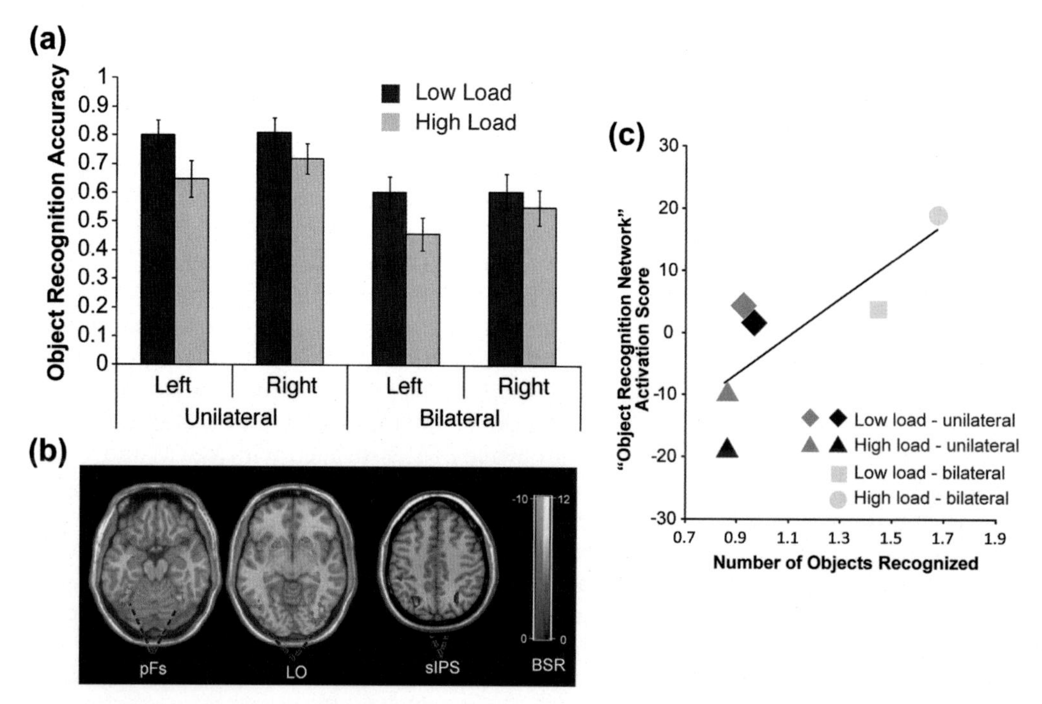

FIGURE 4 **Object recognition/awareness and VSTM load.** (a) Object recognition performance under low- and high-VSTM load. Subjects had to maintain one (low) or three (high) items in VSTM, and then report the identities of objects presented during the delay period. Object performance was worse under a high load, particularly under bilateral presentation. (b) A network of brain areas involved in the "object recognition network" (identified using partial-least squares; PLS). pFs = posterior fusiform sulcus; LO = lateral occipital; sIPS = superior intraparietal sulcus. (c) Activity in this object recognition network, including the sIPS, correlated with the number of objects subjects correctly identified (i.e., the number of objects participants had conscious access to). Note that this network was identified as independent from a VSTM network that also included the inferior and superior IPS. *Adapted from Emrich et al. (2011).*

information must be bound to specific objects and locations. This idea may potentially be considered a combination of the attentional and perceptual hypotheses outlined here, because during both VSTM and perception objects must be individuated (indexed), while also processing the perceptual (featural) information of those items. This proposal may fit well with the findings of Xu and Chun (Xu & Chun, 2006, 2007; Xu, 2007) and similar findings of the SPCN (Wilson et al., 2012) that different areas within the IPS are involved in distinct aspects of perception and memory: the inferior IPS may be involved in indexing a limited number of objects (locations), whereas the superior IPS may be involved in the perceptual processing of the features of those objects. Thus, activity observed in IPS (and the corresponding SPCN) may reflect a combination of the processes necessary to bind perceptual information to object locations.

The binding account is complicated by the finding that IPS activity only increases in response to an increasing number of locations to remember, and not to an increase in the number of object identities (Harrison et al., 2010). Thus, similar to the attentional account, this evidence may suggest that IPS activity simply relates to the number of locations that must be remembered, independent of featural load or binding requirements. However, in our study (Emrich et al., 2013), we observed load-dependent increases in IPS activity even though the moving dot patterns were presented in the same location. Thus, the precise nature of IPS activity remains unclear.

CONCLUSIONS

For most of the past decade, the evidence emerging from the cognitive neuroscience of VSTM converged on an integrated model in which VSTM had a fixed capacity limit of three to four discrete items, mediated by sustained, load-sensitive activity in the IPS. However, in light of findings outlined previously, it is clear that a number of findings no longer fit within this paradigm, requiring a new model that integrates novel findings from psychophysics/behavioral studies of VSTM capacity and new techniques in neuroimaging. A new integrated model likely must be able to incorporate (1) changes in the precision of representations maintained in VSTM, (2) the representation of information encoded in VSTM by the sensory visual cortex, and (3) evidence challenging the role of sustained, delay-period activity in the IPS (including the ERP correlate, the SPCN) as a hallmark of VSTM maintenance.

Although our understanding of the behavioral limitations of VSTM and the neural mechanisms that support them are changing rapidly, it is perhaps possible to make some tentative predictions about what such a model might entail. As outlined previously, there is growing evidence that the representation of information in VSTM occurs in sensory cortex, and that it is these representations that are degraded (i.e., more noisy) when VSTM load increases. Thus, activity previously associated with the maintenance of information in VSTM may reflect other processes associated with VSTM, such as attentional selection, binding, or perceptual awareness. In other words, although sustained activity in the IPS (or the associated ERP correlates) might prove to be a *necessary* for VSTM, it may likely prove not to be a sufficient condition.

Although there remain a number of questions that are not addressed here, a model in which the maintenance and resolution of items in VSTM is encoded in the sensory visual cortex may be parsimonious with a number of findings suggesting that encoding processes may be a limiting factor in VSTM (Keshvari, van den Berg, & Ma, 2013; Van den Berg et al., 2012). That is, if VSTM is mediated by activity in the sensory visual cortex, it is not surprising that perceptual and VSTM tasks may be subject to the same capacity constraints, as both tasks involve competition for the same resources (Emrich et al., 2011; Mitchell & Cusack, 2008; Tsubomi et al., 2013). Competition for sensory resources may also place further limitations on how information is encoded and represented in VSTM (Emrich & Ferber, 2012).

Finally, the possibility that IPS activity could potentially reflect one or more of VSTM, attention, or conscious awareness, speaks to the importance of further demarcating, both ontologically and experimentally, these processes. Although there have been attempts to dissociate attention and consciousness, fewer attempts have been made to try to dissociate VSTM from consciousness (although see Akyürek, Hommel, & Jolicœur, 2007, for one example). Interestingly, the contents of VSTM are known to affect perceptual processes (Gayet, Paffen, & Van der Stigchel, 2013; Kang, Hong, Blake, & Woodman, 2011), and neglect patients exhibit impairments in both conscious awareness and VSTM (Ferber & Danckert, 2006), suggesting an intimate relationship between these processes. There is also some suggestion that things that are outside of the focus of attention (and therefore, also outside of the scope of conscious awareness) can remain active in VSTM (LaRocque, Lewis-Peacock, Drysdale, Oberauer, & Postle, 2013; LaRocque et al., 2013; Larocque, Lewis-Peacock, & Postle, 2014), suggesting that it may be possible to distinguish between these concepts. Ultimately, however, a better understanding of VSTM may require a more complete understanding of the relationships between VSTM, attention, and perception.

References

Akyürek, E. G., Hommel, B., & Jolicœur, P. (2007). Direct evidence for a role of working memory in the attentional blink. *Memory & Cognition, 35*(4), 621–627.

Alvarez, G. A., & Cavanagh, P. (2004). The capacity of visual short-term memory is set both by visual information load and by number of objects. *Psychological Science, 15*(2), 106–111.

Anderson, D. E., Ester, E. F., Serences, J. T., & Awh, E. (2013). Attending multiple items decreases the selectivity of population responses in human primary visual cortex. *The Journal of Neuroscience, 33*(22), 9273–9282. http://dx.doi.org/10.1523/JNEUROSCI.0239-13.2013.

Anderson, D. E., Vogel, E. K., & Awh, E. (2011). Precision in visual working memory reaches a stable plateau when individual item limits are exceeded. *The Journal of Neuroscience, 31*(3), 1128–1138. http://dx.doi.org/10.1523/JNEUROSCI.4125-10.2011.

Bays, P. M., Catalao, R. F. G., & Husain, M. (2009). The precision of visual working memory is set by allocation of a shared resource. *Journal of Vision, 9*(10), 7.1–7.11. http://dx.doi.org/10.1167/9.10.7.

Bays, P. M., & Husain, M. (2008). Dynamic shifts of limited working memory resources in human vision. *Science, 321*(5890), 851–854. http://dx.doi.org/10.1126/science.1158023.

Brady, T. F., Konkle, T., & Alvarez, G. A. (2011). A review of visual memory capacity: beyond individual items and toward structured representations. *Journal of Vision, 11*(5), 4. http://dx.doi.org/10.1167/11.5.4.

Brady, T. F., Konkle, T., Alvarez, G. A., & Oliva, A. (2008). Visual long-term memory has a massive storage capacity for object details. *Proceedings of the National Academy of Sciences of the United States of America, 105*(38), 14325–14329. http://dx.doi.org/10.1073/pnas.0803390105.

Christophel, T. B., & Haynes, J.-D. (2014). Decoding complex flow-field patterns in visual working memory. *NeuroImage, 91*, 43–51. http://dx.doi.org/10.1016/j.neuroimage.2014.01.025.

Christophel, T. B., Hebart, M. N., & Haynes, J.-D. (2012). Decoding the contents of visual short-term memory from human visual and parietal cortex. *The Journal of Neuroscience, 32*(38), 12983–12989. http://dx.doi.org/10.1523/JNEUROSCI.0184-12.2012.

Coutanche, M. N. (2013). Distinguishing multi-voxel patterns and mean activation: why, how, and what does it tell us? *Cognitive, Affective & Behavioral Neuroscience, 13*(3), 667–673. http://dx.doi.org/10.3758/s13415-013-0186-2.

Cowan, N. (2001). The magical number 4 in short-term memory: a reconsideration of mental storage capacity. *Behavioral and Brain Sciences, 24*(1), 87–114; discussion 114–85.

Davis, T., & Poldrack, R. A. (2013). Measuring neural representations with fMRI: practices and pitfalls. *Annals of the New York Academy of Sciences, 1296*(1), 108–134. http://dx.doi.org/10.1111/nyas.12156.

Dell'Acqua, R., Sessa, P., Jolicœur, P., & Robitaille, N. (2006). Spatial attention freezes during the attention blink. *Psychophysiology, 43*(4), 394–400. http://dx.doi.org/10.1111/j.1469-8986.2006.00411.x.

Drew, T., & Vogel, E. K. (2008). Neural measures of individual differences in selecting and tracking multiple moving objects. *The Journal of Neuroscience, 28*(16), 4183–4191. http://dx.doi.org/10.1523/JNEUROSCI.0556-08.2008.

Emrich, S. M., Al-Aidroos, N., Pratt, J., & Ferber, S. (2009). Visual search elicits the electrophysiological marker of visual working memory. *PLoS One, 4*(11), e8042. http://dx.doi.org/10.1371/journal.pone.0008042.

Emrich, S. M., Al-Aidroos, N., Pratt, J., & Ferber, S. (2010). Finding memory in search: the effect of visual working memory load on visual search. *The Quarterly Journal of Experimental Psychology*, *63*(8), 1457. http://dx.doi.org/10.1080/17470218.2010.483768.

Emrich, S. M., Burianová, H., & Ferber, S. (2011). Transient perceptual neglect: visual working memory load affects conscious object processing. *Journal of Cognitive Neuroscience*, *23*(10), 2968–2982. http://dx.doi.org/10.1162/jocn_a_00028.

Emrich, S. M., & Ferber, S. (2012). Competition increases binding errors in visual working memory. *Journal of Vision*, *12* (4). http://dx.doi.org/10.1167/12.4.12.

Emrich, S. M., Riggall, A. C., Larocque, J. J., & Postle, B. R. (2013). Distributed patterns of activity in sensory cortex reflect the precision of multiple items maintained in visual short-term memory. *The Journal of Neuroscience*, *33*(15), 6516–6523. http://dx.doi.org/10.1523/JNEUROSCI.5732-12.2013.

Engle, R. W., Kane, M. J., & Tuholski, S. W. (1999). Individual differences in working memory capacity and what they tell us about controlled attention, general fluid intelligence, and functions of the prefrontal cortex. *Models of Working Memory: Mechanisms of Active Maintenance and Executive Control*, 102–134.

Engle, R. W., Tuholski, S. W., Laughlin, J. E., & Conway, A. R. (1999). Working memory, short-term memory, and general fluid intelligence: a latent-variable approach. *Journal of Experimental Psychology General*, *128*, 309–331.

Ester, E. F., Anderson, D. E., Serences, J. T., & Awh, E. (2013). A neural measure of precision in visual working memory. *Journal of Cognitive Neuroscience*, *25*(5), 754–761. http://dx.doi.org/10.1162/jocn_a_00357.

Ferber, S., & Danckert, J. (2006). Lost in space—the fate of memory representations for non-neglected stimuli. *Neuropsychologia*, *44*(2), 320–325.

Fougnie, D., Suchow, J. W., & Alvarez, G. A. (2012). Variability in the quality of visual working memory. *Nature Communications*, *3*, 1229. http://dx.doi.org/10.1038/ncomms2237.

Gayet, S., Paffen, C. L. E., & Van der Stigchel, S. (2013). Information matching the content of visual working memory is prioritized for conscious access. *Psychological Science*, *24*(12), 2472–2480. http://dx.doi.org/10.1177/0956797613495882.

Harrison, A., Jolicœur, P., & Marois, R. (2010). "What" and "where" in the intraparietal sulcus: an FMRI study of object identity and location in visual short-term memory. *Cerebral Cortex*, *20*(10), 2478–2485. http://dx.doi.org/10.1093/cercor/bhp314.

Harrison, S. A., & Tong, F. (2009). Decoding reveals the contents of visual working memory in early visual areas. *Nature*, *458*(7238), 632–635. http://dx.doi.org/10.1038/nature07832.

Jolicœur, P., Brisson, B., & Robitaille, N. (2008). Dissociation of the N2pc and sustained posterior contralateral negativity in a choice response task. *Brain Research*, *1215*, 160–172. http://dx.doi.org/10.1016/j.brainres.2008.03.059.

Kang, M.-S., Hong, S. W., Blake, R., & Woodman, G. F. (2011). Visual working memory contaminates perception. *Psychonomic Bulletin & Review*, *18*(5), 860–869. http://dx.doi.org/10.3758/s13423-011-0126-5.

Keshvari, S., van den Berg, R., & Ma, W. J. (2013). No evidence for an item limit in change detection. *PLoS Computational Biology*, *9*(2), e1002927. http://dx.doi.org/10.1371/journal.pcbi.1002927.

Kihara, K., Hirose, N., Mima, T., Abe, M., Fukuyama, H., & Osaka, N. (2007). The role of left and right intraparietal sulcus in the attentional blink: a transcranial magnetic stimulation study. *Experimental Brain Research*, *178*(1), 135–140. http://dx.doi.org/10.1007/s00221-007-0896-1.

Klaver, P., Smid, H. G., & Heinze, H. J. (1999). Representations in human visual short-term memory: an event-related brain potential study. *Neuroscience Letters*, *268*(2), 65–68.

Klaver, P., Talsma, D., Wijers, A. A., Heinze, H. J., & Mulder, G. (1999). An event-related brain potential correlate of visual short-term memory. *Neuroreport*, *10*(10), 2001–2005.

Koch, C., & Tsuchiya, N. (2007). Attention and consciousness: two distinct brain processes. *Trends in Cognitive Sciences*, *11*(1), 16–22.

Kundu, B., Sutterer, D. W., Emrich, S. M., & Postle, B. R. (2013). Strengthened effective connectivity underlies transfer of working memory training to tests of short-term memory and attention. *The Journal of Neuroscience*, *33*(20), 8705–8715. http://dx.doi.org/10.1523/JNEUROSCI.5565-12.2013.

Lamme, V. A. F. (2006). Towards a true neural stance on consciousness. *Trends in Cognitive Sciences*, *10*(11), 494–501. http://dx.doi.org/10.1016/j.tics.2006.09.001.

LaRocque, J. J., Lewis-Peacock, J. A., Drysdale, A. T., Oberauer, K., & Postle, B. R. (2013). Decoding attended information in short-term memory: an EEG study. *Journal of Cognitive Neuroscience*, *25*(1), 127–142. http://dx.doi.org/10.1162/jocn_a_00305.

Larocque, J. J., Lewis-Peacock, J. A., & Postle, B. R. (2014). Multiple neural states of representation in short-term memory? It's a matter of attention. *Frontiers in Human Neuroscience*, *8*, 5. http://dx.doi.org/10.3389/fnhum.2014.00005.

Lefebvre, C., Jolicœur, P., & Dell'acqua, R. (2010). Electrophysiological evidence of enhanced cortical activity in the human brain during visual curve tracing. *Vision Research*, *50*(14), 1321–1327. http://dx.doi.org/10.1016/j.visres.2009.12.006.

Linden, D. E., Bittner, R. A., Muckli, L., Waltz, J. A., Kriegeskorte, N., Goebel, R., et al. (2003). Cortical capacity constraints for visual working memory: dissociation of fMRI load effects in a fronto-parietal network. *Neuroimage*, *20*(3), 1518–1530.

Linden, D. E. J., Oosterhof, N. N., Klein, C., & Downing, P. E. (2012). Mapping brain activation and information during category-specific visual working memory. *Journal of Neurophysiology*, *107*(2), 628–639. http://dx.doi.org/10.1152/jn.00105.2011.

Linke, A. C., Vicente-Grabovetsky, A., Mitchell, D. J., & Cusack, R. (2011). Encoding strategy accounts for individual differences in change detection measures of VSTM. *Neuropsychologia*, *49*(6), 1476–1486. http://dx.doi.org/10.1016/j.neuropsychologia.2010.11.034.

Luck, S. J., & Vogel, E. K. (1997). The capacity of visual working memory for features and conjunctions. *Nature*, *390*(6657), 279–281. http://dx.doi.org/10.1038/36846.

Luck, S. J., & Vogel, E. K. (2013). Visual working memory capacity: from psychophysics and neurobiology to individual differences. *Trends in Cognitive Sciences*, *17*(8), 391–400. http://dx.doi.org/10.1016/j.tics.2013.06.006.

Luria, R., & Vogel, E. K. (2011a). Shape and color conjunction stimuli are represented as bound objects in visual working memory. *Neuropsychologia*, *49*(6), 1632–1639. http://dx.doi.org/10.1016/j.neuropsychologia.2010.11.031.

Luria, R., & Vogel, E. K. (2011b). Visual search demands dictate reliance on working memory storage. *The Journal of Neuroscience*, *31*(16), 6199–6207. http://dx.doi.org/10.1523/JNEUROSCI.6453-10.2011.

Ma, W. J., Husain, M., & Bays, P. M. (2014). Changing concepts of working memory. *Nature Neuroscience*, *17*(3), 347–356. http://dx.doi.org/10.1038/nn.3655.

Marois, R., Chun, M. M., & Gore, J. C. (2000). Neural correlates of the attentional blink. *Neuron*, *28*(1), 299–308.

Marois, R., & Ivanoff, J. (2005). Capacity limits of information processing in the brain. *Trends in Cognitive Sciences*, *9*(6), 296–305.

Mitchell, D. J., & Cusack, R. (2008). Flexible, capacity-limited activity of posterior parietal cortex in perceptual as well as visual short-term memory tasks. *Cerebral Cortex*, *18*(8), 1788–1798.

Mitchell, D. J., & Cusack, R. (2011). The temporal evolution of electromagnetic markers sensitive to the capacity limits of visual short-term memory. *Frontiers in Human Neuroscience*, *5*, 18. http://dx.doi.org/10.3389/fnhum.2011.00018.

Munk, M. H. J., Linden, D. E. J., Muckli, L., Lanfermann, H., Zanella, F. E., Singer, W., et al. (2002). Distributed cortical systems in visual short-term memory revealed by event-related functional magnetic resonance imaging. *Cerebral Cortex*, *12*(8), 866–876.

Olson, I. R., & Jiang, Y. (2002). Is visual short-term memory object based? Rejection of the "strong-object" hypothesis. *Perception & Psychophysics*, *64*(7), 1055–1067.

Palmer, J. (1990). Attentional limits on the perception and memory of visual information. *Journal of Experimental Psychology: Human Perception and Performance*, *16*(2), 332–350.

Pashler, H. (1987). Detecting conjunctions of color and form: reassessing the serial search hypothesis. *Perception & Psychophysics*, *41*(3), 191–201.

Phillips, W. A. (1974). On the distinction between sensory storage and short-term visual memory. *Perception & Psychophysics*, *16*, 283–290.

Postle, B. R. (2006). Working memory as an emergent property of the mind and brain. *Neuroscience*, *139*(1), 23–38.

Prime, D. J., & Jolicœur, P. (2010). Mental rotation requires visual short-term memory: evidence from human electric cortical activity. *Journal of Cognitive Neuroscience*, *22*(11), 2437–2446. http://dx.doi.org/10.1162/jocn.2009.21337.

Pun, C., Emrich, S. M., Wilson, K. E., Stergiopoulos, E., & Ferber, S. (2012). In and out of consciousness: sustained electrophysiological activity reflects individual differences in perceptual awareness. *Psychonomic Bulletin & Review*, *19*(3), 429–435. http://dx.doi.org/10.3758/s13423-012-0220-3.

Pylyshyn, Z. (1989). The role of location indexes in spatial perception: a sketch of the FINST spatial-index model. *Cognition*, *32*(1), 65–97.

Riggall, A. C., & Postle, B. R. (2012). The relationship between working memory storage and elevated activity as measured with functional magnetic resonance imaging. *The Journal of Neuroscience*, *32*(38), 12990–12998. http://dx.doi.org/10.1523/JNEUROSCI.1892-12.2012.

Robitaille, N., Grimault, S., & Jolicœur, P. (2009). Bilateral parietal and contralateral responses during maintenance of unilaterally encoded objects in visual short-term memory: evidence from magnetoencephalography. *Psychophysiology*, *46*(5), 1090–1099. http://dx.doi.org/10.1111/j.1469-8986.2009.00837.x.

Robitaille, N., & Jolicœur, P. (2006). Fundamental properties of the N2pc as an index of spatial attention: effects of masking. *Canadian Journal of Experimental Psychology*, *60*(2), 101–111.

Robitaille, N., Marois, R., Todd, J., Grimault, S., Cheyne, D., & Jolicœur, P. (2010). Distinguishing between lateralized and nonlateralized brain activity associated with visual short-term memory: fMRI, MEG, and EEG evidence from the same observers. *Neuroimage*, *53*(4), 1334–1345. http://dx.doi.org/10.1016/j.neuroimage.2010.07.027.

Ruchkin, D. S., Canoune, H. L., Johnson, R., Jr., & Ritter, W. (1995). Working memory and preparation elicit different patterns of slow wave event-related brain potentials. *Psychophysiology*, *32*(4), 399–410.

Ruchkin, D. S., Johnson, R., Jr., Canoune, H., & Ritter, W. (1990). Short-term memory storage and retention: an event-related brain potential study. *Electroencephalography and Clinical Neurophysiology*, *76*(5), 419–439.

Sander, M. C., Lindenberger, U., & Werkle-Bergner, M. (2012). Lifespan age differences in working memory: a two-component framework. *Neuroscience and Biobehavioral Reviews*, *36*(9), 2007–2033. http://dx.doi.org/10.1016/j.neubiorev.2012.06.004.

Serences, J. T., Ester, E. F., Vogel, E. K., & Awh, E. (2009). Stimulus-specific delay activity in human primary visual cortex. *Psychological Science*, *20*(2), 207–214. http://dx.doi.org/10.1111/j.1467-9280.2009.02276.x.

Serences, J. T., & Saproo, S. (2012). Computational advances towards linking BOLD and behavior. *Neuropsychologia*, *50*(4), 435–446. http://dx.doi.org/10.1016/j.neuropsychologia.2011.07.013.

Sperling, G. (1960). The information available in brief visual presentations. *Psychology Monographs*, *74*(11, Whole No. 498), 1–19.

Todd, J. J., Fougnie, D., & Marois, R. (2005). Visual short-term memory load suppresses temporo-parietal junction activity and induces inattentional blindness. *Psychological Science*, *16*(12), 965.

Todd, J. J., & Marois, R. (2004). Capacity limit of visual short-term memory in human posterior parietal cortex. *Nature*, *428*(6984), 751–754. http://dx.doi.org/10.1038/nature02466.

Tsubomi, H., Fukuda, K., Watanabe, K., & Vogel, E. K. (2013). Neural limits to representing objects still within view. *The Journal of Neuroscience*, *33*(19), 8257–8263. http://dx.doi.org/10.1523/JNEUROSCI.5348-12.2013.

Van den Berg, R., Awh, E., & Ma, W. J. (2014). Factorial comparison of working memory models. *Psychological Review*, *121*(1), 124–149. http://dx.doi.org/10.1037/a0035234.

Van den Berg, R., & Ma, W. J. (2014). "Plateau"-related summary statistics are uninformative for comparing working memory models. *Attention, Perception & Psychophysics*, *76*(7), 2117–2135. http://dx.doi.org/10.3758/s13414-013-0618-7.

Van den Berg, R., Shin, H., Chou, W.-C., George, R., & Ma, W. J. (2012). Variability in encoding precision accounts for visual short-term memory limitations. *Proceedings of the National Academy of Sciences of the United States of America*, *109*(22), 8780–8785. http://dx.doi.org/10.1073/pnas.1117465109.

Vogel, E. K., & Machizawa, M. G. (2004). Neural activity predicts individual differences in visual working memory capacity. *Nature*, *428*(6984), 748–751.

Vogel, E. K., Woodman, G. F., & Luck, S. J. (2001). Storage of features, conjunctions and objects in visual working memory. *Journal of Experimental Psychology: Human Perception and Performance*, *27*(1), 92–114.

Wilken, P., & Ma, W. J. (2004). A detection theory account of change detection. *Journal of Vision*, *4*(12), 1120–1135 doi:10:1167/4.12.11.

Wilson, K. E., Adamo, M., Barense, M. D., & Ferber, S. (2012). To bind or not to bind: addressing the question of object representation in visual short-term memory. *Journal of Vision*, *12*(8), 14. http://dx.doi.org/10.1167/12.8.14.

Xu, Y. (2007). The role of the superior intraparietal sulcus in supporting visual short-term memory for multifeature objects. *Journal of Neuroscience*, *27*(43), 11676–11686.

Xu, Y., & Chun, M. M. (2006). Dissociable neural mechanisms supporting visual short-term memory for objects. *Nature*, *440*(7080), 91–95.

Xu, Y., & Chun, M. M. (2007). Visual grouping in human parietal cortex. *Proceedings of the National Academy of Sciences of the United States of America*, *104*(47), 18766–18771.

Zaretskaya, N., Thielscher, A., Logothetis, N. K., & Bartels, A. (2010). Disrupting parietal function prolongs dominance durations in binocular rivalry. *Current Biology*, *20*(23), 2106–2111. http://dx.doi.org/10.1016/j.cub.2010.10.046.

Zhang, W., & Luck, S. J. (2008). Discrete fixed-resolution representations in visual working memory. *Nature*, *453*(7192), 233–235. http://dx.doi.org/10.1038/nature06860.

CHAPTER

7

Hemispheric Organization of Visual Memory: Analyzing Visual Working Memory With Brain Measures

Gabriele Gratton[1], *Eunsam Shin*[2], *Monica Fabiani*[1]

[1]Department of Psychology, University of Illinois at Urbana-Champaign, Illinois, USA;
[2]Department of Psychology, Yonsei University, Yonsei, Korea

INTRODUCTION

The term "working memory" is used to describe a set of processes by which information is kept active for use in an ongoing task. When originally introduced by Baddeley and Hitch (1974), the concept of working memory was intended to replace that of short-term memory to emphasize the dynamic aspects of these memory processes. In other words, working memory is strictly linked to online attention deployment rather than being based on a memory storage system of a duration intermediate to that of other forms of memory storage, such as sensory or long-term memory (as proposed by Atkinson & Shiffrin, 1968). In fact, in current views of working memory, this term is used to reflect the temporary activation or maintenance of mental representations that are parts of other memory systems, such as long-term memory (see Cowan, 1995). In this context, working memory and sensory memory (or, for that matter, long-term memory) are no longer viewed as separate storage systems: rather, they indicate different forms of activations of existing memory representations.

This view leads to the hypothesis that common sets of neural processes underlie all these forms of memory, and that "working memory" tasks (such as the Sternberg's memory search task, Sternberg, 1966; or the change-detection task, Luck & Vogel, 1997) will involve the same neural representations that are used by stimuli that have to be analyzed in isolation, as in "adaptation" or "priming" paradigms (e.g., Reber, Stark, & Squire, 1998). In these latter paradigms, the interest is in investigating the processing "cost savings" that occur after a stimulus that has already been recently processed: the savings, indexed by reduced brain activations to the second stimulus presentation, are believed to reflect the fact that a neural representation of the stimulus is already active, albeit at a reduced level (Dehaene et al., 2001; Schendan & Kutas, 2003; Vuilleumier, Henson, Driver, & Dolan, 2002).

The research from our laboratory reviewed here is based on this general view, and focuses on methods designed to reveal the reactivation of the latent neural representations triggered by visual stimuli (see Roediger, Weldon, & Challis, 1989 for a similar view from a behavioral perspective), affording processing and/or behavioral advantages when the stimulus is presented a second time. Critically, we show that both processing savings and behavioral advantages are most evident when the stimuli are presented in the same visual hemifield (and therefore to the *same* brain hemisphere) during their first and second presentations. This supports the idea that representations embodied within specific anatomical structures (the two hemispheres of the brain) may underlie the facilitation phenomenon.

In the research reviewed in the following sections, processing savings and behavioral advantages are revealed in experimental paradigms that require comparisons between stimuli presented at different moments in time. Depending on time delay and on the number of different stimuli to memorize and compare, these paradigms are classified as either "working memory" or "recognition" paradigms.

Mechanisms of Sensory Working Memory
http://dx.doi.org/10.1016/B978-0-12-811042-3.00007-4

The experiments we review here address the following questions about the latent persistence of neural representations, which we will label *visual memory*, reflecting that all experiments employed the visual modality.

1. Does visual memory reflect latent activation of the original representations that were generated for processing the stimuli in visual cortex?
2. Does this lead to a hemispheric organization of visual memories, depending on the hemifield of initial stimulus presentation?
3. Are there multiple representations for the same stimuli? And, if yes, are they accessed at different latencies, reflecting the flow of information processing?
4. Can the presence of separate representations held in the two hemispheres be used to improve performance in working memory tasks?

GENERAL METHODS

Measuring Brain Activity

To reveal the presence of latent neural representations, we flanked behavioral measures (increased accuracy and reduced reaction times in recognition tasks) with measures of brain activation in response to stimuli. In choosing these measures, we wished to emphasize bottom-up (or feed-forward) processes occurring during initial stimulus evaluation over top-down (or feed-back) processes that might occur as a result of the recognition process itself because the latter may be redundant with respect to overt behavioral measures. For this reason, we selected brain measures with high temporal resolution: event-related brain potentials (ERPs; for a review, see Fabiani, Gratton, & Federmeier, 2007) and the event-related optical signal (EROS; Gratton, Corballis, Cho, Fabiani, & Hood, 1995; for a review, see Gratton & Fabiani, 2010). Both are direct measures of neural activity, the former related to the depolarization (or hyperpolarization) of cortical dendrites involved in information processing and the latter related to the swelling (or shrinking) of the dendrites (with consequent changes in light scattering properties) that cooccurs with the electrical phenomena (Foust & Rector, 2007). Because of their high temporal resolution, these measures can be used to selectively examine brain activity occurring in the first few hundred milliseconds after stimulus presentation. In fact, responses occurring during this early period are more likely to reflect the bottom-up or "feed-forward" processes accompanying initial stimulus evaluation. With this term, we mean to indicate processes occurring in sensory regions that are directly elicited by an external stimulus. As the latency of the brain activity becomes longer, top-down or feedback processes may become more and more prevalent (with this term, we mean the recurrent processes that are generated by nonsensory regions of the brain in reaction to bottom-up processes). This terminology also corresponds to the traditional differentiation used in ERP research between early "exogenous" components (whose parameters are mostly linked to stimulus characteristics) and later "endogenous" components (whose parameters are mostly linked to task-related and strategic factors; see Donchin et al., 1978).

The Encoding-Related Lateralizations. To further focus on the brain processes related to the reactivation of memory representations in response the second presentation of the stimuli, we employed a measurement approach labeled the Encoding-Related Lateralization (ERL, see Figure 1; Gratton, 1998; Gratton, Corballis, & Jain, 1997; see also Talsma, Wijers, Klaver, & Mulder, 2001), a form of event-related lateralization. The main purpose of this procedure is to eliminate brain activity that is *generically* related to stimulus processing rather than *specifically* to response adaptation associated with previous stimulus presentation. This is particularly a problem for ERP measures because they confound phenomena occurring at different locations within the brain because of the propagation of electric fields across the skull and scalp. To reduce the impact of volume conduction, it is therefore useful to subtract activity that is *not* related to the adaptation process per se. This is achieved through a double-subtraction method (see Gratton, Coles, Sirevaag, Eriksen, & Donchin, 1988). A prerequisite for the derivation of the ERL is to generate different memory traces in the two hemispheres. This is accomplished by using a divided-field approach in which stimuli to be memorized are presented to the left or right visual hemifield [left visual field (right hemisphere of the brain) = LVF/RH; right visual field (left hemisphere of the brain) = RVF/LH]. ERPs are then recorded from electrodes located at homologous locations over the left and right hemispheres, in response to the central (full field, LH + RH) presentation of a test stimulus. It is expected that when the test stimulus is presented, the adaptation process will elicit ERP activity that is lateralized to one side of the head if the stimulus matches a neural representation observed in the left hemisphere and to the other side of the head if the stimulus matches a neural representation in the right hemisphere. This is because *reduced* activation should be obtained in the hemisphere where the previous presentation of the same stimulus had induced the adaptation process. As

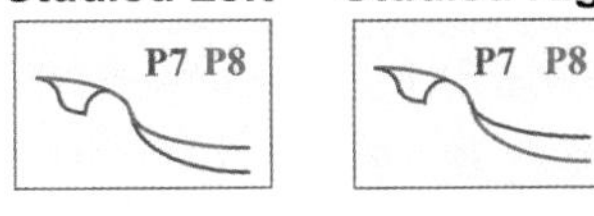
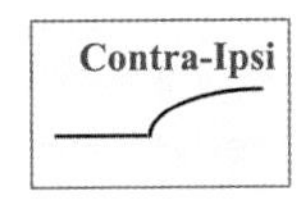

FIGURE 1 Steps in the computation of encoding related lateralizations (ERLs). For this figure, negative polarity is assumed to be plotted upward.

a consequence, by computing lateralization potentials (i.e., differences between activities recorded at electrodes placed at homologous locations over the two hemispheres) and determining which of these activities switches side systematically as a function of the hemifield at which the stimulus was presented during encoding, it is possible to isolate brain activity related to the adaptation process. Truly, this activity may be related to other types of memory-related processes (i.e., not only adaptation, but also recognition, etc.) and therefore may be complex and multiphasic in nature. However, phenomena related to response activation, consistent hemispheric asymmetries in stimulus processing (i.e., lateralized always on the same side regardless of encoding side), or any bilateral process associated with stimulus evaluation will all be removed by the double-subtraction method (Gratton, 1998). This approach therefore can be used to isolate brain activity related to the memory representation left by the previous exposure to the stimulus from all other brain activity occurring during the processing of the test stimulus.

It is important to note that the ERL approach is similar to that employed by other investigators to isolate brain activity in other context, such as motor preparation (lateralized readiness potential, Gratton et al., 1988), selective attention (N2pc, Luck & Hillyard, 1994), and working memory load (contralateral delayed activity; see Eimer & Kiss, 2010; Klaver, Talsma, Wijers, Heinze, & Mulder, 1999; Vogel & Machizawa, 2004; sustained posterior contralateral negativity, or SPCN, Jolicœur, Sessa, Dell'Acqua, & Robitaille, 2006; Robitaille, Grimault, & Jolicœur, 2009).

A prerequisite for all of these procedures to be applicable to the analysis of the brain activity associated with a particular psychological construct is that the psychological construct is implemented in the brain according to a hemispheric organization. In other words, it is important that the relevant neural phenomena occur in a relatively similar fashion in each of the two hemispheres and that the hemisphere in which the phenomenon is likely to be most prominent can be varied using experimental manipulations (such as divided field stimulus presentation, direction of attention, or response requirements). Not all psychological constructs are necessarily linked with brain processes possessing a hemispheric organization. For instance, many phenomena associated with language processes may not be carried out in similar fashion by the two hemispheres and therefore may not possess a type of hemispheric organization that makes this procedure feasible (but see Federmeier, 2007 for a different view on at least some language-related processes). However, to the extent that systematic ERL effects are actually observed, we can conclude that the type of memory underlying the ERL effects is hemispherically organized.

RESULTS AND DISCUSSION

Demonstrating the Hemispheric Organization of Visual Memories

Our initial demonstration that visual memories are hemispherically organized was obtained using a recognition paradigm (Gratton et al., 1997). We presented subjects with lists of unique, novel, line-based, visual stimuli flashed to the left or right of a fixation cross. Subjects were given a superficial orienting task (indicating whether the stimuli were symmetrical along the vertical or horizontal axis). In three different behavioral studies, the same stimuli were then presented in another location of the *same* hemifield (varying in distance depending on the experiment) or of the *opposite* hemifield, intermixed with an equal number of new stimuli, and subjects were asked to provide an old/new judgment. In all the experiments, there was an accuracy advantage when the stimuli were re-presented to the same hemifield with respect to when they were re-presented to the opposite hemifield (see Figure 2). This advantage persisted even when the physical distance between the first and second presentations of the stimuli were matched for

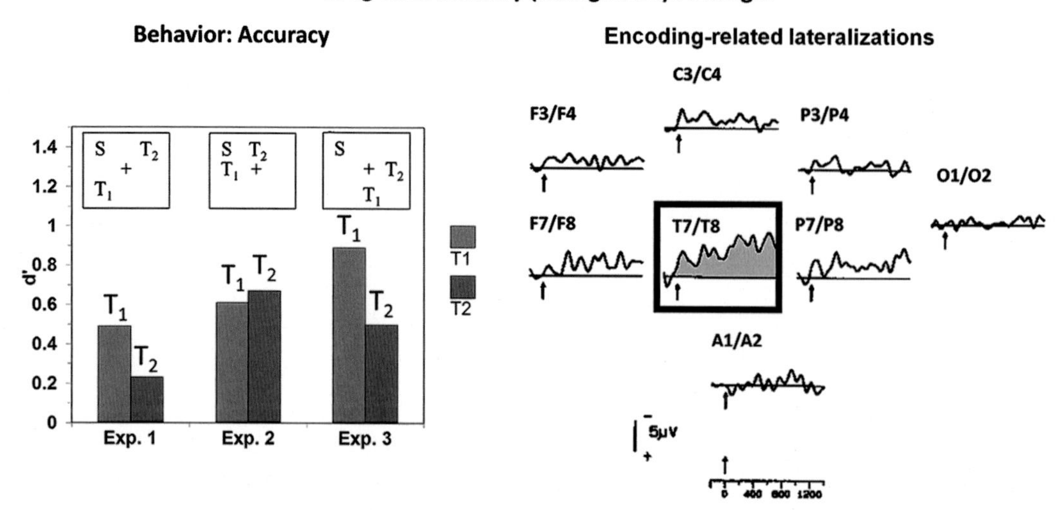

FIGURE 2 Behavioral (left) and ERP (right) results from Gratton et al. (1997). In the left figure, accuracy (d′) measure for a different testing-encoding relationship is presented. S represents the position of the stimulus during encoding (any of the four corners were possible). T_1 and T_2 represent possible test positions, all of which are equidistant to the encoding position. T_1 always presents at test in the same hemifield used for encoding; in Experiments 1 and 3, T_2 presented in the opposite hemifield, whereas in Experiment 2 it also presented in the same hemifield. Notice the higher accuracy for same vs. opposite hemifield presentations. In the right figure, ERL waveforms are presented, with the greatest effect visible at the T7/T8 electrode pair. *Reprinted with permission from the Journal of Cognitive Neuroscience.*

the within- and across-hemifield conditions. In a fourth experiment, we presented the test stimuli at the center (fixation), recorded ERPs, and computed ERL waveforms. The waveforms (also shown in Figure 2) indicated the presence of systematic ERL effects, maximum over temporal scalp locations, consistent with the idea that visual memories are indeed hemispherically organized. Interestingly, when subjects were asked to indicate whether each stimulus had been presented initially to the left or right of fixation, their performance was at chance. This suggests that subjects did not use a simple mnemonic strategy associating a stimulus with a side of presentation and that they were not aware of where the stimuli were originally presented.

Fabiani, Stadler, and Wessels (2000) used a similar paradigm to investigate whether, in the Deese-Roediger-McDermott (DRM) paradigm (Deese, 1959; Roediger & McDermott, 1995), real and illusory memories could be distinguished at retrieval on the basis of their associated brain activity. Specifically, they modified the DRM paradigm by pairing DRM word lists, each associated with a different "lure."[1] Words from each of the two lists were interspersed randomly with words from the other list, with the constraint that words from one list were presented systematically to the left of fixation and words from the other list to the right of fixation. These authors were interested in determining whether correctly identified target (i.e., previously presented) words, but *not* falsely recognized lures (i.e., words that had not been presented previously but were strongly associated with words on the list), would elicit an ERL effect (i.e., ERP activity that was lateralized as a function of list presentation side at encoding) during the recognition/test phase of the DRM paradigm (at which all stimuli were presented at the center of the screen). This was in fact the case, indicating that sensory memory representations (maintaining a trace of the lateralized encoding) existed for the targets but not the lures. As in the Gratton et al. (1997) work, the ERL was most evident in temporal regions. In this paradigm subjects showed very high levels of "illusory" memory (i.e., false alarms to lures), which were identical to the hit levels for words that had been presented during the encoding phase. This indicates that subjects did not make use of (or were not able to access) the lateralized sensory memory representations to make their old/new judgments. Instead, they based their responses on some other form of memory (presumably some form of semantic representation), which does not necessarily have a similar hemispheric organization (i.e., it is not reflected in an ERL). This implies that: (1) stimuli may induce multiple representations within the brain and (2) having a hemispheric organization is not necessarily a property of all memory representations: semantic networks are most likely better developed in one hemisphere than in the other (e.g., in the left in right-handed subjects).

[1] As an example, the "window" and "sleep" DRM lists include words that are associated with window and sleep, respectively, but not the words "sleep" and "window," which are considered lures because they are likely to elicit an erroneous "old" judgment (false memory) at test.

The ERL waveforms obtained in both the Gratton et al. (1997) and the Fabiani et al. (2000) studies were small (<5 μV). Because the number of stimuli that could be garnered in these long-term memory paradigms is limited (generally <100 per condition × subject), the ERL waveforms were inherently noisy. This prevented a more in-depth analysis of the effects, which was instead possible in the working-memory paradigms reviewed in the next section.

The ERL in Working-Memory Paradigms

In divided-field, long-term memory paradigms such as those reviewed previously, each particular stimulus is presented *once* to only one hemisphere during the entire experiment, until it is tested centrally. This allows for representations that possess a sufficiently long decay function to be maintained to show systematic differences as a function of encoding at the moment of test. In a typical working memory paradigm, instead, the same stimuli may be repeated, sometimes hundreds of times. A question may then arise as to whether differences in the latent level of activation (and therefore in the adaptation process) may still be visible after such a large number of stimulus repetitions. This is particularly an issue when the same stimulus is allowed to be presented to either hemisphere across trials. In this case, only memory representations with very rapid decay functions may show a lateralized activation, whereas those with long decay functions may lose this differentiation across trials. It is therefore possible, or even likely, that ERL effects in this case should be different from those seen in long-term memory paradigms.

To determine whether ERL effects could be obtained in working memory paradigms, we (Fabiani, Ho, Stinard, & Gratton, 2003; Gratton, Fabiani, Goodman-Wood, & DeSoto, 1998) employed a classic memory-search paradigm (Sternberg, 1966; Sternberg, 1969), using variable mapping (i.e., the same stimulus could serve as target and distractor on different trials) and a set size of two, with a memory set item presented to the left side of fixation and the other to the right side of fixation. The memory set was presented very briefly (for 200 ms), to avoid saccadic eye movements. The test stimulus was presented to the center of the screen 2 s after the memory set. This paradigm has several advantages. First, using a memory-search task implies that both items had to be encoded, with no particular advantage in shifting attention from one side to the other. This eliminates the possibility that attention shifts, such as those signaled by the N2pc (Luck & Hillyard, 1994), could confound the results. Second, the response requested of the subjects (whether or not the item belonged to the memory set) was orthogonal to the manipulation of interest (whether the test stimulus was previously presented to the left and right of fixation), eliminating possible confounds resulting from the motor response side. Third, this paradigm can be easily adapted to include a large number of trials, which is very useful to obtain stable and reliable brain data.

To observe a behavioral effect in this paradigm, instead of being presented centrally the test stimulus must be presented laterally to match or mismatch the visual hemifield in which it was first encoded. This was done in a separate experiment (Fabiani et al., 2003). The data are presented in Figure 3 and show a clear advantage for the same-hemifield presentation in terms of both accuracy and reaction time.

The ERL data obtained in this paradigm (also presented in Figure 3) show a multiphasic function, suggesting that the sensory memory phenomenon may be quite complex, and involve several different brain regions, or at least different processes that vary over time. Most critically, the data show a positive peak with a latency of approximately 250–300 ms, followed by a negative peak with a latency of 300–400 ms, most visible at posterior lateral electrodes. Both responses were reliable across subjects.

In addition to ERPs, we also recorded EROS from occipital areas in this paradigm (Fabiani et al., 2003; Gratton et al., 1998). An advantage of EROS with respect of ERPs is that signals are much more localized. As a consequence, the responses of individual brain regions can be analyzed separately. This means that it is possible to isolate the activity elicited by test stimuli from brain regions ipsilateral and contralateral to the encoding side. It is also possible to compare these responses with those obtained when the test stimulus was new (i.e., it was not part of the memory set). These data are presented in Figure 4. They indicate that early (80-ms latency) responses in medial occipital cortex were visible in both hemispheres for new items and in the hemisphere ipsilateral to side of encoding for items that were part of the memory set. In other words, responses are reduced in the region of the brain that had just been exposed to the same stimulus. Further, analyses at longer latencies and in different brain regions revealed additional responses, including enhanced activity for "old" items. This suggests that adaptation may not be the only process associated with sensory working memory: template matching or recognition may also occur.

Permanence versus Top-Down Creation of Expectations: Masking Studies

Taken together, these results indicate that the lateralized presentation of visual stimuli leaves "traces" in the hemisphere contralateral to the encoding side. This gives rise to an "adaptation" effect by which subsequent presentations

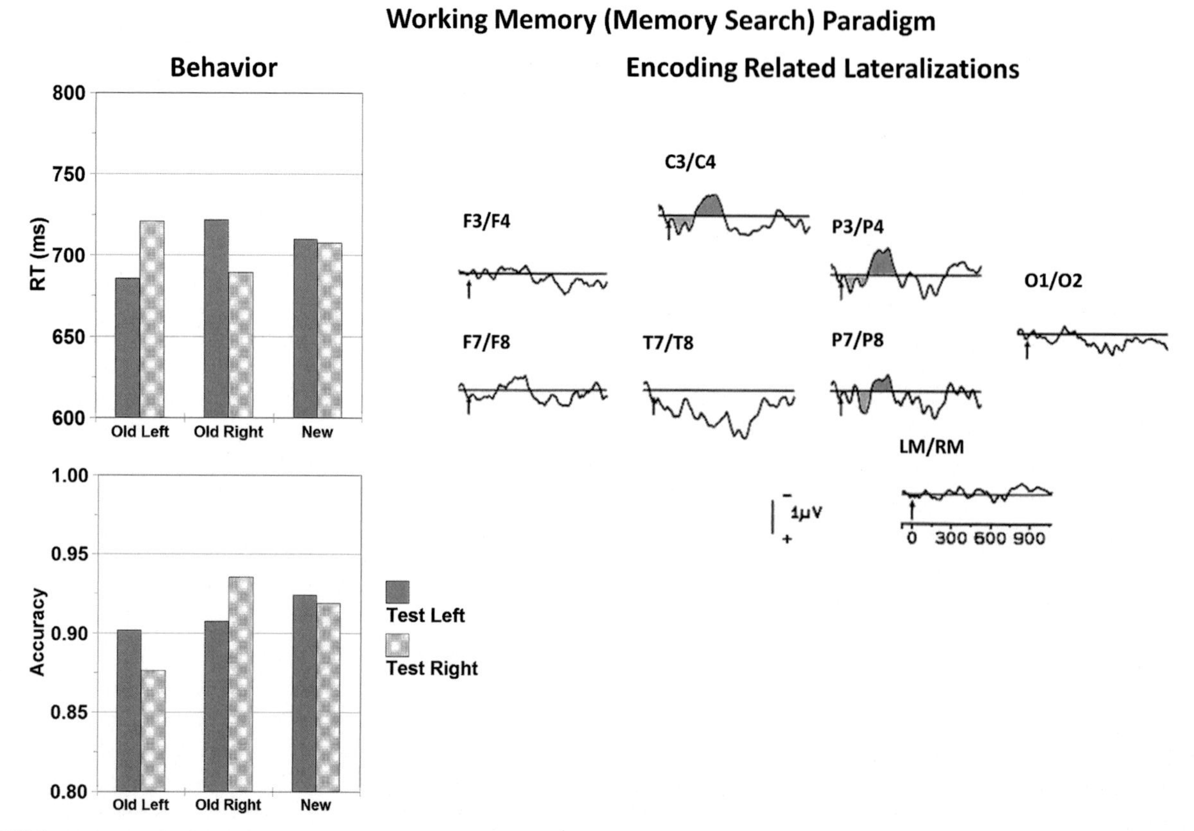

FIGURE 3 Behavior (left) and ERP (right) results from Fabiani et al. (2003). Reaction time and accuracy data for conditions in which the test stimulus matched the stimulus previously presented to the left ("Old Left"), to the right ("Old Right"), or did not match any of the memory set stimuli ("New"). The data reveal an advantage for stimuli presented to the same hemifield at encoding and at test. The ERL results indicate a multiphasic response most evident at parietal locations. *Reprinted with permission from Psychophysiology.*

FIGURE 4 Event-related optical signal (EROS) results from Gratton et al. (1998). Shown are maps of optical (EROS) responses (latency between 50 and 150ms) to test stimuli for conditions in which the test stimulus matched the stimulus previously presented to the left ("Old Left"), to the right ("Old Right"), or did not match any of the memory set stimuli ("New"). The area explored by the optical measures is shown in the left diagram (as viewed from the back of the head). *Reprinted with permission from Psychophysiology.*

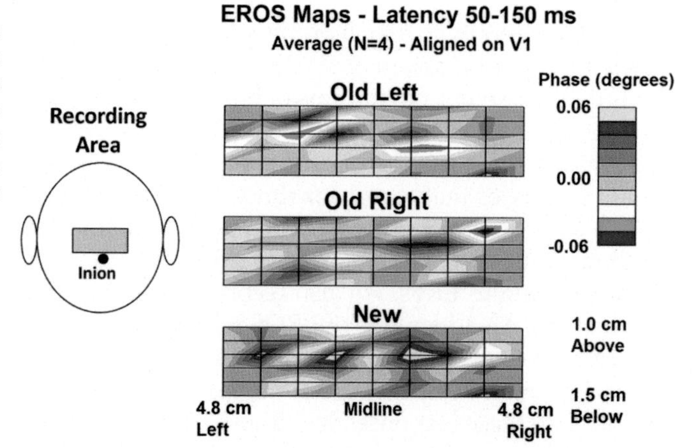

of the same stimulus to the same hemisphere require less processing (and are associated with reduced brain activity). This adaptation phenomenon may reflect the permanence of latent activation of the neural circuitry in the aftermath of the initial processing. Alternatively, it may reflect top-down activation of the sensory circuitry, exerted by control processing areas higher up in the information processing hierarchy, and in preparation for the upcoming test stimulus.

An additional question is the role of adaptation in influencing the subjects' behavior. From the illusory memory experiment summarized previously, it is clear that subjects' performance may be driven by factors other than the sensory memory of the initial stimulus presentation. In the case of illusory memory, activation of semantic nodes may be critical. Are the same representations important for brain and behavioral effects? Are there correlations between

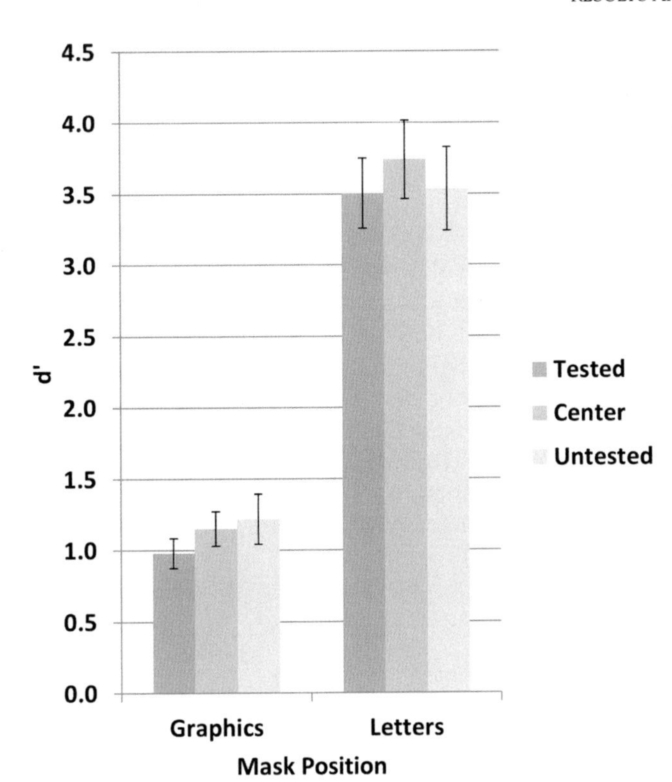

FIGURE 5 Masking studies of visual memory. Left panel: Accuracy (d′) as a function of the location of the mask for graphics stimuli. The data indicate lower accuracy, in particular when the mask was displayed at the same location as the to-be-tested item. Right panel: Accuracy (d′) as a function of the location of the mask for letter symbols. The data indicate high accuracy and no effect of mask location.

brain activity and performance? These questions were difficult to answer in the Fabiani et al. (2003) study previously described because of high accuracy, which did not allow us to compare conditions in which the memory process was successful with those in which it was not. It is possible, however, that the high level of performance was partly from the use of stimuli (letters) that have strong representations in long-term memory—presumably because of their association with language. Subjects, in that case, may rely on phonological representations (the "articulatory loop" in Baddeley and Hitch's 1974 view). Such representations may not be visual and not necessarily hemispherically organized.

To address these issues, in a series of experiments we used backward visual masks displayed 200ms after the presentation of the memory set to prevent residual activation of the processing nodes to be maintained between the onset of the memory set and the response to the test stimulus. In the first of these experiments, we varied the stimulus material by comparing the memory for letters (which are very easy to verbalize) with graphics symbols (which are very difficult to verbalize). The masks were presented either at the same location as the stimulus that was later tested, at the location of the other stimulus to be memorized, or not at all. The results (Figure 5) showed that accuracy was significantly higher when the stimulus material was made up of letters than when it was made up of graphics symbols ($t(15)=11.35$, $p<.0001$). In addition, the masks significantly reduced performance only when presented at the same location as the stimulus to be tested, relative to the contralateral location, and only when graphics symbols ($t(15)=3.48$, $p<.01$) but not letters were used ($t(15)=0.30$, not significant).

These data strongly indicate that different types of representations are used to perform the task in case of letters and graphics symbols. For letters, performance depended on a form of representation that yielded high accuracy and was resistant to the visual mask. The obvious candidate is a phonological representation. For graphics symbols, however, a phonological representation was not readily available and subjects had to rely on a visual representation, which yielded much lower performance and was interfered with by the presentation of the visual mask. These data also provide some indication that residual activation from bottom-up rather than top-down feedback processes might be involved, in that the mask did have an effect on memory performance, even though it was presented very soon after the memory set and well in advance of the test stimulus, thus leaving ample time for top-down expectation processes to take place. However, an alternative interpretation is that the masks interfered with the encoding process because they were presented at the same location as the stimulus to be memorized. This also prevented us from determining whether the phenomenon was specific to a particular location or was a more general phenomenon occurring at the hemispheric level.

To evaluate these different hypotheses, we ran two additional experiments (Experiments 2 and 3) in which we varied the location of the masks. In these experiments, the masks could be presented at the same location as the to-be-tested item, at the location of the alternative item, at other locations within the same hemifield as the to-be-tested item, or as the alternative item (varying the distance from the tested item between the two experiments). The masks could also be presented at the center of the display, where information would be transmitted to and processed in both hemispheres. The results of these studies are also presented in Figure 6 and show three accuracy levels. The lowest accuracy was obtained when the mask overlapped the location of the to-be-tested item ("Ipsil"). An intermediate level of accuracy was achieved when the mask was presented at any other location within the *same* hemifield as the to-be-tested item ("IpsAlt") or at the center of the display ("Center"). Finally, the highest performance was reached when the mask was presented at any other position ("ConAlt" and "Contral") or when no mask at all was presented ("No Mask").

These results indicate that the mask may exert two types of effects: one that was specific to the target location and one that generalized across the hemisphere in which the target was presented. Whereas the first effect may be due to interference with the formation of the stimulus' percept, the second is more likely from a more general disruption of the memory representation to be held in memory, but still confined to only one hemifield (i.e., to one hemisphere). Taken together, these data support the claim that masks may disrupt performance in the working memory task by interfering with the latent activation triggered by the bottom-up processing of the memory set items. This disruption occurred, however, only when alternative and more effective means of maintaining a representation of the to-be-tested item (such as verbalization and use of the phonological loop) were not available.

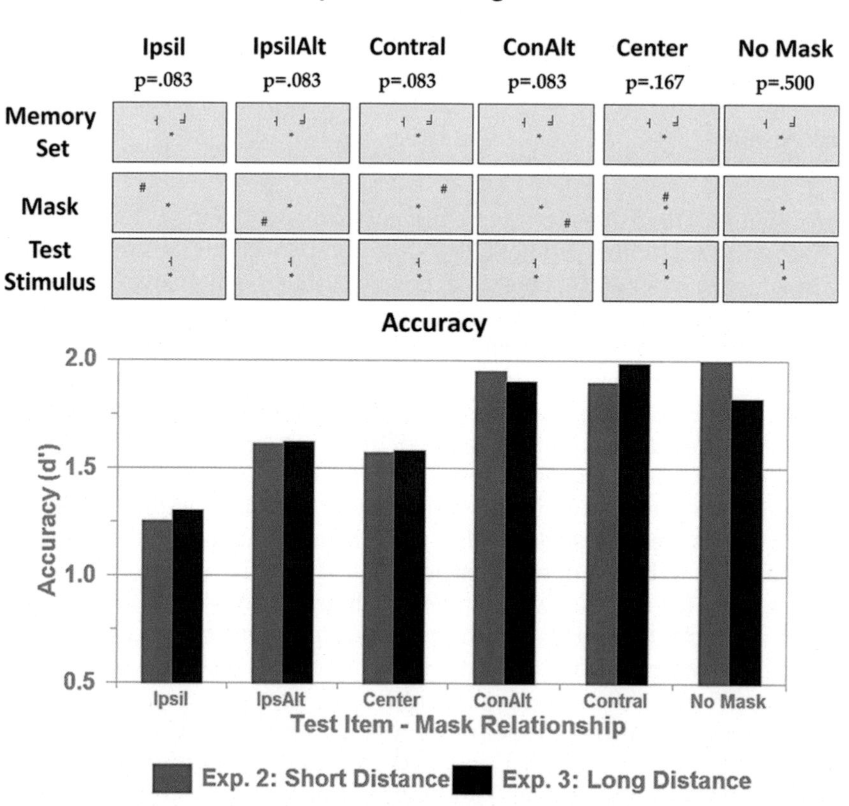

FIGURE 6 Masking studies of visual memory. Top: Examples of different experimental conditions in which the left memory-set stimulus was used as the target stimulus. In an equal number of trials (both corresponding to 25% of the total trials), the right memory-set stimulus was used as target. For half of the trials, the test stimulus did not match any of the memory-set stimuli. Bottom: accuracy (d′) levels reached for "old" test stimuli as a function of the location of the backward mask, relative to the encoding stimulus that was subsequently tested. The encoding stimuli were always presented on two symmetric corners of the display, one in the left and one in the right hemifield. In the "Ipsil" condition, the mask was presented at the encoding location. In the "IpsAlt" condition, the mask was presented at a different location in the same hemifield (at different distances from the encoding location in Experiments 2 and 3). In the "Center" condition, the mask was presented at the center of the display. In the "Contral" condition, the mask was presented over the memory set stimulus that was not the tested hemifield. In the "ConAlt" condition, the mask was presented in the hemifield that was not tested, but at a different location from the alternate memory set stimulus (at different distances from the location of the alternate stimulus in Experiments 2 and 3). In the "NoMask" condition, no mask was presented. Note that three levels of accuracy are reached. Post-hoc Tukey tests showed that the "Ipsil" condition differed from all other conditions, and that the "IpsAlt and "Center" conditions differed from the other four conditions, but not from each other. The remaining three conditions did not differ from each other. No significant differences or interaction were found between the two experiments.

In a fourth experiment, we investigated more directly whether the mask interfered with the neural adaptation process. As a reminder, adaptation is revealed by a *reduced* early-latency (80-ms) response in occipital regions in the hemisphere that was just exposed to the target. The crucial question is whether the mask, when presented at the location of the to-be-tested item, restores the normal bilateral response to the test stimulus, effectively eliminating adaptation. In other words, when the appropriate location is masked, is the repeated target stimulus treated as a new item? We investigated this issue with an EROS experiment, whose results are presented in Figure 7. In the top part of this figure, we show the bilateral brain response to correct rejections (i.e., "new" items) and to hits ("old" items) that were not masked. In the latter case, the response was suppressed in the hemisphere that has just been exposed to the items (replicating the results of our previous study, Gratton et al., 1998). In the bottom part of the figure, we present the effect of the mask on the processing of the test item for "old" trials (hits). When the mask was presented at the location of the item that was not tested, the response was similar to the no-mask condition. However, when the mask was presented at the location of the tested item, a bilateral response was observed (just as for "new" items), indicating that the suppression resulting from adaptation had been removed. These data provide further support to the idea that the processing suppression seen in adaptation is due to residual activation of the stimulus representations left behind by the original bottom-up process rather than by top-down activation in preparation for the upcoming test stimulus, which would occur much later during the memory set-test stimulus interval.

Application of the ERL Methodology to Study the Flow of Activation of Visual Representations

So far, we have reviewed evidence that visual memory representations are hemispherically organized, and, to some extent, rely on the latent activation of neural representations triggered by bottom-up processing. Our data also indicate that several distinct representations may be activated during this processing at different levels within

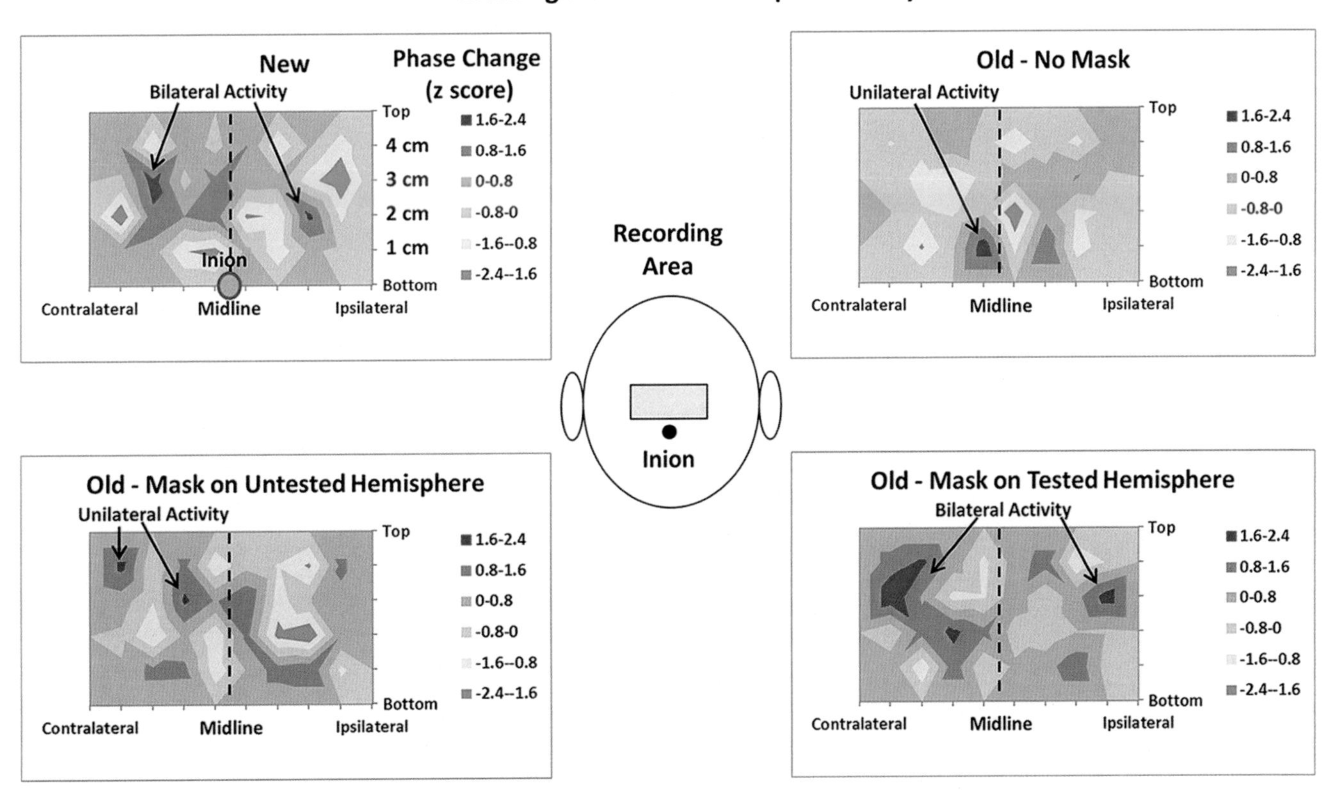

FIGURE 7 Masking studies of visual memory. Shown are maps of the activity elicited by the test stimulus (latency 80 ms) in the brain area indicated in the central diagram. The four maps show brain activity in different masking conditions in response to old (positive) items and to new (negative) items. The color scale refers to "z scores" computed across subjects, based on the phase delay EROS data. Values in red indicate activation levels significant at $p<.05$, uncorrected for multiple comparisons. Note the bilateral activation for new items and the unilateral activation for the old "no mask" condition, replicating the results presented in Figure 4. The presentation of a mask reinstated the bilateral response when presented in the same hemifield as the to-be-tested item, but not when presented in the other hemifield. The purple ball indicates the position of the inion. Each tick mark on the bottom and side of the maps denotes 1 cm.

the information processing system. This idea is consistent with the general concept that visual processes are carried out by a network of distinct areas of cortex, for which visual information is represented according to different rules, which may vary in complexity (see Tse, Low, Fabiani, & Gratton, 2012). If this is the case, then it should be possible to take advantage of the high temporal resolution of ERPs to determine the time at which each of these representations is accessed during information processing. This can be determined by generating different types of overlap between the memory set and test item: When a representation coding the particular feature or dimension over which the memory set and test item overlap is accessed, then at that latency we should observe an adaptation effect in the ERL waveform. In our practice, we compare the ERL waveforms obtained for different types of overlap and determine whether a particular representation is accessed. In the experiments described so far, the exact same stimulus was used for encoding and test, thus affording complete overlap. However, it is possible to make the two stimuli similar on one dimension and not another.

We used this logic in two studies investigating the timing of access of different visual stimulus dimensions. In the first study (Shin, Fabiani, & Gratton, 2006), there were three dimensions of overlap: (1) a physical dimension (i.e., the stimuli were presented either in the same or different letter-case in the memory set and test displays); (2) a semantic/symbolic dimension (i.e., the same or a different letter was presented in the encoding and test displays); and (3) a numerosity overlap (i.e., the same or a different number of letters were presented to each hemisphere in the encoding and test displays). The ERL effects are shown in Figure 8. They indicate three major "bumps" in the ERL waveform: the first (latency 150–200 ms) was sensitive to physical overlap, the second (latency 200–400 ms) was sensitive to symbolic/semantic overlap (i.e., the same ERL effect was obtained at this latency irrespective of the physical properties and numerosity of the memory set letters), and the third (latency 400–600 ms) was sensitive to numerosity overlap. This is consistent with the idea that the representations coding these dimensions are accessed in succession, so that the adaptation processes associated with each of them also occur in succession.

In another experiment (Basak et al., in preparation), we investigated whether the adaptation process would occur only for individual features or could also occur for the conjunction of features. To evaluate this possibility, at encoding, we presented sets of multidimensional stimuli distributed across the two hemifields/hemispheres (see Figure 9). At test, we then presented a stimulus that could completely or partially overlap *one* of the encoded stimuli or that could contain features from *multiple* encoded stimuli (i.e., features from either stimuli encoded in the same or opposite hemifield). We also used new stimuli composed of completely new features (i.e., with no overlap with the stimuli encoded in other hemifield). The critical comparison was between the two types of recombined stimuli. Notice that these recombined stimuli generated a situation of conflict, in that they contained "old" features but had to be responded to as "new" because they did not match any of the stimuli in the memory set. Under these conditions, accuracy was lower for recombined stimuli whose component features were presented in the same hemifield than for those that were

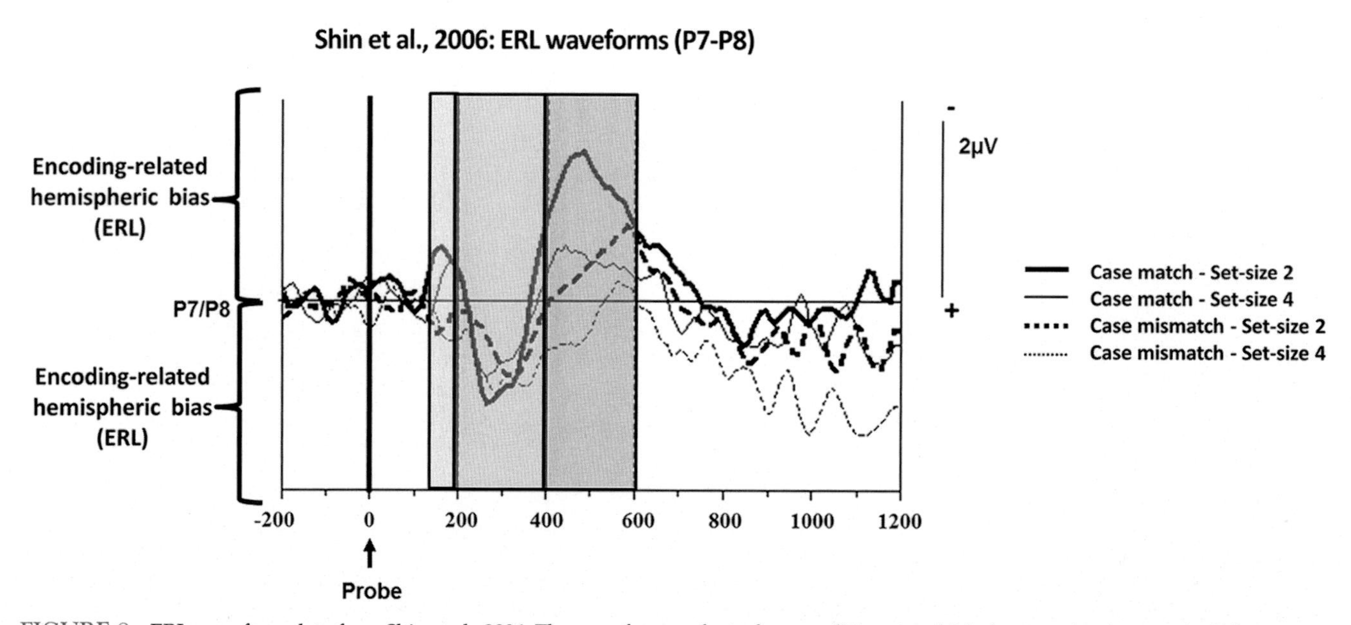

FIGURE 8 ERL waveform data from Shin et al., 2006. The waveforms refer to four conditions in which the test stimulus matched the memory set presented in the contralateral hemifield on the basis of case and numerosity (for all four conditions, it matched it in terms of the symbol presented). *Reprinted with permission from the Journal of Cognitive Neuroscience.*

presented in the opposite hemifield (see Figure 9; note that the distance between the stimuli containing these recombined features was the same in the two cases). Because the incorrect responses to these recombined stimuli were faster than the correct response, we interpreted these incorrect responses as resulting from the formation of illusory conjunctions (Treisman & Gelade, 1980). According to this interpretation, illusory conjunctions are therefore more likely to occur if the individual features were presented to the same than to different hemispheres, suggesting that the process of conjunction is performed within each hemisphere. This idea is further supported by the observation that within-hemifield illusory conjunctions were associated with a very pronounced ERL effect, yielding waveforms that are very similar to conditions of complete match between memory-set and test stimuli (see Figure 9). This match extends not only to the section of the ERL waveform related to physical match, but also to that related to symbolic match. In summary, these findings support the idea that feature conjunction may operate at a hemispheric level.

Advantages of Organizing Information Across Hemispheres

A final question we have addressed in our work is whether subdividing information across different hemifields (and therefore across hemispheres) may generate a computational advantage with respect to when information is all in the same hemifield (or hemisphere). This is a question that has been addressed by several studies in the past

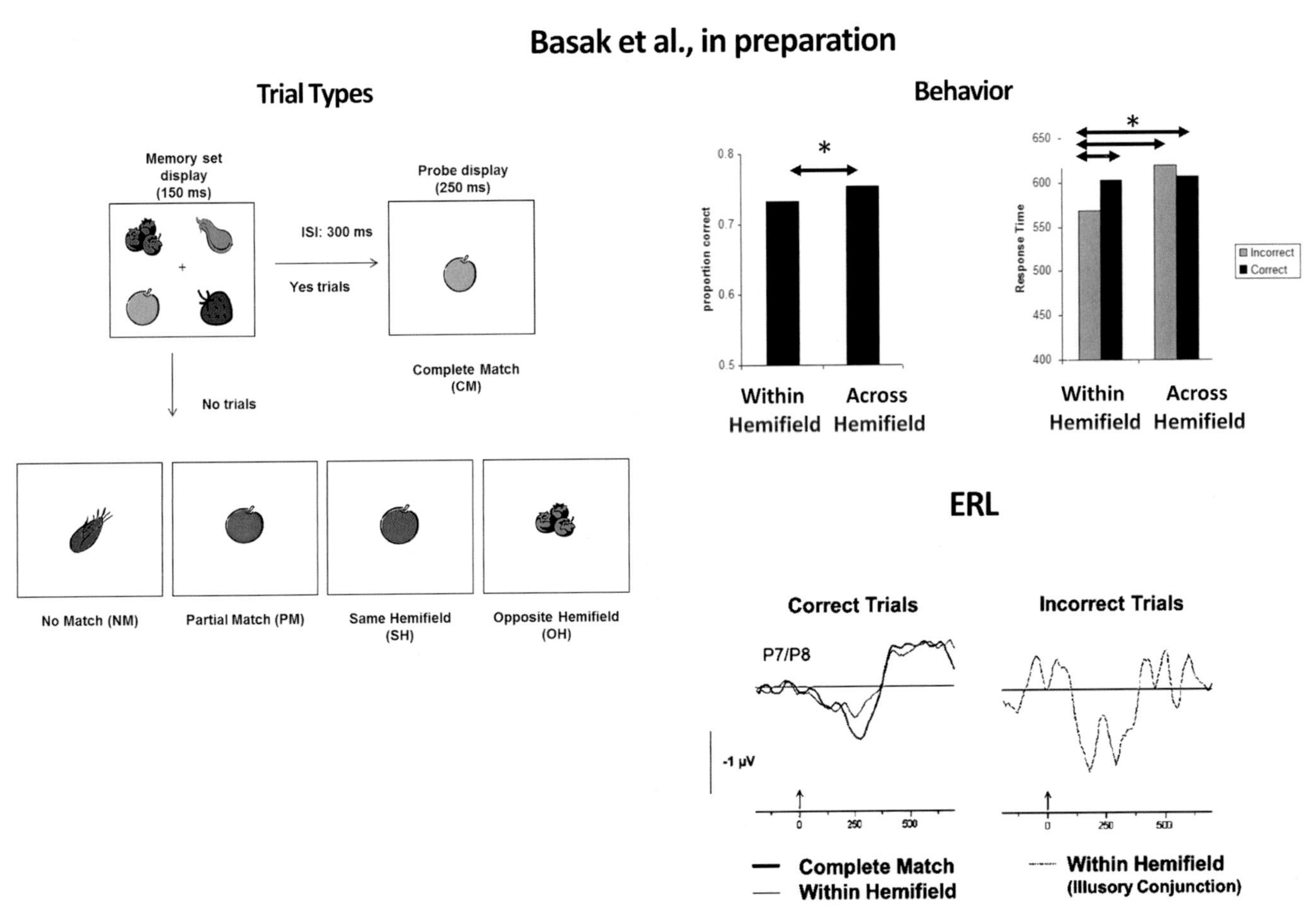

FIGURE 9 Trial types, behavioral data, and ERL data from Basak, Low, Fabiani, & Gratton, in preparation. Trial types are presented to the left. The behavioral data refer to accuracy and reaction time for trials in which the features of the test stimulus were presented already in the memory set items, but belonged to different items. In this case, the correct response should have been "new." However, subjects did commit a relatively large number of errors ("old" responses), indicating the presence of "illusory conjunctions" (Treisman & Gelade, 1980): the errors were significantly ($p<.05$) more frequent when the features composing the test item had been presented in two different stimuli displayed within the same hemifield ("within hemifield" condition) than in two different stimuli presented to different hemifields ("across hemifield" condition). The "within hemifield" incorrect responses were also significantly faster than all other types of responses to these items. Significant comparisons are indicated by arrows and asterisks. The ERL waveforms presented to the left showed lateralizations that occurred when the test stimulus was one of the old stimuli ("complete match," thick line) compared with the correct responses in the "within hemifield" condition (thin line to the left) and the incorrect responses in the "within hemifield" condition (i.e., "illusory conjunctions," dashed line to the right). The ERL deflection was much larger for the "complete match" and "illusory conjunction" conditions than for the correct rejections.

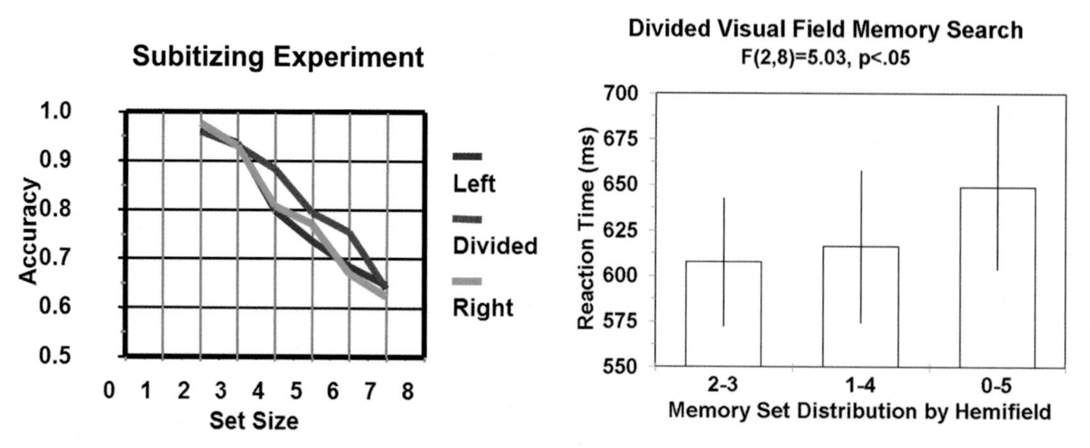

FIGURE 10 Performance data as a function of whether information was split across hemifields in a subitizing task (left) and memory search task (right), from Gratton et al., 2001. In the right panel, vertical bars indicate the standard error computed across subjects. The analysis of variance indicates a difference among the three conditions. *Reprinted with permission from Brain and Behavioral Science.*

few years, some of which involve the computation of memory span using the contralateral delayed activity/SPCN (Vogel & Machizawa, 2004). We (Gratton, Fabiani, & Corballis, 2001) carried out two experiments addressing this issue, one based on numerosity estimation and one based on a memory search task (Figure 10). The numerosity estimation task was based on conditions in which a set of items (varying in number) was rapidly presented (for less than 200 ms to prevent eye movements) within a small constant-size region of the visual field. In different trials, this region encompassed different portions of the visual field, being either located all within the left hemifield, all within the right hemifield, or across the two hemifields. The subjects were asked to report the number of items presented. Performance was higher in the divided hemifield condition than in either of the two "within-hemifield" conditions. In the memory search task, subjects were presented with a memory set comprising five items (different across trials). All five items were presented simultaneously in different locations of the visual field for 200 ms, again to prevent eye movements. However, the number of items in the memory set that were presented to the left and right of fixation varied across trials. Subjects had higher old/new recognition performance for a centrally presented test item (presented 2 s later) when the memory set items were distributed more evenly across the two hemispheres (three to one side and two to the other) than when the distributions were more uneven (four or five to one side and one or none to the other). Thus, both studies showed performance advantages for processing information distributed across the two hemifields/hemispheres. However, these data did not directly measure changes in memory span as a result of exploiting the hemispherically distributed of the visual memory system.

CONCLUSIONS

The studies reviewed in this chapter clearly demonstrate that the divided field/adaptation paradigm coupled with the recording of brain activity is a useful tool for studying visual memory representations in the brain. In particular, among its advantages is the possibility of identifying ERP activity that is specifically related to the access of memory representations. This may provide useful information about the dynamics of information processing.

This paradigm takes advantage of the observation that visual memory is hemispherically organized. This view is shared by other research, such as that focusing on the SPCN (see the work of Eimer & Kiss, 2010; Robitaille et al., 2009; Vogel & Machizawa, 2004). This work has typically focused on the processes that occur during the interval between two displays that need to be compared and shows that the amplitude of the SPCN is monotonically related to the number of items that needs to be maintained in memory, up to the individual's memory span (Vogel & Machizawa, 2004). In this sense, the SPCN may be interpreted as the result of the "attentional effort" spent to keep information alive during the maintenance interval.

In contrast, the ERL effects (and the other data presented in this chapter) are observed after the presentation of the *test* display: They do not provide information about working memory span, but rather about the access of preexisting representations. They also provide information about the nature and mechanisms of visual sensory memory—the target of the effortful processes measured by the SPCN. In this sense, the questions addressed by the two procedures are distinct and complementary.

In this chapter, we have considered the ERL as a tool for revealing latent representations that were left behind during the feed-forward processing of stimuli and we identified this latent activation as "sensory memory." This interpretation is largely based on the results of the masking studies reviewed here. These studies show that this residual latent activation can be interfered with by masks presented to the same hemifield as the to-be-tested stimulus, soon after the presentation of the memory set but well in advance of the upcoming test stimulus. It is also possible to postulate that subjects may "preactivate" stimulus features that are likely to be useful for performance in advance of testing. This is likely mediated by top-down preparatory processes that may be set in place by expectations about the test stimulus. Evidence for this form of reactivation comes from studies of relational memory (e.g., Walker, Low, Cohen, Fabiani, & Gratton, 2014). Both forms of preactivation (feed-forward or bottom-up and feed-back or top-down) are likely to cooccur. However, only the residual latent activation generated by previous bottom-up processes is likely to preserve a lateralized form capable of generating an ERL, as suggested by the "illusory memory" study (Fabiani et al., 2000).

In this chapter, we employed behavioral, ERP, and EROS measures. We see the combination of these measures as more powerful than each of them alone. Behavioral measures are indispensable to understanding the actual functional role of the various brain activities we measure. In some cases, however, we observe brain differences for which the final behavioral outcome is identical. A glaring example is that of illusory memory: in that case, the subject is unable to distinguish between true and illusory memories (Fabiani et al., 2000). Nevertheless, our brain measures show that they are dissociable, and in a fashion that is logically predictable: only true, but not false, memories contain sensory memory signatures, suggesting that the brain contains information that the subject is not accessing to improve performance. In other words, it is the comparison between the behavioral and brain measures that makes this distinction interesting.

ERPs possess exquisite temporal resolution and a reasonably high signal-to-noise ratio. When used with appropriately designed paradigms, they may provide an interesting description of the flow of information across a number of different processing stages in which different representations are accessed. EROS measures are useful because they add spatial precision. They also do not require the complex double subtraction methods used for ERPs, with only a single contralateral control condition required, therefore affording absolute, rather than differential, measures of brain activity. This combined approach could be enriched with yet other modalities to study interactions with attention, brain states, and variations as a function of age, lesions, and so on. Within the context of the hemispheric organization of visual working memory, this approach supports the existence of multiple representations accessed in succession, depending on the relevant feature overlap between the encoded and tested stimulus. The presence of these active representations induces adaptation, which can be interfered with by masks operating at the hemispheric, rather than local level.

Acknowledgments

This chapter is based on research presented at the Attention & Performance XXV meeting, Saint-Hippolyte, Quebec, Canada. We would like to acknowledge the help of many people who were involved in the various studies reviewed in this chapter, including: Paul Corballis, Shamini Jain, Marsha Goodman-Wood, Mike Stadler, Cathy DeSoto, Anita Jurkowsky, and Chandramallika Basak. We would also like to acknowledge the support of a McDonnell-Pew Foundation grant to Dr Fabiani.

References

Atkinson, R. C., & Shiffrin, R. M. (1968). Human memory: a proposed system and its control processes. In K. W. Spence, & J. T. Spence (Eds.), *The psychology of learning and motivation: Advances in research and theory* (Vol. 2) (pp. 89–195). New York: Academic Press.

Baddeley, A. D., & Hitch, G. J. (1974). Working memory. In G. H. Bower (Ed.), *The psychology of learning and motivation* (Vol. 8) (pp. 47–90). New York: Academic Press.

Basak, C., Low, K.A., Fabiani, M. & Gratton, G. (in preparation). An ERP correlate of illusory conjunction. Manuscript in preparation.

Cowan, N. (1995). *Attention and memory: An integrated framework*. New York, NY: Oxford University Press.

Deese, J. (1959). On the prediction of occurrence of particular verbal intrusions in immediate recall. *Journal of Experimental Psychology, 58*, 17–22.

Dehaene, S., Naccache, L., Cohen, L., Bihan, D. L., Mangin, J.-F., Poline, et al. (2001). Cerebral mechanisms of word masking and unconscious repetition priming. *Nature Neuroscience, 4*(7), 752–758.

Donchin, E., Ritter, W., & McCallum, C. (1978). Cognitive psychophysiology: the endogenous components of the ERP. In E. Callaway, P. Tueting, & S. H. Koslow (Eds.), *Event-related brain potentials in man* (pp. 349–411). New York: Academic Press.

Eimer, M., & Kiss, M. (2010). An electrophysiological measure of access to representations in visual working memory. *Psychophysiology, 47*, 197–200.

Fabiani, M., Gratton, G., & Federmeier, K. (2007). Event related brain potentials. In J. Cacioppo, L. Tassinary, & G. Berntson (Eds.), *Handbook of psychophysiology* (3rd ed.) (pp. 85–119). New York, NY: Cambridge University Press.

Fabiani, M., Ho, J., Stinard, A., & Gratton, G. (2003). Multiple visual memory phenomena in a memory search task. *Psychophysiology, 40*, 472–485.

Fabiani, M., Stadler, M. A., & Wessels, P. M. (2000). True but not false memories produce a sensory signature in human lateralized brain potentials. *Journal of Cognitive Neuroscience, 12*(6), 941–949.

Federmeier, K. D. (2007). Thinking ahead: the role and roots of prediction in language comprehension. *Psychophysiology*, *44*(4), 491–505.

Foust, A. J., & Rector, D. M. (2007). Optically teasing apart neural swelling and depolarization. *Neuroscience*, *145*(3), 887–899.

Gratton, G. (1998). The contralateral organization of visual memory: a theoretical concept and a research tool. *Psychophysiology*, *35*, 638–647.

Gratton, G., Coles, M. G. H., Sirevaag, E. J., Eriksen, C. W., & Donchin, E. (1988). Pre- and poststimulus activation of response channels: a psychophysiological analysis. *Journal of Experimental Psychology: Human Perception and Performance*, *14*(3), 331–344.

Gratton, G., Corballis, P. M., Cho, E., Fabiani, M., & Hood, D. (1995). Shades of gray matter: noninvasive optical images of human brain responses during visual stimulation. *Psychophysiology*, *32*, 505–509.

Gratton, G., Corballis, P. M., & Jain, S. (1997). Hemispheric organization of visual memories. *Journal of Cognitive Neuroscience*, *9*(1), 92–104.

Gratton, G., & Fabiani, M. (2010). Fast optical imaging of human brain function. *Frontiers in Human Neuroscience*, *4*(Art. 52), 1–9.

Gratton, G., Fabiani, M., & Corballis, P. M. (2001). Working memory capacity and the hemispheric organization of the brain: a commentary to Cowan, 2001. *Behavioral and Brain Sciences*, *24*(1), 121–122.

Gratton, G., Fabiani, M., Goodman-Wood, M. R., & DeSoto, M. C. (1998). Memory-driven processing in human medial occipital cortex: an event-related optical signal (EROS) study. *Psychophysiology*, *35*, 348–351.

Jolicœur, P., Sessa, P., Dell'Acqua, R., & Robitaille, N. (2006). On the control of visual spatial attention: evidence from human electrophysiology. *Psychological Research*, *70*, 414–424.

Klaver, P., Talsma, D., Wijers, A. A., Heinze, H. J., & Mulder, G. (1999). An event-related brain potential correlate of visual short-term memory. *Neuroreport*, *13*, 2001–2005.

Luck, S. J., & Hillyard, S. A. (1994). Spatial filtering during visual search: Evidence from human electrophysiology. *Journal of Experimental Psychology: Human Perception and Performance*, *20*(5), 1000–1014.

Luck, S. J., & Vogel, E. K. (1997). The capacity of visual working memory for features and conjunctions. *Nature*, *309*, 279–281.

Reber, P. J., Stark, C. E. L., & Squire, L. R. (1998). Cortical areas supporting category learning identified using functional MRI. *Proceedings of the National Academy of Sciences USA*, *95*(2), 747–750.

Robitaille, N., Grimault, S., & Jolicœur, P. (2009). Bilateral parietal and contralateral responses during maintenance of unilaterally encoded objects in visual short-term memory: evidence from magnetoencephalography. *Psychophysiology*, *46*(5), 1090–1099.

Roediger, H. L., & McDermott, K. B. (1995). Creating false memories: remembering words not presented in lists. *Journal of Experimental Psychology: Learning, Memory, & Cognition*, *21*(4), 803–814.

Roediger, H. L., III, Weldon, M. S., & Challis, B. H. (1989). Explaining dissociations between implicit and explicit measures of retention: a processing account. In H. L. Roediger, III, & F. I. M. Craik (Eds.), *Varieties of memory and consciousness: Essays in honor of Endel Tulving* (pp. 3–41). Hillsdale, NJ: Erlbaum.

Schendan, H. E., & Kutas, M. (2003). Time course of processes and representations supporting visual object identification and memory. *Journal of Cognitive Neuroscience*, *15*(1), 111–135.

Shin, E., Fabiani, M., & Gratton, G. (2006). Multiple levels of stimulus representation in visual working memory. *Journal of Cognitive Neuroscience*, *18*, 844–858.

Sternberg, S. (1966). High-speed scanning in human memory. *Science*, *153*, 652–654.

Sternberg, S. (1969). The discovery of processing stages: Extensions of Donders' method. In W. G. Koster (Ed.), *Attention and Performance II* (pp. 276–315). Amsterdam, The Netherlands: North Holland.

Talsma, D., Wijers, A. A., Klaver, P., & Mulder, G. (2001). Working memory processes show different degrees of lateralization: evidence from event-related potentials. *Psychophysiology*, *38*, 425–439.

Treisman, A. M., & Gelade, G. (1980). A feature-integration theory of attention. *Cognitive Psychology*, *12*, 97–136.

Tse, C.-Y., Low, K. A., Fabiani, M., & Gratton, G. (2012). Rules rule! Brain activity dissociates the representations of stimulus contingencies with varying levels of complexity. *Journal of Cognitive Neuroscience*, *24*, 1941–1959.

Vogel, E. K., & Machizawa, M. G. (2004). Neural activity predicts individual differences in working memory capacity. *Nature*, *428*, 748–751.

Vuilleumier, P., Henson, R. N., Driver, J., & Dolan, R. J. (2002). Multiple levels of visual object constancy revealed by event-related fMRI of repetition priming. *Nature Neuroscience*, *5*(5), 491–499.

Walker, J. A., Low, K. A., Cohen, N. J. C., Fabiani, M., & Gratton, G. (2014). When memory leads the brain to take scenes at face value: face areas are reactivated at test by scenes that were paired with faces at study. *Frontiers in Human Neuroscience*, *8*, 18.

CHAPTER

8

Visual Working Memory and Attentional Object Selection

Martin Eimer

Department of Psychological Sciences, Birkbeck College, University of London, London, UK

The processes involved in attention and working memory are assumed to be closely linked. Attention and memory are both selective—they prioritize task-relevant information at the expense of other information that is less important for the adaptive control of current behavior. Both are subject to important capacity limitations—attention can be directed to only a small number of perceptual objects at any time, and the number of items that can be stored and maintained in working memory is also strictly limited. Furthermore, attention and working memory interact at several perceptual and postperceptual stages of information processing (see Awh, Vogel, & Oh, 2006; for a systematic review). These similarities are emphasized in models of working memory that postulate a central role for selective attention. According to Cowan (1995), working memory is best understood as including those representations from long-term memory that are activated because they are within the current focus of attention (see also Oberauer, 2002, 2009, for extensions of Cowan's original model). What is less clear is what the similarities and links between attention and working memory imply for our understanding of the underlying cognitive architecture. Are attention and working memory best conceptualized as two functionally and anatomically separate cognitive subsystems that operate and interact in a coordinated and integrated way but remain essentially distinct? Or should they be regarded as reflections of a unitary mechanism that processes perceptual representations of sensory input and memorized representations of visual objects and events that are no longer physically present in essentially the same selective and goal-directed fashion?

The aim of this chapter is to shed some new light on these questions by reviewing and discussing recent research into links between selective visual attention and visual working memory. The focus of this chapter will be primarily but not exclusively on studies that used event-related brain potential (ERP) measures of attention and working memory. The first section discusses the spatial organization of visual information in working memory. Are objects in visual working memory represented and maintained in a position-dependent fashion or are these representations organized in a more abstract, location-independent way? There is strong experimental support for the hypothesis that representations of visual objects in working memory retain the spatial layout in which these objects were encountered during encoding, and recent findings suggest that the same may also apply to the maintenance of tactile events in somatosensory working memory. However, there is also initial evidence for the existence of spatially global working memory representations. In the second section, the role of spatial attention for visual working memory representations will be discussed. It will be argued that spatially selective visual processing is critical for the individuation and maintenance of visual objects in working memory. This central role of spatial attention may be one main reason why visual working memory contains position-dependent representations. Although selective attention undoubtedly plays a major part in the operation of working memory, the reverse relationship is also important, as working memory is critically involved in the top-down control of visual attention. In the final section of the chapter, the nature of attentional templates in working memory and the role of these templates in the guidance of visual attention toward target objects will be discussed. It will be argued that attentional guidance in visual search requires spatially global attentional templates. Overall, the findings discussed in this chapter suggest that visual attention and visual working memory may be too closely linked to be regarded as functionally and anatomically distinct cognitive

Mechanisms of Sensory Working Memory
http://dx.doi.org/10.1016/B978-0-12-811042-3.00008-6

domains. It might be more appropriate to consider them as different manifestations of a single integrated system, where the same principles and mechanisms of selective visual processing are applied to incoming online sensory information and to stored off-line representations of memorized objects.

ARE REPRESENTATIONS IN VISUAL WORKING MEMORY POSITION-DEPENDENT?

Many neurocognitive models of working memory assume that lateral prefrontal cortex (PFC) plays a critical role in the storage and maintenance of visual information. Classic findings from monkey single-neuron recordings have provided strong support for this view. During the retention period of working memory tasks, many PFC neurons show sustained delay activity (e.g., Fuster & Alexander, 1971), which was interpreted as a direct neural correlate of working memory maintenance. In line with this interpretation, several neuroimaging studies in humans have also found delay activity in lateral PFC during working memory tasks (see Curtis & D'Esposito, 2003; for a review). Such findings suggest that PFC might be a primary locus for the temporary storage of task-relevant information for future use. They are also consistent with Baddeley's cognitive model (Baddeley, 1992; see also Baddeley, 2012 for a recent update), which postulates a central executive system and separate verbal and visual-spatial buffers as the core components of working memory. In fact, Goldman-Rakic (1990) explicitly suggested that the delay activity observed in monkey PFC may be a direct neural correlate of the maintenance of information in the visual storage buffers postulated by Baddeley. Until recently, a commonly held view was that all components of Baddeley's working memory model (i.e., the central executive as well as the specialized storage buffers) are localized in PFC (see Postle, 2006 for a systematic description and critical review of this "standard model" of working memory). In other words, areas in PFC perform both the control and the maintenance functions of working memory that were described by Baddeley and colleagues.

In the past decade, this received view of the neural basis of working memory has faced substantial challenges. In particular, there are now numerous findings from neuroimaging, electrophysiology, and patient studies that appear inconsistent with the hypothesis that PFC is the primary locus for working memory storage (see D'Esposito, 2007; Postle, 2006 for reviews). For example, many human neuroimaging studies have observed delay activity during visual working memory tasks in brain regions outside PFC, and in particular in higher-level visual areas in inferior temporal cortex. Ranganath, Cohen, Dam, and D'Esposito (2004) demonstrated that face-selective and object-selective regions in inferior temporal cortex were sensitive to the type of visual information that had to be actively maintained in working memory. These authors found enhanced activity in the fusiform gyrus in a delayed face-matching task and in the parahippocampal gyrus during a house-matching task. Such findings suggest that posterior cortical regions that are activated in a category-selective fashion during the perception of specific types of visual objects are also involved in the active maintenance of these objects. This emerging view is now known as the "sensory recruitment" hypothesis of visual working memory (Awh & Jonides, 2001; D'Esposito, 2007; Harrison & Tong, 2009; Jonides, Lacey, & Lee, 2005; Postle, 2006; Sreenivasan, Curtis, & D'Esposito, 2014), which proposes that the same posterior cortical brain areas that are responsible for the perception and recognition of visual stimuli are also the primary locus for the temporary maintenance of these stimuli in working memory. Where does this leave the role of the PFC for working memory? That PFC delay activity is a ubiquitous finding in single unit and neuroimaging studies strongly suggests that it does play an important role during working memory tasks. It is likely that this delay activity is primarily related to the top-down control aspects of working memory, such as the activation of goals or task sets, the inhibition of distracting information, or response preparation (see Postle, 2006; for further discussion). In other words, even though the storage and maintenance of task-relevant information (Baddeley's buffer systems) may be delegated to dedicated sensory-perceptual regions of the cortex, PFC may still be the primary locus of the central executive system in Baddeley's model of working memory.

If the storage and maintenance of information in visual working memory is primarily implemented by posterior visual regions that are also responsible for the perceptual analysis of incoming visual signals, the question immediately arises how storage in visual working memory is organized. Early stages of visual information processing in striate and extrastriate visual cortex represent simple features in a small spatially restricted region of the visual field in a retinotopic fashion. Although it is often assumed that higher-level visual object representations are largely independent of the retinal position of an object, this view may not be entirely correct. Although receptive field size and the complexity of stimulus selectivity certainly increases along the occipitotemporal visual pathway (e.g., Gross, Rocha-Miranda, & Bender, 1972; Tsao, Freiwald, Tootell, & Livingstone, 2006), the question whether these changes are accompanied by a corresponding decrease in the retinotopic biases of visual neurons has remained controversial (see Kravitz, Saleem, Baker, Ungerleider, & Mishkin, 2013; for a review). For example, single-unit data have demonstrated a strong preference toward the contralateral visual field in macaque anterior inferotemporal cortex (Desimone & Gross, 1979; Op De Beeck & Vogels, 2000). Although some human functional neuroimaging studies have suggested that high-level visual object representations are not strongly affected by changes in retinotopic position (e.g., Grill-Spector et al., 1999), others have found strong evidence for position-dependence (e.g., Kravitz, Kriegeskorte, & Baker, 2010). If visual working memory storage is located in the same areas that are also responsible for online visual perception, analogous questions can be raised for visual representations in working memory. Do these representations maintain the spatial layout of visual information as it was encountered during perceptual encoding? Or is visual working memory more abstract in the sense that these spatial properties are not always explicitly and obligatorily represented? In other words, are visual working memory representations position-dependent or position-invariant?

ERP studies of visual working memory have provided apparently clear-cut evidence that the maintenance of visual objects in working memory is strongly position-dependent. Many of these studies have used variations of a change detection task where participants are initially presented with a memory array of visual objects (e.g., colored shapes) that they have to remember (Luck & Vogel, 1997). After a retention period, a test array is presented, which is either identical to the first display, or contains one differently colored object, and participants have to make a same–different judgment. By varying the number of to-be-remembered objects in the memory arrays, this task can be used to measure individual working memory capacity. In tasks with simple visual objects, typical capacity estimates range between three or four items (Luck & Vogel, 1997; Cowan, 2001). To measure ERP correlates of visual working memory maintenance, a modified version of the change detection task has been employed (Vogel & Machizawa, 2004). A bilateral memory array containing colored objects in the left and right visual hemifield is presented, and participants are cued to remember the objects in one hemifield. After a retention interval, a test array is shown that is either identical to the memory array or contains one different color on the to-be-remembered side. Again, participants make same–different judgments on each trial. ERPs recorded at lateral posterior electrodes during the retention interval provide strong evidence for the position-dependence of visual working memory maintenance. A sustained enhanced negativity is triggered at electrodes contralateral to the to-be-remembered display side, which starts about 250 ms after memory array onset and persists throughout the retention interval (Vogel & Machizawa, 2004). This contralateral delay activity (CDA) that is also known as the contralateral negative slow wave (Klaver, Talsma, Wijers, Heinze, & Mulder, 1999) or the sustained posterior contralateral negativity (Jolicœur, Sessa, Dell'Acqua, & Robitaille, 2006) is now regarded as an electrophysiological correlate of maintenance in visual working memory. The most direct evidence for this interpretation is provided by the close link between the CDA and working memory load. When the number of objects that participants have to remember is increased, the amplitude of the CDA component increases as well, up to the point where individual working memory capacity is exceeded.

The original motivation for using the CDA component as a marker of visual working memory storage was primarily methodological. Employing lateralized ERP measures of specific cognitive processes has the advantage that possible contributions of other general processes such as arousal, effort, or expectations to ERP waveforms can be readily excluded. By computing ERPs recorded contralateral versus ipsilateral to a set of memorized visual objects, all non-lateralized ERP activity triggered by processes that are simultaneously active in both hemispheres is subtracted. But in addition to this methodological advantage, the presence of CDA components during visual working memory maintenance also has important theoretical implications: it demonstrates that the spatial layout of to-be-remembered visual information is retained when this information is stored and maintained in working memory. Because the CDA is computed by comparing contralateral and ipsilateral ERP waveforms, it cannot reveal the existence of any visual working memory traces that are position-invariant. For the same reason, the CDA will only reflect memory representations that are hemispherically organized, but cannot provide any more detailed insights into the spatiotopic layout of these representations for specific object locations within each hemifield. Nevertheless, the existence of the CDA component demonstrates that memorized visual information is at least to some degree encoded and maintained in a position-dependent fashion.

Further evidence for this conclusion comes from ERP experiments that investigated not memory maintenance as such, but the selective access to information that is already stored in working memory (Kuo, Rao, Lepsien, & Nobre, 2009; Eimer & Kiss, 2010). In these studies, participants first had to memorize all objects in a bilateral stimulus array. After a delay, a retro-cue was presented to instruct participants which of the memorized objects had to be reported. Figure 1 (top panel) shows the procedure employed in a study conducted in our laboratory (Eimer & Kiss, 2010). A bilateral memory array with a shape singleton object (a diamond) on either side was presented first. All objects on one side were red, all objects on the other side were green. After a brief interval (150 ms) or a longer interval (700–1000 ms), a red or green retro-cue was presented at fixation, and this cue informed participants that either red or green objects in the memory array were task-relevant for this trial. Participants then had to report whether the diamond on the cued relevant side was cut at the top or bottom. In this task, objects on both sides had to be encoded, before objects

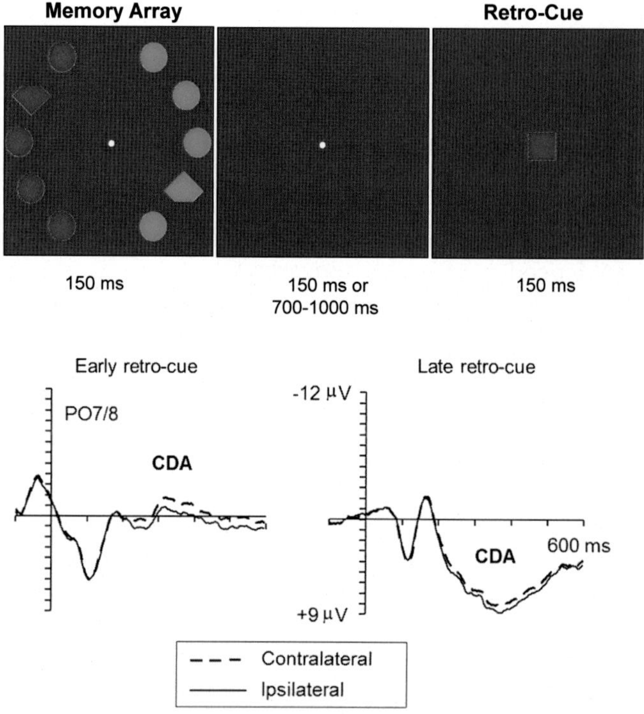

FIGURE 1 Top panel: Working memory task employed by Eimer and Kiss (2010). A memory array with objects in two different colors on the left and right side was followed after a short or a longer interval by a retro-cue, which indicated the task-relevant side for this trial. Bottom panel: event-related brain potentials (ERPs) measured in this study at lateral posterior electrodes PO7/8 contralateral and ipsilateral to the side of the memory array indicated by the retro-cue. Contralateral delay activity (CDA) components are triggered both by early and late retro-cues. ERPs are time-locked to the onset of the retro-cues. *Data from Eimer and Kiss (2010), reproduced with permission. Note that negativity is plotted upward in this figure and in all other figures in this chapter.*

on the cued side could then be accessed in a spatially selective fashion. Figure 1 (bottom panel) shows ERPs triggered in response to the retro-cue at lateral posterior electrodes PO7/8 ipsilateral and contralateral to the relevant memorized display side, as indicated by this cue. Regardless of whether the interval between the array and the retro-cue was short or longer, a sustained negativity was elicited contralateral to the task-relevant side of the memorized array, analogous to the CDA observed during the retention period of change detection tasks (Vogel & Machizawa, 2004). Because the retro-cue itself was a nonlateralized object at fixation, this contralateral response has to be attributed to neural processes involved in the selective access to representations of target objects in visual working memory. If these representations had been stored in a position-independent fashion, no systematic lateralized ERP modulations should have been observed during this selective access phase. The presence of a sustained contralateral negativity in response to the retro-cue (see also Kuo et al., 2009 for similar findings) therefore provides additional support for the view that the storage of visual information in working memory is position-dependent.

However, there is also evidence that visual working memory representations do not always reflect the layout of the encoded information in a spatially corresponding fashion, as was suggested by the presence of CDA components. For example, results from a functional neuroimaging study by Williams et al. (2008) suggested that category-specific information about visual objects presented in the periphery of the visual field is represented in foveal but not in peripheral retinotopic cortex. It is possible that foveal visual areas play an important role in the position-invariant representation of information from different locations in the visual field. Because the objects employed by Williams et al. (2008) appeared repeatedly in rapid succession and did not need to be memorized, this study might have primarily measured online perceptual rather than working memory representations. However, an investigation by Ester, Serences, and Awh (2009) suggested that even in a standard working memory task, task-relevant visual information may be represented in a spatially global fashion during memory maintenance. In this study, participants had to maintain the memorized orientation of a grating presented in the left or right visual field during a long retention period, in order to match it with a test grating. Multivoxel pattern analysis was applied to functional magnetic resonance imaging data obtained during the retention interval to identify where stimulus-specific information about the memorized pattern was stored. Unsurprisingly, Ester et al. (2009) found that activity in contralateral primary visual cortex that retinotopically matched the location of the memorized grating was sensitive to its orientation. However, and more importantly, they found that spatially corresponding areas of ipsilateral primary visual cortex were equally sensitive to the stored orientation of a grating. According to Ester et al. (2009), this apparent spatially global activation of sensory areas during working memory maintenance may help to improve the robustness of the underlying object representations through the recruitment of additional visual areas that are not directly activated during the perceptual encoding of this object. What remains unclear is how this evidence for position-invariant coding in visual working memory can be reconciled with the strong electrophysiological support for position-dependent working memory representations that was discussed earlier.

From a computational perspective, the existence of position-invariant visual working memory representations could offer distinct advantages for memory-based object recognition. The position occupied by a memorized object when it is initially encoded often differs from the position of the same object during a later perceptual episode where it is encountered again and has to be recognized. Because visual object recognition is based on a successful match between incoming perceptual information and stored representations of the same object, this match should be more straightforward if it could be based on position-independent (i.e., spatially global) working memory representations. In contrast, if such representations were strongly position-dependent, performance in object recognition tasks might be impaired, in particular under conditions where a memorized object that appeared in one hemifield during encoding then has to be matched with a perceptual object that is located in the opposite hemifield. There is indeed some evidence for such hemifield-based position switch costs during object recognition. Hornak, Duncan, and Gaffan (2002) found that participants were less accurate in identifying previously seen objects when these objects appeared in opposite visual hemifields during study and test phases. In contrast, vertical position changes in the same hemifield had no detrimental effect. These observations suggest that visual object memories are stored in a hemifield-specific fashion (see also Gratton, Corballis, & Jain, 1997 for similar results and analogous interpretations). However, given the long intervals between study and test phases in the studies by Hornak et al. (2002) and Gratton et al. (1997), their conclusions might apply primarily to long-term memory storage.

To investigate whether visual working memory performance shows hemifield switch costs similar to those observed for long-term memory, Woodman, Vogel, and Luck (2012) employed variations of the standard change detection task. In some blocks, memory and test arrays were presented on the same side. In other blocks, they appeared in opposite visual hemifields. Task performance was essentially unaffected by horizontal translations between memory and test arrays. This was also the case when the relative position of individual objects was changed between memory and test arrays. Even when these two spatial transformations were combined (i.e., test objects appeared on the opposite side and in different positions relative to memory arrays), performance was not different from the baseline condition where side of presentation and relative positions of objects remained unchanged between memory and test arrays. According to Woodman et al. (2012), these results suggest that performance in change detection tasks is based on abstract visual working memory representations that can deal with spatial transformations between memory and test arrays in a flexible and adaptive fashion. The question remains how this apparent flexibility is implemented. One possibility is that visual working memory representations that are active during change detection tasks are genuinely position-invariant. This would again raise the question why the lateralized CDA component is so prominently elicited in this type of task. Another possibility is that these representations are position-dependent, but can be flexibly shifted in mental space in preparation for a comparison with expected perceptual input at a known location, analogous to the mental rotation of visual objects (Shepard & Metzler, 1971). Yet another possibility is that visual working memory always represents stored objects in a position-dependent fashion that reflects their location during encoding, but that the matching process between working memory and perceptual representations is sufficiently powerful and flexible to bridge spatial transformations between memory and test arrays.

If represented object locations in visual working memory can be flexibly shifted to match the anticipated location of test stimuli in the same versus the opposite hemifield, this should be reflected by the CDA component recorded during memory maintenance. To investigate this possibility, we have recently conducted a study in which CDA components were recorded while participants performed two versions of the change detection task (Grubert & Eimer, 2014). All memory and test arrays contained a set of three differently colored objects in the left visual field and another set in the right visual field. The two arrays were separated by a 900 ms interstimulus interval, and an arrow cue presented at the start of each trial signaled that either the left or the right set of objects in the memory array had to be maintained (Figure 2, top panel). In the standard no-shift condition of this task, participants had to compare the memorized objects with the objects that appeared on the same side in the test array. In the new horizontal shift condition, they had to compare the same memorized objects with the set of objects that was presented on the contralateral side of the test array. Analogous to the behavioral study by Woodman et al. (2012), these two conditions were presented in different blocks, so that participants always knew in advance whether they had to compare the stored memory array with a test array in the same hemifield or in the opposite hemifield. Figure 2 (bottom panel) shows

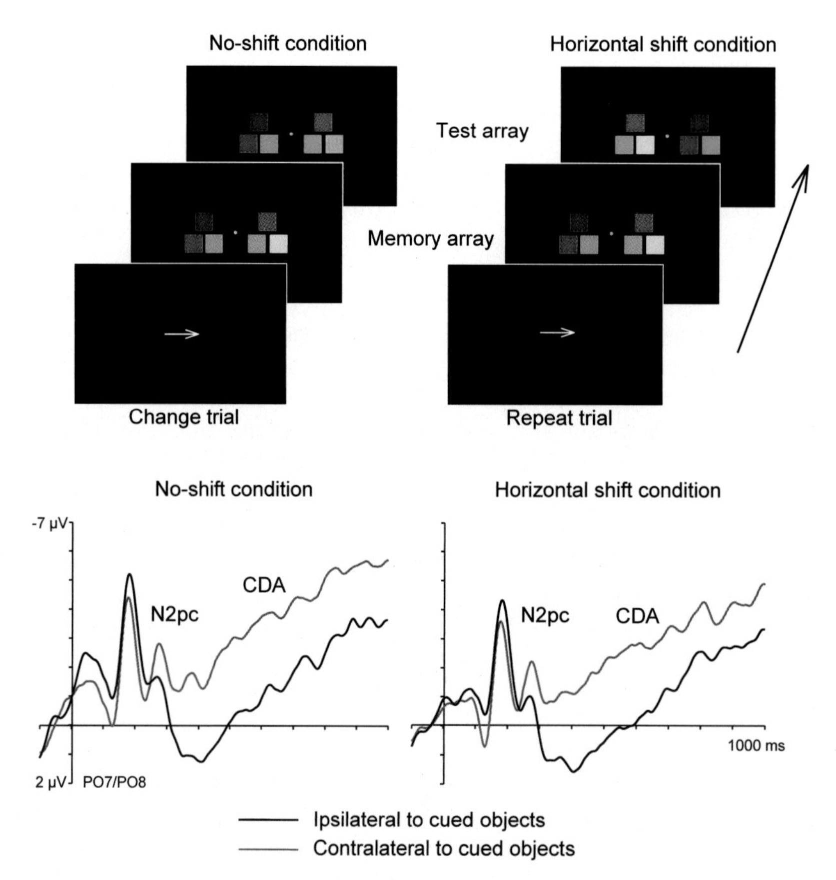

FIGURE 2 Top panel: Change detection task employed by Grubert and Eimer (2014). Arrow cues indicated the side of the to-be-remembered objects in the memory array. In the no-shift condition, these objects had to be compared with objects on the same side in the test array. In the horizontal shift condition, a comparison to objects on the opposite side of the test array was required. Bottom panel: N2pc and contralateral delay activity (CDA) components triggered at electrodes PO7/8 contralateral to the side of the cued objects in the memory array, shown separately for the no-shift and shift conditions. Event-related brain potentials are time-locked to the onset of the memory array. *Data from Grubert and Eimer (2014).*

ERPs triggered at electrodes PO7/8 contralateral and ipsilateral to the side of the cued objects in the memory array in both task conditions. As expected, a sizable CDA component was triggered in the no-shift condition. Critically, this component was equally present and did not differ in amplitude in the horizontal shift condition where the memorized objects had to be compared with test array objects on the opposite side. Even though the later phase of the CDA tended to be slightly smaller in the horizontal shift as compared with the no-shift condition (see Figure 2), this difference was not statistically significant. If participants had moved the represented position of the visual objects in working memory to the opposite side in preparation for memory comparison processes in the horizontal shift condition, this should have been reflected by a polarity reversal of the CDA component in this condition. The observation that the CDA amplitudes were virtually identical for the no-shift and horizontal shift conditions is inconsistent with this scenario. It suggests that object representations in visual working memory retained the spatial layout of these objects as it was encountered during encoding, and that represented locations were not shifted in preparation for an upcoming memory comparison process. Analogous to Woodman et al. (2012), change detection performance did not differ significantly between the no-shift and horizontal shift conditions of our ERP experiment (Grubert & Eimer, 2014), with mean accuracy rates of 86% and 84% in these two conditions, respectively. Instead of anticipatory shifts of represented object locations in working memory, the efficiency of change detection performance with horizontal shifts between memory and test arrays may reflect the operation of a flexible comparison mechanism that can match perceptual and working memory representations across spatial translations. Figure 2 also shows that CDA components in the horizontal shift and no-shift conditions were preceded by transient N2pc components, which reflect the initial spatial selection of the relevant objects on the left or right side of the memory array. Links between the N2pc component, spatial attention, and visual working memory will be discussed in the next section.

In summary, the emerging sensory recruitment view of visual working memory storage assumes that cortical areas that are responsible for visual perception and recognition processes are also the primary locus for the short-term maintenance of visual information. This raises the question of whether these visual working memory representations are position-dependent or position-invariant. On the one hand, the presence of CDA components during the retention and retrieval phases of working memory tasks strongly suggests that these representations retain the spatial layout of visual information that was present during encoding. On the other hand, there is also initial evidence that at least some aspects of visual working memory might operate in a spatially global fashion. It is of course possible that position-dependent and position-invariant visual working memory representations coexist, and that these two types of representations have different functional roles. This possibility will be considered in the final section of this chapter. The next section will discuss links between spatial attention and visual working memory representations. If spatially selective attentional processing is critically involved in the encoding and maintenance of visual objects in working memory, this might explain why at least some aspects of visual working memory appear to be strongly position-dependent.

ATTENTIONAL OBJECT SELECTION AND WORKING MEMORY MAINTENANCE

The idea that selective spatial attention plays an important role in the maintenance of visual information in working memory was first suggested two decades ago (Smyth & Scholey, 1994). Since then, evidence from different sources suggests that there are indeed important links between spatial attention and visual working memory maintenance (see Awh et al., 2006 for a detailed review). For example, there is substantial overlap in the activity of frontoparietal control areas observed during attention and working memory tasks (e.g., Corbetta, Kincade, & Shulman, 2002). Functional imaging and ERP studies have demonstrated that visual processing is enhanced at locations that are currently being maintained in working memory, and these effects bear striking resemblance to the enhancements observed as a result of shifts of spatial attention (e.g., Awh, Jonides, & Reuter-Lorenz, 1998; Awh, Anllo-Vento, & Hillyard, 2000). These observations strongly suggest that selective spatial attention is critically involved in the maintenance of task-relevant locations in spatial working memory, but do not reveal the exact role that attention plays in this context. Another important question is whether spatial attention is similarly engaged in change detection tasks where nonspatial information such as the color of objects has to be maintained and object locations are not explicitly task-relevant. This section will discuss recent findings from ERP studies that may shed new light on these questions.

As mentioned in the previous section, the CDA component is sensitive to the number of objects that are currently being maintained in working memory (Vogel & Machizawa, 2004). In change-detection tasks, CDA amplitudes increase as memory set size is increased from one to four items. Adding further items to the memory set typically does not produce a further increase of the CDA component, suggesting that working memory capacity is exceeded. Importantly, the function linking CDA amplitude and memory set size depends on individual differences in working memory capacity. For individuals with low capacity, CDA amplitudes reach asymptote with smaller memory set sizes than for high-capacity individuals, who can hold more objects simultaneously in working memory. Although these observations provide strong evidence that the CDA is indeed a valid electrophysiological marker for visual working memory maintenance, they do not offer direct insight into the underlying functional architecture of working memory. Which mechanisms are reflected by the CDA and its sensitivity to working memory load and capacity, and what role does spatial attention play in this context?

To answer these questions, it is important to consider another ERP component that is known to be directly linked to attentional object selection. The N2pc component is an enhanced negativity in response to candidate target objects that are presented among nontarget distractors in multistimulus visual displays (Luck & Hillyard, 1994; Eimer, 1996; Girelli & Luck, 1997). This component emerges around 200 ms after stimulus onset at occipito-temporal electrodes contralateral to the hemifield where a potential target object is presented. The N2pc is primarily generated in retinotopic extrastriate occipito-temporal cortex (Hopf et al., 2000), and reflects the spatial selection of a target object among distractors in multistimulus visual arrays. Because it is generally elicited during the attentional selection of task-relevant visual stimuli, the N2pc can also be observed in ERP experiments that use the change detection task to measure the CDA component. In these experiments, the location of the to-be-remembered stimuli on the left or right side of the memory array is usually indicated by precues at the start of each trial. The attentional selection of memorized stimuli triggers a transient N2pc component at posterior electrodes contralateral to these stimuli, before the subsequent sustained CDA component. This is illustrated in Figure 2, which shows ERPs measured in response to memory arrays in blocks where participants had to maintain the objects on the cued side to compare them to test array objects in the same or the opposite hemifield (Grubert & Eimer, 2014; see previous section). In both horizontal shift and no-shift conditions, N2pc components of similar size emerged about 220 ms after stimulus onset at posterior electrodes contralateral to the set of objects that had to be maintained in working memory. N2pc and CDA components generally have a similar scalp distribution. Although there is some evidence that the CDA may have a slightly more dorsal focus relative to the N2pc (McCollough, Machizawa, & Vogel, 2007), this difference has not always been observed. The question whether these two components have dissociable neural generators still needs to be resolved.

What is the functional relationship between the attentional object selection processes reflected by the N2pc and the memory maintenance processes that give rise to the CDA component? A study by Anderson, Vogel, and Awh (2011) has revealed striking similarities in the sensitivity of both components to memory set size and to individual differences in memory capacity. Confirming previous observations (Vogel & Machizawa, 2004), this study found that the CDA was sensitive to memory set size and that individual memory capacity limits predicted when CDA amplitudes reached asymptote. Importantly, essentially the same pattern of results was observed for N2pc components that were triggered in response to the memory arrays. N2pc amplitudes increased linearly as the number of to-be-remembered items on the cued side of the memory array was increased from one to three, and remained at a constant level for larger memory set sizes. Furthermore, the memory set size where N2pc amplitude reached asymptote for individual participants was reliably correlated with individual working memory capacity. These remarkable similarities between N2pc and CDA components in terms of their sensitivity to manipulations of memory load strongly suggest that these two components reflect functionally linked mechanisms.

To understand the nature of this link, it is useful to consider how the effects of memory set size on N2pc amplitudes observed by Anderson et al. (2011) might be interpreted, taking into account the generally accepted view that the N2pc is a marker of the spatially selective processing of target objects. Several other studies have also found links between N2pc amplitudes and the number of simultaneously presented task-relevant visual objects. For example, Mazza and Caramazza (2011) presented zero, one, two, or three uniquely colored target objects among uniform distractors (e.g., red targets among green distractors) in the left or right hemifield and asked participants to report the number of targets on each trial. Similar to Anderson et al. (2011), N2pc amplitudes increased with the number of targets. Notably, no such increase was observed in another experiment of the same study where participants simply had to report the presence versus absence of target-color objects, regardless of how many targets were presented simultaneously. Mazza and Caramazza (2011) interpreted these results as evidence that the N2pc component reflects a mechanism by which task-relevant visual objects are individuated and distinguished from other objects in the visual field on the basis of their spatial location, and suggested that this spatially selective object individuation process may operate in parallel for at least three different objects.

Using a quite different experimental paradigm, Drew and Vogel (2008) observed effects of target set size on N2pc amplitudes that were very similar to those reported by Mazza and Caramazza (2011). In this study, participants had to simultaneously track multiple moving objects in the visual field for 1500 ms. The objects that had to be tracked were indicated by their color at the start of each trial in a stationary cue array. N2pc amplitudes to these cue arrays increased as the number of to-be-tracked items increased. During the subsequent tracking period, CDA components were elicited, and CDA amplitudes also increased as a function of the number of tracked objects. Both N2pc and CDA amplitudes reached asymptote for three tracked objects and both components were highly sensitive to individual limitations in tracking capacity. In other words, even though Drew and Vogel (2008) employed a multiple object tracking procedure, their results were very similar to the pattern of N2pc and CDA modulations observed by Anderson et al. (2011) in a change detection task, suggesting that both tasks may recruit analogous attentional object selection and working memory maintenance mechanisms. Some models of multiple object tracking assume that a separate focus of spatial attention is established for each tracked target object, and that these parallel attentional foci then follow each targets' movement in the visual field (e.g., Cavanagh & Alvarez, 2005). In line with these assumptions, Drew and Vogel (2008) interpreted the observed link between N2pc amplitudes and the number of objects to be tracked as evidence for the parallel attentional selection of multiple target objects at the start of each trial. Along similar lines, the CDA and its sensitivity to the number of tracked objects was interpreted as reflecting sustained spatial attention on one versus more target object locations.

In short, the N2pc and CDA components observed by Anderson et al. (2011) in a change detection task and by Drew and Vogel (2008) in an object tracking task, and their sensitivity to target set size, might reflect the parallel and independent attentional selection of multiple target objects, which is followed by the subsequent maintenance of sustained spatial attention directed towards these objects. If this interpretation is correct, the observation that the N2pc and CDA amplitude modulations observed in these studies were closely linked to individual limits in working memory or tracking capacity is particularly important. It suggests that individual differences in the ability to attend simultaneously to multiple object locations determines performance in multiple object tracking tasks, and also has fundamental consequences for the maintenance of objects in visual working memory. If individual working memory capacity limitations reflect differences in the ability to establish and maintain multiple foci of spatial attention, such individual capacity differences should also predict performance in visual search tasks where target objects have to be selected among multiple distractors. Anderson, Vogel, and Awh (2013) have recently demonstrated such a link between working memory capacity and visual search performance. They employed a difficult search task where targets were presented among perceptually heterogeneous distractor objects, so that each item in the display had to be individuated to determine its target or nontarget status. Search slopes were shallower for individuals with high working memory capacity, demonstrating more efficient search performance. In contrast, no such link between memory capacity and search efficiency was observed in another task of the Anderson et al. (2013) study, in which all distractor objects were identical and therefore could be grouped and rejected together as nontargets (Duncan & Humphreys, 1992). This suggests that performance in difficult visual search tasks and working memory capacity may both be determined by a common underlying factor—individual differences in the ability to simultaneously select and maintain multiple representations of individuated visual objects. Not all studies have observed an increase in N2pc amplitudes with increased memory set sizes (see Jolicœur, Brisson, & Robitaille, 2008; Perron et al., 2009). The reasons for these discrepancies will need to be clarified in future research.

Electrophysiological research into the role of spatial attention in working memory discussed has so far almost exclusively focused on the visual modality and on visual N2pc and CDA components. It is important to know whether similar mechanisms might also be involved in the encoding and maintenance of information in other sensory modalities, such as touch. This question was addressed in a recent tactile working memory study in our laboratory (Katus, Grubert, & Eimer, 2014). This study used a task that was closely modeled on the visual change detection procedures employed by Vogel and Machizawa (2004). On each trial, tactile sample stimuli were followed after a 2000-ms delay period by tactile test stimuli. Sample and test stimulus sets were always delivered simultaneously to both hands, but only one hand was relevant for the tactile memory task. The relevant hand was swapped between successive experimental blocks. Participants' task was to compare the locations of the tactile sample stimuli on the task-relevant hand with subsequent test stimuli on the same hand to make a same/different judgment on each trial. On low-load trials, a single tactile sample stimulus had to be maintained and compared. On high-load trials, two tactile sample pulses presented to two different fingers of the task-relevant hand had to be memorized.

Figure 3 shows grand-averaged ERP waveforms obtained in low-load and high-load trials during the 2000 ms delay period after the onset of the sample stimuli. These ERPs were recorded from lateral central electrodes (Figure 3, top left insert) contralateral and ipsilateral to the currently task-relevant hand. Between 180 ms and 260 ms after sample stimulus onset, a transient central contralateral negativity (N2cc component) emerged. The N2cc was followed by

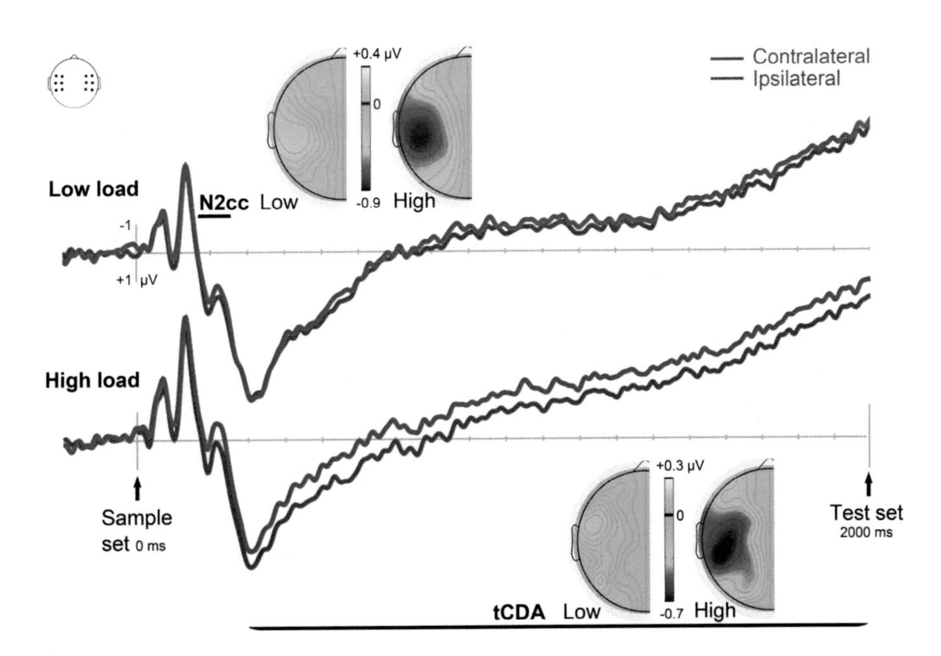

FIGURE 3 Event-related brain potential (ERP) results obtained in a tactile working memory study by Katus et al. (2014), in which participants memorized one tactile event (low load) or two tactile events (high load) presented to the currently task-relevant hand. These sample stimuli had to be compared with a tactile test stimulus set on the same hand that was presented after a 2000 ms delay period. ERP waveforms during this delay period are shown for a lateral central electrode cluster (as indicated in the insert, top left) for electrodes contralateral and ipsilateral to the task-relevant hand, separately for low-load and high-load trials. An early transient central contralateral negativity (N2cc component) was followed by a sustained tactile contralateral delay activity (tCDA component). Both N2cc and tCDA amplitudes were larger for high-load as compared to low-load trials, and both had a modality-specific centroparietal scalp distribution (as shown in the topographical maps). *Data reproduced from Katus et al. (2014), in a different format.*

a sustained tactile contralateral delay activity (tCDA component) that remained present throughout the delay period. As can be seen from the topographical maps in Figure 3, the N2cc and the subsequent tCDA components were both characterized by distinctive modality-specific scalp distributions over somatosensory areas. Importantly, both components were load-dependent, with larger amplitudes in high-load relative to low-load trials. The tCDA component was also found to be sensitive to individual differences in tactile working memory capacity, and tCDA amplitudes predicted the speed and accuracy of working memory performance on individual trials. These findings strongly suggest that analogous to the visual CDA component, the tCDA is an electrophysiological correlate of the maintenance of somatosensory information in working memory. The earlier N2cc component is likely to reflect the spatial selection of task-relevant tactile information, and thus represents the somatosensory equivalent of the visual N2pc component. Overall, the results from this study (Katus et al., 2014) reveal striking similarities between the mechanisms that underlie the spatial selection and selective maintenance of sensory stimuli in vision and touch. During both visual and tactile working memory tasks, two successive contralateral ERP components are triggered that are very similar in terms of their timing, but show distinct modality-specific scalp distributions, with a centroparietal focus for the tactile N2cc and tCDA components, and an occipitotemporal focus for the visual N2pc and CDA components.

Overall, the research reviewed in this section provides strong evidence that spatial attention plays a critical role for visual (and possibly also tactile) working memory. Our ability to store and maintain multiple objects in working memory may be determined by our capacity to activate multiple simultaneous foci of attention. In other words, object representations in working memory are individuated and sustained by directing and maintaining spatial attention at the represented location of these objects. At the neural level, this mechanism may be implemented by spatially selective activations of modality-specific sensory-perceptual cortical areas that represent objects and their features at specific locations. If attentional object selection and working memory encoding and maintenance are linked in this way, this may also provide an answer to the question why working memory representations should be strongly position-dependent. If space-based attentional selectivity is the means through which object representations are established and maintained in working memory, these objects will necessarily be represented in a location-dependent fashion.

The question remains why spatial attention should play such a critical role for the maintenance of visual objects in working memory. According to Feature Integration Theory (e.g., Treisman & Gelade, 1980), focal spatial attention is required to combine visual features that are initially represented independently in visual cortex into integrated perceptual objects. Essentially the same requirement might apply to visual object representations in working memory. To maintain representations of individual memorized objects, attention has to be selectively allocated to the location of these objects. Behavioral evidence for attention-dependent feature binding in visual working memory was provided by Wheeler and Treisman (2002), who demonstrated that the maintenance of integrated object representations is more difficult than the storage of simple visual features from different dimensions, and more vulnerable to interference. Based on these results, they argued that focal attention is required to maintain object representations where features are integrated. It is plausible to assume that the attention-dependent binding of object features postulated by Wheeler and Treisman (2002) and the object individuation processes through focal attention suggested by Drew and Vogel (2008) or Mazza and Caramazza (2011) refer to a similar mechanism—the spatially selective modulation of visual feature processing at a specific position in visual space, which generates a stable visual representation of a particular visual object at this location.

THE GUIDANCE OF ATTENTIONAL OBJECT SELECTION BY VISUAL WORKING MEMORY

The previous section has discussed the role of selective attention for the maintenance of visual objects in working memory. In this final section, the reverse relationship will be considered. What role does working memory play for the control of selective attention? In a visual search, where visual targets have to be detected among task-irrelevant distractors, attentional target selection is guided by observers' knowledge about the target and its features. It is assumed that this information is represented in working memory as an attentional template (e.g., Duncan & Humphreys, 1992) or top-down task set (e.g., Folk, Remington, & Johnston, 1992). Attentional templates play a critical role in the top-down control of visual search, because they can guide spatial attention toward the location of likely target stimuli (e.g., Wolfe, 1994, 2007). In other words, visual working memory representations of currently task-relevant features or objects control the allocation of spatial attention in the visual field.

If working memory guides attentional object selection, the properties of visual working memory that were discussed in the two previous sections should have important implications for the top-down control of visual attention. For example, it is generally accepted that visual working memory has a storage capacity of approximately three to four objects (Luck & Vogel, 1997; Cowan, 2001). If attentional templates are working memory representations of currently task-relevant features or objects, the capacity of working memory should make it possible to simultaneously activate multiple attentional templates and to search efficiently for more than one target object at the same time. However, there is compelling evidence that this is not the case. Searching for two or more objects or features is often much less efficient than a search for one particular object or feature. For example, Houtkamp and Roelfsema (2009) demonstrated that the detection of targets in a rapid serial visual presentation stream was strongly impaired when observers searched for one of two possible target objects relative to a task where they searched for a single object. Modeling of these behavioral results suggested that exactly one attentional template was active at any given time. Additional evidence for severely capacity-limited attentional templates was reported by Stroud, Menneer, Cave, Donnelly, and Rayner (2011), who used visual search tasks designed to resemble airport security checking procedures. Search for a single object or for two different objects that were defined by the same color was much more efficient than search for two different objects in different colors (see also Meneer, Cave, & Donnelly, 2009 for similar observations). What these observations seem to suggest is that attentional object selection can be guided only by a single target template at a time.

In a recent study in our laboratory (Grubert & Eimer, 2013), we employed the N2pc component to demonstrate the impaired efficiency of attentional target selection under conditions in which observers have to search simultaneously for two possible target colors. As shown in Figure 4 (top panels), each search array contained a colored digit on one side and a gray distractor digit on the other. In the One-Color task, participants had to report the numerical value of digits in one specific target color and to ignore all displays that contained a digit in any other color. In the Two-Color task, they had to attend to two possible target colors (e.g., "report all red or green digits") and to ignore displays with a digit in a different color. Figure 4 (bottom left) displays N2pc components (shown as difference waveforms obtained by subtracting ipsilateral from contralateral ERPs) in response to the color-defined target items, separately for both tasks. The N2pc was delayed by more than 30 ms in the Two-Color task as compared with the One-Color task, and this N2pc onset latency difference matched the observed target reaction time difference between both tasks.

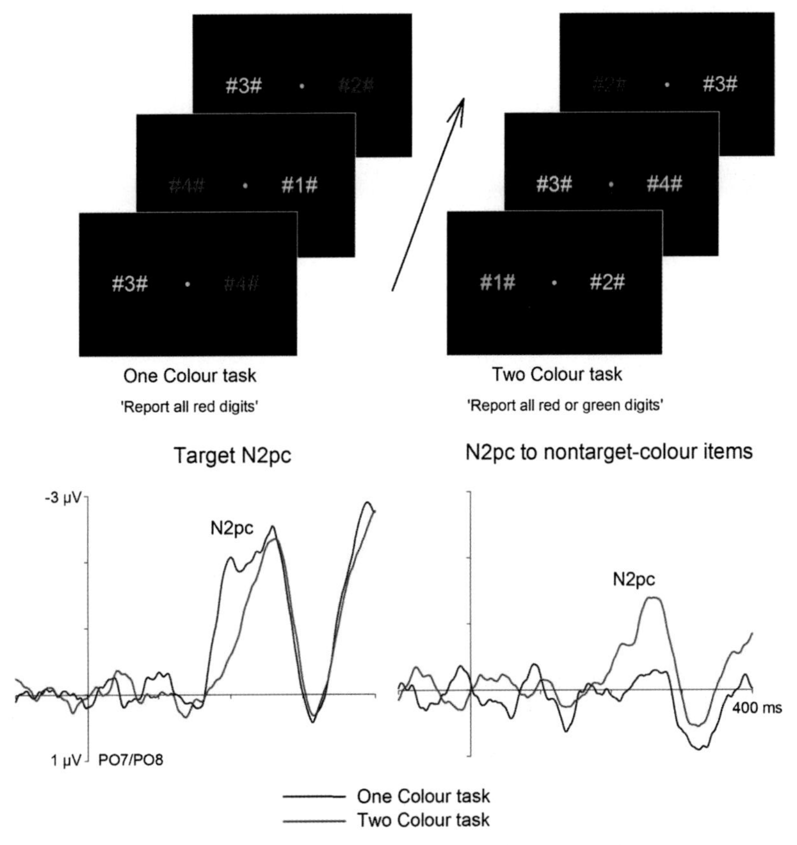

FIGURE 4 Top panel: Color selection task employed by Grubert and Eimer (2013). A single color was task-relevant in the One-Color task, and two different colors were relevant in the Two-Color Task. Bottom panel: N2pc components measured at lateral posterior electrodes PO7/8 contralateral to the side of a target-color digit (left), and contralateral to the side of a nontarget-color digit (right), shown separately for the One-Color and Two-Color tasks. *Data reproduced from Grubert and Eimer (2013).*

This result shows that when observers attempt to attend simultaneously to two possible target colors, attentional target selection is slower than when only one color is attended. There is yet another important cost of multiple-color search, and this was revealed by N2pc components measured on trials where an irrelevant nontarget-color object was presented. In the One-Color task, these color nontargets did not trigger an N2pc (Figure 4, bottom right), demonstrating that when a top-down attentional template for a single target color is active, colored objects in a different nonmatching color are unable to capture attention. In marked contrast, these nontarget-color objects triggered a substantial N2pc in the Two-Color task, which demonstrates that they attracted attention in spite of the fact that they were task-irrelevant. This result illustrates the difficulty of guiding attention toward more than one possible target color at the same time. During multicolor search, the selectivity of color-based attention is strongly reduced, so that nontarget-color objects can no longer be efficiently excluded from attentional processing.

If attentional templates are representations of task-relevant objects in visual working memory, it is not immediately obvious why there should be such strict limitations in the number of objects or features that can be used simultaneously in the top-down guidance of attention. If working memory has a capacity of three to four objects, why should the capacity of attentional templates be restricted to one object or feature at a time? In a recent review article, Olivers, Peters, Houtkamp, and Roelfsema (2011) suggested that object representations in visual working memory can occupy different states. According to these authors, exactly one of these representations can attain a privileged status as a target template for a visual search task, which enables it to bias focal attentional processing in lower-level visual areas towards perceptually matching objects. Even though other object representations may be simultaneously maintained in working memory, these representations (accessory memory items; see Olivers et al., 2011) are actively inhibited and therefore unable to affect spatial attention. This model assumes that attentional templates and other visual working memory representations are functionally and structurally equivalent, except for the fact that the representation that currently acts as attentional template is in a highly active state, while other representations are temporarily suppressed.

The model proposed by Olivers et al. (2011) provides a plausible account for the fact that the capacity limitations for template-guided attentional target selection in visual search are more severe than the general capacity limits of visual working memory. However, the research discussed in the previous sections of this chapter suggests a different possible explanation for this discrepancy. If object representations in visual working memory are position-dependent, and if their position-dependence is a consequence of the fact that focal spatial attention is required for their individuation and maintenance, these representations might not be suitable to act as attentional templates and to guide attentional object selection in visual search tasks. When observers search for a target object with known visual features, the key difficulty is that the position of this target object in the visual field among other irrelevant distractor objects is unknown. Under such conditions of spatial uncertainty, top-down attentional guidance will be most useful if it operates in a spatially global fashion. Attentional processing has to be prioritized in favor of possible target objects, irrespective of their location in the visual field.

How could this type of spatially global attentional guidance be implemented? According to the biased competition model of selective attention (Desimone & Duncan, 1995), multiple simultaneously active object representations compete for neural processing resources and the control of behavior, and attentional processes resolve this competition in favor of one particular object. Critically, this competition is biased in favor of currently task-relevant objects, and the features of these objects are represented in attentional templates. A plausible neural correlate of such preparatory attentional templates has been observed in several single-unit recording studies with primates (e.g., Chelazzi, Duncan, Miller, & Desimone, 1998; Haenny, Maunsell, & Schiller, 1988) during the delay period of cued attention tasks. When a cue signals that a particular object or feature is relevant for an upcoming attentional selection task, visual neurons that respond preferentially to this object or feature remain activated in a sustained fashion throughout a delay period, until task-relevant objects have been processed. Activity modulations analogous to such target-selective sustained "baseline shifts" have also been observed in human neuroimaging studies (Chawla, Rees, & Friston, 1999; Giesbrecht, Woldorff, Song, & Mangun, 2003). According to Desimone and Duncan (1995), selective modulations in the baseline activity of visual neurons may provide a competitive advantage for the subsequent processing of target versus distractor objects, and thus bias the competition between multiple visual objects in favor of those that are currently task-relevant. Because the position of targets is usually unknown in visual search tasks, any feature- or object-specific attentional baseline shifts need to be activated in a spatially global fashion across the visual field to bias visual processing in favor of possible target objects regardless of their location.

Is there any evidence that such spatially global attentional templates do in fact exist? Results from single-unit and human neuroimaging studies have shown that when attention is directed to a specific task-relevant feature, the neural processing of this feature is enhanced at the expense of the processing of other features in the same dimension.

Crucially, these feature-based attentional modulations of neural activity appear to be elicited in a spatially global fashion across the visual field. For example, Martinez-Trujillo and Treue (2004) presented two sets of moving dots in the left and right visual field that both moved in the same direction, and trained monkeys to detect small changes in the speed and direction of one of these sets of dots. They recorded the activity of movement-selective MT neurons with receptive fields that covered the movement in the currently unattended visual field. Neuronal activity triggered by these unattended stimuli was strongly modulated by the direction of movement that the monkey was attending on the other side. The activity of neurons that preferred the currently task-relevant movement direction was enhanced, while the activity of neurons that preferred the opposite direction of movement was suppressed. That these feature-selective activity modulations of movement-sensitive neurons were observed in response to visual stimuli outside the current focus of spatial attention provides strong evidence that feature-based attention operates in a spatially global fashion across the visual field (see also Bichot, Rossi, & Desimone, 2005 for evidence from single-unit recordings in behaving monkeys for the independence of feature-based and space-based attention). Human neuroimaging studies have provided converging support for the spatially global nature of feature-based attention (e.g., Saenz, Buracas, & Boynton, 2002). It has also been demonstrated that this global spread of feature-based attention across the visual field even includes retinal locations that do not contain a visual stimulus (Serences & Boynton, 2007). If feature-based attention operates in a spatially global fashion, its utility for the top-down guidance of attention in visual search tasks is obvious. Feature-based attention can bias perceptual processing in favor of candidate target objects, irrespective of the location that these objects occupy in the visual field, and can thus guide spatial attention towards objects that match one or more currently task-relevant features. In other words, these feature-specific and spatially global neuronal activity modulations can act as attentional templates in visual search.

These considerations may have implications for our understanding of how visual working memory representations control the allocation of attention in visual search. Working memory representations of visual objects are position-dependent (see the first section) because focal attention is required to individuate objects and maintain object representations across time (see the second section). Position-dependent representations are of limited use for the guidance of attention in visual search tasks where the position of targets among distractors is unknown. Because object representations in visual working memory rely on focal spatial attention, they cannot be employed to guide attention to potentially task-relevant locations in the visual field. To find targets at unpredictable locations, spatially global position-independent attentional templates are required (see also Eimer, 2014, for further discussion of this issue). But this does not necessarily imply that visual working memory plays no role in the guidance of visual attention. As discussed in the first section, not all visual working memory representations are strictly position-dependent. Ester et al. (2009) have demonstrated that simple visual features (such as the memorized orientation of grating patterns) may be represented and maintained in a spatially global fashion. This type of representation could be a working memory analogue of the spatially global operation of feature-based attention (e.g., Martinez-Trujillo & Treue, 2004; Serences & Boynton, 2007). Critically, position-independent working memory representations of the type suggested by the results of Ester et al. (2009) have the potential to act as attentional templates during visual search for targets at unpredictable locations. Because of their spatially global nature, they can bias visual processing in parallel across the whole visual field in favor of features that match the description of current target objects.

If the arguments developed in this section are valid, they may provide an explanation for the puzzling fact that capacity limits of attentional templates appear to be more severe than the known capacity limitations of visual working memory. There are marked efficiency costs in the top-down guidance of attention when observers search for two target objects or features simultaneously (e.g., Houtkamp & Roelfsema, 2009; Grubert & Eimer, 2013), even though such tasks are unlikely to exceed working memory capacity. Olivers et al. (2011) have attributed this to quantitative differences in the activation level of visual working memory representations, with a strongly activated attentional target template and additional temporarily suppressed accessory memory items. Alternatively, there may be fundamental qualitative differences between object representations in working memory and attentional templates that are involved in the top-down control of visual attention. Working memory for visual objects is based on position-dependent representations of individuated visual objects. This type of object memory depends on focal attention, and its capacity limitations reflect the fact that only a small number of separate attentional foci (typically two or three) can be maintained simultaneously. In contrast, attentional templates are spatially global representations of currently task-relevant features that can be activated across the visual field, independently of the current focus of spatial attention.

If visual search is guided by spatially global attentional templates for target features, why is the efficiency of attentional target selection strongly impaired when observers search for more than one feature or object? It is possible that the spatially global nature of feature-based attention comes at a cost. While feature-selective visual neurons can be activated in parallel across the visual field when looking for a target defined by a specific feature at an unknown location, it may only be possible to maintain such a spatially global bias for simple features (e.g., color, orientation) and, importantly, only for one feature in a given dimension at a time. The costs for attentional selectivity that arise when attending simultaneously to two possible target colors (Grubert & Eimer, 2013) may reflect the basic fact that feature-based attention cannot be set in parallel for two or more colors. In contrast, it may be possible to simultaneously activate two or more spatially global attentional templates for features from different dimension, such as color and orientation. This hypothesis that top-down attentional guidance can only be based on a single feature for any given dimension is of course a central component of the Guided Search model of visual search (e.g., Wolfe, 2007). What is proposed here is that this limitation may reflect a fundamental constraint for top-down attentional templates and more generally for the operation of spatially global feature-based attention.

SUMMARY

This review started with the question of whether working memory and selective attention should be regarded as functionally and anatomically separate cognitive domains or as a unitary system for the selective goal-directed processing of sensory-perceptual and memorized visual information. The research discussed in this chapter suggests that many aspects of visual working memory can be conceptualized as selective attention directed in a sustained and location-dependent fashion towards perceptual representations of visual objects that are no longer physically present. This account is similar to previous suggestions that working memory includes those long-term visual representations that are within the current focus of attention (e.g., Cowan, 1995; Oberauer, 2002, 2009). The results from recent electrophysiological and brain imaging studies that were reviewed in this chapter suggest that such a model may indeed provide a coherent and neuroscientifically plausible account of the encoding and maintenance of visual object representations in working memory. They also suggest that the selective attentional processing of online sensory information and of stored representations of visual objects may be based on very similar processes. In this sense, the operation of selective attention in perception and in working memory may indeed reflect a single unitary mechanism. However, not all aspects of visual working memory are necessarily linked to focal spatial attention because some information about task-relevant visual features may be maintained in a location-unspecific spatially global fashion. This second type of visual working memory maintenance may rely on the same mechanisms that implement feature-based attention in the perceptual domain, and may be particularly important for the top-down control of attentional object selection in visual search (see Eimer, 2014, for more details).

Terms such as selective attention and working memory are psychological concepts that refer to a set of specific cognitive functions, rather than to the functions of particular dedicated brain areas such as PFC or modality-specific sensory areas. Although this review chapter has focused on the role of sensory-perceptual areas, the successful attentional selection, maintenance, and subsequent retrieval of task-relevant information will always involve the coordinated activation of interconnected neural structures in different brain regions (e.g., Sreenivasan et al., 2014). At a general level, attentional mechanisms can be described as the set of all processes that contribute to the selective activation of currently task-relevant perceptual representations, whereas working memory refers to processes that are responsible for maintaining these representations in an activated state for longer periods, even when the represented objects are no longer physically present. When considered from this perspective, the question whether attention and working memory are unitary or functionally distinct may become increasingly moot, as there is unlikely to be a distinct temporal boundary that could separate the "attentional" and "working memory" aspects of task-dependent selective information processing.

Acknowledgments

Thanks to Katie Fisher, Anna Grubert, Christoph Huber–Huber, Tobias Katus, Rebecca Nako, Joanna Parketny, John Towler, and an anonymous reviewer for comments and contributions. This research was supported by a grant from the Economic and Social Sciences Research Council (ESRC), UK.

References

Anderson, D. E., Vogel, E. K., & Awh, E. (2011). Precision in visual working memory reaches a stable plateau when individual item limits are exceeded. *Journal of Neuroscience, 31*, 1128–1138.

Anderson, D. E., Vogel, E. K., & Awh, E. (2013). A common discrete resource for visual working memory and visual search. *Psychological Science, 24*, 929–938.

Awh, E., Anllo-Vento, L., & Hillyard, S. A. (2000). The role of spatial selective attention in working memory for locations: evidence from event-related potentials. *Journal of Cognitive Neuroscience, 12*, 840–847.
Awh, E., & Jonides, J. (2001). Overlapping mechanisms of attention and spatial working memory. *Trends in Cognitive Sciences, 5*, 119–126.
Awh, E., Jonides, J., & Reuter-Lorenz, P. A. (1998). Rehearsal in spatial working memory. *Journal of Experimental Psychology: Human Perception and Performance, 24*, 780–790.
Awh, E., Vogel, E. K., & Oh, S. H. (2006). Interactions between attention and working memory. *Neuroscience, 139*, 201–208.
Baddeley, A. (1992). Working memory. *Science, 255*, 556–559.
Baddeley, A. (2012). Working memory: theories, models, and controversies. *Annual Review of Psychology, 63*, 1–29.
Bichot, N. P., Rossi, A. F., & Desimone, R. (2005). Parallel and serial neural mechanisms for visual search in macaque area V4. *Science, 308*, 529–534.
Cavanagh, P., & Alvarez, G. A. (2005). Tracking multiple targets with multifocal attention. *Trends in Cognitive Sciences, 9*, 349–354.
Chawla, D., Rees, G., & Friston, K. J. (1999). The physiological basis of attentional modulation in extrastriate visual areas. *Nature Neuroscience, 2*, 671–676.
Chelazzi, L., Duncan, J., Miller, E. K., & Desimone, R. (1998). Responses of neurons in inferior temporal cortex during memory-guided visual search. *Journal of Neurophysiology, 80*, 2918–2940.
Corbetta, M., Kincade, J. M., & Shulman, G. L. (2002). Neural systems for visual orienting and their relationships to spatial working memory. *Journal of Cognitive Neuroscience, 14*, 508–523.
Cowan, N. (1995). *Attention and memory: An integrated framework*. Oxford ; New York: Oxford University Press.
Cowan, N. (2001). The magical number 4 in short-term memory: a reconsideration of mental storage capacity. *Behavioral and Brain Sciences, 24*, 87–114; discussion 114–185.
Curtis, C. E., & D'Esposito, M. (2003). Persistent activity in the prefrontal cortex during working memory. *Trends in Cognitive Sciences, 7*, 415–423.
D'Esposito, M. (2007). From cognitive to neural models of working memory. *Philosophical Transactions of the Royal Society of London. Series B: Biological Sciences, 362*, 761–772.
Desimone, R., & Duncan, J. (1995). Neural mechanisms of selective visual attention. *Annual Review of Neuroscience, 18*, 193–222.
Desimone, R., & Gross, C. G. (1979). Visual areas in the temporal cortex of the macaque. *Brain Research, 178*, 363–380.
Drew, T., & Vogel, E. K. (2008). Neural measures of individual differences in selecting and tracking multiple moving objects. *Journal of Neuroscience, 28*, 4183–4191.
Duncan, J., & Humphreys, G. (1992). Beyond the search surface: Visual search and attentional engagement. *Journal of Experimental Psychology: Human Perception and Performance, 18*, 578–588.
Eimer, M. (1996). The N2pc component as an indicator of attentional selectivity. *Electroencephalography and Clinical Neurophysiology, 99*, 225–234.
Eimer, M. (2014). The neural basis of attentional control in visual search. *Trends in Cognitive Sciences, 18*, 526–535.
Eimer, M., & Kiss, M. (2010). An electrophysiological measure of access to representations in visual working memory. *Psychophysiology, 47*, 197–200.
Ester, E. F., Serences, J. T., & Awh, E. (2009). Spatially global representations in human primary visual cortex during working memory maintenance. *Journal of Neuroscience, 29*, 15258–15265.
Folk, C. L., Remington, R. W., & Johnston, J. C. (1992). Involuntary covert orienting is contingent on attentional control settings. *Journal of Experimental Psychology: Human Perception and Performance, 18*, 1030–1044.
Fuster, J. M., & Alexander, G. E. (1971). Neuron activity related to short-term memory. *Science, 173*, 652–654.
Giesbrecht, B., Woldorff, M. G., Song, A. W., & Mangun, G. R. (2003). Neural mechanisms of top-down control during spatial and feature attention. *Neuroimage, 19*, 496–512.
Girelli, M., & Luck, S. J. (1997). Are the same attentional mechanisms used to detect visual search targets defined by color, orientation, and motion? *Journal of Cognitive Neuroscience, 9*, 238–253.
Goldman-Rakic, P. S. (1990). Cellular and circuit basis of working memory in prefrontal cortex of nonhuman primates. *Progress in Brain Research, 85*, 325–335; discussion 335–326.
Gratton, G., Corballis, P. M., & Jain, S. (1997). Hemispheric organization of visual memories. *Journal of Cognitive Neuroscience, 9*, 92–104.
Grill-Spector, K., Kushnir, T., Edelman, S., Avidan, G., Itzchak, Y., & Malach, R. (1999). Differential processing of objects under various viewing conditions in the human lateral occipital complex. *Neuron, 24*, 187–203.
Gross, C. G., Rocha-Miranda, C. E., & Bender, D. B. (1972). Visual properties of neurons in inferotemporal cortex of the Macaque. *Journal of Neurophysiology, 35*, 96–111.
Grubert, A., & Eimer, M. (2013). Qualitative differences in the guidance of attention during single-color and multiple-color visual search: Behavioral and electrophysiological evidence. *Journal of Experimental Psychology: Human Perception and Performance, 39*, 1433–1442.
Grubert, A., & Eimer, M. (2014). Does visual working memory represent the predicted locations of future target objects? An event-related brain potential study. *Brain Research* http://dx.doi.org/10.1016/j.brainres.2014.10.011.
Haenny, P. E., Maunsell, J. H., & Schiller, P. H. (1988). State dependent activity in monkey visual cortex. II. Retinal and extraretinal factors in V4. *Experimental Brain Research, 69*, 245–259.
Harrison, S. A., & Tong, F. (2009). Decoding reveals the contents of visual working memory in early visual areas. *Nature, 458*, 632–635.
Hopf, J.-M., Luck, S. J., Girelli, M., Hagner, T., Mangun, G. R., Scheich, H., et al. (2000). Neural sources of focused attention in visual search. *Cerebral Cortex, 10*, 1233–1241.
Hornak, J., Duncan, J., & Gaffan, D. (2002). The role of the vertical meridian in visual memory for objects. *Neuropsychologia, 40*, 1873–1880.
Houtkamp, R., & Roelfsema, P. R. (2009). Matching of visual input to only one item at any one time. *Psychological Research, 73*, 317–326.
Jolicœur, P., Brisson, B., & Robitaille, N. (2008). Dissociation of the N2pc and sustained posterior contralateral negativity in a choice response task. *Brain Research, 1215*, 160–172.
Jolicœur, P., Sessa, P., Dell'Acqua, R., & Robitaille, N. (2006). On the control of visual spatial attention: evidence from human electrophysiology. *Psychological Research, 70*, 414–424.
Jonides, J., Lacey, S. C., & Nee, D. E. (2005). Processes of working memory in mind and brain. *Current Directions in Psychological Science, 14*, 2–5.
Katus, T., Grubert, A., & Eimer, M. (2014). Electrophysiological evidence for a sensory recruitment model of somatosensory working memory. *Cerebral Cortex*. http://dx.doi.org/10.1093/cercor/bhu153 Online publication.

Klaver, P., Talsma, D., Wijers, A. A., Heinze, H.-J., & Mulder, G. (1999). An event-related brain potential correlate of visual short-term memory. *NeuroReport*, *10*, 2001–2005.

Kravitz, D. J., Kriegeskorte, N., & Baker, C. I. (2010). High-level visual object representations are constrained by position. *Cerebral Cortex*, *20*, 2916–2925.

Kravitz, D. J., Saleem, K. S., Baker, C. I., Ungerleider, L. G., & Mishkin, M. (2013). The ventral visual pathway: an expanded neural framework for the processing of object quality. *Trends in Cognitive Sciences*, *17*, 26–49.

Kuo, B. C., Rao, A., Lepsien, J., & Nobre, A. C. (2009). Searching for targets within the spatial layout of visual short-term memory. *Journal of Neuroscience*, *29*, 8032–8038.

Luck, S. J., & Hillyard, S. A. (1994). Spatial filtering during visual search: evidence from human electrophysiology. *Journal of Experimental Psychology: Human Perception and Performance*, *20*, 1000–1014.

Luck, S. J., & Vogel, E. K. (1997). The capacity of visual working memory for features and conjunctions. *Nature*, *390*, 279–281.

Martinez-Trujillo, J. C., & Treue, S. (2004). Feature-based attention increases the selectivity of population responses in primate visual cortex. *Current Biology*, *14*, 744–751.

Mazza, V., & Caramazza, A. (2011). Temporal brain dynamics of multiple object processing: the flexibility of individuation. *PLoS One*, *6*, e17453.

McCollough, A. W., Machizawa, M. G., & Vogel, E. K. (2007). Electrophysiological measures of maintaining representations in visual working memory. *Cortex*, *43*, 77–94.

Menneer, T., Cave, K. R., & Donnelly, N. (2009). The cost of search for multiple targets: effects of practice and target similarity. *Journal of Experimental Psychology: Applied*, *15*, 125–139.

Oberauer, K. (2002). Access to information in working memory: exploring the focus of attention. *Journal of Experimental Psychology: Learning, Memory, and Cognition*, *28*, 411–421.

Oberauer, K. (2009). Design for a working memory. *Psychology of Learning and Motivation*, *51*, 45–100.

Olivers, C. N. L., Peters, J., Houtkamp, R., & Roelfsema, P. R. (2011). Different states in visual working memory: when it guides attention and when it does not. *Trends in Cognitive Sciences*, *15*, 327–334.

Op De Beeck, H., & Vogels, R. (2000). Spatial sensitivity of macaque inferior temporal neurons. *Journal of Comparative Neurology*, *426*, 505–518.

Perron, R., Lefebvre, C., Robitaille, N., Brisson, B., Gosselin, F., Arguin, M., et al. (2009). Attentional and anatomical considerations for the representation of simple stimuli in visual short-term memory: evidence from human electrophysiology. *Psychological Research*, *73*, 222–232.

Postle, B. R. (2006). Working memory as an emergent property of the mind and brain. *Neuroscience*, *139*, 23–38.

Ranganath, C., Cohen, M. X., Dam, C., & D'Esposito, M. (2004). Inferior temporal, prefrontal, and hippocampal contributions to visual working memory maintenance and associative memory retrieval. *Journal of Neuroscience*, *24*, 3917–3925.

Saenz, M., Buracas, G. T., & Boynton, G. M. (2002). Global effects of feature-based attention in human visual cortex. *Nature Neuroscience*, *5*, 631–632.

Serences, J. T., & Boynton, G. M. (2007). Feature-based attentional modulations in the absence of direct visual stimulation. *Neuron*, *55*, 301–312.

Shepard, R. N., & Metzler, J. (1971). Mental rotation of three-dimensional objects. *Science*, *171*, 701–703.

Smyth, M. M., & Scholey, K. A. (1994). Interference in immediate spatial memory. *Memory & Cognition*, *22*, 1–13.

Sreenivasan, K. K., Curtis, C. E., & D'Esposito, M. (2014). Revisiting the role of persistent neural activity during working memory. *Trends in Cognitive Sciences*, *18*, 82–89.

Stroud, M. J., Menneer, T., Cave, K. R., Donnelly, N., & Rayner, K. (2011). Search for multiple targets of different colours: misguided eye movements reveal a reduction of colour selectivity. *Applied Cognitive Psychology*, *25*, 971–982.

Treisman, A., & Gelade, G. (1980). A feature-integration theory of attention. *Cognitive Psychology*, *12*, 97–136.

Tsao, D. Y., Freiwald, W. A., Tootell, R. B., & Livingstone, M. S. (2006). A cortical region consisting entirely of face-selective cells. *Science*, *311*, 670–674.

Vogel, E. K., & Machizawa, M. G. (2004). Neural activity predicts individual differences in visual working memory capacity. *Nature*, *428*, 748–751.

Wheeler, M. E., & Treisman, A. M. (2002). Binding in short-term visual memory. *Journal of Experimental Psychology: General*, *131*, 48–64.

Williams, M. A., Baker, C. I., Op de Beeck, H. P., Shim, W. M., Dang, S., Triantafyllou, C., et al. (2008). Feedback of visual object information to foveal retinotopic cortex. *Nature Neuroscience*, *11*, 1439–1445.

Wolfe, J. M. (1994). Guided Search 2.0: a revised model of visual search. *Psychonomic Bulletin and Review*, *1*, 202–238.

Wolfe, J. M. (2007). Guided search 4.0: current progress with a model of visual search. In W. Gray (Ed.), *Integrated models of cognitive systems* (pp. 99–119). New York: Oxford.

Woodman, G. F., Vogel, E. K., & Luck, S. J. (2012). Flexibility in visual working memory: accurate change detection in the face of irrelevant variations in position. *Visual Cognition*, *20*, 1–28.

C H A P T E R

9

Individual Differences in Visual Working Memory Capacity: Contributions of Attentional Control to Storage

Keisuke Fukuda[1], *Geoffrey F. Woodman*[1], *Edward K. Vogel*[2]

[1]Department of Psychological Sciences, Vanderbilt Vision Research Center, Center for Integrative and Cognitive Neuroscience, Vanderbilt University, Nashville, TN, USA; [2]Department of Psychology, University of Oregon, Eugene, OR, USA

THE CAPACITY OF VISUAL WORKING MEMORY

The amount of visual information an individual can actively represent is severely limited; this limitation is known as the capacity of visual working memory (VWM). To estimate an individual's VWM capacity, we present an observer with a memory array containing multiple visual stimuli to remember over a relatively short delay interval, after which we test their memory (Figure 1(a)). When we examine memory performance as a function of the number of stimuli to remember (i.e., set size), we find that healthy young adults can easily retain up to three simple objects (Figure 1(b)), with their performance declining thereafter as the set size increases. Further, when memory performance for a set size larger than three is transformed to an estimate of the number of stimuli retained in VWM, the average VWM capacity estimate (K) is consistently computed as around three objects across a wide range of paradigms (Cowan, 2001; Luck & Vogel, 1997; Zhang & Luck, 2008). Based on such robust findings, the average VWM capacity for healthy young adults is thought to be around three simple objects. To derive a single metric of VWM capacity, we often average the K estimate across set sizes above four guided by the logic that across these large set sizes individuals' VWM stores are equally filled to capacity (e.g., Fukuda & Vogel, 2009; Fukuda, Vogel, Mayr, & Awh, 2010; Vogel & Machizawa, 2004). In this chapter, we call this estimate Kave for the sake of clarity.

Individual Differences in VWM Capacity

Despite this robust estimate of the average capacity in healthy young adults, it is well known that K estimates differ substantially across different subject populations. VWM capacity is known to develop throughout childhood, and it reaches its peak at young adulthood followed by a gradual decline as we age (Brockmole, Parra, & Sala, 2008; Brockmole & Logie, 2013; Cowan, Naveh-Benjamin, Kilb, & Saults, 2006; Riggs, McTaggart, Simpson, & Freeman, 2006). Furthermore, an impaired ability to store information in VWM, inferred by differences in K estimates, is associated with a wide variety of cognitive disorders, such as schizophrenia (Goldman-Rakic, 1994; Lee & Park, 2005), and Parkinson's disorder (Gabrieli, Singh, Stebbins, & Goetz, 1996; Lee et al., 2010). Even in the healthy young adult population, researchers find sizable and reliable individual differences in K estimates (Awh, Barton, & Vogel, 2007; Cowan et al., 2005; Fukuda, Awh, & Vogel, 2010). Some individuals show Kave of four or above, whereas others show Kave of two or less. Interestingly, these individual differences are robustly correlated with a variety of higher cognitive functions such as fluid intelligence (Cowan, Fristoe, Elliott, & Brunner, 2006; Fukuda, Vogel, et al., 2010; Shipstead, Redick, Hicks, & Engle, 2012). Therefore, it is very important to understand what contributes to the individual differences in this capacity estimate.

Mechanisms of Sensory Working Memory
http://dx.doi.org/10.1016/B978-0-12-811042-3.00009-8

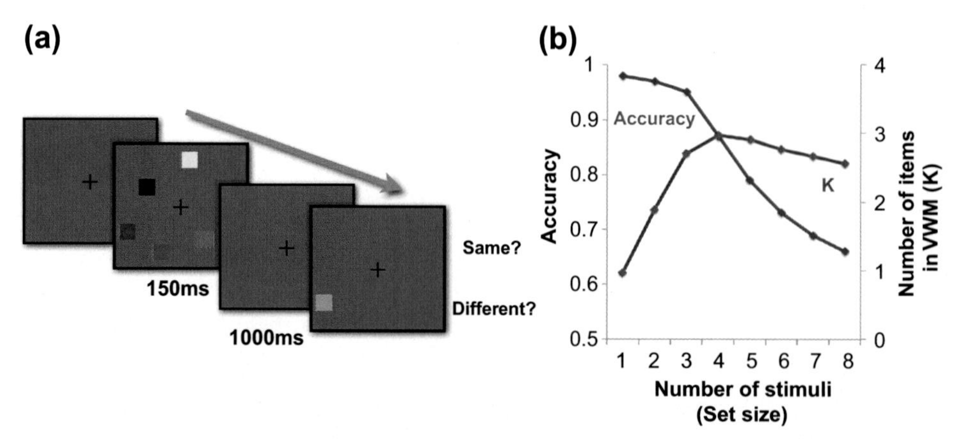

FIGURE 1 (a) A schematic of a standard change-detection task used to estimate individuals' visual working memory capacity. A memory array consisting of multiple colored squares are presented for a brief time, and participants are asked to hold as many squares as possible in mind. After a short retention interval, one item is presented on a test array, and participants have to indicate whether the test item is the same or different from the memory item presented at the same location. (b) Typical performance during this type of change-detection task. The accuracy is near perfect up to set size 3, and it drops monotonically as a function of a set size. When performance is converted to visual working memory capacity estimate (K) using a standard formula (Cowan, 2001), it increases up to three and stops increasing for higher set sizes.

Why do some individuals have higher Kave than others? One simple explanation would be that the variation in Kave reflects the variation in the size of mental storage space. That is, those with high Kave have enough storage capacity to hold four or more objects in VWM, and others with low Kave have less storage space and can only hold two or fewer objects. However, here we also consider the alternative hypothesis that individuals might have roughly the same amount of storage capacity, but what differs across individuals is the ability to consistently use their capacity to its maximum.

Evidence supporting this alternative hypothesis comes from studies that have investigated the relationship between individuals' VWM capacity and attentional-control ability (Fukuda & Vogel, 2009, 2011; McNab & Klingberg, 2008; Vogel, McCollough, & Machizawa, 2005). A well-established consensus is that individuals with low Kave are poor at exerting attentional control over what gets encoded and maintained in VWM. For example, Vogel and colleagues have shown that low-Kave individuals cannot help but orient their attention to distractors (Fukuda & Vogel, 2009), and thus they end up storing more distractors in their limited-capacity VWM than high Kave individuals (Vogel et al., 2005). Though this individual-differences approach has been successful in highlighting the important link between VWM capacity and attentional control, it still is unclear how attentional-control ability manifests itself in a conventional VWM task in which participants simply remember as many objects as possible.

How would attentional control become relevant in the absence of distractors? The attentional-capture literature suggests a plausible mechanism through which the onset of a large number of task-relevant objects induces an overwhelming competition for the limited VWM resource. It is well known that the objects sharing critical features with the target items serve as potent distractors (i.e., contingent-capture distractors), and their onset automatically demands deployment of attentional resources (Folk, Leber, & Egeth, 2002; Folk, Remington, & Johnston, 1992). If we extend this perspective to VWM tasks, all target items in a memory array would automatically claim their share of VWM resources with their onset. If the set size is equal to or below an individual's VWM capacity, then all items would get a sufficient share of the resources to be encoded into VWM. However, if the set size exceeds one's VWM capacity, then the competition for the resources becomes overwhelming, and VWM storage would require the exertion of attentional selection so that a subset of items can be successfully encoded into VWM. If an individual is poor at exerting the attentional control needed to adequately allocate VWM resources, then the number of items successfully retained in VWM could be less than they can actually hold.

Our goal here was to better characterize the nature of individual differences in VWM capacity observed in a standard VWM task (i.e., the change-detection task). We collected a large pool of data (n=495), and analyzed it with the following questions in mind. First, does the capacity estimate systematically change as a function of set size? If K estimates simply reflect the amount of mental storage space, then they should not change as a function of set size so long as the set size exceeds an individual's VWM capacity (e.g., larger than set size 3). On the other hand, if the capacity estimate also reflects an individual's ability to exert attentional control over encoding into VWM, then it might show a decrement at supra-capacity set sizes in which the competition among stimuli for storage overwhelms attentional selection mechanisms. More importantly, if this attentional-control ability to resolve the competition for encoding is a the key factor in determining an individual's VWM capacity estimate, then we should observe that the low-capacity individuals show a disproportionate drop in K estimates at supra-capacity set sizes. To test this hypothesis, we computed a traditional single metric for VWM capacity for each individual by averaging K estimates across set sizes larger than three (Kave). Based on Kave, we split the participants into high- and low-capacity groups using a median split, and examined how the high- and low-capacity groups performed across set sizes. If the storage-space account holds true, performance differences between groups will emerge as soon as the to-be-remembered set size surpasses their VWM capacity and stay constant across larger set sizes. On the other hand, if attentional control is what differentiates high- and low-capacity individuals, performance differences between groups will grow larger as the set size increases.[1]

EXPERIMENT 1

To test the competing theoretical accounts of the individual differences in Kave that we described previously, we accumulated a large pool of data (n=495) using a standard VWM task (i.e., the change-detection task). In this task, participants were asked to remember a briefly presented memory array that consisted of either four or eight colored squares. After a 1000-ms retention interval, participants had to judge if a single item matched the original item presented at the same location in the previous memory array. Using this large sample size, we tested the following predictions. If the attentional-control account is true, then we should first observe that the K estimates drop from set size 4 to set size 8 (i.e., the 4–8 drop). Next, and more importantly, the distribution of K estimates should be much tighter for set size 4 than for set size 8. The attentional-control account predicts that the increased spread of the distribution should be mainly due to those who show a decrement in K estimates from set size 4 to set size 8. In contrast, if the storage-space account is correct, we should observe stable capacity estimates across set sizes, and the distribution of the K estimates should not show any difference for low-Kave individuals.

Method

Participants

After providing informed consent of procedures approved by the University of Oregon Institutional Review Board, 495 young adults with normal or corrected-to-normal vision participated in the study in return for either course credit in psychology classes or monetary compensation ($8/h).

Stimuli and Procedure

Participants performed a standard change-detection task. In this task, participants were presented with a memory array that consisted of either four (set size 4) or eight (set size 8) colored squares for 150ms. The memory array was created from nine highly discriminable colors (red, green, blue, yellow, magenta, cyan, orange, white, and black) that were randomly chosen without replacement. Participants were asked to remember as many colored squares as possible across a retention interval of 1000ms during which the screen remained blank. After the retention interval, a single colored square was presented, and the participants indicated whether or not the test square had the same color as the original one at the same location. Participants used the "z" key on the keyboard to indicate that the color of the test item was the same and the "/" key to indicate that the color of the test item was different in an unspeeded manner. Half of the trials were same trials, and the others were different trials. The sequence of the trials was pseudorandomly determined, and participants completed 60 trials at each set size. After instructions were given and the subjects were given the opportunity to ask questions the experimental trials began.

Results

Attentional-Control Account of Individual Differences in Kave

The accuracy for each set size was converted to VWM capacity estimate (K4 for set size 4, and K8 for set size 8) using the standard formula (Cowan, 2001; Rouder, Morey, Morey, & Cowan, 2011). Then, each individual's

[1] Importantly, we are splitting individuals based on a metric derived by the dependent measure of interest. Therefore, the main effect of group will be meaningless (i.e., high-capacity individuals have higher capacity estimate than low-capacity individuals by definition). However, the two models of individual differences in visual working memory capacity make qualitatively different predictions for the interaction between set sizes and capacity groups, and this justifies our approach.

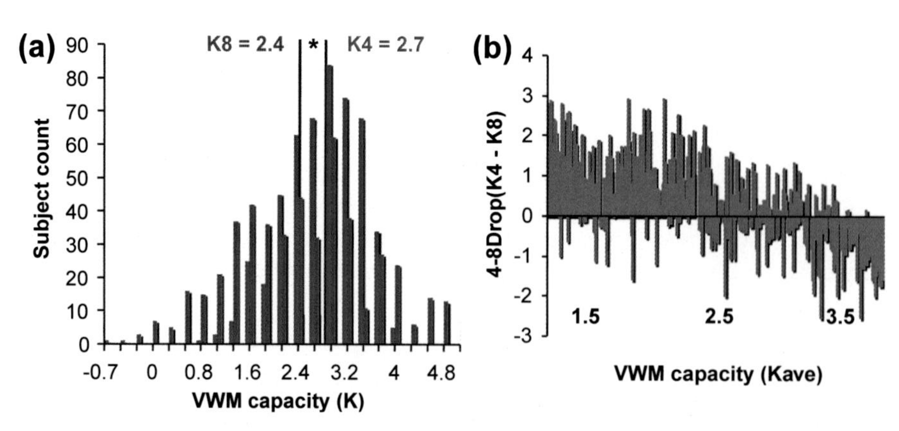

FIGURE 2 (a) A histogram of visual working memory capacity estimate (K) for set size 4 and set size 8. The blue bars depict the distribution of K estimate for set size 4 (K4) and red bars for set size 8 (K8). The black bar represents the mean K4 and K8 scores. As can be seen, the distribution for K8 is wider compared with that for K4. Particularly, there is a substantial increase in the number of individuals toward the lower end of distribution. (b) 4–8 drop for every individual sorted by their traditional visual working memory capacity measure (Kave). For each individual, the difference between K4 and K8 (4–8 drop) was calculated as a measure of the decrement in the capacity estimate from the capacity overload. The figure shows the sorted 4–8 drop for all the participants from the lowest Kave estimate to the largest Kave estimate. Clearly, it was the low-capacity individuals who showed the larger 4–8 drop.

traditional VWM capacity (Kave) was computed by averaging K4 and K8. Based on Kave, individuals were grouped into high- and low-capacity groups by a median split (median Kave = 2.53). The resultant data were subjected to a mixed-design repeated-measures analysis of variance (ANOVA) with the between-subject factor of group (high vs low Kave) and the within-subject factor of set size (4 vs 8). There was a main effect of set size ($F(1, 492) = 60.15$, $p < .0001$). K4 (mean = 2.7) was significantly higher than K8 (mean = 2.4) (Figure 2). This finding is consistent with a previous observation by Cusack and colleagues (Linke, Vicente-Grabovetsky, Mitchell, & Cusack, 2011) that K decreases when estimated with larger set sizes. Further, the spread of distribution nearly doubled from set size 4 (SD = 0.61) to set size 8 (SD = 1.18). This increase in the spread is not entirely driven by the limited range of K estimates at set size 4 because the largest capacity estimate for set size 8 was 5.06 compared with 4 for set size 4. Rather, the greater variability across set sizes was primarily driven by the lower end of the distributions. For set size 4, the number of the individuals with K4 less than 1.5 was 22. However, this number showed nearly a five-fold increase (106 individuals) at set size 8. Although we do see a small portion of individuals who showed a K estimate greater than 4 in set size 8 (8% of the sample), the pattern of the data is largely consistent with the attentional-control account as opposed to the storage-space account in which K estimates should remain fixed once the capacity is reached for an individual.

Our next observation was that there was a significant interaction between set size and capacity group ($F(1, 492) = 213.06$, $p < .0001$), suggesting that the drop in the capacity estimate for the supra-capacity set size was primarily driven by the low-capacity individuals. To further decompose this interaction, we calculated the drop in the K estimates at the supra-capacity set size (i.e., K4–K8, or the 4–8 drop). If this is the main source of individual differences in the standard VWM capacity estimates (Kave = the average of K4 and K8), we should expect that the individuals with low Kave should show a larger 4–8 drop than those with high Kave. When low- and high-Kave groups were compared, high-Kave individuals showed a modest increase from K4 to K8 (M = +0.25), whereas low-Kave individuals showed a sizable 4–8 drop (M = −0.83). This resulted in a larger group difference in K8 than in K4. A correlational analysis buttressed this observation by showing a strong negative correlation between individuals' Kave and the size of the 4–8 drop ($r = -0.64$, $p < .0001$). Taken together, these findings support the attentional-control account of individual differences in VWM capacity. That is, our findings suggest that the individual differences in Kave are heavily affected by how good individuals are at regulating the competition induced by more task-relevant information than they can represent in their limited VWM capacity. More specifically, it is the low-Kave individuals who are negatively affected by an excessive information load, and as a result, they store less information than their VWM could actually hold.

Time Invariant Nature of the Visual Working Memory Capacity Estimate within a Session

The previous findings show that estimated capacity changes with the set size of the to-be-remembered array, and this change strongly drives the individual differences in VWM capacity estimates. One unexplored hypothesis is that these individual differences change their structure over time. One plausible scenario is that early on in the experiment, high-capacity individuals might also show a similar K deficit for the supra-capacity set size as exhibited by the low-capacity individuals. However, high-capacity individuals learn to counteract this memory overload in the course of the experiment, thus leading to stable K estimates across set sizes. This scenario would predict an increase in the individual differences in the K estimate for a supra-capacity set size over time. Another plausible scenario is that the K deficit resulting from memory overload for low-capacity individuals might be only observable in the early part of the experiment, and as they learn how to attentionally select the manageable set of items, this deficit might disappear. This scenario, on the other hand, would predict a decrease in the individual differences in the K estimate for a supra-capacity set size over the course of the experiment. Or alternatively, the individual differences in K estimates are very robust and its structure might not change at all over the course of the experiment.

Recently, Luck and colleagues examined a similar question using a single set size to measure VWM storage as the experiment unfolded. They found that the VWM capacity estimate measured in a standard change-detection task is not influenced by proactive interference, as evidenced by the fact that average performance did not change across the experiment or across trials with a specific type of stimulus (Lin & Luck, 2012). They interpreted these observations as indicating that the standard change-detection task is insensitive to the accumulation of representations in long-term memory over the course of the experiment, and thus K provides a pure measure of the amount of information actively held in VWM at a given moment, uninfluenced by storage in other memory systems. One thing that Luck and colleagues did not directly point out in this study is the time course of the individual differences. Thus, we examined the metric of individual capacity across the experiment in the following analysis. To address the question of the stability of individual differences in K measures, we calculated K estimates for every temporal order for each set size by pooling single trial data from all individuals in each capacity group (e.g., performance of all low-capacity individuals on the first trial of set size 4 condition, the second trial, the third, etc.). We first separately sorted set size 4 and 8 trials in the temporal order that they occurred for each individual. This resulted in trial one through 60 for each set size for each individual. Then, individuals were divided into high- and low-capacity groups by median split based on the standard VWM capacity estimate (i.e., Kave = the average of K4 and K8). Last, performance for each trial order was pooled within high- and low-capacity group to calculate the group capacity estimate for each trial order. Figure 3 shows the temporal fluctuation of the K estimate for each set size. As can be seen, there was no systematic change in the K estimate over time for both set sizes for both groups (absolute rs < 0.14, not significant). This clearly

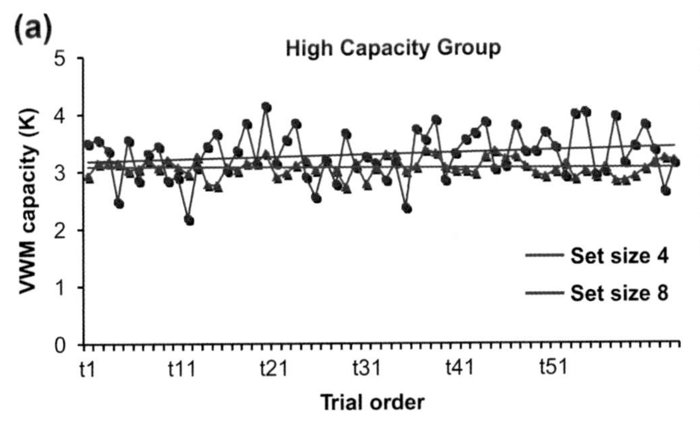

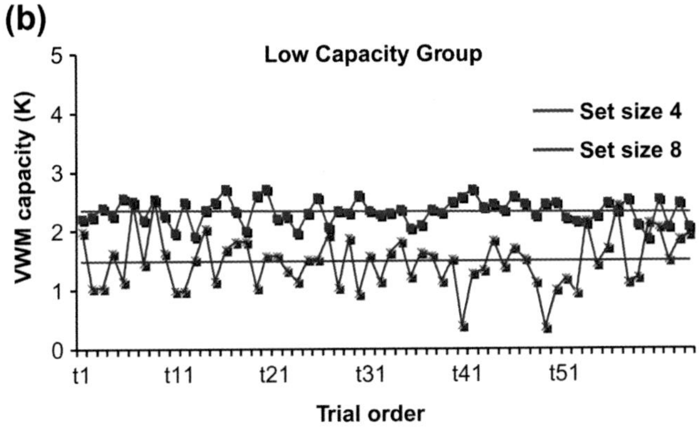

FIGURE 3 **The fluctuations of K estimates across the experiment.** (a) The temporal fluctuation of K estimates for the high-capacity group. The blue line shows the temporal fluctuation of the K estimate for set size 4 (K4); the red line shows that for set size 8 (K8). K4 and K8 are nearly identical throughout the time course of the experiment. (b) The temporal fluctuation of K estimates for the low-capacity group. The blue and red lines represent the fluctuation of K4 and K8, respectively. K4 is consistently higher than K8 throughout the experiment, and the difference does not change over time.

indicates that the individual differences reflected in this capacity estimate are not malleable over the course of the experiment and that K is a stable and reliable metric of individual differences in information processing.

Discussion

The results from Experiment 1 strongly supported the attentional-control account of the individual differences in VWM capacity estimates. Unlike the prediction of the space account, the capacity estimate was significantly smaller for a supra-capacity set size (i.e., set size 8) than for a near capacity set size (i.e., set size 4). This suggests that the onset of excessive task-relevant information causes an overwhelming competition for the limited VWM capacity. Additionally, the decrement in the capacity estimates with the supra-capacity set size was strongly driven by low-capacity individuals, and in turn, it showed a significant contribution to the standard VWM capacity estimates. Further, our large dataset uniquely enabled us to estimate the fluctuation of VWM capacity estimates from trial to trial. The time-course analysis of the capacity estimates revealed that the individual differences are time invariant and can be reliably measured at any point in time. This finding also may suggest that the contribution of the attentional control to memory storage is rather constant over the course of the experiment.

However, one glaring limitation in Experiment 1 was that we drew our inferences about the number of items held in VWM based solely on participants' behavioral report of a match between the memoranda and the test item. We know that VWM representations are not perfect, and their precision worsens as the number of items increases (Anderson, Vogel, & Awh, 2011; Bays & Husain, 2008; Zhang & Luck, 2008). Further, if the representations are not precise enough, individuals are prone to committing errors when comparing the test item with the memoranda. These comparison errors alone can be responsible for a pronounced decrement in the behavioral capacity estimate (Alvarez & Cavanagh, 2004; Awh et al., 2007). It could have been these comparison errors that induced the decrement in the K estimate for the supra-capacity set size. Thus, behavioral data alone do not provide definitive evidence for a decrease in the number of items represented in VWM.

EXPERIMENT 2

In Experiment 2, we used a neural measure to more directly test the predictions of the competing storage-space account and attentional-control account of VWM capacity limitations. Neural measures obtained during VWM maintenance have the distinct advantage of measuring storage without contamination by comparison errors, output interference, and simple breakdowns in late-stage response selection. The contralateral delay activity (CDA) is an electrophysiological measure of the number of items actively represented in VWM (Vogel & Machizawa, 2004). That is, the amplitude of CDA provides a direct measure of the number of items actively maintained in VWM without the complications inherent in basing our conclusions on behavioral output. The CDA is an event-related potential (ERP) component defined as a sustained negative voltage over the parieto-occipital channels that are contralateral to the hemifield in which memory items are presented. It onsets approximately 300 ms after the onset of memory items, and it lasts until the end of the retention interval. In previous experiments that parametrically manipulated memory set size, the amplitude of the CDA showed a monotonic increase until the set size reached an individual's VWM capacity, with no further increase after set size reached K (Anderson et al., 2011). Furthermore, the differences in the CDA amplitudes between a subcapacity set size (e.g., set size 2) and a supra-capacity set size (e.g., set size 4 or more) was shown to strongly correlate with VWM capacity estimates (Anderson et al., 2011; Tsubomi, Fukuda, & Vogel, 2012; Vogel & Machizawa, 2004). In the first CDA article that showed the link between the CDA amplitude and VWM capacity (Vogel & Machizawa, 2004), higher asymptotic CDA amplitudes were found in the high-capacity individuals. This finding was interpreted as evidence that these individuals have more storage space than low-capacity individuals. In the present experiment, we return to this issue and used this neural signature of VWM capacity to directly distinguish between the originally assumed storage-space account (Vogel & Machizawa, 2004) and the attentional-control account of individual differences in VWM capacity.

In this experiment, we had participants perform an ERP version of the change-detection task while we manipulated the set size between 1 and 8 items across trials. In this version, we controlled for the electrophysiological responses to the stimulus onset per se by presenting the same number of items on both sides of the screen. As in the typical paradigm used to measure the CDA, we instructed participants to remember the items presented on the precued side only (e.g., Vogel & Machizawa, 2004). The CDA amplitude was characterized as the difference in the parieto-occipital ERP amplitude between the contralateral and ipsilateral channels relative to the hemifield where the to-be-remembered stimuli were presented. If the space account of individual differences in VWM capacity is correct, then we should observe that the differences in the CDA amplitudes between high- and low-capacity individuals stay constant once the set size surpasses their VWM capacity (i.e., set sizes higher than 3). Alternatively, if the attentional control-account is correct, then we should observe the following. First, all the individuals should show a rise in the CDA amplitude up to their capacity limit. However, when the set size surpasses their capacity, low-capacity individuals should show a decrease in the CDA amplitude, whereas high-capacity individuals should show sustained amplitudes.

Method

Participants

After providing informed consent of procedures approved by the University of Oregon Institutional Review Board, 36 neurologically normal young adults with normal or corrected-to-normal vision participated in the study in return for monetary compensation ($10/h).

Stimuli and Procedure

An example trial is shown in Figure 4(a). Participants were first presented with a central arrow cue that indicated which hemifield (i.e., left or right) to remember. Participants were instructed to remember the colored squares in the cued hemifield. Five hundred milliseconds after the cue presentation, a bilateral memory array consisting of one through eight colored squares (1° × 1° each) on each side was presented for 100 ms. The minimum distance between

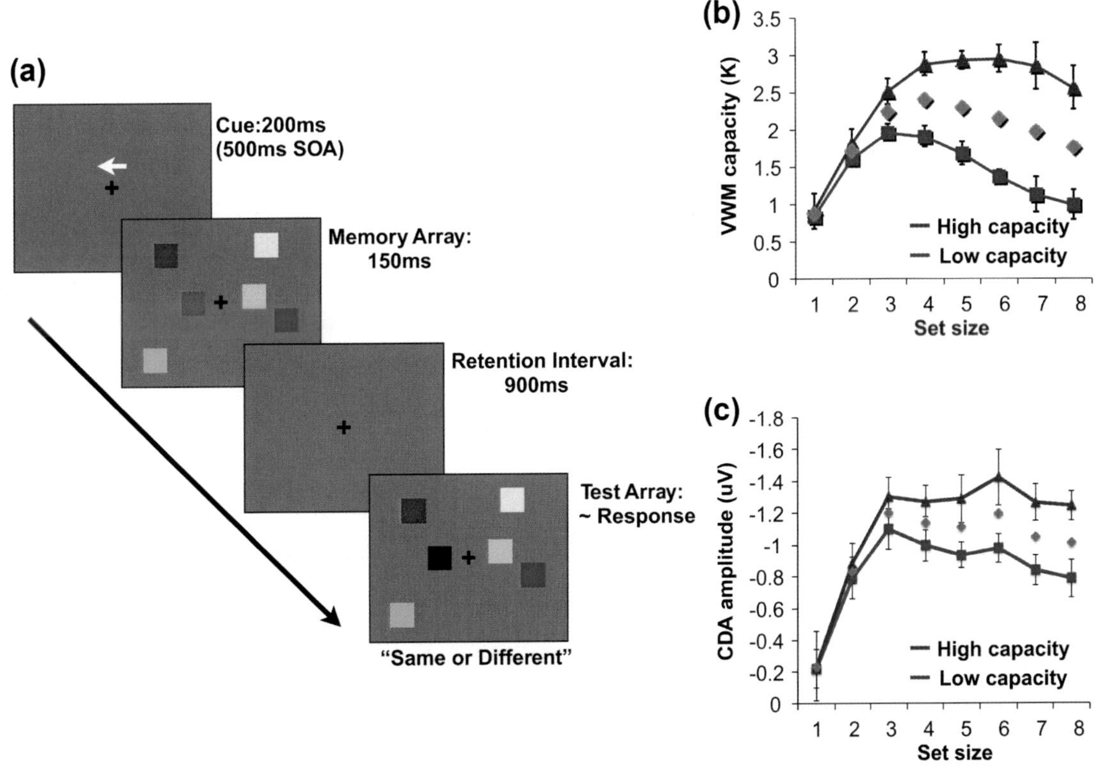

FIGURE 4 (a) A schematic of the bilateral change-detection task. In this task, participants were first presented with a central arrow cue indicating which side they should shift their attention to while holding a central fixation. Five hundred milliseconds after the onset of the cue, a memory array consisted of the same number of items on both sides were presented for 150 ms, and participants remembered items presented on the cued side while ignoring those on the other side. After a 900-ms retention interval, a test array that is either identical or different by one item on the cued side was presented, and participants indicated whether it was the same or different compared with the memory array by a button press. (b) Set size functions of K estimates. The red and blue lines represent the set size functions of K estimates for low- and high-capacity groups, respectively. The error bars indicate the within-subject 95% confidence intervals. The green diamonds show the mean K estimate across all individuals. Consistent with the results of Experiment 1, K estimates for the low-capacity group showed a continuous decline above set size 4, whereas those for the high-capacity group remained constant. (c) Set size function of the CDA amplitudes. The blue and red lines represent the set size functions of the CDA amplitudes for low- and high-capacity groups, respectively. The error bars indicate the within-subject 95% confidence intervals. The green diamonds show the mean CDA amplitudes across all individuals. Much like the K estimates, the low-capacity group showed a monotonic decline in the CDA amplitude after set size 4, whereas the high-capacity group showed stable amplitudes.

squares was 2°, each as at least 2.5° from fixation, but no more than 6.5° from fixation. The colors were randomly chosen from six highly distinguishable colors (i.e., red, blue, yellow, green, magenta, and black), allowing one repetition of each color. After a 900-ms blank retention interval, the test array was presented. The test array was either identical to the memory array or different by one colored square, and participants had to press the key indicating same or different ("z" or "/" on the keyboard, respectively). The test array was available until participants made a response. After responding, participants were allowed to blink or make eye movements for 1 s before the next trial started. The set size was pseudo-randomly changed throughout the experiment, and each participant performed 200 trials with each of the 8 set sizes.

Electroencephalogram Recording

ERPs were recorded in each experiment using our standard recording and analysis procedures, including rejection of trials contaminated by blinks or large (>1°) eye movements, movement artifacts, or amplifier saturation (Vogel & Machizawa, 2004). We recorded from 22 standard electrode sites spanning the scalp, including International 10/20 sites F3, F4, C3, C4, P3, P4, O1, O2, PO3, PO4, P7, P8, PO7, and PO8.[2] Trials containing ocular artifacts, movement artifacts, or amplifier saturation were excluded from the averaged ERP waveforms. Participants who had more than 20% of trials rejected in any condition were replaced (five subjects replaced).

Measuring the Contralateral Delay Activity

As is now standard procedure for measuring the CDA (Vogel & Machizawa, 2004; Vogel et al., 2005), ERPs recorded at posterior parietal, lateral occipital, and posterior temporal electrode sites (PO3, PO4, T5, T6, OL, and OR) were first binned as either contralateral side or ipsilateral side with respect to the memorized hemifield. Because each pair of electrode sites showed the CDA, we maximized the signal-to-noise ratio of our measurements by averaging the channels for each bin to make a single pair of the contralateral and the ipsilateral channels. The CDA amplitude was calculated as the difference between the mean amplitude for the contralateral and the ipsilateral activity in 300–1000 ms time window after the onset of the memory array.

Results

Behavioral Results

As shown in Figure 4(b), the accuracy for each set size was first transformed into the standard capacity estimate, K. To obtain a single metric of VWM capacity (Kave), we averaged the estimates from set size 4 through set size 8 for each individual. Then based on Kave, we divided the participants into high-Kave (M = 2.8) and low-Kave (M = 1.4) groups using a median split. A mixed-model repeated-measures ANOVA with the between-subject factor of capacity group (high vs low) and the within-subject factor of set size (1 through 8) on the accuracy data revealed that there was a main effect of set size ($F(7, 28) = 47.73$, $p < .0001$). As typical in this design, the estimates increased monotonically up to set size 4 (K4 = 2.4), and showed no further increase. More importantly, there was a significant interaction between capacity group and set size ($F(7, 196) = 22.34$, $p < .001$). In contrast to high-Kave individuals who showed relatively stable capacity estimates across set size 4 through set size 8, low-Kave individuals showed a monotonic decrease in the capacity estimates as the set size increased. This interaction was further supported by the strong correlation between individuals' VWM capacity estimate (Kave) and the size of drop in the capacity estimates from set size 4 to set size 8 ($r = -0.83$, $p < .001$). Thus, the behavioral findings of Experiment 2 replicated the essential findings of Experiment 1 even while using different memory and test arrays and requiring fixation.

ERP Analyses

The CDA Amplitude

First, we analyzed the standard CDA waveforms across set sizes, as shown in Figure 4(c). A mixed-model repeated-measures ANOVA was run with capacity group as the between-subject factor and set size as the within-subject factor. Replicating the previous observations, the CDA amplitude increased monotonically up to set size 3

[2] P7, P8 PO7, and PO8 are identical to T5, T6, OL, and OR, respectively, in the previous CDA literature.

and showed no further increase as a function of the set size ($F(7, 28) = 43.21$, $p < .0001$). We also replicated the tight relationship between individuals' Kave and the increase in the CDA amplitude from set size 2 to set size 4 ($r = -0.56$, $p < .01$).

Most critically for the current hypothesis, there was a significant interaction between Kave and the set size ($F(7, 196) = 2.91$, $p < .01$). In contrast to high-Kave individuals who showed stable CDA amplitudes across set size 3 through set size 8, low-Kave individuals showed a monotonic decrease in the CDA amplitudes as the set size increased beyond set size 3. This interaction was supported by a robust correlation between individuals' Kave and the size of drop in the CDA amplitude from set size 3 to set size 8 ($r = -0.50$, $p < .01$).

Separating Contralateral and Ipsilateral Activity

The set size function of the CDA amplitudes supported the attentional-control account of the individual differences in VWM capacity estimate (Kave). If attentional control is the key factor that determines whether an individual will have a high or low VWM capacity, then the activity elicited by items presented on the to-be-ignored side should contribute to the relationship between the CDA amplitudes and the behavioral capacity estimates. The paradigm we used in Experiment 2 involved presenting an entire hemifield full of items that need to be filtered out so as not to induce a sensory confound in the ERPs (Drew, McCollough, & Vogel, 2006; Woodman, 2010). It is easy to imagine that the presence of to-be-ignored items further taxed the attentional selection of the manageable subset of items because they shared critical features with the target items in the cued hemifield. If low-capacity individuals are unable to do this, then we should see that the waveforms recorded contralateral to the irrelevant items (i.e., ipsilateral to the to-be-remembered items) would be more negative relative to the waveforms from high-capacity individuals as the CDA is elicited by these irrelevant items. To test this idea, we separately analyzed the contralateral and the ipsilateral waveforms (relative to the to-be-remembered hemifield) across set sizes for the high- and low-capacity individuals. There was a monotonic increase in the negativity of the contralateral potential up to set size 3 and showed no further increase for larger set sizes (see Figures 5(a,b)). This resulted in a significant main effect of set size on these contralateral waveforms ($F(7, 28) = 25.86$, $p < .00001$), but the effect of capacity group was not significant ($F(1, 28) < 1.0$, not significant). Interestingly, the ipsilateral activity revealed a significant interaction between set size and capacity group ($F(7, 196) = 2.38$, $p < .03$). More specifically, low-Kave individuals showed a monotonic rise in ipsilateral negativity as a function of a set size, whereas the high-Kave group showed constant amplitudes across all set sizes. In fact, when the difference in the ipsilateral negativity between subcapacity set sizes (i.e., average of set size 1 and set size 2) and supra-capacity set sizes (the average of set sizes above 3) were correlated with the capacity estimates, low-Kave individuals showed a larger difference in the ipsilateral negativity than high-Kave individuals ($r = -0.39$, $p = .03$). These analyses suggest that the onset of to-be-ignored items also overload low-capacity individuals' selection mechanisms and they cannot help but allocate VWM resources to them.

Discussion

In Experiment 2, we found evidence from both behavioral and ERP measures consistent with our predictions that low-capacity individuals showed a drop in the amount of task-relevant information represented in VWM as the set size increased beyond their storage capacity. It appears that it is this drop that magnifies the differences between high- and low-Kave individuals. This finding contradicts with the space account of individual differences in VWM capacity, and again supported the attentional-control account by showing that the difference between high- and low-Kave individuals is largely driven by the ability to use their VWM to the fullest when there are more objects than they can maintain in their VWM. We have observed supportive ERP evidence for this interpretation in the literature. The CDA amplitudes showed a similar trend in previous studies (Anderson et al., 2011; Vogel & Machizawa, 2004). One study that employed a task that heavily taxes attentional selection to gate access to VWM capacity (i.e., multiple-object tracking) even reported that the drop in the CDA amplitudes was associated with lower performance (Drew & Vogel, 2008). Thus, it suggests that strong correlations previously observed between the CDA and the behavioral capacity estimates might have been reflecting the underutilization of VWM capacity by low-capacity individuals' when attentional selection was overloaded.

Another interesting finding came from a separate analysis of the ipsilateral activities. If attentional control against the automatic deployment of attention to the onset of an overwhelming large number of targets is what impedes low-Kave individual's performance, we might expect that the neural response to the to-be-ignored side would also differ between high- and low-Kave individuals. It was indeed the case. Low-Kave individuals showed significantly greater ipsilateral negativities for supra-capacity set sizes, which contributed to the decrease in the

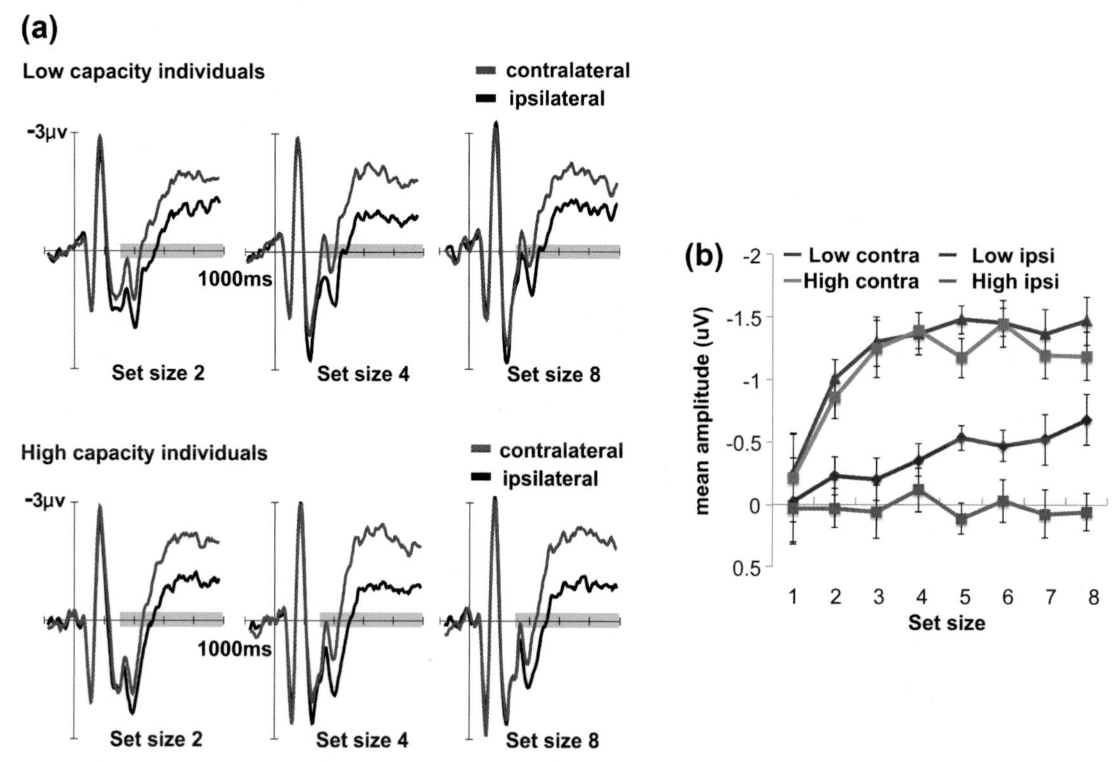

FIGURE 5 **Findings of the separate analyses of the waveforms measured contralateral to the memoranda versus ipsilateral to the memoranda (and contralateral to the to-be-ignored items).** (a) Contralateral and ipsilateral ERP responses for set size 2, 4, and 8. The figures show the contralateral (purple) and ipsilateral (black) response to the onset of the memory arrays. The tick marks on the x axis indicate every 200 ms after the onset of the memory array. The difference between the contralateral responses and the ipsilateral responses in the 300–1000 ms (highlighted by the green bars) time window defines the CDA. The top and bottom rows show the response for set size 2, 4 and 8 for low- and high-capacity groups, respectively. (b) Set size functions of contralateral and ipsilateral responses. The blue and green lines represent the set size functions of the mean amplitudes of the contralateral responses in the CDA time window for low- and high-capacity groups, respectively. The red and magenta lines represent the set size functions of the mean amplitudes of the ipsilateral responses in the CDA time window for low- and high-capacity groups, respectively. The error bars indicate the within-subject 95% confidence intervals. The contralateral responses show similar patterns across capacity groups, but the ipsilateral responses clearly dissociate the capacity groups. The ipsilateral responses for low-capacity group showed monotonic increase in negativity, consistent with what we predicted if these individuals were attempting to store the task-irrelevant items, whereas those for high-capacity group remained constant across set sizes.

CDA amplitudes for supra-capacity set sizes. This nicely matches with our attentional-control account of individual differences in VWM capacity. Our findings extend the potential implications of measuring CDA amplitudes as the marker of the number of items held in VWM. That is, our findings indicate that there is utility in measuring the amplitude of activity at one hemisphere as task demands are manipulated, instead of being constrained to measuring the CDA as a difference between hemispheres to understand the nature of capacity limitations in the brain.

EXPERIMENT 3

The results from Experiments 1 and 2 suggest that the VWM capacity estimate obtained from the standard change-detection task is also sensitive to individuals' ability to use their storage space in the face of an overload of task-relevant information. Particularly, it is the low-capacity individuals who suffer more from this overload and end up storing fewer usable task-relevant items than their storage space allows. Our current working hypothesis argues that this deficit is triggered by the onset of the supra-capacity load of items that demand attentional selection. Recently, Shapiro and colleagues have found evidence supporting this explanation by manipulating the presentation of memory items (Ihssen, Linden, & Shapiro, 2010). In their study, they presented eight memory items, either simultaneously or sequentially in two four-item groups. Arguably, in the sequential-presentation condition, the amount of the competition among items in each memory array is substantially reduced. According to our hypothesis, this should lead to the elimination of the decrement in the capacity estimate associated with supra-capacity set sizes. Indeed, the capacity estimate was higher for the sequential condition than that for the simultaneous condition. More important to our hypothesis, the benefit of the sequential presentation was larger for low-capacity individuals. However, one limitation in this study by Shapiro and colleagues was that sequential presentation could have potentially altered how participants performed the task. For instance, participants could have used different mnemonic strategies (e.g., verbal rehearsal) to retain the initially presented items, and then devoted VWM exclusively to the information in the second display. Alternatively, they could have encoded the initially presented items into offline memory storage (i.e., long-term memory) before the onset of the second display. To eliminate such alternative hypotheses, we sought to diminish the capacity decrement associated with the supra-capacity set size while preserving the simultaneous onset of memory items.

According to attentional-capture literature, resolving involuntary competition for resource deployment requires time for attentional control to be exerted (Folk et al., 2002). More specifically, our recent study demonstrated that it is the low-capacity individuals who take significantly longer to exert proper attentional control to disengage their attention from task-irrelevant distractors that have target properties (Fukuda & Vogel, 2011). If this is the case, we should find that the decrement in the capacity estimates for a supra-capacity set size diminishes when target items are presented long enough for the attentional control mechanisms to resolve the competition. Thus, in Experiment 3, we parametrically manipulated the exposure duration of the memory items and examined the decrement in the capacity estimates for a supra-capacity set size. Our predictions were straightforward. First, with a typical exposure duration (i.e., 150 ms) we should observe that the capacity estimates for a supra-capacity set size (i.e., set size 8) are lower than those for a near-capacity set size (i.e., set size 4). Of note, this decrement in the capacity estimates (i.e., the 4–8 drop) should be observed more for low-capacity individuals than high-capacity individuals. Most critically, this 4–8 drop should decrease at longer exposure durations.

Method

Participants

After providing informed consent of procedures approved by the University of Oregon Institutional Review Board, 36 young adults with normal or corrected-to-normal vision participated in the study in return for course credit in psychology classes.

Stimuli and Procedure

Participants performed a variant of the standard change-detection task used in Experiment 1. In this task, participants were presented with a memory array that consisted of either four (set size 4), or eight (set size 8) colored squares. These memory arrays were presented for 150 ms (short), 300 ms (medium), or 450 ms (long). The other aspects were identical to Experiment 1. The types of trials were pseudo-randomly determined, and participants completed 60 trials for each set size and exposure duration combination.

Results

Similar to Experiments 1 and 2, performance at each set size and exposure duration was transformed to K estimate using the standard formula (Figure 6). Then, individuals were classified as high- or low-capacity using a median split based on the mean K estimate across set size 4 and 8 (i.e., Kave = the average of K4 and K8) using the short exposure duration condition. A mixed-design repeated-measures ANOVA with three factors (exposure duration × set size × capacity group) revealed the following. First, there were main effects of set size and exposure duration. The K estimates were significantly smaller for set size 8 than for set size 4 (mean K4 = 2.7, mean K8 = 2.3, $F(1, 34) = 7.68$, $p < .01$). Also, the K estimates increased as the exposure duration increased (short Kave = 2.49, medium Kave = 2.67, long Kave = 2.80; $F(2, 68) = 5.11$, $p < .01$) In addition, there were two significant two-way interactions, namely between set size and capacity group ($F(1, 34) = 15.60$, $p < .001$) and between exposure duration and capacity group ($F(2, 68) = 5.36$, $p < .01$). The first interaction was driven by a larger difference in K estimates for set size 8 than for set size 4 between high- and low-Kave groups. The second interaction was driven by the low-Kave group showing a larger improvement in Kave at longer exposure durations. Critically, we observed a significant three-way interaction across capacity group, set size, and exposure duration ($F(2, 68) = 3.39$, $p < .05$). This shows that the selective improvement of the traditional capacity estimate due to increased exposure duration was primarily driven by the increase in low-capacity individuals' K8.

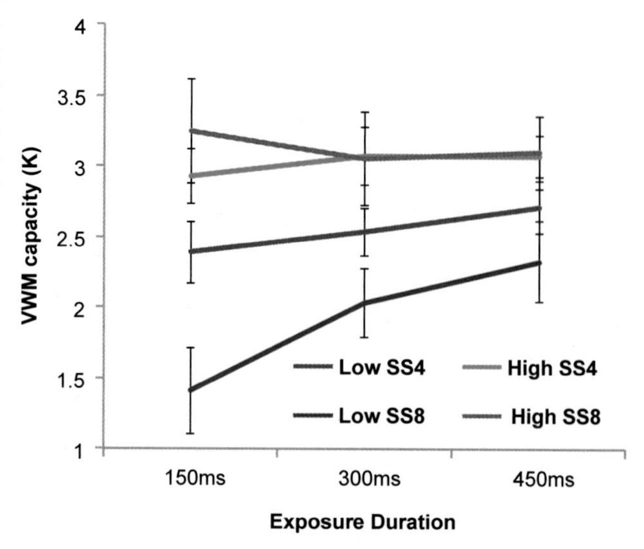

FIGURE 6 **The effect of exposure duration on visual working memory capacity estimate (K).** The blue and red line indicates the K estimate for set sizes 4 and 8 for the low-capacity group, respectively. The green and magenta line indicates the K estimate for set sizes 4 and 8 for the high-capacity group, respectively. The error bars represent the within-subject 95% confidence interval. In a stark contrast to the high-capacity group that showed a stable capacity estimate across set sizes and exposure durations, the low-capacity group's K8 was significantly smaller than K4 at a 150-ms exposure. However, the difference diminished as exposure duration increased mainly from a monotonic increase in K8.

Discussion

In Experiment 3, we sought to investigate the nature of the decrement in the capacity estimates for supra-capacity set sizes. If the decrement is due to the failure in exerting attentional control over the overwhelming competition amongst an excessive number of task-relevant items, we should be able to decrease or eliminate such a decrement by providing enough time for attentional control to resolve the competition. To directly test this hypothesis, we manipulated the interval in which memory stimuli were available for encoding. Consistent with this prediction, we found that increasing the exposure duration of the memory arrays systematically decreased the drop in K to the point that it did not occur at the longest exposure duration. One important observation to note is that the extended exposure only benefitted the performance for low-capacity individuals, particularly in the supra-capacity set size. This critically argues against the alternative hypothesis that the extended exposure encouraged individuals to engage in additional strategies (e.g., verbal re-encoding of stimuli) because it should lead to performance improvement across all set sizes for both high- and low-capacity individuals. In addition, it is also inconsistent with an alternative hypothesis that low-capacity individuals are simply slower at perceiving color information because this hypothesis would also predict an equal, if not larger, amount of increase in the K estimate for the near-capacity set size (i.e., set size 4). Our finding that low-capacity individuals are as quick as high-capacity individuals at identifying a colored target embedded in a rapid-serial-visual presentation is also inconsistent with this alternative account (Fukuda & Vogel, 2009; Experiment 4). Thus, the results from Experiment 3 confirm that a large portion of individual differences in the standard VWM capacity estimate is due to differences in the individuals' ability to exert attentional control in the face of overwhelming competition for the limited representational space in VWM.

GENERAL DISCUSSION

Our ability to actively hold multiple objects in VWM is critically involved in many aspects of our visual cognition (see Luck & Vogel, 2013 for a recent review). At the same time, it is well known that the capacity of VWM reliably and significantly varies among individuals. Numerous studies have investigated why individuals differ in this fundamental cognitive ability, and it has been shown that low-capacity individuals tend to have poor attentional control in allocating VWM to a manageable set of task-relevant objects. However, such observations have been established by examining the correlations among a set of tasks, and it has been unclear how the attentional control component manifests itself in the measure of VWM capacity.

In this study, we sought to examine the contribution of an individual's attentional-control ability to their VWM capacity estimate itself. In Experiment 1, we analyzed a data set based on a sample size of 495 subjects, which enabled a detailed examination of the nature of individual differences in visual short-term memory performance. We found that the capacity estimate systematically changed as a function of a set size. More precisely, the capacity estimate was significantly lower for a supra-capacity set size (i.e., set size 8) than a near-capacity set size (i.e., set size 4). Critically, this decrement in the capacity estimate was primarily driven by low-capacity individuals. These findings support the view that the traditional VWM capacity estimate is not only a measure of an individual's maximum storage space in VWM, but is sensitive to an individual's ability to exert attentional control when faced with overwhelming competition induced by an excessive number of task-relevant objects. In Experiment 2, we used the CDA, a well-established neural marker of VWM capacity, to confirm that the decrement in the behavioral capacity estimates for supra-capacity set sizes is not caused by a failure in the postmaintenance comparison processes, but caused by a failure in maintaining manageable amounts of the correct information. Analyses of the waveforms contralateral to the to-be-ignored items revealed that low-capacity individuals processed the task-irrelevant information so that it interfered with the usable task-relevant information as measured with the CDA amplitude. In Experiment 3, we attempted to obtain further evidence for the contribution of the attentional-control component in memory tasks by selectively eliminating the capacity decrement induced by a supra-capacity set size (i.e., the 4–8 drop). We hypothesized that given enough time low-capacity individuals should also be able to exert proper attentional control to resolve the overwhelming competition (Fukuda & Vogel, 2011). Indeed, we successfully eliminated the 4–8 drop for low-capacity individuals by simply extending the exposure duration. Consistent with our hypothesis, the benefit of extended exposure was primarily observed with a supra-capacity set size in low-capacity individuals. This observation rules out the alternative hypothesis that extended exposure led individuals to engage in other mnemonic strategy (e.g., verbal coding). Given these findings, we conclude that the VWM capacity estimate derived from a standard change-detection task is not a pure measure of the size of the storage space, but rather, it is also a measure of how well individuals can attentionally regulate the allocation of the limited storage space in VWM in the face of overwhelming competition.

One favored mechanism of representing multiple items in VWM is that each item is represented by recurring synchronous firing of feature-coding neurons for that item. The recurring neuronal firing for each representation is desynchronized from one another to avoid the confusion of representations that leads to reduction in the number of successfully stored representations using the phase of a low-frequency carrier wave (i.e., theta ~ alpha wave) (Lee, Simpson, Logothetis, & Rainer, 2005; Liebe, Hoerzer, Logothetis, & Rainer, 2012; Lisman & Idiart, 1995; Luck & Vogel, 1997, 2013; Raffone & Wolters, 2001; Sauseng et al., 2009; Siegel, Warden, & Miller, 2009). In light of this account, the attentional-control mechanism might be playing a critical role in selecting a manageable subset of items so that each representation can be desynchronized from one another in the limited phase space of the carrier wave.

The results of our study have broad implications for the interpretation of individual differences in VWM capacity. First, our results show why VWM capacity measures are strongly correlated with other attentional-control measures. It has been proposed that variability in VWM capacity is the causal factor in individual variations in attentional control (see Luck & Vogel, 2013 for raising a similar argument). Given that the standard capacity measure is heavily influenced by individuals' attentional-control ability as we show here, it is no surprise that such a measure is strongly predicted by other attentional-control measures. Our findings indicate that we need further refinement of our understanding of the relationship between VWM capacity measures and high-level cognitive constructs such as fluid intelligence and reasoning ability. Is it our amount of storage space, our attentional control abilities, or both, that determine our ability to carry out high-level cognitive functions?

Gaining a better understanding of the individual differences in VWM capacity across the population also has important practical implications. As discussed earlier, reduced VWM capacity is one of the most prevalent cognitive deficits found in a wide variety of mental disorders (e.g., schizophrenia, Alzheimer's disease). Also, it is known that VWM capacity estimates change dramatically over the course of normal development and aging. Indeed, there have been multiple attempts at characterizing the nature of VWM capacity differences in such populations (Cashdollar et al., 2013; Jost, Bryck, Vogel, & Mayr, 2011; Lee et al., 2010; Leonard et al., 2013; Mayer, Fukuda, Vogel, & Park, 2012), but these approaches have exclusively relied on examining the correlations between the performance of different tasks. We believe that this work would benefit from taking an approach like that of this study to manipulate the structure of the VWM task itself to understand the bases of these changes across development and aging. Studying the fluctuations of VWM capacity estimate across set sizes would provide the cleanest method for studying how VWM capacity deficits come about in each population.

Last, knowing the nature of the mechanisms underlying VWM capacity deficits is integral to developing an effective training regimen to reduce the capacity limits of human cognition. Training working memory abilities has been a recent focus of investigation given the intimate relationship between this capacity limit and a variety of cognitive abilities. Though its trainability and the transferability of the training effect is still unclear (Jaeggi, Buschkuehl, Jonides, & Perrig, 2008; Jaeggi, Buschkuehl, Jonides, & Shah, 2012; Shipstead, Redick, & Engle, 2012; Shipstead, Hicks, & Engle, 2012), a part of the issue may well be that the training methods are not tailored to the specific type of problem that a given individual faces when trying to encode and maintain multiple items in VWM. By identifying individuals with low capacities because of breakdowns in attentional control, it may be possible to train these individuals to be more selective in what they try to store in VWM.

References

Alvarez, G. A., & Cavanagh, P. (2004). The capacity of visual short-term memory is set both by visual information load and by number of objects. *Psychological Science*, *15*(2), 106–111.

Anderson, D. E., Vogel, E. K., & Awh, E. (2011). Precision in visual working memory reaches a stable plateau when individual item limits are exceeded. *The Journal of Neuroscience*, *31*(3), 1128–1138.

Awh, E., Barton, B., & Vogel, E. K. (2007). Visual working memory represents a fixed number of items regardless of complexity. *Psychological Science*, *18*(7), 622–628.

Bays, P. M., & Husain, M. (2008). Dynamic shifts of limited working memory resources in human vision. *Science*, *321*(5890), 851–854.

Brockmole, J. R., & Logie, R. H. (2013). Age-related change in visual working memory: a study of 55,753 participants aged 8–75. *Frontiers in Psychology*. http://dx.doi.org/10.3389/fpsyg.2013.00012.

Brockmole, J. R., Parra, M. A., & Sala, Della, S. (2008). Do binding deficits account for age-related decline in visual working memory? *Psychonomic Bulletin & Review*, *15*(3), 543–547.

Cashdollar, N., Fukuda, K., Bocklage, A., Aurtenetxe, S., Vogel, E. K., & Gazzaley, A. (2013). Prolonged disengagement from attentional capture in normal aging. *Psychology and Aging*, *28*(1), 77–86.

Cowan, N. (2001). The magical number 4 in short-term memory: a reconsideration of mental storage capacity. *Behavioral and Brain Sciences*, *24*(1), 87–185.

Cowan, N., Elliott, E. M., Scott Saults, J., Morey, C. C., Mattox, S., Hismjatullina, A., et al. (2005). On the capacity of attention: its estimation and its role in working memory and cognitive aptitudes. *Cognitive Psychology*, *51*(1), 42–100.

Cowan, N., Fristoe, N. M., Elliott, E. M., & Brunner, R. P. (2006). Scope of attention, control of attention, and intelligence in children and adults. *Memory & Cognition*, *34*(8), 1754–1768.

Cowan, N., Naveh-Benjamin, M., Kilb, A., & Saults, J. S. (2006). Life-span development of visual working memory: when is feature binding difficult? *Developmental Psychology*, *42*(6), 1089–1102.

Drew, T., McCollough, A. W., & Vogel, E. K. (2006). Event-related potential measures of visual working memory. *Clinical EEG and Neuroscience*, *37*(4), 286–291.

Drew, T., & Vogel, E. K. (2008). Neural measures of individual differences in selecting and tracking multiple moving objects. *The Journal of Neuroscience*, *28*(16), 4183–4191.

Folk, C. L., Leber, A. B., & Egeth, H. E. (2002). Made you blink! contingent attentional capture produces a spatial blink. *Perception & Psychophysics*, *64*(5), 741–753.

Folk, C. L., Remington, R. W., & Johnston, J. C. (1992). Involuntary covert orienting is contingent on attentional control settings. *Journal of Experimental Psychology: Human Perception and Performance*, *18*(4), 1030–1044.

Fukuda, K., Awh, E., & Vogel, E. K. (2010). Discrete capacity limits in visual working memory. *Current Opinion in Neurobiology*, *20*, 177–182.

Fukuda, K., & Vogel, E. K. (2009). Human variation in overriding attentional capture. *The Journal of Neuroscience*, *29*(27), 8726–8733.

Fukuda, K., & Vogel, E. K. (2011). Individual differences in recovery time from attentional capture. *Psychological Science*, *22*(3), 361–368.

Fukuda, K., Vogel, E., Mayr, U., & Awh, E. (2010). Quantity, not quality: the relationship between fluid intelligence and working memory capacity. *Psychonomic Bulletin & Review*, *17*(5), 673–679.

Gabrieli, J., Singh, J., Stebbins, G. T., & Goetz, C. G. (1996). Reduced working memory span in Parkinson's disease: evidence for the role of frontostriatal system in working and strategic memory. *Neuropsychology*, *10*(3), 322–332.

Goldman-Rakic, P. S. (1994). Working memory dysfunction in schizophrenia. *The Journal of Neuropsychiatry and Clinical Neurosciences*, *6*(4), 348–357.

Ihssen, N., Linden, D. E. J., & Shapiro, K. L. (2010). Improving visual short-term memory by sequencing the stimulus array. *Psychonomic Bulletin & Review*, *17*(5), 680–686.

Jaeggi, S. M., Buschkuehl, M., Jonides, J., & Perrig, W. J. (2008). Improving fluid intelligence with training on working memory. *Proceedings of the National Academy of Sciences of the United States of America*, *105*(19), 6829–6833.

Jaeggi, S. M., Buschkuehl, M., Jonides, J., & Shah, P. (2012). Cogmed and working memory training—current challenges and the search for underlying mechanisms. *Journal of Applied Research in Memory and Cognition*, *1*, 211–215.

Jost, K., Bryck, R. L., Vogel, E. K., & Mayr, U. (2011). Are old adults just like low working memory young adults? filtering efficiency and age differences in visual working memory. *Cerebral Cortex*, *21*(5), 1147–1154.

Lee, E.-Y., Cowan, N., Vogel, E. K., Rolan, T., Valle-Inclán, F., & Hackley, S. A. (2010). Visual working memory deficits in patients with Parkinson's disease are due to both reduced storage capacity and impaired ability to filter out irrelevant information. *Brain*, *133*(9), 2677–2689.

Lee, J., & Park, S. (2005). Working memory impairments in schizophrenia: a meta-analysis. *Journal of Abnormal Psychology*, *114*(4), 599–611.

Lee, H., Simpson, G. V., Logothetis, N. K., & Rainer, G. (2005). Phase locking of single neuron activity to theta oscillations during working memory in monkey extrastriate visual cortex. *Neuron*, *45*(1), 147–156.

Leonard, C. J., Kaiser, S. T., Robinson, B. M., Kappenman, E. S., Hahn, B., Gold, J. M., et al. (2013). Toward the neural mechanisms of reduced working memory capacity in schizophrenia. *Cerebral Cortex*, *23*(7), 1582–1592.

Liebe, S., Hoerzer, G. M., Logothetis, N. K., & Rainer, G. (2012). Theta coupling between V4 and prefrontal cortex predicts visual short-term memory performance. *Nature Neuroscience*, *15*(3), 456–462.

Linke, A. C., Vicente-Grabovetsky, A., Mitchell, D. J., & Cusack, R. (2011). Encoding strategy accounts for individual differences in change detection measures of VSTM. *Neuropsychologia*, *49*(6), 1476–1486.

Lin, P. H., & Luck, S. J. (2012). Proactive interference does not meaningfully distort visual working memory capacity estimates in the canonical change detection task. *Frontiers in Psychology*. http://dx.doi.org/10.3389/fpsyg.2012.00042.

Lisman, J. E., & Idiart, M. (1995). Storage of 7+/−2 short-term memories in oscillatory subcycles. *Science*, *267*(5203), 1512–1515.

Luck, S. J., & Vogel, E. K. (1997). The capacity of visual working memory for features and conjunctions. *Nature*, *390*(6657), 279–281.

Luck, S. J., & Vogel, E. K. (2013). Visual working memory capacity: from psychophysics and neurobiology to individual differences. *Trends in Cognitive Sciences*, *17*(8), 391–400.

Mayer, J. S., Fukuda, K., Vogel, E. K., & Park, S. (2012). Impaired contingent attentional capture predicts reduced working memory capacity in schizophrenia. *PloS One*, *7*(11), e48586.

McNab, F., & Klingberg, T. (2008). Prefrontal cortex and basal ganglia control access to working memory. *Nature Neuroscience*, *11*(1), 103–107.

Raffone, A., & Wolters, G. (2001). A cortical mechanism for binding in visual working memory. *Journal of Cognitive Neuroscience*, *13*(6), 766–785.

Riggs, K. J., McTaggart, J., Simpson, A., & Freeman, R. P. J. (2006). Changes in the capacity of visual working memory in 5- to 10-year-olds. *Journal of Experimental Child Psychology*, *95*(1), 18–26.

Rouder, J. N., Morey, R. D., Morey, C. C., & Cowan, N. (2011). How to measure working memory capacity in the change detection paradigm. *Psychonomic Bulletin & Review*, *18*(2), 324–330.

Sauseng, P., Klimesch, W., Heise, K. F., Gruber, W. R., Holz, E., Karim, A. A., et al. (2009). Brain oscillatory substrates of visual short-term memory capacity. *Current Biology*, *19*(21), 1846–1852.

Shipstead, Z., Hicks, K. L., & Engle, R. W. (2012). Cogmed working memory training: does the evidence support the claims? *Journal of Applied Research in Memory and Cognition*, *1*, 185–193.

Shipstead, Z., Redick, T. S., & Engle, R. W. (2012). Is working memory training effective? *Psychological Bulletin*, *138*(4), 628–654.

Shipstead, Z., Redick, T. S., Hicks, K. L., & Engle, R. W. (2012). The scope and control of attention as separate aspects of working memory. *Memory*, *20*(6), 608–628.

Siegel, M., Warden, M. R., & Miller, E. K. (2009). Phase-dependent neuronal coding of objects in short-term memory. *Proceedings of the National Academy of Sciences of the United States of America*, *106*(50), 21341–21346.

Tsubomi, H., Fukuda, K., & Vogel, E. K. (2012). Neural limits to representing objects still within view. *The Journal of Neuroscience*, *33*(19), 8257–8263.

Vogel, E. K., & Machizawa, M. G. (2004). Neural activity predicts individual differences in visual working memory capacity. *Nature*, *428*(6984), 748–751.

Vogel, E. K., McCollough, A. W., & Machizawa, M. G. (2005). Neural measures reveal individual differences in controlling access to working memory. *Nature*, *438*(7067), 500–503.

Woodman, G. F. (2010). A brief introduction to the use of event-related potentials in studies of perception and attention. *Attention, Perception, & Psychophysics*, *72*(8), 2031–2046.

Zhang, W., & Luck, S. J. (2008). Discrete fixed-resolution representations in visual working memory. *Nature*, *453*(7192), 233–235.

CHAPTER

10

Working Memory and Aging: A Review

Monica Fabiani, Benjamin Zimmerman, Gabriele Gratton

Department of Psychology, University of Illinois at Urbana-Champaign, Illinois, USA

INTRODUCTION

Normal aging is characterized by a number of sensory, cognitive, and brain changes that are evidenced in laboratory tasks comparing older and younger adults, but also have practical consequences in everyday life. Front and center among these age-related changes are declines in working memory (see Craik & Byrd, 1982), which is a central node for most cognitive functions, including problem solving, abstract thinking, and creativity. Working memory also provides a bridge between the processing and maintenance of sensory input (which is often somewhat degraded with advancing age due to changes in the sensory organs) and higher-order functions, including other forms of memory. In keeping with the theme of this book, this chapter pays special attention to how sensory and nonsensory aspects of working memory interact with one another as well as with sensory and episodic memory.

The concept of working memory implies the idea of a memory workspace ("working with memory," Moscovitch & Winocur, 1992), where information held in short-term memory can be manipulated. In the classic model of Baddeley and Hitch (1974), which was primarily based on behavioral findings, working memory is not only provided with maintenance "loops" but is also equipped with a "central executive," which has the purpose of directing and allocating working memory resources to task-relevant goals. More recent theories of working memory (e.g., Cowan, 1995, 2000) emphasize this latter attention-control component (see McCabe, Roediger, McDaniel, Balota, & Hambrick, 2010 for an insightful study on the relationship between the concepts of executive function and working memory). Behavioral evidence shows changes in both of these aspects of working memory in aging—holding capacity and attention/executive control (for reviews see Braver & West, 2008; Fabiani & Wee, 2001). Such performance changes are flanked by evidence from brain imaging studies, indicating volumetric and functional activation changes in brain networks associated with working memory and the top-down control of attention (Fabiani & Gratton, 2013; Reuter-Lorenz & Sylvester, 2005).

One of the advantages of examining working memory from a lifespan perspective is that the disruptions inherent to aging help decompose and analyze this process and its biological underpinnings. In addition, an examination of the individual differences that are present in aging allows us to investigate possible factors that may prevent, stave off, or remediate cognitive decline. In this chapter, we review theories of cognitive aging that are relevant to our understanding of working memory function. In doing so, we try to bridge behavioral and neuroimaging data for a more integrated theoretical view, which we discuss at the end of this chapter.

In this respect, it is important to keep in mind that the concept of working memory, however useful a construct for understanding behavior, may not necessarily parallel how the brain actually implements such functions. The concept of working memory became extremely popular in the 1970s from a powerful metaphor assimilating the mind to a computer, and likening working memory to a computer's RAM. Potential disconnections between working memory as a psychological construct and the way it is implemented in the brain will be discussed in the course of this chapter. Among these are the inherent plasticity and adaptability of the nervous system, which, contrary to earlier beliefs, continues throughout the lifespan, and the extreme complexity of the brain and its reshaping in response to maturational and disease pressures.

Mechanisms of Sensory Working Memory
http://dx.doi.org/10.1016/B978-0-12-811042-3.00010-4

Working Memory and Attention

As mentioned previously, current memory theories have begun to explain working memory in terms of its relationship with attention control. For example, Cowan (1995, 2000) pushed this insight forward by describing working memory basically as a long-term memory constrained by focused attention. Because the concept of a central executive was embedded in the Baddeley and Hitch's (1974) model, early attempts to connect working memory with its brain implementation were focused on localizing the central executive to the frontal cortex. For example, Moscovitch and Winocur (1992) pointed at the existence of some neuropsychological performance overlaps between older adults and patients with frontal lobe lesions, in tasks that require to maintain dynamic contextual representations, and to avoid repetitions and confusion errors in the presence of interference. This neuropsychological evidence led to the *frontal aging hypothesis* (for a review, see West, 1996), which suggests that cognitive losses in aging stem from a weakening of the frontal "executive" cortex.

The more recent focus on the ties between the working memory and selective attention systems has allowed investigators to move away from a localization approach and focus instead on the network organization and deployment of this executive control system. Changes in the ability to focus attention and maintain that focus will have broad consequences for all tasks that require some form of top-down control. In fact, a consistent observation is that aging increases difficulty in the three most commonly agreed upon executive functions: (1) in maintaining relevant information, especially in the presence of distractors; (2) in shifting attention between multiple concurrent tasks; and (3) in the inhibition of irrelevant responses (Fisk & Sharp, 2004; Miyake et al., 2000). These difficulties are historically explained from two perspectives. On the one hand it is postulated that top-down control itself becomes less efficient, and that a deficiency in the brain's ability to suppress irrelevant information causes deficits in executive attention tasks (Craik & Byrd, 1982; Gazzaley & D'Esposito, 2007; Hasher & Zacks, 1988). This general impairment in attention control decreases the efficiency of the working memory system, affecting performance in any task that requires it. On the other hand, it can also be hypothesized that there is more noise associated with sensorimotor processing, which requires more top-down intervention to resolve. Within this latter framework, increased top-down control can be seen as compensatory (e.g., Cabeza, 2002; Reuter-Lorenz & Cappell, 2008). It can also be inferred, however, that top-down intervention to maintain the level of noise within an acceptable range will decrease available resources. We describe and discuss these competing theoretical frameworks in the next sections.

Noise from below. A major hypothesis that attempts to explain declines in working memory and attention control holds that increased noise in sensorimotor processing is related to declines in executive attention through some "common cause." The starting point for this hypothesis is that normal aging is accompanied by declines in the sensory, sensorimotor, and executive control systems (including both the sensory and nonsensory aspects of working memory), rather than in executive attention alone (Schäfer, Huxhold, & Lindenberger, 2006).

The major line of evidence in support of this hypothesis is research that has demonstrated increased correlations between sensorimotor and executive function from early adulthood to old age (e.g., Baltes & Lindenberger, 1997). This observation has been called *dedifferentiation* and refers to the fact that the once highly modular cognitive functions seem to overlap more and more throughout adulthood, both in behavior and in neuroimaging results (see Grady, 2012, for a concise functional review of dedifferentiation). The logic behind interpreting this evidence is that if the functional integrity of the brain were generally affected, then one would expect increased covariation in the functioning of the affected brain systems. Alternatively, if specific areas of the brain were affected, then one would expect to see specific functional deficits only (although this may be an oversimplification, as brain regions are highly interconnected; see Gratton, Nomura, Pérez, & D'Esposito, 2012).

Some support for this interpretation of cognitive aging comes from work that shows that controlling for general information processing speed can reduce age-related deficits in executive control tasks to nonsignificance (Salthouse, 1996). For example, it has been demonstrated that processing speed is sufficient to explain age-related declines in reading and computation span (Salthouse, Babcock, & Shaw, 1991), random letter generation (Fisk & Warr, 1996), and performance on the Wisconsin Card Sorting Task (Fristoe, Salthouse, & Woodard, 1997). However, although the data are compelling, it is not clear what would cause the generalized slowing in the first place, because overall slowing is the final outcome of a very large number of processes, involving much of the brain. In addition, there are also data showing brain-processing differences in the *absence* of reaction time (RT) slowing in older adults (e.g., Fabiani & Friedman, 1995).

One possibility is that slowing is due to a reduced "processing bandwidth" (a phenomenon we all experience when computer networks are overloaded). In other words, if the system is overloaded and the signals coming in are noisy, a greater bandwidth is needed to individuate each stimulus (for similar conceptualizations see Holden, Hoebel, Loftis, & Gilbert, 2012; Stark, Yassa, & Stark, 2010). If such larger bandwidth is not available, as it may be the case for aging, slowing will likely occur, and there will be loss of sharpness in the memory representations generated by the incoming stimuli. Consistent with this idea, Lindenberger and Baltes (1994) showed that visual and auditory acuity explained more than 93% of the age-related variance in intelligence in a group of 156 older individuals. This very strong connection between sensory and higher cognitive domains and other similar evidence has led some researchers to make the even stronger hypothesis that sensorimotor decline may precede cognitive decline (Anstey, Lord, & Williams, 1997). It still remains unclear, however, whether the loss of peripheral sensory acuity should be considered as a *direct* cause of cognitive loss (i.e., cognitive processing is degraded because of faulty input or long-term sensory deprivation) or rather as a *correlate* (i.e., those same age-related conditions, such as arterial stiffening, hypertension, diabetes, oxidative stress, etc., leading to loss of peripheral sensory acuity can also operate at the central level, leading to loss of cognitive function).

Consistent with this latter view, neuroimaging research shows that older adults often have more diffuse brain activation patterns during cognitive tasks (for a recent review, see Goh, 2011). Similarly, older adults often show bilateral prefrontal activity associated with a variety of cognitive tasks (including working memory tasks) that typically induce lateralized activity in younger adults (*Hemispheric asymmetry reduction in older adults* or HAROLD model, Cabeza, 2002). Further, older adults show less selective activity than younger adults across a variety of tasks (Grady, 2002; Townsend, Adamo, & Haist, 2006). For example, older adults have been shown to have less selective activity in occipital cortex than young adults during working memory tasks (Carp, Gmeindl, & Reuter-Lorenz, 2010). Similarly, older adults have less selective activity to specific categories of visual stimuli (Park et al., 2004); this lack of specificity is correlated with lowered measures of executive function (Park, Carp, Hebrank, Park, & Polk, 2010). Roski et al. (2013) have provided additional evidence for decreased specificity across functional domains. In their study, they demonstrated dedifferentiation beyond the visual system by showing an age-related decrease in activity in the parietal-operculum (a sensorimotor processing area) and an age-related increase in activity in visual areas during a *motor* task. Likewise, during a *visual* task, there was an age-related increase in pre-motor areas and an age-related decrease in visual areas.

Several authors have also observed that while certain brain areas show reduced activity with aging, others show *increased* activity (see Spreng, Wojtowicz, & Grady, 2010 for a meta-analysis of age-related changes in brain activation patterns). Particularly, observations were made early on in the functional literature that older adults tended to have more activity than younger adults in the prefrontal cortex (PFC) across a number of tasks (Cabeza et al., 1997; Reuter-Lorenz et al., 2000). The consistent observation of increased activity in the PFC alongside of decreased activity in visual processing areas has become known as the *posterior-anterior shift with aging* (PASA; Davis, Dennis, Daselaar, Fleck, & Cabeza, 2008).

Recent work using event-related potentials (ERPs) has further examined age-related differences in the activity of top-down compared with bottom-up attention (Li, Gratton, Fabiani, & Knight, 2013). This study showed that bottom-up, "pop-out" target detection was associated with greater activity at parietal locations than at prefrontal locations in younger participants but had a more equal distribution in older adults. Furthermore, this activation pattern was associated with target detection and was interpreted as evidence that frontal processing was used to help enhance sensory representations.

As mentioned previously, this increased activation in the PFC in older adults is often bilateral. These findings gave rise to the idea of *compensation*, which explained these observations as the PFC ramping up activity to compensate for reduced activity in the visual processing system (Davis et al., 2008; Grady et al., 1994). More recently, similar observations have extended into the temporal domain, suggesting that, in addition to spatial changes in activity, there are also changes in the dynamics of activity. For example, in an executive control task, participants were presented with pairs of cue-target letters with the instructions to respond to the target "X" when it follows the cue "A." Behavioral results in this task show that older adults wait for the target rather than using the cue information, suggesting that they use reactive rather than proactive control strategies (Paxton, Barch, Storandt, & Braver, 2006). Follow-up functional magnetic resonance imaging (fMRI) studies showed that PFC regions had reduced activity in cue-related trials but increased activity during probe-related trials in older compared with younger adults (Paxton, Barch, Racine, & Braver, 2008). These changes in the temporal dynamics of activation in the PFC are referred to as the *early-to-late shift in aging*.

The compensation interpretation is also often used to explain imaging results when either older adults have more activity than younger adults in a brain region related to the task, without showing any difference in behavioral performance, or when increased activity in some brain region is positively correlated with task performance in older adults but not in younger adults (Cabeza, Anderson, Locantore, & McIntosh, 2002; Grady, 2012; Grady, McIntosh, & Craik, 2005; McIntosh et al., 1999). These types of results are often thought of in terms of the *compensation-related utilization of neural circuits hypothesis* (CRUNCH; Reuter-Lorenz & Cappell, 2008), which states that older adults recruit more neural resources at lower levels of cognitive load than younger adults, but less at higher levels of load.

It is important to note that this idea of compensation is quite different from its original instantiations and does not require that older adults rely on different brain regions or networks to make up for some deficit in processing somewhere else. However, the field has moved toward accepting this definition of compensation, which simply requires an increase of activation at lower cognitive loads (easier tasks) in older adults than younger adults. This idea of compensation has been supported by several studies of executive control. For instance, the ability to inhibit otherwise dominant responses is positively correlated with activity in the dorsal PFC and parietal regions (Vallesi, McIntosh, & Stuss, 2011). Another study found that bilateral activity in the PFC was positively associated with performance on a task requiring attention to a specific side of the visual field (Davis, Kragel, Madden, & Cabeza, 2012). Specifically in line with CRUNCH, several studies have also shown that older adults exhibit higher activation at low cognitive loads in the PFC and parietal cortex while maintaining similar performance to young adults, but exhibit lower activation and performance at high cognitive loads than young adults (Cappell, Gmeindl, & Reuter-Lorenz, 2010; Mattay et al., 2006; Schneider-Garces et al., 2010).

Losing control from the top. An alternative to the idea of sensory or sensorimotor noise draining cognitive resources is that the attention control system itself becomes less efficient and may more easily fail. This is typically pitched as either reduced efficiency of the processes responsible for working memory generally (Craik & Byrd, 1982) or more specifically as a diminished inhibitory ability (Hasher & Zacks, 1988; Hasher, Lustig, & Zacks, 2007).

Many early views of working memory efficiency (and its relationship with aging) equated poorer working memory with lower working memory "capacity." That is, a memory workspace that can store and operate on six items will perform better than a memory workspace that can store and operate on three items. There is ample evidence from a number of methodologies that working memory capacity is reduced in aging. This includes analyses based on the computation of Cowan's K (Cowan, 2001; see also Cowan et al., 2005) in memory search tasks (Schneider-Garces et al., 2010) and data from electrophysiological studies (measuring the contralateral delayed activity, Vogel & Machizawa, 2004, also called sustained posterior contralateral negativity, Jolicœur, Sessa, Dell'Acqua, & Robitaille, 2006). However, it is also clear that, although age-related reductions in working memory capacity are small in simple tasks such as the forward digit span, they greatly increase under conditions of load (i.e., when additional processing is needed beyond holding items in memory). An example of a loaded working memory task is the "O-Span" task, in which subjects need to verify simple math problems while memorizing word lists of varying length (Turner & Engle, 1989; see Brumback-Peltz, Gratton, & Fabiani, 2011). Overall, these observations beg the question of what are the basic mechanisms that underlie this reduced working memory span in aging.

The original conceptualization of short-term memory in the Atkinson and Shiffrin's model (1968) emphasized the idea that the limited capacity of working memory (Miller, 1956; Sperling, 1967) reflected a reduced *storage* capacity. However, since then, the field has moved toward the idea that limitations in the capacity of working memory reflect *attentional*, rather than storage, factors. The active maintenance of representations is now typically viewed as an inherent characteristic of executive control (Baddeley, 2003; Braver & West, 2008), with perceptual representations held in perceptual processing areas (Sreenivasan, Curtis, & D'Esposito, 2014; Sreenivasan, Gratton, Vytlacil, & D'Esposito, 2014). According to this view, working memory is the outcome of a process by which currently relevant long-term representations are activated (so as to be readily accessible for the current task) and representations that are no longer relevant (which might have been useful up to a few seconds before) are suppressed, in a goal-oriented fashion. This dynamic view of working memory focuses on the capacity of orchestrating and executing activation and inhibition processes in a very fine-tuned and timely manner, according to the current task requirements. Impaired tuning mechanisms may be reflected in an effective span loss because some of the relevant representations may not be fully activated and some of the irrelevant representations may not be fully inhibited.

This goal-maintenance framework is based on findings that suggest a commonality between executive control tasks and the maintenance of behavioral goals. This view therefore sees age-related declines in executive function as declines in goal-maintenance, or so-called *goal neglect* (Kane & Engle, 2003). Because goal neglect parallels findings from studies on patients with frontal lesions, proponents of this view often focus on the frontal lobe (and especially on the dorsolateral prefrontal cortex) as the crucial brain area responsible for age-related cognitive decline (West, 1996). Goal-maintenance mechanisms act to bias competition between representations in favor of goal-relevant perceptual representations (Braver & West, 2008; Sreenivasan, Curtis, et al., 2014). This view closely parallels Desimone and Duncan's (1995) biased competition theory of attention. The cognitive functions involved in goal maintenance must include an ability to maintain goals over time, to enhance goal-relevant representations, and to inhibit goal-irrelevant representations, especially when those irrelevant representations would otherwise be dominant (Braver & West, 2008). Cognitive aging has been implicated in the decline of all these domains, including decreased response inhibition, and decreased performance in task-switching and dual-task paradigms (Braver & West, 2008).

In this view of executive control, cognitive resources still play a prominent role in describing the efficiency of the executive system, and although inhibition is included as an important facet, it is not regarded as the primary factor in age-related executive function declines. A related view is that the main function of the attention control systems is to *limit* mental representations through inhibition, and that working memory capacity should not, therefore, be thought of in terms of size or resources. This view, originally purported by Hasher and Zacks (1988; see also Lustig, Hasher, & Zacks, 2007) subdivides inhibitory processes into at least three subprocesses that all act to maintain goal representations. These subprocesses include (1) gating the access to the focus of attention; (2) deleting irrelevant items from attention; and (3) restraining prepotent responses to allow evaluation of other response candidates (Hasher, Zacks, & May, 1999).

In general, there is strong evidence that aging includes declines in inhibitory processes, which are especially evident when there is competition between stimuli for mental representation in both the visual (Carlson, Hasher, Zacks, & Connelly, 1995; Connelly, Hasher, & Zacks, 1991; Dywan & Murphy, 1996; Li, Hasher, Jonas, Rahhal, & May, 1998; Phillips & Lesperance, 2003; Plude & Hoyer, 1986; Zacks & Zacks, 1993) and auditory (Tun, O'Kane, & Wingfield, 2002) domains. These findings are based on behavioral data in which inhibition is defined as the capacity to stop or reduce one's tendency to respond to stimuli or stimulus features that should normally be responded to, but that are deemed irrelevant under the specific task conditions. This difficulty in inhibiting otherwise prepotent responses occurs more frequently with age.

Within this framework, several questions arise regarding the underlying mechanisms that lead to behavioral changes in the capacity to inhibit processing or action. In fact, a reduced inhibition of overt responses is likely to derive from complex activations and deactivation of brain networks that may change the level of activation in "response channels" (for a review of this logic, see Gratton, Coles, Sirevaag, Eriksen, & Donchin, 1988). It can be hypothesized that age-related impairments in appropriate response modulation may be due to deficits in neurotransmitter systems involved in neuromodulation (such as dopamine, norepinephrine, acetylcholine, or gamma-aminobutyric acid [GABA]). If this were the case, then a common mechanism could underlie the effective reduction in working memory span observed in older adults and their difficulty inhibiting no-longer-relevant responses. In addition, age-related downregulation of the GABAergic system (Bishop, Lu, & Yankner, 2010; Loerch et al., 2008) can lead to a reduction in lateral inhibition, through which the spreading of neuronal activity is controlled. This may result in less distinctive representations and thus effectively reduce the communication bandwidth between cortical regions. As a consequence, processing speed may also be reduced, providing a unifying thread across a number of theories of cognitive aging.

To show that decline in inhibitory processing can also affect speed of processing, Lustig, Hasher, and Tonev (2006) manipulated extraneous information in a letter comparison task. In this task, two-letter strings (e.g., TRGWH and TRCWH) are evaluated as being the same or different to assess perceptual speed. Including additional, irrelevant letter strings significantly increased older adults', but not younger adults', RTs, suggesting that a decline in inhibitory abilities is sufficient to affect perceptual speed.

An interesting prediction derived from the "decreased inhibition" hypothesis is that the inability to suppress irrelevant representations and processes may in some cases be advantageous—and that therefore there may exist tasks in which older adults outperform younger adults. Evidence in favor of this view comes from a study using a Remote Associates Task, in which three loosely related words (e.g., rat, blue, and cottage) are presented and the participant must find a word (cheese), which relates all the other words together (May, 1999). In this study, on experimental trials target words were presented together with distractors, which were supposed to be ignored, but were chosen to either lead toward the correct answer or away from it. The critical result was that older adults were unable to ignore the distractors, which led to either inferior or superior performance compared to younger adults depending on whether the distractors led toward or away from the correct answer. In this manipulation, it was clear that the inability to inhibit (rather than a generalized speed of processing deficit) was the key to performance differences in older adults.

Some evidence in support of age-related decline in inhibitory processes comes also from studies using ERPs (see Fabiani, Gratton, & Federmeier, 2007). For instance, there is a well-known age-related increase in the Stroop effect, in which incongruence between stimulus features leads to RT slowing. For example, if a participant is instructed to name the ink color in which a word is written, and the word "red" is presented in blue ink, the participant is typically slower and less accurate in naming the ink color than when the color and the word meaning match. The Stroop effect increase in aging has been attributed to either generalized slowing or to decline in inhibitory processes. According to the first view, slowing in stimulus processing increases with conflict; according to the second view, the delay occurs at a motor level, as decline in top-down control allows for greater activation of the incorrect response in conflict conditions. West and Alain (2000) proposed that ERPs could be used to distinguish between these hypotheses because the latency of P3 can be viewed as a "pure" measure of stimulus-processing time independent of response processing (see Magliero, Bashore, Coles, & Donchin, 1984; McCarthy & Donchin, 1981). Therefore, if the latency of P3 in the conflict condition does not change as a function of aging, then the increase in RT should be seen as due to response-related conflict (contrary to the predictions of the generalized slowing hypothesis). This was the result reported in the West and Alain's (2000) study.

Other ERP research has examined decreases in habituation in sensory areas in aging with the assumption that they derive from diminished top-down control. For example, in one study, younger and older adults were presented with trains of auditory tones while they read a book of their choice and were instructed to ignore the tones (Fabiani, Low, Wee, Sable, & Gratton, 2006). N100 amplitudes to the repeated, to-be-ignored tones, along with concurrently recorded event-related optical signals (EROS; Gratton & Fabiani, 2010) showed that older adults, unlike young adults, maintained high N100 responses to the tones all the way through each train for the duration of the experiment. In other words, despite the hundreds of tone repetitions, older adults were unable to stop processing the repeated irrelevant tones occurring in the background to at least some degree (see also Gazzaley, Cooney, Rissman, & D'Esposito, 2005 for a similar example in the visual domain).

In the same study, the lag between tone trains was varied to test whether the decrease in habituation was due to deficits in inhibition or faster decay of sensory memory. The logic behind this manipulation was that if sensory memory representations are decaying faster in older adults, then older adults should have the same N100 as younger adults after a short lag between tone trains, but should have a *larger* N100 compared with younger adults after longer lags between tone trains. In fact, older adults showed no differences in N100 amplitude from younger adults at either lag, supporting the idea that an inability to inhibit was the culprit for the diminished habituation. A similar study showed that in the presence of background tones, older adults performed worse at a digit-matching task, and that the less was the habituation of the auditory N100, the worse was the performance (Kazmerski, Lee, Gratton, & Fabiani, 2005). Results like these support the idea that the inability to inhibit goal-irrelevant perceptual representations may cloud the mental landscape and decrease the signal-to-noise ratio in cognitive processing.

There is further evidence from ERP studies using the "novelty oddball" paradigm that older adults may have problems inhibiting processes that are initially useful, but become redundant and unnecessary in the long run (Fabiani & Friedman, 1995). In this paradigm, subjects are presented with randomized series of three types of stimuli: standard stimuli (frequently presented tones to which the subject does not have to respond), target stimuli (rarely presented tones to which the subject has to respond), and novel stimuli (a collection of unique auditory stimuli, which altogether occur rarely and to which the subject should not respond). Initially, both target and novel stimuli elicit a frontal positive ERP component (frontal P3 or P3a), which is likely generated in inferior frontal cortex and is linked to an orienting response to novel stimuli (see Polich, 2007; see also Knight, 1984). In young adults, this response is maintained over time for the novel stimuli (which keep changing), but is quickly lost for the target stimuli (which are repeated many times in the course of the study). In young adults, the P3a is replaced by a parietal positivity (P300 or P3b), presumably reflecting that these items are critical for the subject's task (Donchin & Coles, 1988; Polich, 2007). In older adults, however, the frontal P3a to the target tones persists over an entire experimental session, suggesting that they have difficulties inhibiting the orienting response to these stimuli even though they are repeated many times. In subsequent work, we showed that the extent to which older adults fail to inhibit this response is correlated with other indices of cognitive deterioration in aging, including performance in the Wisconsin Card Sorting Test (a neuropsychological measure of frontal lobe function; Fabiani, Friedman, & Cheng, 1998) and in the O-Span task (Brumback-Peltz et al., 2011), again linking inhibitory function and working memory.

The inhibition hypothesis suggests that declines in the performance of working memory tasks in aging may be due in part to a reduction in the "efficiency" of working memory, because some of the available capacity or resources are being tied up by no-longer-relevant information. One line of research shows that there are inhibition-related differences in cognitive performance when no working memory capacity differences are present. For example, when older adults read a slightly ambiguous passage, they showed different patterns in the availability of inferences than younger adults (Hamm & Hasher, 1992). If *subsequent* information made it clear that one of the two inferences was incorrect, older adults continued to consider both competing inferences, whereas young adults were able to suppress the incorrect inference and adopt the other. This age-related effect could not be explained by differences in working memory capacity.

Some researchers have also attempted to explain working memory capacity through inhibition. One aspect of inhibitory processing is the "deletion" of no-longer-relevant information. If deletion is inefficient, then prior representations will corrupt current representations as proactive interference builds up. May, Hasher, and Kane (1999) implemented a simple manipulation on a reading-span task, which is used to assess working memory span, by reversing the order of presentation of items so that the most difficult, largest trials occurred first. Usually in working memory span tasks small trials are presented first, and then over time they include more and more information until a capacity limit in memorization is reached. Reversing the administration of the items significantly improved the performance of older adults to the point where there were no differences in working memory span from younger adults, suggesting that age-related differences in reading span resulted from a buildup of interference (from inefficient suppression of previously presented material) rather than to a smaller working memory capacity.

Taken together, results like these demonstrate that differences in inhibitory abilities cannot be completely explained by working memory capacity, and that inhibitory processes may explain at least some of the declines in working memory capacity. Expanding on this, a recent study examined the effect of the three major attention-control functions (shifting, inhibition, and updating) according to the goal-maintenance view of executive control, and found that after controlling for processing speed only inhibition remained impaired in aging (Sylvain-Roy, Lungu, & Belleville, 2014). Intriguingly, other research has examined how the phenomenon of inattentional blindness changes with aging (Graham & Burke, 2011). Inattentional blindness is the failure to notice an unexpected event. Therefore, the authors argued, the inhibition hypothesis would predict that older adults would be more likely to be distracted by an unexpected stimulus. In fact, in that study older adults were *more* likely to experience inattentional blindness than younger adults. This result is surprising in the context of data showing enhanced distractibility in older adults. However, it should be taken into account that this experiment (noticing or failing to note a gorilla crossing a group of people passing a ball to each other while counting ball passes) is based on a single trial, in which fatigue and interference may not have had a chance to set in.

Another study using dichotic listening by Naveh-Benjamin et al. (2014) also found that older adults had more difficulty detecting that their names were presented to the unattended ear than younger adults. This may appear paradoxical because this phenomenon is also observed in young adults with high attention span. Naveh-Benjamin and colleagues argue that this result indicates that older adults may have more limited attention capacity compared with younger adults, and therefore cannot process the information presented to the unattended ear (differently from the high-capacity young adults, who process this information, but actively suppress it). Thus, not only inhibitory processes may decline with age, but also attention span. This account could also be used to explain the "inattention blindness" phenomenon described by Graham and Burke (2011).

Comparing bottom-up and top-down views of cognitive brain aging. As described previously, theories of cognitive brain aging may be classified into those that emphasize deficits in bottom-up processing, and those that emphasize deficits in top-down processing. The bottom-up views consider that a central deficit in aging is in feed-forward processes, either at the peripheral level (Baltes & Lindenberger, 1997) or at the central level (Cabeza, 2002; Grady, 2012; Parks & Reuter-Lorenz, 2009), resulting in "fuzzy" representations. To compensate for this "noise," additional processing resources (signaled by increased brain activity) are recruited. This compensation can be seen as an automatic servo-mechanism that comes into play to increase stimulus individuation when representations are noisy, or as a strategic behavior used by the older adults to deal with a less efficient system.

As an example, CRUNCH (Reuter-Lorenz & Cappell, 2008) is used to explain both over- and under-activations in the brains of older adults: at low task loads, older adults compensate by recruiting additional processing areas; conversely, at high loads they can no longer compensate and they show under-activation and lower performance compared with younger adults. We used this view to interpret changes in brain activity during a memory-search task in which we manipulated memory load in a parametric fashion from sub- to supra-span values (Schneider-Garces et al., 2010). We found increased brain activity in a set of cortical regions, largely corresponding to the frontoparietal dorsal attention network (Corbetta & Shulman, 2002) as a function of increasing set size. However, the increase was not linear. In younger adults, the amplitude of brain activity did not begin to increase until the set size exceeded four, whereas in older adults the increase occurred at low set sizes (fewer than four), with the brain activation curve saturating at higher set sizes. This apparent activation difference between the two age groups, however, disappeared if instead of considering the curves as a function of the objective load, we considered them as a function of the load relative to the subjects' span (measured using Cowan's K). In that case, the shapes of the curves for younger and older adults were identical. This is consistent with CRUNCH, in that it shows that brain activity increases with task demands until a subject-specific limit is reached, at which point brain activity does not increment any further. This limit is reached at lower objective loads in older adults because older adults have a lower working memory span.

A possible interpretation for the increase in activity is that it is compensatory, just as proposed by several theorists of cognitive brain aging (Cabeza, 2002; Parks & Reuter-Lorenz, 2009; Reuter-Lorenz & Cappell, 2008). However, there are also alternative ways of considering the increase in activity: It may merely represent a greater response to a more demanding task condition—just as a larger response in a sensory processing area could be expected to be elicited by a stronger stimulus. Indeed, this is the way we would typically interpret the increase in response as a function of set size observed in the young adults. There are other examples in which increased activity can hardly be considered as signaling compensation. These include the ERPs studies showing lack of suppression to repeated stimulation reviewed earlier (Fabiani et al., 2006; see also Fabiani & Friedman, 1995; Fabiani et al., 1998; see also Gazzaley et al., 2005), in which increased activity signaled lack of habituation to repeated stimuli and was also correlated with neuropsychological indices of frontal dysfunction.

In contrast to the bottom-up views, the top-down views of cognitive aging consider that the most critical deficit is in feedback and control processes, including attention control (Braver & Barch, 2002; Rypma & D'Esposito, 2000) and inhibition (Hasher & Zacks, 1988). Diminished top-down control, perhaps mediated by an age-related decrease in the efficiency of lateral-inhibition mechanisms, will lead to less differentiated representations of *both* goals and percepts.

Although diverging on several aspects, these two major explanations for the age-related decline in working memory and executive function (dedifferentiation and decreased inhibition) do agree that there is some loss in the brain's ability to generate sharp representations (see Holden et al., 2012; Li, Lindenberger, & Sikström, 2001; Stark et al., 2010; Yassa, Mattfeld, Stark, & Stark, 2011; Yassa & Stark, 2011; see also Velanova, Lustig, Jacoby, & Buckner, 2007). Obtaining sharp representations is critical for increasing the bandwidth of communication between brain areas. One way of measuring how sharp representations are in the brain is to assess the size of receptive fields in various cortical areas. Recently, Serences, Saproo, Scolari, Ho, and Muftuler (2009; see also Gratton, Sreenivasan, Silver, & D'Esposito, 2013) developed a methodology for studying receptive field sizes in humans from neuroimaging data. Both of these studies showed that receptive fields become smaller and better defined under attention-demanding conditions. It is accepted that receptive field size is largely mediated by lateral inhibition and is influenced by ascending neuromodulatory circuits based on acetylcholine and dopamine, which have also been implicated in cognitive aging (Wallace & Bertrand, 2013). It is likely, therefore, that the brain's ability to create and maintain sharp representations requires some combination of efficient sensorimotor and perceptual processing circuits that enhance the signal-to-noise ratio in a bottom-up, automatic fashion as well as proper top-down, executive control mechanisms that bias competition between representations in favor of the most goal-relevant ones. The presence of such common mechanism makes it difficult to distinguish between these two hypotheses, because in many cases they may lead to similar predictions. This is particularly the case when only brain imaging data are considered, because these types of data are inherently correlational.

Methodological considerations. Studies using neuroimaging techniques have greatly facilitated our understanding of how the executive control system functions and how functional activations change across the lifespan. Much of the work involves fMRI based on the blood oxygen level–dependent (BOLD) signal. However, in a discussion of aging, it is important to keep in mind that there may be age-related changes in the cerebral vascular system and that these changes may impact the BOLD signal in various ways. For example, we recently conducted a study to investigate neurovascular coupling—the relationship between hemodynamic and neuronal responses—in younger and older adults (Fabiani et al., 2014). Although younger and older participants showed similar coupling functions in visual cortex when BOLD fMRI and deoxyhemoglobin functional near-infrared spectroscopic measures were used, there were changes in the coupling of oxy- and deoxyhemoglobin, especially in low-fit individuals (see also D'Esposito et al., 1999, 2003; Safonova et al., 2004). Low-fit individuals also typically show steeper declines in cognitive functions with aging (e.g., Colcombe & Kramer, 2003). Similarly, in a recent study, we showed a relationship between arterial stiffening and other physiological and cognitive indices of brain health (Fabiani et al., 2014). Additional research will further elucidate the cognitive consequences of these physiological changes.

Aging and brain networks. Another major area of functional imaging research has examined changes in brain networks and functional connectivity between different brain regions. One finding that may help explain some of the age-related declines in executive control is the presence of weaker structural and functional connections between the PFC and other areas of the brain (see Rykhlevskaia, Fabiani, & Gratton, 2008 for a discussion). For example, in the context of functional connectivity, one study found that older adults exhibited weaker connectivity between the premotor cortex and the left dorsolateral PFC than younger adults in a parametric letter n-back task (Nagel et al., 2011). Another recent study found that in a delayed recognition working memory task, after affecting the functional connectivity between the PFC and the parahippocampal place area with a task designed to produce interference, older adults were unable to regain connectivity to preinterference levels, unlike younger adults who could (Clapp, Rubens, Sabharwal, & Gazzaley, 2011). Likewise, other studies have found reduced functional connectivity between frontal and parietal regions in older adults compared to younger adults (Bollinger, Rubens, Masangkay, Kalkstein, & Gazzaley, 2011; Madden et al., 2010).

In the context of structural connectivity, white matter health is known to decline in older adults (Kennedy & Raz, 2009; Pfefferbaum, Adalsteinsson, & Sullivan, 2005), likely affecting the propagation of signals across brain areas. As an example, we investigated the role of the corpus callosum in task switching in younger and older adults (Gratton, Wee, Rykhlevskaia, Leaver, & Fabiani, 2009). We found that older adults have more difficulties switching from verbal to spatial tasks (much more than vice versa). Using EROS, we showed that switching from a verbal to a spatial task involves the activation of areas in both hemispheres, differently from switching from a spatial to a verbal task, which only involves structures in the left hemisphere. We determined that the anterior part of the corpus callosum, the major white matter structure connecting the two frontal lobes, exerts an important role in the switch-to-spatial condition: Subjects with a large corpus callosum (both young and old) had significantly lower task-switching costs in this condition than subjects with a small corpus callosum. Further, older adults did have, on average, a smaller corpus callosum than younger adults, which accounted for their larger performance deficit.

A core approach in the study of functional connectivity involves an examination of connections in large-scale brain networks (Sporns, 2011, 2013a, 2013b). Using this approach, age-related changes in connectivity have been observed, especially within the default-mode network (DMN; Raichle et al., 2001; Raichle & Snyder, 2007). The DMN tends to be active when people are at rest or their mind is wandering, and is suppressed when they are engaged in attention-demanding, stimulus-driven tasks (Andrews-Hanna et al., 2007; Damoiseaux et al., 2008; Sambataro et al., 2010). However, the reduction of DMN activity is much less robust in older adults compared with younger adults (Andrews-Hanna et al., 2007; Damoiseaux et al., 2008; Gordon et al., 2014; Sambataro et al., 2010). The connectivity of this network has been shown to be a predictor of performance in many different tasks (Damoiseaux et al., 2008; Sambataro et al., 2010). A reduced ability to suppress activity in the DMN and dynamically switch between functional brain networks is another possible factor mediating age-related declines in executive control. In a recent study (Gordon et al., 2014), we examined whether age-related differences in DMN suppression may be mediated by the way signals propagate in the DMN and frontoparietal network. During a memory-search task, we found that older adults showed a more widespread signal around areas of activation in the frontoparietal network than younger adults (consistent with a decrease in the sharpness of representations). Conversely, older adults also showed a much less diffused *deactivation* signal in the DMN. This latter result suggests a problem with inhibitory function because the DMN is supposed to be suppressed during the memory-search task.

Another recent study investigated changes in functional connectivity in a wider range of functional networks using fMRI during a visual oddball task (Geerligs, Maurits, Renken, & Lorist, 2014). Older adults showed decreased functional connectivity compared with younger adults *within* the DMN and somatomotor networks. In addition to this reduced functional connectivity within networks, however, older adults also showed increased functional connectivity between *different* functional networks compared with younger adults. Again, decreased connectivity within functional networks correlated with worse cognitive functioning. This evidence supports the idea that the brain's ability to both process efficiently within pre-existing functional networks and to separate processing into distinct functional networks declines with age and is related to decrements in cognitive functioning. As suggested by the Gordon et al. (2014) study reviewed previously, this may be due to reduced capacity to spread inhibitory influences in an efficient manner with advancing age.

GOLDEN aging? A fundamental question that has arisen from age-related imaging studies on executive control is whether the functional changes that we discussed previously should be viewed as a continuation of ongoing lifespan development in cognitive processes, or if completely new patterns of activation arise in the brain, perhaps out of necessity, as described in the original conception of compensation. Some recent studies have attempted to investigate this question by taking advantage of naturally occurring individual differences to understand the extent of overlap between younger and older adults. In one study, younger and older adults varying in O-Span were required to respond to each stimulus in a series of Xs or Os, presented randomly on a computer screen with equal probability while ERPs were recorded (Brumback-Peltz et al., 2011; see also Brumback, Low, Gratton, & Fabiani, 2004; Brumback, Low, Gratton, & Fabiani, 2005). Stimuli presented in such "oddball-like" series elicit a P300, whose amplitude reflects the subjective probability of the stimulus (i.e., how rare or unexpected it is within the context of the surrounding stimuli; Squires, Wickens, Squires, & Donchin, 1976). Because the P300 is thought to reflect the updating of representations in working memory, it was expected that individuals with lower working memory would have larger P300s to a sequence change (e.g., OOX) than individuals with higher working memory, presumably because a representation of the current stimulus (X) would still be active in individuals with higher working memory (e.g., XOOX). The results supported our predictions. In addition, although the older adults had significantly lower O-Span scores on average than the younger adults, within each age group the relationship between O-Span and P300 sequential effects was the same, supporting the idea that the neural underpinnings of individual differences in working memory capacity in older adults are similar to those in younger adults. Similar evidence of continuity over the lifespan comes from the Schneider-Garces et al. (2010) study reviewed earlier in this chapter. There we showed that by equalizing the subject-specific working memory load (Cowan's K) between older and younger adults we also equalized their brain patterns of activation (which appeared different when they were averaged according to an objective load based on set size). Along the same lines, the Gratton et al. (2009) study reviewed earlier in this chapter showed that switch-costs in a spatial Stroop task were explained by individual differences in callosum size irrespective of age.

These and other similar studies have led to the hypothesis that the *growing lifelong differences explain normal aging* (GOLDEN aging framework, Fabiani, 2012; see also Hedden & Gabrieli, 2005), which suggests that many of the age-related functional changes in executive control or working memory tasks may represent continuations of individual-difference trajectories that started earlier in life. In other words, the difference between activation patterns in older and younger adults can often be explained in terms of how the cognitive load of a task compares with the cognitive capacity of the individual. The term "capacity" here is vague, and could either refer to declines in the efficiency of the executive control systems including inhibitory processes or to declines in early perceptual processing that reduce the overall clarity of perceptual representations in the brain. Most likely, some combination of both of these influences the overall cognitive capacity of an individual. Note that the GOLDEN aging framework does not address the causes of age-related cognitive decline directly, but rather focuses on contextualizing those declines in terms of individual behavioral and biological differences that occur throughout the entire lifespan, rather than positing that functional differences in the aging brain represent new, perhaps compensatory, patterns of activation.

Anatomical Underpinnings

As mentioned previously, the weakening of frontal lobe function and connectivity is often invoked as a hallmark of cognitive aging (West, 1996), which fits in nicely with views that look to reductions in the efficiency of the executive control system to explain cognitive declines. Support for this view comes from both cross-sectional and longitudinal anatomical studies, which show more pronounced reductions in the volume of prefrontal regions and more cortical thinning than in the rest of the neocortex (Raz & Rodrigue, 2006; Gordon et al., 2008; Raz, Ghisletta, Rodrigue, Kennedy, & Lindenberger, 2010). Other researchers have proposed extending this fronto-centric view to include the cerebellum (*frontocerebellar aging hypothesis*), which also undergoes substantial age-related volumetric decline and could help explain more of the changes in overall processing speed and automaticity (Hogan, 2004). Changes in white matter, particularly in the anterior dorsal portion of the corpus callosum, is also affected by aging, and may be related to the efficiency of information transfer between the two hemispheres' dorsolateral prefrontal cortices (Gratton et al., 2009; Rykhlevskaia, Fabiani, & Gratton, 2006). Better integrity of the corpus callosum also correlates with faster response times in older adults (Grady et al., 2005).

Much ongoing research is attempting to elucidate the causes of these macro-level structural changes in both gray and white matter in the aging brain, and to understand smaller scale changes that may impact cognitive decline. The ventral–striatal dopamine system is often implicated to explain some of the loss in sharp neural representations (Bäckman, Lindenberger, Li, & Nyberg, 2010; Bäckman et al., 2011; Klosterman, Braskie, Landau, O'Neil, & Jagust, 2012). In fact, both the dopamine and norepinephrine systems seem to be essential for the "gating" of information, an idea very reminiscent of the inhibition hypothesis of aging (see reviews by Arnsten, 2011; Braver & Barch, 2002). For example, studies in animals have demonstrated that dysregulation of the tonic levels of either dopamine or norepinephrine increase phasic firing to distractors in target detection tasks, which accompany behavioral false alarms (Aston-Jones, Rajkowski, & Cohen, 1999). This evidence is in line with the Gordon et al. (2014) study from our laboratory, which shows that the spread of inhibition is reduced in aging, especially in the DMN.

Other researchers have been inspired by the positive impact of aerobic exercise on measures of executive function and on brain measures (see Colcombe & Kramer, 2003 for a meta-analytic review of this literature on cognition, and Colcombe et al., 2006; Erickson et al., 2009 for the positive effects of fitness intervention on the brain). These studies predict that increases in vascularization, vascular health, neurogenesis, and neurotrophic growth factors may all help increase the overall health of the brain in aging. In particular, fitness level seems to play a major role in the health of the brain's vasculature. For example, Zimmerman et al. (2014) recently showed that in older adults (aged 55–87) cardiorespiratory fitness fully mediated the age effects on cerebrovascular blood flow, which declines with age. Similarly, optically derived measures of arterial elasticity, pulse pressure, and pulse wave velocity within the brain also showed sensitivity to age and fitness level and were predictors of cognitive performance (Fabiani et al., 2014).

GENERAL DISCUSSION AND CONCLUSIONS

To optimally function, the brain needs to generate, maintain, and access sharp, goal-relevant representations within the focus of attention. Limitations in the ability to switch between possible representations effectively, to dynamically inhibit irrelevant representations, or to efficiently overcome incoming noise from the bottom-up perceptual processing stream may all serve to reduce the sharpness of attentional focus in older adults. Although it seems clear that younger and older adults balance perceptual and control processes differently, it is still unclear whether this difference is driven largely by diminished bottom-up/sensory processing or by deficits in the actual control processes. Some recent research has attempted to answer this question. For example, Miller et al. (2011) used transcranial magnetic stimulation (TMS) on the prefrontal cortex and found decreased posterior perceptual tuning, favoring an interpretation of frontal, top-down decline as responsible for declines in perception.

However, as the Spanish proverb goes, "the two horns of a dilemma are usually on the same bull." In other words, similar age-related changes in physiology are likely responsible for overall processing declines in the brain, affecting both top-down control and bottom-up representations. For example, if lateral inhibitory processes are globally affected in aging, this may explain a degree of "fuzziness" in both goal-directed, top-down control (because goal representations would lose sharpness) and in perceptual bottom-up representations (which may become less sharp from both a lack of top-down control and local inhibitory constraints; see Sander, Linderberger, & Werkle-Bergner, 2012, for a similar perspective).

An intriguing candidate for this sort of global impact is in the declining coherence of oscillatory activity in the brain during the course of aging. In the working memory literature, oscillations have been recently hypothesized to be playing a role in both coding representations (gamma oscillations; Roux & Uhlhaas, 2014), in inhibiting irrelevant sensory information (alpha oscillations; Mathewson, Gratton, Fabiani, Beck, & Ro, 2009; Mathewson et al., 2011), and in holding both sequential (theta/gamma oscillations) and spatial information (alpha/gamma oscillations) in short-term memory through cross-frequency coupling between frequency ranges (Roux & Uhlhaas, 2014).

Oscillations are a promising candidate in part because of some physiological observations about the nature of working memory. Particularly, a specific representation is not stored in the brain by means of a fixed subpopulation of neurons, and yet there must be some mechanism that is sufficiently stable to retain a memory trace (Barak & Tsodyks, 2014). In addition, there are no major synaptic or receptor alterations hypothesized to take place in the brain while using online representations in working memory (Barak & Tsodyks, 2014). In younger adults, suppression of visual processing is typically correlated with a greater alpha power in the ipsilateral visual cortex (Vaden, Hutcheson, McCollum, Kentros, & Visscher, 2012; see also Mathewson et al., 2009, 2011). In one study, a cue was used to indicate a task relevant hemifield (Sauseng et al., 2009). The results showed that alpha power increased with the number of distractors and was lateralized to the irrelevant hemifield. In addition, repetitive TMS at an alpha frequency over the task-irrelevant hemisphere *improved* working memory capacity. However, older adults do not modulate alpha power in this way to suppress visual processing (Vaden et al., 2012). Sander, Werkle-Bergner, and Lindenberger (2012) studied lifespan differences in alpha power using a cued change-detection paradigm. They observed lateralized increases in alpha power for conditions of low and medium load in all age groups. However, in older adults, alpha power decreased for high load conditions. This might be interpreted as a tradeoff between maintaining the strength of a distinct representation (requiring high inhibition), which may be most important for performing some mental operation with the representation, and the number of representations that could be held at once (requiring less inhibition and greater activation).

GABAergic interneurons are critical for producing gamma oscillations, which have been shown to relate to sensory representations (Roux & Uhlhaas, 2014). In a similar vein, some GABAergic inputs are modulated by cholinergic inputs, which have been implicated in the generation of theta oscillations (Pignatelli, Beyeler, & Leinekugel, 2012). Finally, long-range neuronal communication of information is likely to be carried by the cross-frequency coupling between the fast gamma oscillations, which are capable of holding sensory representations, and the slower alpha and theta oscillations, thought to play a role in the short-term memory of visual/spatial and sequential representations, respectively (Roux & Uhlhaas, 2014). According to this view, different representations could be maintained simultaneously and separately (i.e., eliminating cross-talk) by "phase-multiplexing" the associated gamma bursts during the excitatory phase of low-frequency rhythms (theta or alpha). The two "magical numbers" of storage capacity in the working memory literature (7: Miller, 1956; 4: Cowan, 2001), would therefore result from the constraint of how many gamma-related representations may fit within either the theta or alpha temporal windows of excitation (Roux & Uhlhaas, 2014). In fact, as shown by Mathewson et al. (2009), during periods of high alpha power, different phases of the alpha wave represent a pulsating inhibition mechanism. Gamma bursts would ride on the excitatory part of this biphasic rhythm, enabling the maintenance of recent representations.

It is well known that the alpha rhythm slows down with aging (e.g., Gratton et al., 1992; Roubicek, 1977). Given this slowing in the carrier alpha frequency, there should also be a slowing of the gamma burst frequency to account for the typically lower working memory capacity in aging. Interestingly, a study on rats does show a slowing of their spontaneous gamma frequency activity (measured using local field potentials) in the frontal cortex (Insel et al., 2012). This may result in a more limited capacity for phase multiplexing. Alternatively, the deterioration of white matter tracks, which is conceivably important for keeping cross-frequency coupling well timed, may impact the storage capacity of working memory through a loss of oscillatory timing, in addition to affecting task-switching (Gratton et al., 2009; see also Rykhlevskaia et al., 2006) and overall processing speed in the brain, as others have proposed (Kerchner et al., 2012).

Other mechanisms have been proposed to contribute to working memory function. The dopamine-modulated *N*-methyl-D-aspartate receptors in the prefrontal cortex may also play a role in regulating neuronal activity during working memory processing (Durstewitz, 2009). In addition, short-term synaptic plasticity may play a role in stabilizing representations in working memory (Barak & Tsodyks, 2014). Finally, the ability of certain areas of the hippocampus (CA3/dentate gyrus) to maintain separate stimulus representations may decrease with age (Stark et al., 2010) and undermine not only long-term memory but also working-memory (as the same representations may be used for both). Although there is much still to learn about the functional mechanisms of working memory, these advances may help to integrate ideas about the mechanisms underlying working memory with the changes in behavioral and functional outcomes in working memory that often accompany aging.

This chapter has also highlighted how naturally occurring individual differences likely develop into the observed age-related changes in functional activity. Certainly the brain is a plastic organ, capable of both functional and structural modulation. However, it is important to discern which aspects of the functional data that are typically observed with aging are due to true network reorganization and which functional activations represent responses that are consistent throughout the lifespan, but are perhaps more common in older adults. For example, the bilateral frontal activation observed with high working memory loads in older adults could be interpreted as network reorganization to "compensate" for the age-related declines in working memory or it could be interpreted as the "normal" brain response to high working-memory loads, where the *subjective* load of the working memory task is what influences the functional response (Schneider-Garces et al., 2010). In addition, some functional activation patterns may be the result of breakdowns in some brain processing (e.g., a loss of inhibitory function) without necessarily reflecting or causing a change in behavior. This idea is related to the concept of cognitive reserve (e.g., Barulli & Stern, 2013), which postulates that factors affecting cognitive and brain health across the lifespan may influence the point at which overt cognitive decline begins.

In this chapter, we have considered working memory in a broader context, as the set of processes used for maintaining sensory (and long-term) memory representations active and available for the performance of ongoing tasks. As such, changes in working memory function occurring with aging can either result from degraded sensory representations (a bottom-up phenomenon) or from degraded attentional processes (a top-down phenomenon). In other words, working memory is the product of the interaction between these two classes of processes. In this context, aging research can be very useful to understand the different challenges that may interfere with the normal operation of working memory function and its sensory and attentional components.

Moving toward the future, to improve our understanding of age-related declines in working memory, it will be critical to combine data encompassing both the temporal dynamics of the activity and its localization in the brain. For example, learning more about the order of activation of brain areas, within networks, along with how this relates to functional and anatomical connectivity, will provide a more complete picture of the changes that accompany lifespan development and give deeper insights into the biology behind how sensory working memory interacts with the executive control system to maintain and fulfill goals.

Acknowledgments

We wish to acknowledge the support of National Institute of Aging grant 1RC1AG035927 to M. Fabiani and Ben Zimmerman's support on a National Science Foundation Integrative Graduate Education and Research Traineeship training grant 0903622. Address all correspondence to: Monica Fabiani, University of Illinois, Beckman Institute, 405N. Mathews Ave., Urbana, IL 61801. E-mail: mfabiani@illinois.edu.

References

Andrews-Hanna, J. R., Snyder, A. Z., Vincent, J. L., Lustig, C., Head, D., Raichle, M. E., et al. (2007). Disruption of large-scale brain systems in advanced aging. *Neuron, 56*(5), 924–935. http://dx.doi.org/10.1016/j.neuron.2007.10.038.

Anstey, K. J., Lord, S. R., & Williams, P. (1997). Strength in the lower limbs, visual contrast sensitivity, and simple reaction time predict cognition in older women. *Psychology and Aging, 12*(1), 137. http://dx.doi.org/10.1037/0882-7974.12.1.137.

Arnsten, A. F. (2011). Catecholamine influences on dorsolateral prefrontal cortical networks. *Biological Psychiatry, 69*(12), e89–e99. http://dx.doi.org/10.1016/j.biopsych.2011.01.027.

Aston-Jones, G., Rajkowski, J., & Cohen, J. (1999). Role of locus coeruleus in attention and behavioral flexibility. *Biological Psychiatry, 46*(9), 1309–1320. http://dx.doi.org/10.1016/S0006-3223(99)00140-7.

Atkinson, R. C., & Shiffrin, R. M. (1968). Human memory: a proposed system and its control processes. In K. W. Spence, & J. T. Spence (Eds.), *The psychology of learning and motivation* (Vol. 2) (pp. 89–195). New York: Academic Press.

Bäckman, L., Karlsson, S., Fischer, H., Karlsson, P., Brehmer, Y., Rieckmann, A., et al. (2011). Dopamine D(1) receptors and age differences in brain activation during working memory. *Neurobiology of Aging*, *32*(10), 1849–1856. http://dx.doi.org/10.1016/j.neurobiolaging.2009.10.018.

Bäckman, L., Lindenberger, U., Li, S.-C., & Nyberg, L. (2010). Linking cognitive aging to alterations in dopamine neurotransmitter functioning: recent data and future avenues. *Neuroscience and Biobehavioral Reviews*, *34*(5), 670–677. http://dx.doi.org/10.1016/j.neubiorev.2009.12.008.

Baddeley, A. (2003). Working memory: looking back and looking forward. *Nature Reviews Neuroscience*, *4*(10), 829–839. http://dx.doi.org/10.1038/nrn1201.

Baddeley, A. D., & Hitch, G. J. (1974). Working memory. *The Psychology of Learning and Motivation*, *8*, 47–89. http://dx.doi.org/10.1016/S0079-7421(08)60452-1.

Baltes, P. B., & Lindenberger, U. (1997). Emergence of a powerful connection between sensory and cognitive functions across the adult life span: a new window to the study of cognitive aging? *Psychology and Aging*, *12*(1), 12–21. http://dx.doi.org/10.1037//0882-7974.12.1.12.

Barak, O., & Tsodyks, M. (2014). Working models of working memory. *Current Opinion in Neurobiology*, *25*, 20–24. http://dx.doi.org/10.1016/j.conb.2013.10.008.

Barulli, D., & Stern, Y. (2013). Efficiency, capacity, compensation, maintenance, plasticity: emerging concepts in cognitive reserve. *Trends in Cognitive Sciences*, *17*(10), 502–509. http://dx.doi.org/10.1016/j.tics.2013.08.012.

Bishop, N. A., Lu, T., & Yankner, B. A. (2010). Neural mechanisms of ageing and cognitive decline. *Nature*, *464*(7288), 529–535. http://dx.doi.org/10.1038/nature08983.

Bollinger, J., Rubens, M. T., Masangkay, E., Kalkstein, J., & Gazzaley, A. (2011). An expectation-based memory deficit in aging. *Neuropsychologia*, *49*(6), 1466–1475. http://dx.doi.org/10.1016/j.neuropsychologia.2010.12.021.

Braver, T. S., & Barch, D. M. (2002). A theory of cognitive control, aging cognition, and neuromodulation. *Neuroscience and Biobehavioral Reviews*, *26*(7), 809–817. http://dx.doi.org/10.1016/S0149-7634(02)00067-2.

Braver, T. S., & West, R. (2008). Working memory, executive control, and aging. In F. I. M. Craik, & T. A. Salthouse (Eds.), *The handbook of aging and cognition* (3rd ed.) (pp. 311–372). Brighton, UK: Psychology Press.

Brumback-Peltz, C. R., Gratton, G., & Fabiani, M. (2011). Age-related changes in electrophysiological and neuropsychological indices of working memory, attention control, and cognitive flexibility. *Frontiers in Cognition*, *2*. (Article 190). http://dx.doi.org/10.3389/fpsyg.2011.00190.

Brumback, C. R., Low, K. A., Gratton, G., & Fabiani, M. (2004). Sensory brain responses predict individual differences in working memory span and fluid intelligence. *NeuroReport*, *15*(2), 373–376. http://dx.doi.org/10.1097/00001756-200402090-00032.

Brumback, C. R., Low, K., Gratton, G., & Fabiani, M. (2005). Putting things into perspective: differences in working memory span and the integration of information. *Experimental Psychology*, *52*(1), 21–30. http://dx.doi.org/10.1027/1618-3169.52.1.21.

Cabeza, R. (2002). Hemispheric asymmetry reduction in older adults: the HAROLD model. *Psychology and Aging*, *17*(1), 85. http://dx.doi.org/10.1037/0882-7974.17.1.85.

Cabeza, R., Anderson, N. D., Locantore, J. K., & McIntosh, A. R. (2002). Aging gracefully: compensatory brain activity in high-performing older adults. *NeuroImage*, *17*(3), 1394–1402. http://dx.doi.org/10.1006/nimg.2002.1280.

Cabeza, R., Grady, C. L., Nyberg, L., McIntosh, A. R., Tulving, E., Kapur, S., et al. (1997). Age-related differences in neural activity during memory encoding and retrieval: a positron emission tomography study. *The Journal of Neuroscience: The Official Journal of the Society for Neuroscience*, *17*(1), 391–400.

Cappell, K. A., Gmeindl, L., & Reuter-Lorenz, P. A. (2010). Age differences in prefontal recruitment during verbal working memory maintenance depend on memory load. *Cortex; A Journal Devoted to the Study of the Nervous System and Behavior*, *46*(4), 462–473. http://dx.doi.org/10.1016/j.cortex.2009.11.009.

Carlson, M. C., Hasher, L., Zacks, R. T., & Connelly, S. L. (1995). Aging, distraction, and the benefits of predictable location. *Psychology and Aging*, *10*(3), 427–436. http://dx.doi.org/10.1037//0882-7974.10.3.427.

Carp, J., Gmeindl, L., & Reuter-Lorenz, P. A. (2010). Age differences in the neural representation of working memory revealed by multi-voxel pattern analysis. *Frontiers in Human Neuroscience*, *4*, 217. http://dx.doi.org/10.3389/fnhum.2010.00217.

Clapp, W. C., Rubens, M. T., Sabharwal, J., & Gazzaley, A. (2011). Deficit in switching between functional brain networks underlies the impact of multitasking on working memory in older adults. *Proceedings of the National Academy of Sciences*, *108*(17), 7212–7217. http://dx.doi.org/10.1073/pnas.1015297108.

Colcombe, S. J., Erickson, K. I., Scalf, P. E., Kim, J. S., Prakash, R., McAuley, E., et al. (2006). Aerobic exercise training increases brain volume in aging humans. *The Journals of Gerontology Series A: Biological Sciences and Medical Sciences*, *61*(11), 1166–1170. http://dx.doi.org/10.1093/gerona/61.11.1166.

Colcombe, S., & Kramer, A. F. (2003). Fitness effects on the cognitive function of older adults a meta-analytic study. *Psychological Science*, *14*(2), 125–130. http://dx.doi.org/10.1111/1467-9280.t01-1-01430.

Connelly, S. L., Hasher, L., & Zacks, R. T. (1991). Age and reading: the impact of distraction. *Psychology and Aging*, *6*(4), 533–541. http://dx.doi.org/10.1037//0882-7974.6.4.533.

Corbetta, M., & Shulman, G. L. (2002). Control of goal-directed and stimulus-driven attention in the brain. *Nature Reviews Neuroscience*, *3*(3), 201–215. http://dx.doi.org/10.1038/nrn755.

Cowan, N. (1995). *Attention and memory*. Oxford University Press.

Cowan, N. (2000). Processing limits of selective attention and working memory: potential implications for interpreting. *Interpreting*, *5*(2), 117–146. http://dx.doi.org/10.1075/intp.5.2.05cow.

Cowan, N. (2001). The magical number 4 in short-term memory: a reconsideration of mental storage capacity. *Behavioral and Brain Sciences*, *24*, 87–185. http://dx.doi.org/10.1017/S0140525X01003922.

Cowan, N., Elliott, E. M., Scott Saults, J., Morey, C. C., Mattox, S., Hismjatullina, A., et al. (2005). On the capacity of attention: its estimation and its role in working memory and cognitive aptitudes. *Cognitive Psychology*, *51*(1), 42–100. http://dx.doi.org/10.1016/j.cogpsych.2004.12.001.

Craik, F. I., & Byrd, M. (1982). Aging and cognitive deficits. In *Aging and cognitive processes* (pp. 191–211). Springer US. http://dx.doi.org/10.1007/978-1-4684-4178-9_11.

D'Esposito, M., Deouell, L. Y., & Gazzaley, A. (2003). Alterations in the BOLD fMRI signal with ageing and disease: a challenge for neuroimaging. *Nature Reviews Neuroscience*, *4*(11), 863–872. http://dx.doi.org/10.1038/nrn1246.

D'Esposito, M., Zarahn, E., Aguirre, G. K., & Rypma, B. (1999). The effect of normal aging on the coupling of neural activity to the bold hemodynamic response. *NeuroImage*, *10*(1), 6–14. http://dx.doi.org/10.1006/nimg.1999.0444.

Damoiseaux, J. S., Beckmann, C. F., Arigita, E. J. S., Barkhof, F., Scheltens, P., Stam, C. J., et al. (2008). Reduced resting-state brain activity in the "default network" in normal aging. *Cerebral Cortex (New York, N.Y.: 1991)*, *18*(8), 1856–1864. http://dx.doi.org/10.1093/cercor/bhm207.

Davis, S. W., Dennis, N. A., Daselaar, S. M., Fleck, M. S., & Cabeza, R. (2008). Que' PASA? the posterior-anterior shift in aging. *Cerebral Cortex*, *18*(5), 1201–1209. http://dx.doi.org/10.1093/cercor/bhm155.

Davis, S. W., Kragel, J. E., Madden, D. J., & Cabeza, R. (2012). The architecture of cross-hemispheric communication in the aging brain: linking behavior to functional and structural connectivity. *Cerebral Cortex*, 232–242. http://dx.doi.org/10.1093/cercor/bhr123.

Desimone, R., & Duncan, J. (1995). Neural mechanisms of selective visual attention. *Annual Review of Neuroscience*, *18*(1), 193–222. http://dx.doi.org/10.1146/annurev.ne.18.030195.001205.

Donchin, E., & Coles, M. G. (1988). Is the P300 component a manifestation of context updating? *Behavioral and Brain Sciences*, *11*(03), 357–374. http://dx.doi.org/10.1017/S0140525X00058027.

Durstewitz, D. (2009). Implications of synaptic biophysics for recurrent network dynamics and active memory. *Neural Networks: The Official Journal of the International Neural Network Society*, *22*(8), 1189–1200. http://dx.doi.org/10.1016/j.neunet.2009.07.016.

Dywan, J., & Murphy, W. E. (1996). Aging and inhibitory control in text comprehension. *Psychology and Aging*, *11*(2), 199. http://dx.doi.org/10.1037/0882-7974.11.2.199.

Erickson, K. I., Prakash, R. S., Voss, M. W., Chaddock, L., Hu, L., Morris, K. S., et al. (2009). Aerobic fitness is associated with hippocampal volume in elderly humans. *Hippocampus*, *19*(10), 1030–1039. http://dx.doi.org/10.1002/hipo.20547.

Fabiani, M. (2012). It was the best of times, it was the worst of times: a psychophysiologist's view of cognitive aging. *Psychophysiology*, *49*(3), 283–304. http://dx.doi.org/10.1111/j.1469-8986.2011.01331.x.

Fabiani, M., & Friedman, D. (1995). Changes in brain activity patterns in aging: the novelty oddball. *Psychophysiology*, *32*, 579–594. http://dx.doi.org/10.1111/j.1469-8986.1995.tb01234.x.

Fabiani, M., Friedman, D., & Cheng, J. C. (1998). Individual differences in P3 scalp distribution in older adults, and their relationship to frontal lobe function. *Psychophysiology*, *35*, 698–708. http://dx.doi.org/10.1111/1469-8986.3560698.

Fabiani, M., Gordon, B. A., Maclin, E. L., Pearson, M. A., Brumback-Peltz, C. R., Low, K. A., et al. (2014). Neurovascular coupling in normal aging: a combined optical, ERP and fMRI study. *NeuroImage*, *85*(Pt 1), 592–607. http://dx.doi.org/10.1016/j.neuroimage.2013.04.113.

Fabiani, M., & Gratton, G. (2013). Aging, working memory and attention control: a tale of two processing streams? In D. T. Stuss, & R. T. Knight (Eds.), *Principles of frontal lobe function: Second edition* (pp. 582–592). New York, NY: Oxford University Press.

Fabiani, M., Gratton, G., & Federmeier, K. (2007). Event related brain potentials. In J. Cacioppo, L. Tassinary, & G. Berntson (Eds.), *Handbook of psychophysiology* (3rd ed.) (pp. 85–119). New York, NY: Cambridge University Press.

Fabiani, M., Low, K. A., Tan, C. -H., Fletcher, M., Zimmerman, B., Schneider-Garces, N., Maclin, E., Chiarelli, A. M., Sutton, B. P., & Gratton, G. (2014). Taking the pulse of aging: Mapping pulse pressure and elasticity in cerebral arteries with diffuse optical methods. *Psychophysiology*, *51*, 1072–1088. http://dx.doi.org/10.1111/psyp.12288.

Fabiani, M., Low, K. A., Wee, E., Sable, J. J., & Gratton, G. (2006). Reduced suppression or labile memory? Mechanisms of inefficient filtering of irrelevant information in older adults. *Journal of Cognitive Neuroscience*, *18*(4), 637–650. http://dx.doi.org/10.1162/jocn.2006.18.4.637.

Fabiani, M., & Wee, E. (2001). Age-related changes in working memory function: a review. In C. Nelson, & M. Luciana (Eds.), *Handbook of developmental cognitive neuroscience* (pp. 473–488). Cambridge, MA: MIT press.

Fisk, J. E., & Sharp, C. A. (2004). Age-related impairment in executive functioning: updating, inhibition, shifting, and access. *Journal of Clinical and Experimental Neuropsychology*, *26*(7), 874–890. http://dx.doi.org/10.1080/13803390490510680.

Fisk, J. E., & Warr, P. (1996). Age and working memory: the role of perceptual speed, the central executive, and the phonological loop. *Psychology and Aging*, *11*(2), 316. http://dx.doi.org/10.1037/0882-7974.11.2.316.

Fristoe, N. M., Salthouse, T. A., & Woodard, J. L. (1997). Examination of age-related deficits on the Wisconsin Card Sorting Test. *Neuropsychology*, *11*(3), 428. http://dx.doi.org/10.1037/0894-4105.11.3.428.

Gazzaley, A., Cooney, J. W., Rissman, J., & D'Esposito, M. (2005). Top down suppression deficit underlies working memory impairment in normal aging. *Nature Neuroscience*, *8*, 1298–1300. http://dx.doi.org/10.1038/nn1543.

Gazzaley, A., & D'Esposito, M. (2007). Top-down modulation and normal aging. *Annals of the New York Academy of Sciences*, *1097*, 67–83. http://dx.doi.org/10.1196/annals.1379.010.

Geerligs, L., Maurits, N. M., Renken, R. J., & Lorist, M. M. (2014). Reduced specificity of functional connectivity in the aging brain during task performance. *Human Brain Mapping*, *35*(1), 319–330. http://dx.doi.org/10.1002/hbm.22175.

Goh, J. O. S. (2011). Functional dedifferentiation and altered connectivity in older adults: neural accounts of cognitive aging. *Aging and Disease*, *2*(1), 30–48.

Gordon, B. A., Tse, C., Gratton, G., & Fabiani, M. (2014). Spread of activation and deactivation in the brain: Does age matter? *Frontiers in Aging Neuroscience*, *6*, 288. http://dx.doi.org/10.3389/fnagi.2014.00288.

Gordon, B., Rykhlevskaia, E., Brumback, C. R., Lee, Y., Elavsky, S., Konopack, J. F., McAuley, E., Kramer, A. F., Colcombe, S., Gratton, G., & Fabiani, M. (2008). Anatomical correlates of aging, cardiopulmonary fitness level, and education. *Psychophysiology*, *45*(5), 825–838. http://dx.doi.org/10.1111/j.1469-8986.2008.00676.x.

Grady, C. L. (2002). Age-related differences in face processing: a meta-analysis of three functional neuroimaging experiments. *Canadian Journal of Experimental Psychology = Revue Canadienne de Psychologie Expérimentale*, *56*(3), 208–220. http://dx.doi.org/10.1037/h0087398.

Grady, C. (2012). The cognitive neuroscience of ageing. *Nature Reviews Neuroscience*, *13*(7), 491–505. http://dx.doi.org/10.1038/nrn3256.

Grady, C. L., Maisog, J. M., Horwitz, B., Ungerleider, L. G., Mentis, M. J., Salerno, J. A., et al. (1994). Age-related changes in cortical blood flow activation during visual processing of faces and location. *The Journal of Neuroscience: The Official Journal of the Society for Neuroscience*, *14*(3 Pt 2), 1450–1462.

Grady, C. L., McIntosh, A. R., & Craik, F. I. M. (2005). Task-related activity in prefrontal cortex and its relation to recognition memory performance in young and old adults. *Neuropsychologia*, *43*(10), 1466–1481. http://dx.doi.org/10.1016/j.neuropsychologia.2004.12.016.

Graham, E. R., & Burke, D. M. (2011). Aging increases inattentional blindness to the gorilla in our midst. *Psychology and Aging, 26*(1), 162–166. http://dx.doi.org/10.1037/a0020647.

Gratton, G., Coles, M. G. H., Sirevaag, E., Eriksen, C. W., & Donchin, E. (1988). Pre and poststimulus activation of response channels: a psychophysiological analysis. *Journal of Experimental Psychology: Human Perception and Performance, 11*, 331–344. http://dx.doi.org/10.1037//0096-1523.14.3.331.

Gratton, G., & Fabiani, M. (2010). Fast optical imaging of human brain function. *Frontiers in Human Neuroscience, 4*. http://dx.doi.org/10.3389/fnhum.2010.00052.

Gratton, C., Nomura, E. M., Pérez, F., & D'Esposito, M. (2012). Focal brain lesions to critical locations cause widespread disruption of the modular organization of the brain. *Journal of Cognitive Neuroscience, 24*(6), 1275–1285. http://dx.doi.org/10.1162/jocn.

Gratton, C., Sreenivasan, K. K., Silver, M. A., & D'Esposito, M. (2013). Attention selectively modifies the representation of individual faces in the human brain. *The Journal of Neuroscience, 33*(16), 6979–6989. http://dx.doi.org/10.1523/JNEUROSCI.4142-12.2013.

Gratton, G., Villa, A. E. P., Fabiani, M., Colombis, G., Palin, E., Bolcioni, G., et al. (1992). Functional correlates of a three-component spatial model of the alpha rhythm. *Brain Research, 582*, 159–162. http://dx.doi.org/10.1016/0006-8993(92)90332-4.

Gratton, G., Wee, E., Rykhlevskaia, E. I., Leaver, E. E., & Fabiani, M. (2009). Does white matter matter? Spatio-temporal dynamics of task switching in aging. *Journal of Cognitive Neuroscience, 21*(7), 1380–1395. http://dx.doi.org/10.1162/jocn.2009.21093.

Hamm, V. P., & Hasher, L. (1992). Age and the availability of inferences. *Psychology and Aging, 7*(1), 56. http://dx.doi.org/10.1037/0882-7974.7.1.56.

Hasher, L., Lustig, C., & Zacks, R. T. (2007). Inhibitory mechanisms and the control of attention. In A. Conway, C. Jarrold, M. Kane, A. Miyake, & J. Towse (Eds.), *Variation in working memory* (pp. 227–249). New York: Oxford University Press.

Hasher, L., & Zacks, R. T. (1988). Working memory, comprehension, and aging: a review and a new view. In G. H. Bower (Ed.), *The psychology of learning and motivation* (Vol. 22) (pp. 193–225). New York, NY: Academic Press.

Hasher, L., Zacks, R. T., & May, C. P. (1999). Inhibitory control, circadian arousal, and age. In D. Gopher, & A. Koriat (Eds.), *Attention and performance XVII. Cognitive regulation of performance: Interaction of theory and application* (pp. 653–675). Cambridge, MA: MIT Press.

Hedden, T., & Gabrieli, J. D. E. (2005). Healthy and pathological processes in adult development: new evidence from neuroimaging of the aging brain. *Current Opinion in Neurology, 18*(6), 740–747. http://dx.doi.org/10.1097/01.wco.0000189875.29852.48.

Hogan, M. J. (2004). The cerebellum in thought and action: a fronto-cerebellar aging hypothesis. *New Ideas in Psychology, 22*(2), 97–125. http://dx.doi.org/10.1016/j.newideapsych.2004.09.002.

Holden, H. M., Hoebel, C., Loftis, K., & Gilbert, P. E. (2012). Spatial pattern separation in cognitively normal young and older adults. *Hippocampus, 22*(9), 1826–1832. http://dx.doi.org/10.1002/hipo.22017.

Insel, N., Patron, L. A., Hoang, L. T., Nematollahi, S., Schimanski, L. A., Lipa, P., et al. (2012). Reduced gamma frequency in the medial frontal cortex of aged rats during behavior and rest: implications for age-related behavioral slowing. *The Journal of Neuroscience, 32*(46), 16331–16344. http://dx.doi.org/10.1523/JNEUROSCI.1577-12.2012.

Jolicœur, P., Sessa, P., Dell'Acqua, R., & Robitaille, N. (2006). On the control of visual spatial attention: evidence from human electrophysiology. *Psychological Research, 70*(6), 414–424. http://dx.doi.org/10.1007/s00426-005-0008-4.

Kane, M. J., & Engle, R. W. (2003). Working-memory capacity and the control of attention: the contributions of goal neglect, response competition, and task set to Stroop interference. *Journal of Experimental Psychology: General, 132*(1), 47. http://dx.doi.org/10.1037/0096-3445.132.1.47.

Kazmerski, V. A., Lee, Y., Gratton, G., & Fabiani, M. (2005). Evidence for inefficient sensory filtering mechanisms in aging. *Journal of Cognitive Neuroscience Supplement*, 89–90.

Kennedy, K. M., & Raz, N. (2009). Aging white matter and cognition: differential effects of regional variations in diffusion properties on memory, executive functions, and speed. *Neuropsychologia, 47*(3), 916–927. http://dx.doi.org/10.1016/j.neuropsychologia.2009.01.001.

Kerchner, G. A., Racine, C. A., Hale, S., Wilheim, R., Laluz, V., Miller, B. L., et al. (2012). Cognitive processing speed in older adults: relationship with white matter integrity. *PLoS ONE, 7*(11), e50425. http://dx.doi.org/10.1371/journal.pone.0050425.

Klosterman, E. C., Braskie, M. N., Landau, S. M., O'Neil, J. P., & Jagust, W. J. (2012). Dopamine and frontostriatal networks in cognitive aging. *Neurobiology of Aging, 33*(3), 623.e15–623.e24. http://dx.doi.org/10.1016/j.neurobiolaging.2011.03.002.

Knight, R. T. (1984). Decreased response to novel stimuli after prefrontal lesions in man. *Electroencephalography and Clinical Neurophysiology/Evoked Potentials Section, 59*(1), 9–20. http://dx.doi.org/10.1016/0168-5597(84)90016-9.

Li, L., Gratton, C., Fabiani, M., & Knight, R. T. (2013). Age-related frontoparietal changes during the control of bottom-up and top-down attention: an ERP study. *Neurobiology of Aging, 34*(2), 477–488. http://dx.doi.org/10.1016/j.neurobiolaging.2012.02.025.

Li, K. Z., Hasher, L., Jonas, D., Rahhal, T. A., & May, C. P. (1998). Distractibility, circadian arousal, and aging: a boundary condition? *Psychology and Aging, 13*(4), 574–583. http://dx.doi.org/10.1037//0882-7974.13.4.574.

Li, S. C., Lindenberger, U., & Sikström, S. (2001). Aging cognition: from neuromodulation to representation. *Trends in Cognitive Science, 5*(11), 479–486. http://dx.doi.org/10.1016/S1364-6613(00)01769-1.

Lindenberger, U., & Baltes, P. B. (1994). Sensory functioning and intelligence in old age: a strong connection. *Psychology and Aging, 9*(3), 339. http://dx.doi.org/10.1037//0882-7974.9.3.339.

Loerch, P. M., Lu, T., Dakin, K. A., Vann, J. M., Isaacs, A., Geula, C., et al. (2008). Evolution of the aging brain transcriptome and synaptic regulation. *PloS One, 3*(10), e3329. http://dx.doi.org/10.1371/journal.pone.0003329.

Lustig, C., Hasher, L., & Tonev, S. T. (2006). Distraction as a determinant of processing speed. *Psychonomic Bulletin & Review, 13*(4), 619–625. http://dx.doi.org/10.3758/BF03193972.

Lustig, C., Hasher, L., & Zacks, R. T. (2007). Inhibitory deficit theory: recent developments in a "new view."*Inhibition in Cognition*, 145–162. http://dx.doi.org/10.1037/11587-008.

Madden, D. J., Costello, M. C., Dennis, N. A., Davis, S. W., Shepler, A. M., Spaniol, J., et al. (2010). Adult age differences in functional connectivity during executive control. *NeuroImage, 52*(2), 643–657. http://dx.doi.org/10.1016/j.neuroimage.2010.04.249.

Magliero, A., Bashore, T. R., Coles, M. G., & Donchin, E. (1984). On the dependence of P300 latency on stimulus evaluation processes. *Psychophysiology, 21*(2), 171–186. http://dx.doi.org/10.1111/j.1469-8986.1984.tb00201.x.

Mathewson, K., Gratton, G., Fabiani, M., Beck, D., & Ro, A. (2009). To see or not to see: pre-stimulus alpha phase predicts visual awareness. *The Journal of Neuroscience, 29*(8), 2725–2732. http://dx.doi.org/10.1523/JNEUROSCI.3963-08.2009.

Mathewson, K. E., Lleras, A., Beck, D. M., Fabiani, M., Ro, T., & Gratton, G. (2011). Pulsed out of awareness: EEG alpha oscillations represent a pulsed inhibition of ongoing cortical processing. *Frontiers in Perception Science*, 2. (Article 99). http://dx.doi.org/10.3389/fpsyg.2011.00099.

Mattay, V. S., Fera, F., Tessitore, A., Hariri, A. R., Berman, K. F., Das, S., et al. (2006). Neurophysiological correlates of age-related changes in working memory capacity. *Neuroscience Letters*, *392*(1–2), 32–37. http://dx.doi.org/10.1016/j.neulet.2005.09.025.

May, C. P. (1999). Synchrony effects in cognition: the costs and a benefit. *Psychonomic Bulletin & Review*, *6*(1), 142–147. http://dx.doi.org/10.3758/BF03210822.

May, C. P., Hasher, L., & Kane, M. J. (1999). The role of interference in memory span. *Memory & Cognition*, *27*(5), 759–767. http://dx.doi.org/10.3758/BF03198529.

McCabe, D. P., Roediger, H. L., III, McDaniel, M. A., Balota, D. A., & Hambrick, D. Z. (2010). The relationship between working memory capacity and executive functioning: evidence for a common executive attention construct. *Neuropsychology*, *24*(2), 222. http://dx.doi.org/10.1037/a0017619.

McCarthy, G., & Donchin, E. (1981). A metric for thought: a comparison of P300 latency and reaction time. *Science*, *211*(4477), 77–80. http://dx.doi.org/10.1126/science.7444452.

McIntosh, A. R., Sekuler, A. B., Penpeci, C., Rajah, M. N., Grady, C. L., Sekuler, R., et al. (1999). Recruitment of unique neural systems to support visual memory in normal aging. *Current Biology: CB*, *9*(21), 1275–1278. http://dx.doi.org/10.1016/S0960-9822(99)80512-0.

Miller, B. T., Vytlacil, J., Fegen, D., Pradhan, S., & D'Esposito, M. (2011). The prefrontal cortex modulates category selectivity in human extrastriate cortex. *Journal of Cognitive Neuroscience*, *23*, 1–10. http://dx.doi.org/10.1162/jocn.2010.21516.

Miller, G. A. (1956). The magical number seven, plus or minus two: some limits on our capacity for processing information. *Psychological Review*, *63*(2), 81. http://dx.doi.org/10.1037/h0043158.

Miyake, A., Friedman, N. P., Emerson, M. J., Witzki, A. H., Howerter, A., & Wager, T. D. (2000). The unity and diversity of executive functions and their contributions to complex "frontal lobe" tasks: a latent variable analysis. *Cognitive Psychology*, *41*(1), 49–100.

Moscovitch, M., & Winocur, G. (1992). The neuropsychology of memory and aging. In T. A. Salthouse, & F. I. M. Craik (Eds.), *The handbook of aging and cognition* (pp. 315–372). Hillsdale, NJ: Erlbaum.

Nagel, I. E., Preuschhof, C., Li, S.-C., Nyberg, L., Bäckman, L., Lindenberger, U., et al. (2011). Load modulation of BOLD response and connectivity predicts working memory performance in younger and older adults. *Journal of Cognitive Neuroscience*, *23*(8), 2030–2045. http://dx.doi.org/10.1162/jocn.2010.21560.

Naveh-Benjamin, M., Kilb, A., Maddox, G., Thomas, J., Fine, H., Chen, T., et al. (2014). Older adults don't notice their names: a new twist to a classic attention task. *Journal of Experimental Psychology: Learning, Memory, and Cognition*, *40*(6), 1540–1550. http://dx.doi.org/10.1037/xlm0000020.

Park, J., Carp, J., Hebrank, A., Park, D. C., & Polk, T. A. (2010). Neural specificity predicts fluid processing ability in older adults. *The Journal of Neuroscience: The Official Journal of the Society for Neuroscience*, *30*(27), 9253–9259. http://dx.doi.org/10.1523/JNEUROSCI.0853-10.2010.

Park, D. C., Polk, T. A., Park, R., Minear, M., Savage, A., & Smith, M. R. (2004). Aging reduces neural specialization in ventral visual cortex. *Proceedings of the National Academy of Sciences of the United States of America*, *101*(35), 13091–13095. http://dx.doi.org/10.1073/pnas.0405148101.

Parks, D. C., & Reuter-Lorenz, P. (2009). The adaptive brain: aging and neurocognitive scaffolding. *Annual Review of Psychology*, *60*, 173–196. http://dx.doi.org/10.1146/annurev.psych.59.103006.093656.

Paxton, J. L., Barch, D. M., Racine, C. A., & Braver, T. S. (2008). Cognitive control, goal maintenance, and prefrontal function in healthy aging. *Cerebral Cortex (New York, N.Y.: 1991)*, *18*(5), 1010–1028. http://dx.doi.org/10.1093/cercor/bhm135.

Paxton, J. L., Barch, D. M., Storandt, M., & Braver, T. S. (2006). Effects of environmental support and strategy training on older adults' use of context. *Psychology and Aging*, *21*(3), 499–509. http://dx.doi.org/10.1037/0882-7974.21.3.499.

Pfefferbaum, A., Adalsteinsson, E., & Sullivan, E. V. (2005). Frontal circuitry degradation marks healthy adult aging: evidence from diffusion tensor imaging. *NeuroImage*, *26*(3), 891–899. http://dx.doi.org/10.1016/j.neuroimage.2005.02.034.

Phillips, N. A., & Lesperance, D. (2003). Breaking the waves: age differences in electrical brain activity when reading text with distractors. *Psychology and Aging*, *18*(1), 126. http://dx.doi.org/10.1037/0882-7974.18.1.126.

Pignatelli, M., Beyeler, A., & Leinekugel, X. (2012). Neural circuits underlying the generation of theta oscillations. *Journal of Physiology, Paris*, *106*(3–4), 81–92. http://dx.doi.org/10.1016/j.jphysparis.2011.09.007.

Plude, D. J., & Hoyer, W. J. (1986). Age and the selectivity of visual information processing. *Psychology and Aging*, *1*(1), 4. http://dx.doi.org/10.1037/0882-7974.1.1.4.

Polich, J. (2007). Updating P300: an integrative theory of P3a and P3b. *Clinical Neurophysiology*, *118*(10), 2128–2148. http://dx.doi.org/10.1016/j.clinph.2007.04.019.

Raichle, M. E., MacLeod, A. M., Snyder, A. Z., Powers, W. J., Gusnard, D. A., & Shulman, G. L. (2001). A default mode of brain function. *Proceedings of the National Academy of Sciences*, *98*(2), 676–682. http://dx.doi.org/10.1073/pnas.98.2.676.

Raichle, M. E., & Snyder, A. Z. (2007). A default mode of brain function: a brief history of an evolving idea. *NeuroImage*, *37*(4), 1083–1090. http://dx.doi.org/10.1016/j.neuroimage.2007.02.041.

Raz, N., Ghisletta, P., Rodrigue, K. M., Kennedy, K. M., & Lindenberger, U. (2010). Trajectories of brain aging in middle-aged and older adults: regional and individual differences. *NeuroImage*, *51*(2), 501–511. http://dx.doi.org/10.1016/j.neuroimage.2010.03.020.

Raz, N., & Rodrigue, K. M. (2006). Differential aging of the brain: patterns, cognitive correlates and modifiers. *Neuroscience & Biobehavioral Reviews*, *30*(6), 730–748. http://dx.doi.org/10.1016/j.neubiorev.2006.07.001.

Reuter-Lorenz, P. A., & Cappell, K. A. (2008). Neurocognitive aging and the compensation hypothesis. *Current Directions in Psychological Science*, *17*(3), 177–182. http://dx.doi.org/10.1111/j.1467-8721.2008.00570.x.

Reuter-Lorenz, P. A., Jonides, J., Smith, E. E., Hartley, A., Miller, A., Marshuetz, C., et al. (2000). Age differences in the frontal lateralization of verbal and spatial working memory revealed by PET. *Journal of Cognitive Neuroscience*, *12*(1), 174–187. http://dx.doi.org/10.1162/089892900561814.

Reuter-Lorenz, P. A., & Sylvester, C. Y. (2005). The cognitive neuroscience of aging and working memory. In D. Park, R. Cabeza, & L. Backman (Eds.), *The cognitive neuroscience of aging* (pp. 186–217). Oxford University Press.

Roski, C., Caspers, S., Lux, S., Hoffstaedter, F., Bergs, R., Amunts, K., et al. (2013). Activation shift in elderly subjects across functional systems: an fMRI study. *Brain Structure & Function*. http://dx.doi.org/10.1007/s00429-013-0530-x.

Roubicek, J. (1977). The electroencephalogram in the middle-aged and the elderly. *Journal of the American Geriatric Society*, *25*(4), 145–152.

Roux, F., & Uhlhaas, P. J. (2014). Working memory and neural oscillations: alpha–gamma versus theta–gamma codes for distinct WM information? *Trends in Cognitive Sciences*, *18*(1), 16–25. http://dx.doi.org/10.1016/j.tics.2013.10.010.

Rykhlevskaia, E., Fabiani, M., & Gratton, G. (2006). Lagged covariance structure models for studying functional connectivity in the brain. *NeuroImage*, *30*(4), 1203–1218. http://dx.doi.org/10.1016/j.neuroimage.2005.11.019.

Rykhlevskaia, E. I., Fabiani, M., & Gratton, G. (2008). Combining structural and functional neuroimaging data for studying brain connectivity: a review. *Psychophysiology*, *45*, 173–187. http://dx.doi.org/10.1111/j.1469-8986.2007.00621.x.

Rypma, B., & D'Esposito, M. (2000). Isolating the neural mechanisms of age-related changes in human working memory. *Nature Neuroscience*, *3*(5), 509–515. http://dx.doi.org/10.1038/74889.

Safonova, L. P., Michalos, A., Wolf, U., Wolf, M., Hueber, D. M., Choi, J. H., et al. (2004). Age-correlated changes in cerebral hemodynamics assessed by near-infrared spectroscopy. *Archives of Gerontology and Geriatrics*, *39*(3), 207–225. http://dx.doi.org/10.1016/j.archger.2004.03.007.

Salthouse, T. A. (1996). The processing-speed theory of adult age differences in cognition. *Psychological Review*, *103*(3), 403. http://dx.doi.org/10.1037/0033-295X.103.3.403.

Salthouse, T. A., Babcock, R. L., & Shaw, R. J. (1991). Effects of adult age on structural and operational capacities in working memory. *Psychology and Aging*, *6*(1), 118. http://dx.doi.org/10.1037/0882-7974.6.1.118.

Sambataro, F., Murty, V. P., Callicott, J. H., Tan, H.-Y., Das, S., Weinberger, D. R., et al. (2010). Age-related alterations in default mode network: impact on working memory performance. *Neurobiology of Aging*, *31*(5), 839–852. http://dx.doi.org/10.1016/j.neurobiolaging.2008.05.022.

Sander, M. C., Linderberger, U., & Werkle-Bergner, M. (2012a). Lifespan age differences in working memory: a two-component framework. *Neuroscience and Biobehavioral Reviews*, *36*, 2007–2033. http://dx.doi.org/10.1016/j.neubiorev.2012.06.004.

Sander, M. C., Werkle-Bergner, M., & Lindenberger, U. (2012b). Amplitude modulations and inter-trial phase stability of alpha-oscillations differentially reflect working memory constraints across the lifespan. *NeuroImage*, *59*(1), 646–654. http://dx.doi.org/10.1016/j.neuroimage.2011.06.092.

Sauseng, P., Klimesch, W., Heise, K. F., Gruber, W. R., Holz, E., Karim, A. A., et al. (2009). Brain oscillatory substrates of visual short-term memory capacity. *Current Biology: CB*, *19*(21), 1846–1852. http://dx.doi.org/10.1016/j.cub.2009.08.062.

Schäfer, S., Huxhold, O., & Lindenberger, U. (2006). Healthy mind in healthy body? A review of sensorimotor–cognitive interdependencies in old age. *European Review of Aging and Physical Activity*, *3*(2), 45–54. http://dx.doi.org/10.1007/s11556-006-0007-5.

Schneider-Garces, N. J., Gordon, B. A., Brumback-Peltz, C. R., Shin, E., Lee, Y., Sutton, B. P., et al. (2010). Span, CRUNCH, and beyond: working memory capacity and the aging brain. *Journal of Cognitive Neuroscience*, *22*(4), 655–669. http://dx.doi.org/10.1162/jocn.2009.21230.

Serences, J. T., Saproo, S., Scolari, M., Ho, T., & Muftuler, L. T. (2009). Estimating the influence of attention on population codes in human visual cortex using voxel-based tuning functions. *NeuroImage*, *44*, 223–231. http://dx.doi.org/10.1016/j.neuroimage.2008.07.043.

Sperling, G. (1967). Successive approximations to a model for short term memory. *Acta Psychologica*, *27*, 285–292. http://dx.doi.org/10.1016/0001-6918(67)90070-4.

Sporns, O. (2011). The non-random brain: efficiency, economy, and complex dynamics. *Frontiers in Computational Neuroscience*, *5*, 5. http://dx.doi.org/10.3389/fncom.2011.00005.

Sporns, O. (2013a). Structure and function of complex brain networks. *Dialogues in Clinical Neuroscience*, *15*(3), 247–262.

Sporns, O. (2013b). The human connectome: origins and challenges. *NeuroImage*, *80*, 53–61. http://dx.doi.org/10.1016/j.neuroimage.2013.03.023.

Spreng, R. N., Wojtowicz, M., & Grady, C. L. (2010). Reliable differences in brain activity between young and old adults: a quantitative meta-analysis across multiple cognitive domains. *Neuroscience & Biobehavioral Reviews*, *34*(8), 1178–1194. http://dx.doi.org/10.1016/j.neubiorev.2010.01.009.

Squires, K. C., Wickens, C., Squires, N. K., & Donchin, E. (1976). The effect of stimulus sequence on the waveform of the cortical event-related potential. *Science*, *193*(4258), 1142–1146. http://dx.doi.org/10.1126/science.959831.

Sreenivasan, K. K., Curtis, C. E., & D'Esposito, M. (2014). Revisiting the role of persistent neural activity during working memory. *Trends in Cognitive Sciences* Retrieved from http://www.sciencedirect.com/science/article/pii/S1364661313002726.

Sreenivasan, K. K., Gratton, C., Vytlacil, J., & D'Esposito, M. (2014). Evidence for working memory storage operations in perceptual cortex. *Cognitive, Affective, & Behavioral Neuroscience*, 1–12. http://dx.doi.org/10.3758/s13415-013-0246-7.

Stark, S. M., Yassa, M. A., & Stark, C. E. L. (2010). Individual differences in spatial pattern separation performance associated with healthy aging in humans. *Learning and Memory*, *17*(6), 284–288. http://dx.doi.org/10.1101/lm.1768110.

Sylvain-Roy, S., Lungu, O., & Belleville, S. (2014). Normal aging of the attentional control functions that underlie working memory. *The Journals of Gerontology. Series B, Psychological Sciences and Social Sciences*. http://dx.doi.org/10.1093/geronb/gbt166.

Townsend, J., Adamo, M., & Haist, F. (2006). Changing channels: an fMRI study of aging and cross-modal attention shifts. *NeuroImage*, *31*(4), 1682–1692. http://dx.doi.org/10.1016/j.neuroimage.2006.01.045.

Tun, P. A., O'Kane, G., & Wingfield, A. (2002). Distraction by competing speech in young and older adult listeners. *Psychology and Aging*, *17*(3), 453. http://dx.doi.org/10.1037/0882-7974.17.3.453.

Turner, M. L., & Engle, R. W. (1989). Is working memory capacity task dependent? *Journal of Memory and Language*, *28*(2), 127–154. http://dx.doi.org/10.1016/0749-596X(89)90040-5.

Vaden, R. J., Hutcheson, N. L., McCollum, L. A., Kentros, J., & Visscher, K. M. (2012). Older adults, unlike younger adults, do not modulate alpha power to suppress irrelevant information. *NeuroImage*, *63*(3), 1127–1133. http://dx.doi.org/10.1016/j.neuroimage.2012.07.050.

Vallesi, A., McIntosh, A. R., & Stuss, D. T. (2011). Overrecruitment in the aging brain as a function of task demands: evidence for a compensatory view. *Journal of Cognitive Neuroscience*, *23*(4), 801–815. http://dx.doi.org/10.1162/jocn.2010.21490.

Velanova, K., Lustig, C., Jacoby, L. L., & Buckner, R. L. (2007). Evidence for frontally mediated controlled processing differences in older adults. *Cerebral Cortex*, *17*(5), 1033–1046. http://dx.doi.org/10.1093/cercor/bhl013.

Vogel, E. K., & Machizawa, M. G. (2004). Neural activity predicts individual differences in working memory capacity. *Nature*, *428*, 748–751. http://dx.doi.org/10.1038/nature02447.

Wallace, T. L., & Bertrand, D. (2013). Importance of the nicotinic acetylcholine receptor system in the prefrontal cortex. *Biochemical Pharmacology*, *85*(12), 1713–1720. http://dx.doi.org/10.1016/j.bcp.2013.04.001.

West, R. L. (1996). An application of prefrontal cortex function theory to cognitive aging. *Psychological Bulletin*, *120*(2), 272. http://dx.doi.org/10.1037/0033-2909.120.2.272.

West, R. L., & Alain, C. (2000). Age–related decline in inhibitory control contributes to the increased Stroop effect observed in older adults. *Psychophysiology*, *37*(2), 179–189. http://dx.doi.org/10.1111/1469-8986.3720179.

Yassa, M. A., Mattfeld, A. T., Stark, S. M., & Stark, C. E. (2011). Age-related memory deficits linked to circuit-specific disruptions in the hippocampus. *Proceedings of the National Academy of Sciences*, *108*(21), 8873–8878. http://dx.doi.org/10.1073/pnas.1101567108.

Yassa, M. A., & Stark, C. E. (2011). Pattern separation in the hippocampus. *Trends in Neurosciences*, *34*(10), 515–525. http://dx.doi.org/10.1016/j.tins.2011.06.006.

Zacks, J. L., & Zacks, R. T. (1993). Visual search times assessed without reaction times: a new method and an application to aging. *Journal of Experimental Psychology. Human Perception and Performance*, *19*(4), 798–813. http://dx.doi.org/10.1037//0096-1523.19.4.798.

Zimmerman, B., Sutton, B. P., Low, K. A., Tan, C. H., Schneider-Garces, N., Fletcher, M. A., Li, Y., Ouyang, C., Maclin, E. L., Gratton, G., & Fabiani, M. (2014). Cardiorespiratory fitness mediates the effects of aging on cerebral blood flow. *Frontiers in Aging Neuroscience*, *8*, 59. http://dx.doi.org/10.3389/fnagi.2014.00059.

C H A P T E R

11

Defining a Role for Lateral Prefrontal Cortex in Memory-Guided Decisions About Visual Motion

Tatiana Pasternak

Departments of Neurobiology & Anatomy, Brain and Cognitive Science, and Center for Visual Science, University of Rochester, Rochester, NY, USA

Active observers are often faced with a ubiquitous task of comparing visual motion stimuli across time and space. To successfully perform such comparisons subjects must be able to not only identify these stimuli but also store them in memory, so they can be retrieved at the time of sensory comparison. Thus, the neural mechanisms underlying such tasks are likely to involve cortical regions subserving processing of visual motion stimuli, their storage as well as areas involved in control of visual attention and decision-making. Two reciprocally interconnected cortical regions likely to be relevant to these operations are motion-processing middle temporal area (MT) and the lateral region of the prefrontal cortex (LPFC) strongly linked to sensory maintenance and executive control (Barbas, 1988; Miller & Cohen, 2001; Petrides & Pandya, 2006). In recent years, the evidence has emerged that area MT not only shows strong selectivity for visual motion, but also represents processes related such cognitive task components as storage, attention, and perceptual decisions (Martinez-Trujillo & Treue, 2004; Pasternak & Greenlee, 2005; Zaksas & Pasternak, 2006; Lui & Pasternak, 2011). These processes are likely to represent top-down modulations, most likely arising in LPFC traditionally associated with storage and decision-making (Miller & Cohen, 2001). Conversely, there is accumulating evidence that LPFC not only carries signals associated with storage and cognitive control but also shows stimulus selectivity, reminiscent of responses recorded in MT (e.g., Kim and Shadlen (1999), Zaksas and Pasternak (2006), Hussar and Pasternak (2009, 2012)). Because LPFC is a likely source of modulation of neurons processing visual information, elucidation of such influences provides important insights into the role its neurons play in the circuitry subserving such tasks.

In this chapter, I will characterize the activity of LPFC neurons during all components of memory-guided motion comparison tasks, focusing on the identity of neurons likely to provide sensory neurons with such influences. For the analysis of neuronal activity during this task, we used waveform durations of action potentials of individual neurons to distinguish between putative pyramidal projection neurons, a likely source of top-down influences LPFC may be exerting on upstream sensory neurons, and putative local interneurons (Rockland, 1997; Markram et al., 2004). In our analysis, we took advantage of differences in the temporal dynamics of action potentials between these two classes of neurons, with pyramidal neurons having broader action potentials compared with the relatively narrow spikes characteristic of inhibitory interneurons (Connors & Gutnick, 1990; Nowak, Azouz, Sanchez-Vives, Gray, & McCormick, 2003). We used these differences to identify the two cells groups as broad-spiking (BS) putative pyramidal and narrow-spiking (NS) putative interneurons (Hussar & Pasternak, 2009).

In our studies, described in detail in several articles (Zaksas & Pasternak, 2006; Hussar & Pasternak, 2009, 2012, 2013), the monkeys compared directions (Figure 1(b)) or speeds (Figure 1(d)) of two moving random dot stimuli, S1 and S2, separated by a delay and reported whether the two stimuli were the same or different by pressing the right or the left button, respectively. On each trial, the two stimuli moved in the same or different

Mechanisms of Sensory Working Memory
http://dx.doi.org/10.1016/B978-0-12-811042-3.00011-6

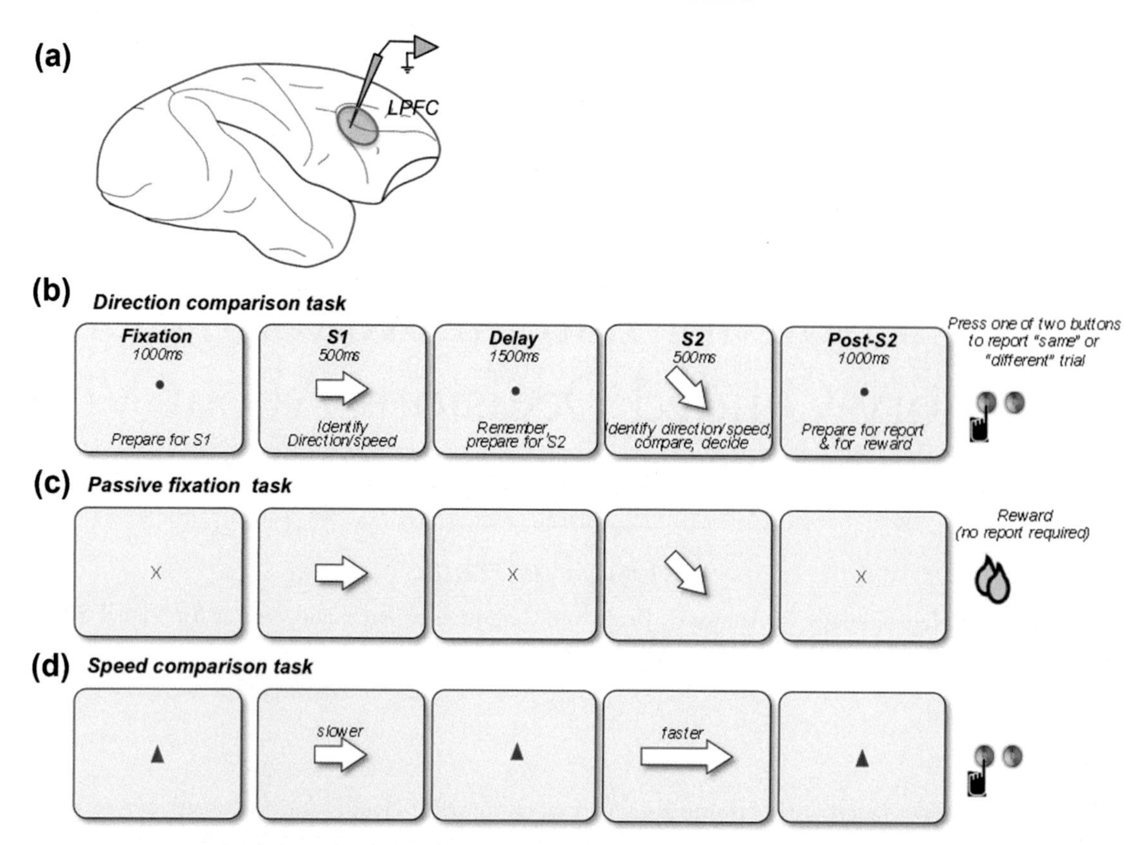

FIGURE 1 (a) Lateral view of the monkey cortex. Recordings discussed in the chapter were aimed at the lateral prefrontal cortex (LPFC) shown in blue. (b, d) Direction/speed comparison tasks. Monkeys reported whether S1 and S2, each lasting 500 ms, moved coherently in the same or different directions (b) or speeds (d) by pressing one of two response buttons. Each task was cued by a different fixation target. Task difficulty was manipulated by decreasing the difference in direction (or speed) between S1 and S2. The figure shows only trials with different directions during S1 and S2. (c) Passive fixation task. Stimulus conditions were identical to those in the direction task. The monkeys maintained fixation throughout the trial on a small cross and were rewarded after the offset of S2.

directions (or speed) and we manipulated task difficulty by varying the difference between S1 and S2. Recordings during this task revealed that many LPFC neurons of both types are selective for motion direction and show some activity during the delay. This activity was more prominent and more likely to exhibit periods of stimulus selectivity in putative pyramidal neurons. These patterns can be seen in the activity of the two example neurons, one of each type, shown in Figure 2. Both neurons showed strong selectivity for motion direction during S1 and S2. They also showed modulation of activity with time in delay, which particularly pronounced for the BS neuron (Figure 2(b)).

Although both neurons showed periods with activity differences following opposite directions of motion, in the NS cell, this period was brief and was dominated by the antipreferred direction, whereas in the BS neuron the period of selectivity lasted a bit longer and it was dominated by the direction identified as preferred during S1.

RESPONSES TO VISUAL MOTION IN LPFC

During our task, the majority of neurons of both types with task-related activity showed significant differences in their responses to opposite directions of motion and this selectivity equally strong for NS and BS cells (Figure 3(a) and (b)) (Hussar & Pasternak, 2009). We asked whether this selectivity persists when the behavioral task does not involve the use of information about motion direction and recorded responses to different directions of motion during a speed comparison task (Figure 1(c)). In that task, cued with a different fixation target, the two comparison stimuli moved at the same or different speeds and the monkeys were rewarded for reporting whether the speeds were the same or different. The two stimuli always moved at the same direction and we compared direction selectivity of the same neurons recorded during the speed and direction comparison task. We found that although BS putative pyramidal neurons maintained their direction selectivity when the task was switched from direction to speed, selectivity for direction of

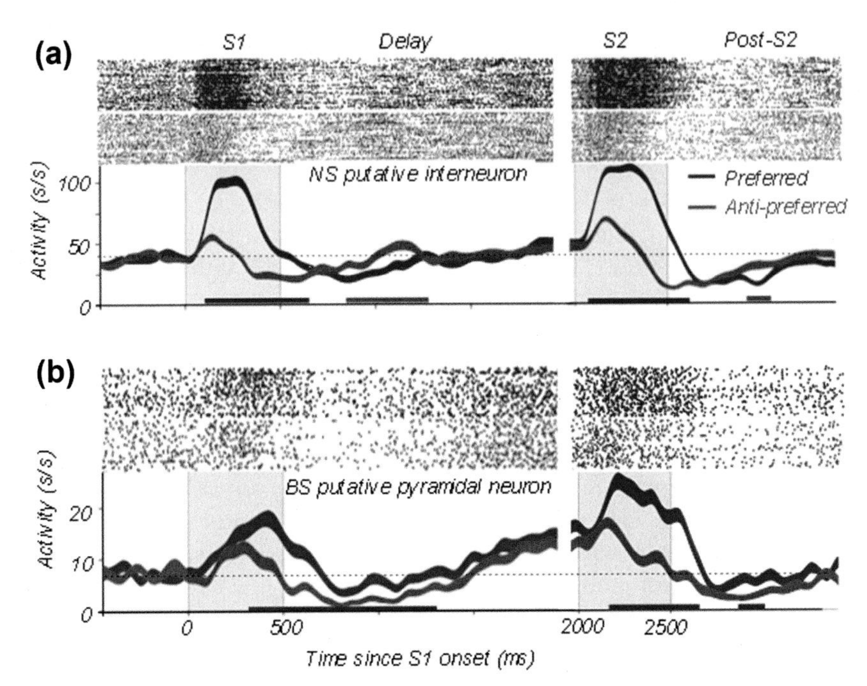

FIGURE 2 Activity of two example neurons during the direction comparison task. Raster plots and average activity for an example NS putative interneuron (a) and BS putative pyramidal neuron (b). Periods of significantly higher activity following S1 moving in preferred and anti-preferred directions are indicated by blue and red lines, respectively, plotted along the x-axis (Wilcoxon signed-rank, $p<.05$). *From Hussar and Pasternak (2012).*

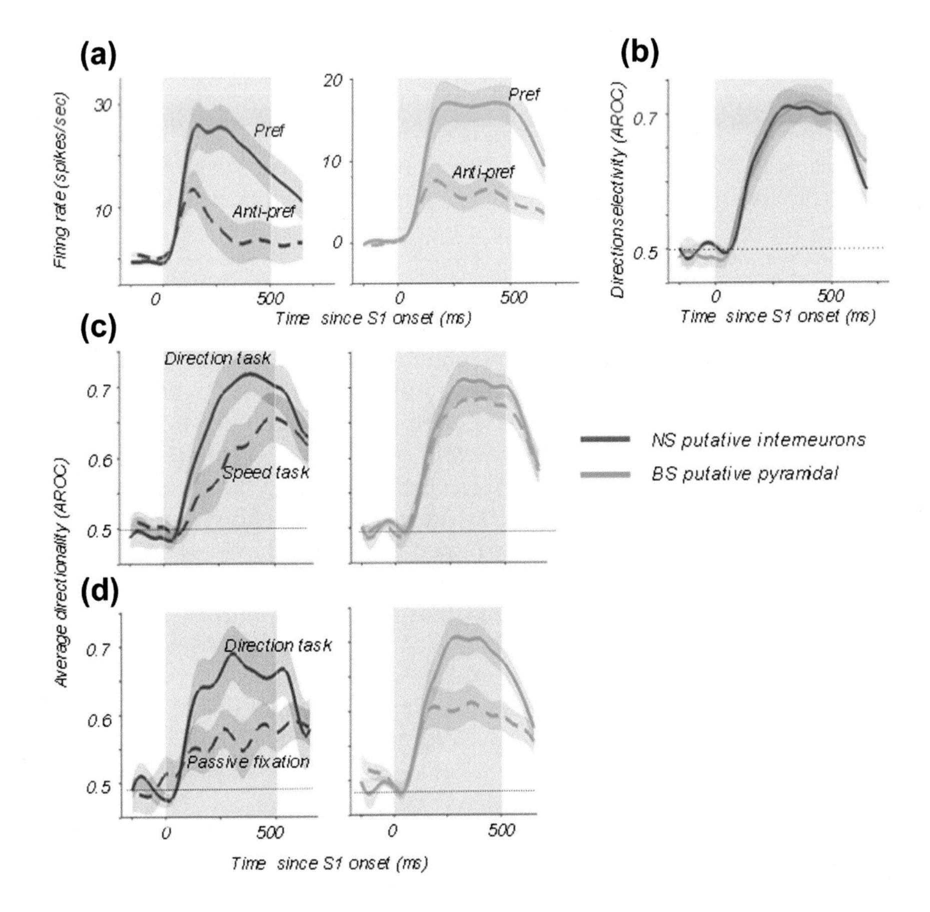

FIGURE 3 (a) Direction selective (DS) responses of NS (left, n=14) and BS (right, n=37) neurons. Both cell groups show DS activity. (b) DS of the two cell groups computed with receiver operating characteristic analysis is nearly identical. (c) Average DS during direction and speed tasks. Only NS neurons showed reduced DS when animals actively ignored direction during the speed task. BS neurons maintained stable representation of direction. (d) Average DS selectivity during direction and passive fixation task. Both cell types showed reduced drastically DS during passive fixation. *From Hussar and Pasternak (2009).*

many NS neurons drastically decreased during the speed task (Figure 3(c)). These results demonstrate the ability of NS putative interneurons to actively reduce their selectivity to the unattended stimulus dimension. Similar selective effects of attentional shifts on responses of NS putative interneurons were observed in visual area V4 (Mitchell, Sundberg, & Reynolds, 2007). These results suggest the existence of a specialized neuronal network, with putative interneurons playing a key role in detecting shifts in attention between stimulus dimensions. We should note that the selective effect of task demands on NS neurons appears to be limited to tasks involving active shifts of attention, because when animals are not required to make perceptual judgments and receive a reward at the end of each trial (Figure 1(c)), both cell types show a dramatic loss of response selectivity (Figure 3(d)). Although direction selectivity characteristic of many LPFC neurons is reminiscent of selectivity recorded in area MT, its strong dependence on behavioral context is very different. Although responses of neurons in MT can be modulated by spatial and feature-based attention (Treue & Maunsell, 1996, 1999; Cook & Maunsell, 2004; Martinez-Trujillo & Treue, 2004; Anton-Erxleben, Stephan, & Treue, 2009), this modulation is relatively modest and direction selectivity is robust in anesthetized or passively fixating monkeys (e.g., Maunsell and Van Essen (1983), DeAngelis and Newsome (1999), Zaksas and Pasternak (2006)).

These data demonstrate that LPFC neurons can faithfully represent fundamental stimulus modalities, supplied by bottom-up inputs from motion processing neurons. However, they also show remarkable flexibility and adaptation to the behavioral context, the property that may be reflected in the nature of top-down influences. This access to fundamental sensory information and sensitivity to task demands allows for specificity and precision with which LPFC neurons can exert executive control on sensory processing.

ACTIVITY DURING THE MEMORY DELAY

During our task, the appearance of S2 was highly predictable because the duration of the preceding delay was constant. This predictability was reflected in gradually increasing or decreasing firing rates with time in delay, the pattern characteristic of nearly 60% of putative pyramidal neurons but only 23% of putative interneurons (for details see Figure 3 in Hussar and Pasternak (2012)). The higher incidence of delay modulation among putative pyramidal cells translates into a significantly larger proportion of these neurons being active in late delay (Figure 4(a)). The reduced delay activity during passive fixation when no sensory comparison is required supports the preparatory nature of delay modulation observed during the active task (see Figure S3 in Hussar and Pasternak (2010)).

We were interested whether the activity recorded during the delay reflected the identity of the remembered direction. In many individual neurons, the firing rates recorded during the delay that followed each of the two directions presented during S1 showed reliable differences. However, these differences were relatively brief (Figure 4(b)), appearing at different times in different neurons, and were independent of directional preferences exhibited during S1 (Zaksas & Pasternak, 2006; Hussar & Pasternak, 2012). Furthermore, these differences were inconsistent across the delay, sometimes dominated by one and sometimes by the other S1 direction (see example neuron in Figure 2(a)). Nevertheless, this stimulus selective activity was reliable and when averaged across cells, persisted throughout the delay, particularly in the more active putative pyramidal neurons (Figure 4(c)), the likely source of the top-down influences exerted by the LPFC neurons on other cortical areas.

The inconsistent and transient nature of this activity raises the question whether such activity is used during the task. Our results suggest that it is, because it becomes substantially weaker when motion direction is not behaviorally relevant during the speed discrimination task (Figure 4(d)) or during the passive fixation task (Figure 4(e)) (also see Hussar and Pasternak (2012)). These results also highlight the differential contribution of the two cell types to different aspects of the direction comparison task. Although the reduction of direction selectivity during the speed comparison task was relatively limited for putative pyramidal neurons, during the delay these same cells showed a significant selectivity loss, further highlighting that direction selectivity of delay activity depends on the behavioral context.

The nature of delay activity appears to be task-dependent. Thus, studies of spatial memory employing paradigms in which monkeys make saccades to previously cued locations report that prefrontal neurons continue to fire throughout the delay separating the cue and the response, and this persistent activity is spatially selective (e.g., Funahashi, Bruce, and Goldman-Rakic (1989)). The discovery of this activity, proposed as the possible substrate of working memory (Goldman-Rakic, 1995), has spurred a large body of computational work that gave rise to biophysically realistic attractor network models as a mechanism for this process (e.g., Compte, Brunel, Goldman-Rakic, and Wang (2000)). With time, such models were generalized to account not only for the maintenance of spatial information but also for the retention of other stimulus features, including visual motion (Durstewitz, Seamans, & Sejnowski, 2000; Machens, Romo, & Brody, 2005; Engel & Wang, 2011), giving rise to the widely accepted notion that individual prefrontal cortex neurons perform sensory maintenance by means of persistent stimulus selective activity.

It is noteworthy that the transient nature of stimulus-selective delay activity in the PFC is not unique to our studies because similar transient delay activity is also detectible in the other data sets (Wallis & Miller, 2003; Shafi et al., 2007; Jun et al., 2010; Cromer, Roy, Buschman, & Miller, 2011). Thus, it is likely that the persistent delay activity is characteristic of spatial tasks in which monkeys saccade to previously cued locations. Wise and his colleagues pointed out that the design of this task makes it difficult to determine whether this activity represents maintenance or attentional selection (Lebedev, Messinger, Kralik, & Wise, 2004). They addressed this question in a study that allowed them to assess the two possibilities and concluded that attentional selection is a more likely mechanism underlying delay activity of many prefrontal neurons during such tasks. This problem does not apply to tasks involving memory-guided comparisons used in our studies. Thus, the question whether short-term maintenance of spatial locations and stimulus features are represented by similar neuronal code will require the use of matched similarly structured behavioral tasks.

Although we did not observe sustained representation of the preceding stimulus in individual neurons, across the population of recorded neurons the information about S1 direction was indeed present throughout the delay (Figure 4(c)), particularly in putative pyramidal cells. Overall, the transient and inconsistent nature of stimulus selective

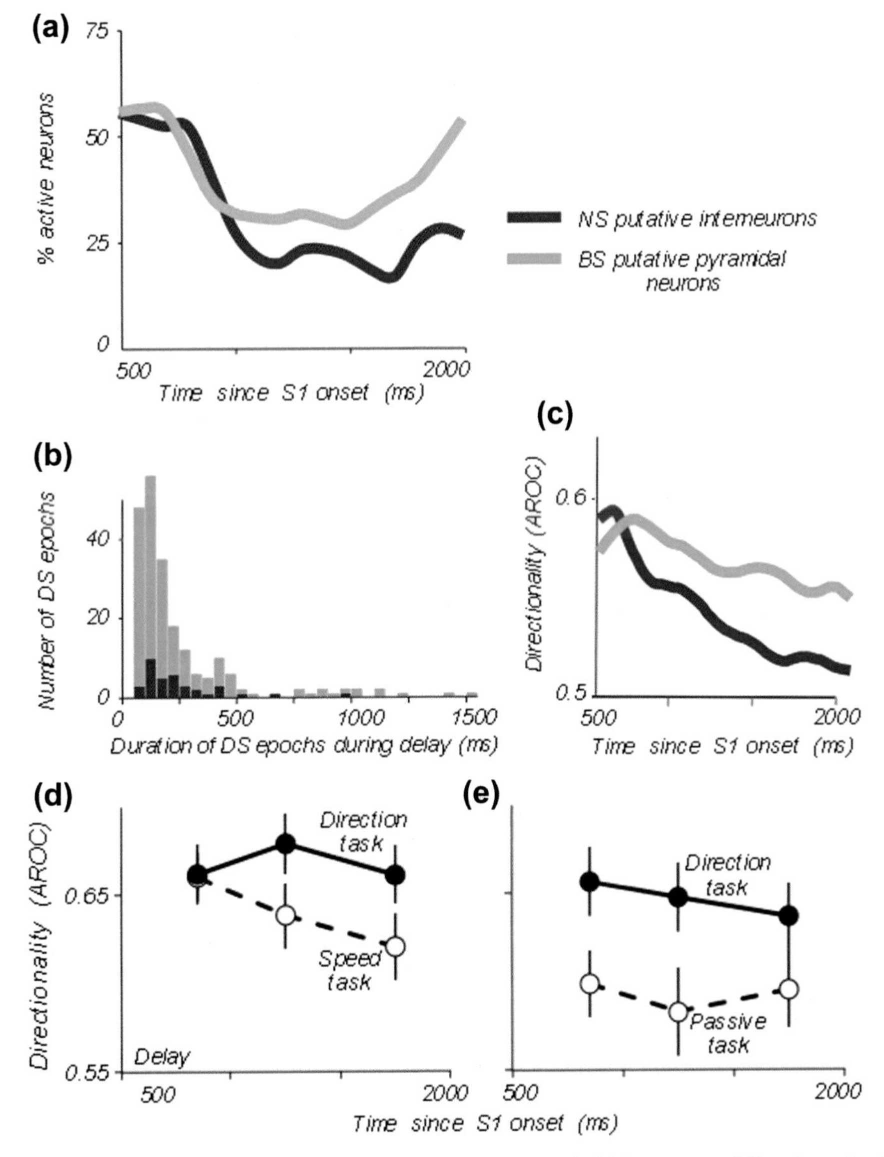

FIGURE 4 (a) Incidence of neurons active during the delay (NS, n=35; BS, n=124). (b) Durations of direction selective (DS) epochs encountered during the delay. The short duration of DS epochs (~245ms) highlights the transient nature of activity reflecting the delay selectivity for BS neurons was higher. (d, e) Comparison of DS delay activity during the direction and speed or passive fixation tasks. The data are shown for three consecutive 500-ms periods of delay. DS was reduced when direction was not relevant to the task, during speed comparisons (d), or during passive fixation. *From Hussar and Pasternak (2012).*

delay activity in individual cells suggests that the representation of the information about the preceding stimulus is likely to be distributed among many neurons. Although it is not yet known whether and how this type of activity is used by the brain to decode stimulus identity, some promising population-based decoding schemes have been proposed (Meyers et al., 2008; Barak et al., 2010).

COMPARISON-RELATED ACTIVITY

Because our task requires that the subjects report whether the directions of S1 and S2 are the same or different, their decision must be based on the stored representation of the preceding stimulus and during or after the appearance of S2. Thus, the neural code underlying the perceptual report required by our task must include the information about both the current and the previous stimulus. To examine this possibility, we compared responses to identical stimuli presented during S2 on trials where the preceding S1 moved in the same direction (S-trials) and on trials where S1 moved in a different direction (D-trials) (Figure 5(a)) (Hussar & Pasternak,

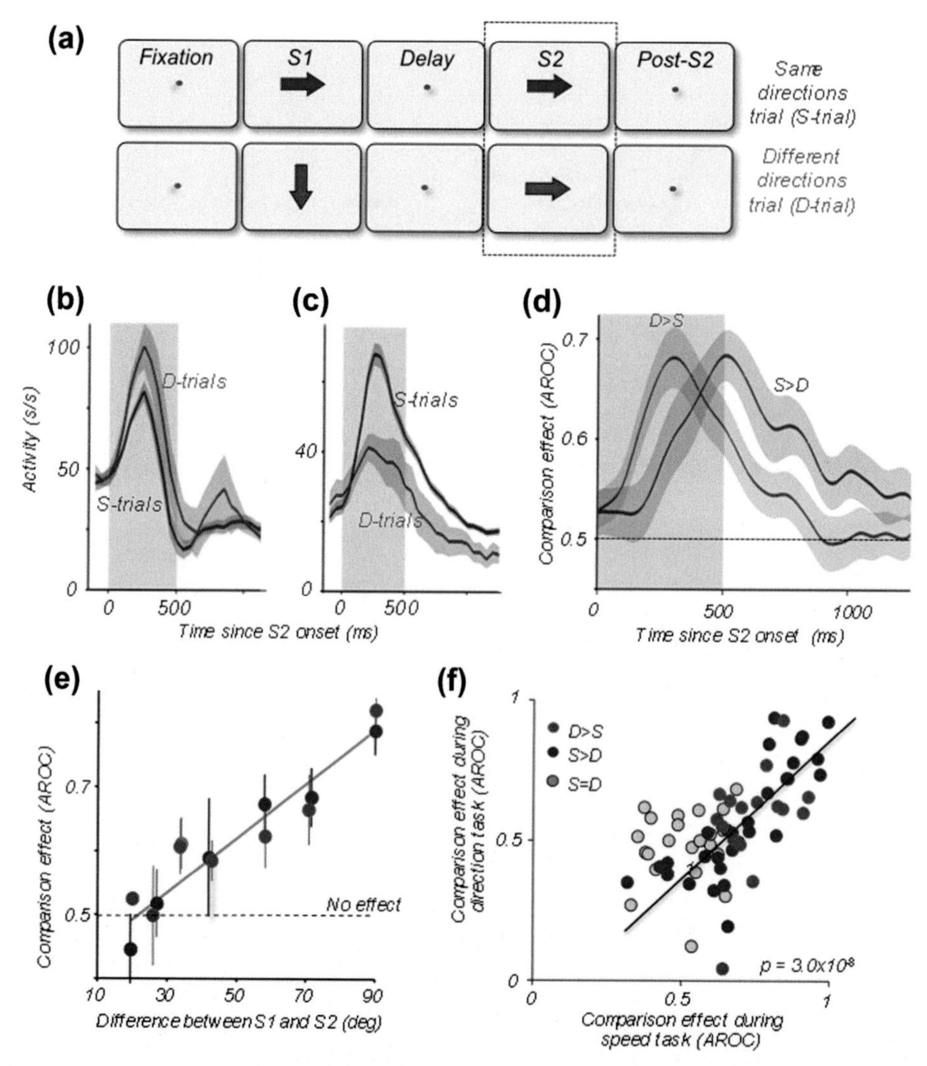

FIGURE 5 Comparison effects (CE) recorded during and after S2. (a) Diagram illustrating the two types of trials used to evaluate CE, S-trials (same), and D-trials (different). (b, c) Example responses during S2 of two neurons showing the two types of modulation by S1 direction. (d) Average CE for S>D cells and D>S cells. (e) Dependence of CE on the difference in direction between S1 and S2. Correlation between CE and direction difference was highly significant ($p < 7.5 \times 10^6$). (f) Correlation between comparison effects computed for each task. Neurons are color-coded for trial-type preference exhibited during the speed discrimination task. Trial type preferences of individual neurons during the two tasks were strongly correlated ($R^2 = 0.604$, Pearson's correlation, $p < 3.0 \times 10^{-8}$). *From Hussar and Pasternak (2012, 2013).*

2012). We found that responses of a substantial proportion of neurons during and after S2 was modulated by S1 direction, an effect indicative of these neurons having access to the stored information about S1. In some neurons, this modulation was in the form of stronger responses on S-trials, whereas in other neurons it consisted of stronger responses on D-trials (Figure 5(b)) and the two types of modulation was equally likely to be carried by NS and BS neurons. We used receiver operating characteristic (ROC) analysis to quantify the differences in responses between the two types of trials for the two groups of cells (for details see Hussar and Pasternak (2012)). The results of this analysis, performed only on trials with large (90°) differences in direction between S1 and S2, showed robust differences in activity between S- and D-trials. The two types of modulation had similar temporal profiles, persisting after the offset of S2, although in neurons preferring different trials (D>S) this effect emerged earlier. Because during our task we used both large and small differences between S1 and S2 directions, we were able to show that the magnitude of this modulation, termed *comparison effect*, decreased as the two directions became more similar (Figure 5(e)). This relationship paralleled behaviorally measured accuracy of direction discrimination, and the associated effect was detectible for pairs of directions that the monkeys were unable to discriminate reliably. In addition, comparison effects were nearly absent during the passive fixation task when the monkeys were not required to make the comparison between the two stimuli. These observations demonstrate that during the comparison phase of the task LPFC neurons carry signals reflecting the difference between directions being compared, the signals highly relevant to the perceptual decision required by the task.

We often recorded from the same neurons during both direction and speed comparison tasks (see Figure 1(b) and (d)). Thus, we were able to obtain data that allowed us to directly compare responses during S- and D-trials in the two tasks and determine whether a given neuron's trial preferences during the two tasks coincide. Such coincidence would be indicative of neurons exhibiting a general preference for "same" or "different" trials, independent of stimulus features being compared. The comparison of trial preferences during the two tasks revealed that more than 70% of LPFC neurons tended to prefer the same type of trial and their comparison effects were highly correlated (Figure 5(f)), suggesting that many LPFC neurons reflect an abstract rule ("same" or "different"), regardless of stimulus conditions or task structure.

DECISION-RELATED ACTIVITY

So far, the analysis here has examined the behavior of LPFC neurons in response to visual motion, whether and how its direction is represented during the delay period and revealed that responses during S2 reflect the comparison process between the current and the stored direction. In this section, we will focus on activity associated with behavioral report that occurred at the end of each trial and determine whether this activity recorded during or after the termination of S2 was predictive of that report. To address this question only activity recorded on S-trials in neurons with sufficient numbers of "same" and "different" reports (for details, see Hussar & Pasternak, 2012). Only trials containing identical stimuli during S1 and S2 (S-trials) were used to rule out sensory conditions affecting the animal's choices. We used ROC analysis to compare activity associated with each of the two reports and computed choice probability, often used to relate neuronal activity to perceptual decision (e.g., Britten, Newsome, Shadlen, Celebrini, and Movshon (1996); also see Hussar and Pasternak (2012, 2013)). We found that more than half of the PFC neurons showed significant choice probability (CP). This decision-related activity was of two types, one associated with "same" report and the other with "different" report, paralleling the two types of comparison effects associated with S- and D-trials. (Figure 6(a)). To explore the relationship between activity reflecting trial type (i.e., S>D or D>S) and activity associated with "same" or "different" reports about that trial, we identified neurons that carried both types of activity and on a cell-by-cell basis compared the time-course of comparison effects and of choice-related signals. We found that while the average time-course of the two types of signals was similar, comparison effects emerged ~150–190ms earlier than choice-related activity (Figure 6(b) and (c)), a pattern consistent with the possibility of comparison effects being used in the decision process. Furthermore, the two types of activity were strongly correlated (Figure 6(d)), with neurons preferring S-trials also showing activity predictive of "same" reports, whereas neurons preferring D-trials showing activity prior predictive of "different" reports. Although this correlation does not prove that the comparison signals provide the basis for the perceptual report, its strength and the consistency in the sign of the two types of signals strongly supports this possibility.

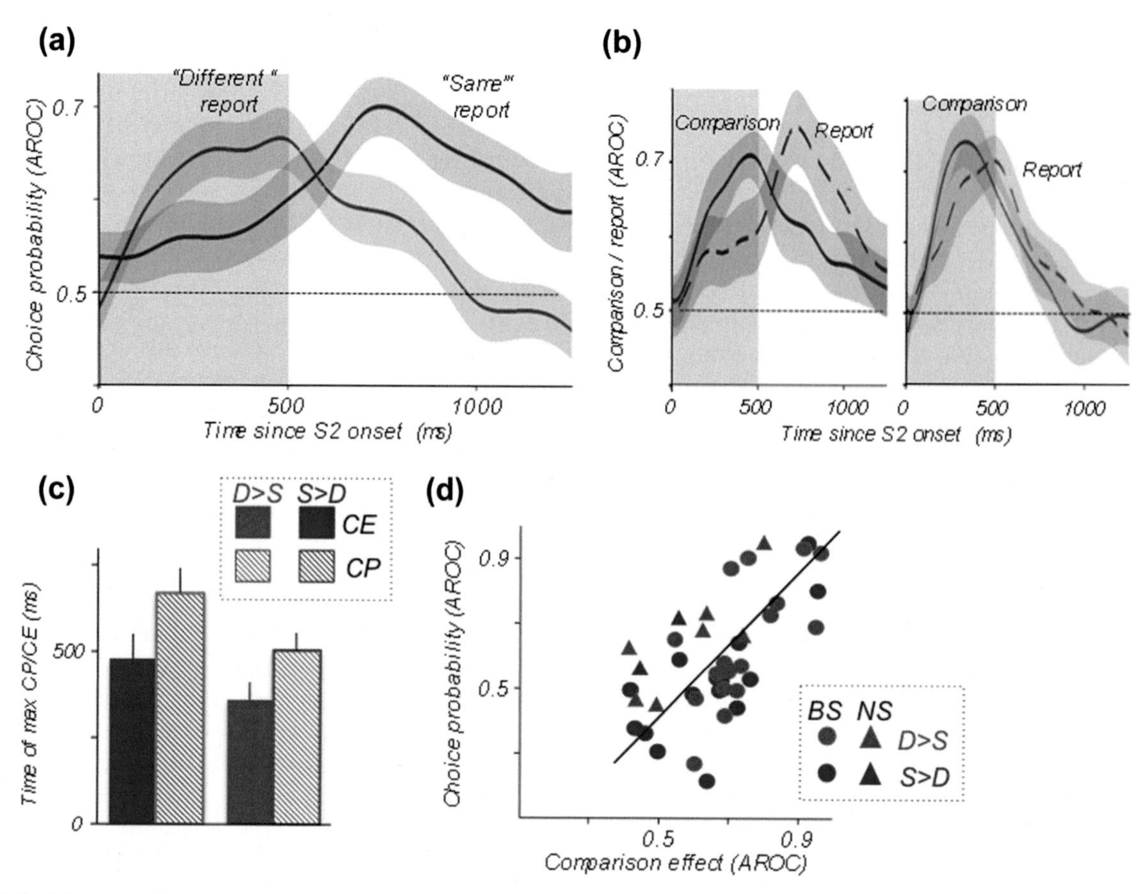

FIGURE 6 (a) Average choice probability (CP) of neurons signaling "different" and "same." (b) Average comparison effects (CE) and CP signals recorded from neurons carrying both types of signals. Note that CP activity lags behind CE. (c) Average times of maximal CP and CE. CE preceded CP by 145ms in D>S cells and by 190ms in S>D cells. (d) Correlation between CE and CP computed for individual S>D and D>S cells was highly significant during S2 ($p=1.3\times10^{-7}$) shown here, and later in the trial, 600–800ms after S2 onset (not shown; $p=1.2\times10^{-4}$). *From Hussar and Pasternak (2012).*

SUMMARY AND CONCLUSIONS

In this chapter, I focused on the way neurons in LPFC represent and use visual stimuli during memory-guided sensory comparison tasks. The work revealed that these neurons faithfully represent behaviorally relevant visual motion and actively participate in *all* aspects of memory-guided decisions. Thus, these neurons represented visual motion in a manner resembling responses of motion processing cortical neurons. However, this representation was flexible, with neurons adjusting their stimulus selectivity to task demands, the process achieved by a specialized neuronal network, with putative inhibitory interneurons playing a key role in enhancing stimulus selectivity, when necessary. During the memory delay, LPFC carried memory-related and anticipatory signals, and this activity was more prevalent in putative pyramidal neurons, a likely source of the top-down influences arising in the PFC. Although during the delay the information about the remembered stimulus was transient and inconsistent in individual neurons, it was rather robust when combined across cells and weakened when the stimuli it represented were not relevant to the task. Finally, LPFC neurons represent signals reflecting similarities and differences between the comparison stimuli and their activity predicts the monkey's decision. The behavior of LPFC neurons during the direction comparison task was strikingly similar to that recorded during a similar comparison task involving a different sensory dimension, stimulus speed (Hussar & Pasternak, 2013). This similarity suggests the implementation of common rules that govern the way LPFC neurons represent and use sensory information during memory-guided sensory comparisons. Overall, our results highlight access of LPFC to fundamental sensory dimensions and provide insights into the nature of top-down influences that the area is likely to provide to upstream neurons processing sensory information.

Acknowledgments

This work was supported National Institutes of Health grants: RO1 EY11749; P30 EY01319.

References

Anton-Erxleben, K., Stephan, V. M., & Treue, S. (2009). Attention reshapes center-surround receptive field structure in macaque cortical area MT. *Cerebral Cortex, 19*, 2466–2478.

Barak, O., Tsodyks, M., & Romo, R. (2010). Neuronal Population Coding of Parametric Working Memory. *Journal of Neuroscience, 30*, 9424–9430.

Barbas, H. (1988). Anatomic organization of basoventral and mediodorsal visual recipient prefrontal regions in the rhesus monkey. *Journal of Comparative Neurology, 276*, 313–342.

Britten, K. H., Newsome, W. T., Shadlen, M. N., Celebrini, S., & Movshon, J. A. (1996). A relationship between behavioral choice and the visual responses of neurons in macaque MT. *Visual Neuroscience, 13*, 87–100.

Compte, A., Brunel, N., Goldman-Rakic, P. S., & Wang, X. J. (2000). Synaptic mechanisms and network dynamics underlying spatial working memory in a cortical network model. *Cerebral Cortex, 10*, 910–923.

Connors, B. W., & Gutnick, M. J. (1990). Intrinsic firing patterns of diverse neocortical neurons. *Trends in Neurosciences, 13*, 99–104.

Cook, E. P., & Maunsell, J. H. R. (2004). Attentional modulation of motion integration of individual neurons in the middle temporal visual area. *Journal of Neuroscience, 24*, 7964–7977.

Cromer, J. A., Roy, J. E., Buschman, T. J., & Miller, E. K. (2011). Comparison of primate prefrontal and premotor cortex neuronal activity during visual categorization. *Journal of Cognitive Neuroscience*, 3355–3365.

DeAngelis, G. C., & Newsome, W. T. (1999). Organization of disparity-selective neurons in macaque area MT. *Journal of Neuroscience, 19*, 1398–1415.

Durstewitz, D., Seamans, J. K., & Sejnowski, T. J. (2000). Neurocomputational models of working memory. *Nature Neuroscience, 3*, 1184–1191.

Engel, T. A., & Wang, X. J. (2011). Same or different? A neural circuit mechanism of similarity based pattern-match decision making. *Journal of Neuroscience, 31*, 6982–6996.

Funahashi, S., Bruce, C. J., & Goldman-Rakic, P. S. (1989). Mnemonic coding of visual space in the monkey's lateral prefrontal cortex. *Journal of Neurophysiology, 61*, 331–349.

Goldman-Rakic, P. S. (1995). Cellular basis of working memory. *Neuron, 14*, 477–485.

Hussar, C. R., & Pasternak, T. (2009). Flexibility of sensory representations in prefrontal cortex depends on cell type. *Neuron, 64*, 730–743.

Hussar, C., & Pasternak, T. (2010). Trial-to-trial variability of the prefrontal neurons reveals the nature of their engagement in a motion discrimination task. *Proceedings of the National Academy of Sciences, 107*, 21842–21847.

Hussar, C. R., & Pasternak, T. (2012). Memory-guided sensory comparisons in the prefrontal cortex: contribution of putative pyramidal cells and interneurons. *The Journal of Neuroscience, 32*, 2747–2761.

Hussar, C. R., & Pasternak, T. (2013). Common rules guide comparisons of speed and direction of motion in the lateral prefrontal cortex. *The Journal of Neuroscience, 33*, 972–986.

Jun, J. K., Miller, P., Hernandez, A., Zainos, A., Lemus, L., Brody, C. D., et al. (2010). Heterogenous population coding of a short-term memory and decision task. *Journal of Neuroscience, 30*, 916–929.

Kim, J. N., & Shadlen, M. N. (1999). Neural correlates of a decision in the lateral prefrontal cortex of the macaque. *Nature Neuroscience, 2*, 176–185.

Lebedev, M. A., Messinger, A., Kralik, J. D., & Wise, S. P. (2004). Representation of attended versus remembered locations in prefrontal cortex. *PLoS Biology, 2*, 1919–1935.

Lui, L. L., & Pasternak, T. (2011). Representation of comparison signals in cortical area MT during a delayed direction discrimination task. *Journal of Neurophysiology, 106*, 1260–1273.

Machens, C. K., Romo, R., & Brody, C. D. (2005). Flexible control of mutual inhibition: a neural model of two-interval discrimination. *Science, 307*, 1121–1124.

Markram, H., Toledo-Rodriguez, M., Wang, Y., Gupta, A., Silberberg, G., & Wu, C. (2004). Interneurons of the neocortical inhibitory system. *Nature Reviews Neuroscience, 5*, 793–807.

Martinez-Trujillo, J. C., & Treue, S. (2004). Feature-based attention increases the selectivity of population responses in primate visual cortex. *Current Biology, 14*, 744–751.

Maunsell, J. H., & Van Essen, D. C. (1983). Functional properties of neurons in middle temporal visual area of the macaque monkey. I. Selectivity for stimulus direction, speed, and orientation. *Journal of Neurophysiology, 49*, 1127–1147.

Meyers, E. M., Freedman, D. J., Kreiman, G., Miller, E. K., & Poggio, T. (2008). Dynamic Population Coding of Category Information in Inferior Temporal and Prefrontal Cortex. *Journal of Neurophysiology, 100*, 1407–1419.

Miller, E. K., & Cohen, J. D. (2001). An integrative theory of prefrontal cortex function. *Annual Review of Neuroscience, 24*, 167–202.

Mitchell, J. F., Sundberg, K. A., & Reynolds, J. H. (2007). Differential attention-dependent response modulation across cell classes in macaque visual area V4. *Neuron, 55*, 131–141.

Nowak, L. G., Azouz, R., Sanchez-Vives, M. V., Gray, C. M., & McCormick, D. A. (2003). Electrophysiological classes of cat primary visual cortical neurons in vivo as revealed by quantitative analyses. *Journal of Neurophysiology, 89*, 1541–1566.

Pasternak, T., & Greenlee, M. (2005). Working memory in primate sensory systems. *Nature Reviews Neuroscience, 6*, 97–107.

Petrides, M., & Pandya, D. N. (2006). Efferent association pathways originating in the caudal prefrontal cortex in the macaque monkey. *The Journal of Comparative Neurology, 498*, 227–251.

Rockland, K. S. (1997). Elements of cortical architecture. Hierarchy revisited. *Cerebral Cortex, 12*, 243–293.

Shafi, M., Zhou, Y., Quintana, J., Chow, C., Fuster, J., & Bodner, M. (2007). Variability in neuronal activity in primate cortex during working memory tasks. *Neuroscience, 146*, 1082–1108.

Treue, S., & Maunsell, J. H. (1996). Attentional modulation of visual motion processing in cortical areas MT and MST. *Nature, 382*, 539–541.

Treue, S., & Maunsell, J. H. R. (1999). Effects of attention on the processing of motion in macaque middle temporal and medial superior temporal visual cortical areas. *Journal of Neuroscience, 19*, 7591–7602.

Wallis, J. D., & Miller, E. K. (2003). Neuronal activity in primate lateral and orbital prefrontal cortex during performance of a reward preference task. *European Journal of Neuroscience, 18*, 2069–2081.

Zaksas, D., & Pasternak, T. (2006). Directional signals in the prefrontal cortex and in area MT during a working memory for visual motion task. *Journal of Neuroscience, 26*, 11726–11742.

CHAPTER

12

Working Memory Representations of Visual Motion along the Primate Dorsal Visual Pathway

Diego Mendoza-Halliday, Santiago Torres, Julio Martinez-Trujillo
Cognitive Neurophysiology Laboratory, Department of Physiology, McGill University, Montreal, QC, Canada

THE SEARCH FOR THE NEURAL CORRELATES OF WORKING MEMORY IN THE PRIMATE BRAIN

Working memory (WM) can be defined as the temporary storage and manipulation of stimuli representations that are no longer available to the senses. It allows us to momentarily use knowledge from past experiences to solve a current task. WM is a form of short-term memory that lasts longer than sensory memories (i.e., iconic memory), but extinguishes usually within seconds or minutes or may be stored in long-term memory (diagram in Figure 1(a)). WM is essential to many other cognitive processes that define human intelligence, such as top-down attention, abstract thinking, decision-making, and action planning (Baddeley, 2012). Importantly, WM is one of the most frequently impaired functions in patients with mental disease, such as schizophrenia (Goldman-Rakic, 1994). Thus, studies regarding the neural substrates of WM are a priority in cognitive neuroscience and associated disciplines, such as psychology and psychiatry.

A milestone in the search for the neural substrates of WM was reported by Fuster and Alexander (1971), who found that neurons in the macaque monkey prefrontal cortex increased and sustained firing rates (i.e., the frequency of action potentials fired in a second) when the animal holds the representation of a visual stimulus or object in WM. This sustained firing, selective for remembered locations, objects, or features, has since been known as the neural correlate of WM. It is important to distinguish these dynamic correlates from those characterizing the long-lasting changes in the strength of synapses underlying the storage of long-term memories (Kandel, Schwartz, Jessell, Siegelbaum, & Hudspeth, 2013). After the finding by Fuster and Alexander, several groups started to search for the neural correlates of WM in different areas of the primate brain. The large majority of these studies used the macaque monkey as a model system because of how similar this animal's brain is to that of a human's. In this chapter, we will focus on studies of WM on areas along the macaque monkey visual dorsal pathway, where neurons are specialized for the processing of motion attributes (Kandel et al., 2013).

Several studies in macaque monkeys have reported that, during visual WM tasks, neurons in high-order association areas, such as the lateral prefrontal (LPFC) (Funahashi, Bruce, & Goldman-Rakic, 1989; Miller, Erickson, & Desimone, 1996), posterior parietal (Andersen, Essick, & Siegel, 1987), and inferotemporal (Miller, Li, & Desimone, 1991) cortices, show elevated and sustained firing rates that encode memorized information (Figure 1(b)). Moreover, it has been proposed that because neurons in high-order association areas lack fine selectivity for single visual features (e.g., direction of motion, color, orientation), WM representations of such features must be encoded by feature-selective neurons in early visual cortical areas (Harrison & Tong, 2009; Pasternak & Greenlee, 2005). However, reports of WM neural correlates in early sensory areas are scarce. Only one study found weak WM selective activity in area V1 (Super, Spekreijse, & Lamme, 2001) and another in MT (Zaksas & Pasternak, 2006). However, the former study only found selective sustained WM activity in V1 when the receptive fields of the neurons under scrutiny

Mechanisms of Sensory Working Memory
http://dx.doi.org/10.1016/B978-0-12-811042-3.00012-8

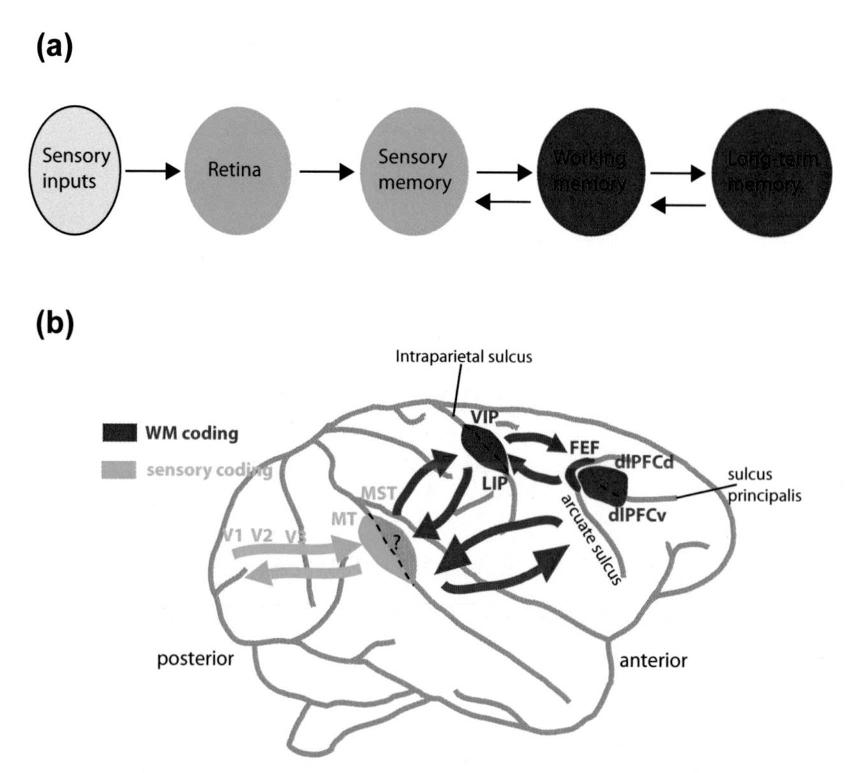

FIGURE 1 (a) Different types of memory and hypothetical interactions in a serial model. (b) Hierarchical organization of brain areas along the dorsal visual pathway. Areas where WM coding has been clearly documented (red); areas where WM coding has not been documented or where the existing evidence is contradictory (green).

were stimulated by a background object. The latter study, in area MT, reported WM-related activity following the presentation of the stimulus to be remembered inside the neuron's receptive fields; however, this activity was short and transient. Thus, it is very likely that, in early visual areas, neuronal firing representing memorized features or locations cannot be robustly sustained in the absence of sensory inputs. Evidence supporting this hypothesis has been reported in several electrophysiological studies in monkeys (Bisley, Zaksas, Droll, & Pasternak, 2004; Ferrera, Rudolph, & Maunsell, 1994; Lee, Simpson, Logothetis, & Rainer, 2005; Super et al., 2001).

A puzzling finding from functional magnetic resonance imaging studies in humans is that, during WM tasks, blood oxygenation level–dependent (BOLD) signals in early visual areas remain at baseline levels, yet the memorized features can be decoded from these signals using pattern classification analysis techniques (Harrison & Tong, 2009; Riggall & Postle, 2012; Sneve, Alnaes, Endestad, Greenlee, & Magnussen, 2012). However, because increases in BOLD signals may not directly reflect the firing rate of individual neurons, these results need to be corroborated by single cell studies in nonhuman primates.

Hence, so far, it remains controversial whether the neural correlates of WM coexist with representations of visual inputs in early visual areas or, alternatively, whether such correlates emerge in association areas further downstream along the visual pathways, whereas early visual areas exclusively encode representations of stimuli available to the eyes. In the next sections, we will summarize the results of a series of experiments aimed at addressing this and other related issues.

NEURAL CORRELATES OF WM FOR MOTION IN THE DORSAL VISUAL PATHWAY

Visual processing in the primate brain takes place along two main pathways, the dorsal and the ventral. In the dorsal pathway, neurons are selective for the direction and speed of moving stimuli. As a general rule, areas along the pathway are serially connected through numerous feed-forward and feedback projections (Figure 1(b); Kandel et al., 2013). For example, area middle temporal (MT) projects and receives signals from area medial superior temporal (MST) (Born & Bradley, 2005). This knowledge of visual processing in the primate brain becomes very useful when searching for the neural correlates of WM. For example, collecting single neuron activity (the firing rate or frequency

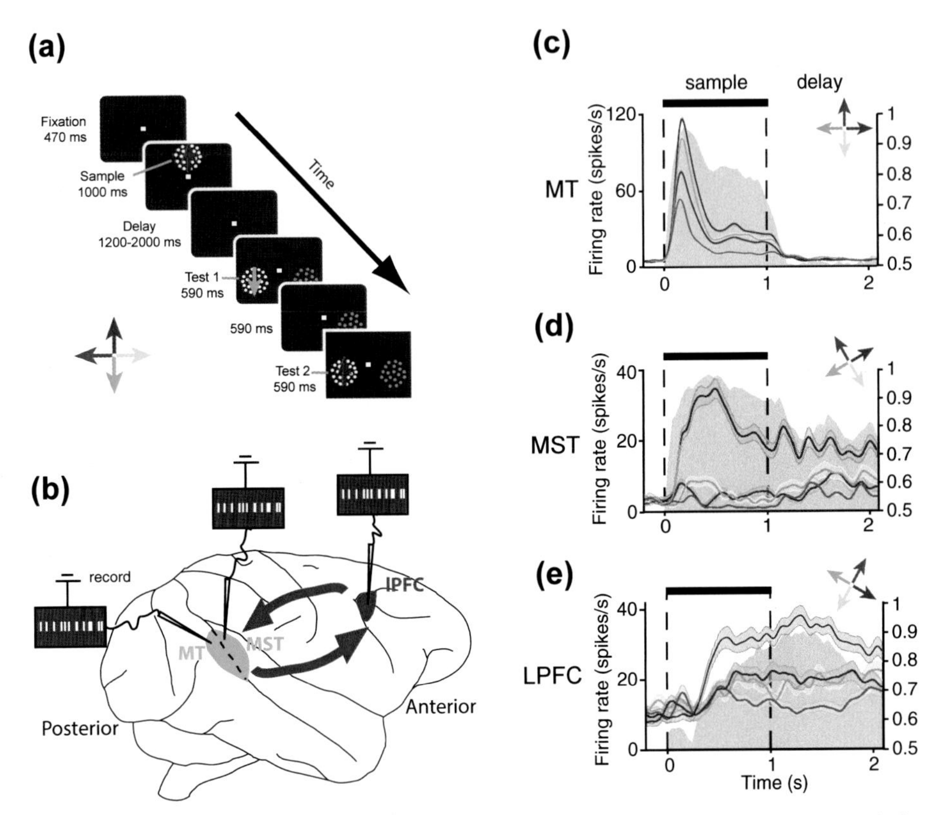

FIGURE 2 (a) Behavioral task. The motion directions of all stimuli were chosen among four orthogonal directions (color-coded arrows) aligned to the neuron's preferred direction (red). The task timing is indicated. (b) Cartoon of the macaque brain illustrating the areas where recordings were conducted. The inserts represent the recording electrodes and hypothetical trains of action potentials. (c), (d) Firing rate of example neurons in MT, MST, and lPFC. Mean firing rate (±standard error) over time, across trials for each of the four sample directions (color-coded arrows) for single neuron examples in MT (c), MST (d), and lPFC (e). Each neuron's preferred direction is shown in red. The gray area corresponds to the auROC over time (right axis label).

of action potentials or spikes that the cell fires in a given time unit, usually 1 s) from the different areas illustrated in Figure 1(b), during a WM task, may help in determining where the neural correlates of WM emerge along the visual pathways. The experiment illustrated in Figure 2(a) follows this line of reasoning.

Two macaque monkeys were trained to perform a delayed match-to-sample task. The animals viewed a sample random dot pattern moving in one of four orthogonal directions for 1000 ms. After a variable delay period (between 1200 and 2000 ms), two test stimuli were sequentially presented for 590 ms each, separated by a 590-ms interval. The animals had to indicate which one of the two test stimuli had the same motion direction as the sample in order to obtain a juice reward. A behaviorally irrelevant stimulus (0% coherent motion) was presented in the opposite hemifield simultaneously with each of the two test stimuli. Because the position of the test and irrelevant stimuli were randomized from trial to trial, the animals could not predict the location of the test. Importantly, to perform the task, the animals were required to maintain a representation of the sample direction in WM throughout the delay period.

The responses of neurons in MT, MST, and lPFC were recorded across many experimental sessions (Figure 2(b)). In all sessions, the performance of both animals was above chance, indicating that they correctly learned and performed the task (chance performance: 50%; mean performance: 79% for monkey M, 76% for monkey S). The responses of prototypical example neurons in the three areas are shown in Figure 2(c)–(e). In the three graphs, the abscissa represents time from sample onset, the left ordinate represents the instantaneous firing rate in spikes per second, and the right ordinate represents the area under the receiver operating characteristic curve (auROC) between the responses to the two most discriminable directions. The ROC analysis produces a measurement of how discriminable or separable two firing rate distributions are. An ROC value of 0.5 means that the distributions fully overlap, whereas values of one or zero mean that they are fully separated. In our data analysis, auROC was taken as a measurement of a neuron's selectivity for a motion direction.

As one can easily appreciate in these graphs, there are noticeable differences in firing rates and auROC values across the different time periods among the different areas. The MT neuron was sensory selective but not memory

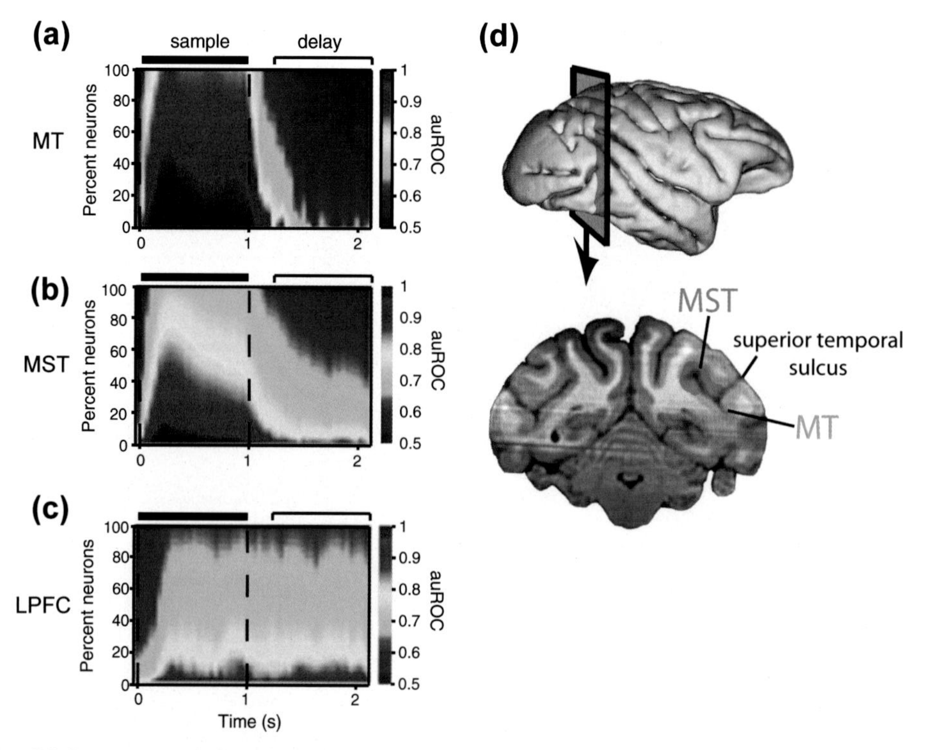

FIGURE 3 Average auROC across recorded MT (a), MST (b), and lPFC (c) neurons over time as a function of the percentage of averaged neurons (organized from maximum to minimum auROC in each time bin). Neurons were included in this analysis if one of the sample/remembered directions gave a significantly different firing rate relative to the other directions in any of the analyzed periods (analysis of variance test). Black horizontal bar and dashed lines delineate the sample period; line segment shows delay period used for analysis; error bars denote standard error of the mean. (d) Reconstruction of one of the animal's brain showing the anatomical proximity between areas MT and MST in the superior temporal sulcus.

selective (Figure 2(c)). Surprisingly, the MST neuron showed both sensory selectivity and memory selectivity (Figure 2(d)). The lPFC neuron also showed sensory and memory selectivity (Figure 2(e)). A total of 109 neurons in MT (97%), 218 in MST (88%), and 86 in lPFC (32%) showed selectivity either during the sample (sensory) or memory period. From these neurons, 100% in MT, 93% in MST, and 70% in lPFC showed sensory selectivity, whereas 8% in MT, 36% in MST, and 55% in lPFC showed memory selectivity (Figure 3(d)). All percentages, except the percentage of memory-selective neurons in MT, were significantly higher than expected by chance (permutation tests, $p<.05$).

As previously mentioned, to further explore how well sensory-selective and/or memory-selective neurons discriminated between motion directions (discriminability), we used signal detection analysis. The auROC between the activity in trials with the sample moving in the neuron's preferred and least-preferred directions was computed across a sliding time window (right ordinate and gray area in Figure 2(c)–(e)) as a measure of how well the neuron distinguished between motion directions (Lennert & Martinez-Trujillo, 2011). Average discriminability across task periods and neurons in the three areas is displayed in Figure 3(a)–(c). The abscissa in these plots represents time from sample onset, and the ordinate measures the percentage of neurons from the recorded sample that show a given level of discriminability (see Figure 3 legend and Mendoza-Halliday, Torres, & Martinez-Trujillo, 2014). After the sample onset, discriminability increased in all three areas (see color scale), but did so more rapidly and reached higher values in MT, followed by MST, and then by lPFC. During the sample period, the discriminability strength (i.e., mean auROC) and duration (i.e., time that the auROC remained above chance), were higher in MT than MST (strength: unpaired t test, $p<.001$; duration, $p<.001$, t tests), and higher in MST than lPFC (strength, $p<0.001$; duration, $p<.01$, t tests). Thus, the motion direction of visual stimuli is strongly represented by neurons in area MT; such representations become weaker downstream in MST and the lPFC.

Immediately after the sample offset, discriminability in MT neurons quickly dropped and remained at chance throughout the delay period (Figure 3(a)). In contrast, discriminability of many MST and lPFC neurons remained high during the delay period (Figure 3(b)–(c)). Remarkably, discriminability in MST was as strong as in lPFC ($p=.13$) but lasted significantly longer ($p=.03$) persisting throughout the entire delay period in 22% of MST and

11% of lPFC memory-selective neurons. Thus, WM representations were more robust over time in MST than in lPFC.

Given the proximity and direct feedforward and feedback connectivity between MT and MST (Boussaoud, Ungerleider, & Desimone, 1990), it is somewhat surprising that neurons from both areas show such dramatic differences in the neural correlates of WM—absent in MT but robustly present in MST (Figure 3(d)). This indicates that along the dorsal visual pathway, WM representations emerge sharply as a de novo property of MST neurons. Moreover, MST neurons displayed heterogeneity of coding: some had sensory selectivity but not memory selectivity, others had memory selectivity but not sensory selectivity, and others had both types of selectivity. This indicates that information about present and past stimuli remains segregated in some neurons and coexists in others, and suggests that along the dorsal visual pathway, the transformation of sensory representations of motion direction into WM representations occurs within the MST circuitry. It remains to be determined whether such a transformation also involves lPFC, since this area's neurons also showed a similar heterogeneity of coding.

It is important to note that WM representations of motion direction were also encoded by neurons in the high-order association area lPFC. Thus, in addition to visual locations and complex visual objects (Miller et al., 1996), lPFC neurons also encode WM representations of simple visual features. This is consistent with observations of WM coding in lPFC for other simple sensory features such as visual motion speed gradients (Hussar & Pasternak, 2013) and vibrotactile stimuli (Hernandez et al., 2010). On the other hand, this seems at odds with results from functional imaging experiments showing the inability to decode the contents of WM recorded in lPFC using multivoxel pattern classification analysis of BOLD signals (Riggall & Postle, 2012). This apparent contradiction may be explained considering lPFC lacks the topographical organization of feature coding at a scale that can be detected with the relative coarse resolution of functional magnetic resonance imaging.

Why are WM signals present in MST and lPFC but not in MT? Based on known functional differences between these three areas, it may be that the characteristic small receptive fields and stable tuning for sensory features of neurons in early sensory areas (such as MT) limit them to play a primary role in encoding sensory inputs. In contrast, WM coding for stimulus features may be an exclusive property of association areas further downstream (such as MST and lPFC), where large and bilateral receptive fields provide position invariance to the representation of visual features and where multiple simple sensory features and/or modalities can be integrated into a unified representation of entire objects. For example, area MST integrates local motion signals from MT into patterns of optical flow. Thus, it may be advantageous for WM representations of such stimuli to be encoded in association areas rather than in early sensory areas, where representations of individual stimulus features remain segregated.

Interestingly, Zaksas and Pasternak (2006) used a task similar to ours and reported no persistent activity encoding WM representations of motion direction in MT. Instead, they observed brief and transient periods of weak motion direction coding during the delay period in individual MT neurons. These periods were mostly restricted to the first few hundred milliseconds after sample offset, rather than being homogenously distributed along the entire delay period. They hypothesized that these transient signals could be part of a mechanism to transfer information from MT to other areas with WM coding properties to "load" information in memory. Although this possibility cannot be fully discarded, our observation that motion direction representations in lPFC and MST are fully present during the first half of the sample presentation period strongly indicates that there is no need for additional transfer of information from MT to these areas after the sample offset.

One major problem with the information transfer hypothesis is that sensory neurons show residuals of the sensory response during the first few hundred milliseconds after stimulus offset, which may also account for Zaksas and Pasternak's observations. This residual activity may be due to a combination of two phenomena: first, it takes a brief period after stimulus offset for the response of stimulated sensory neurons to decay to baseline (Osborne, Bialek, & Lisberger, 2004; Priebe, Churchland, & Lisberger, 2002); second, rapid neural adaptation has been observed to follow brief stimulation (Glasser, Tsui, Pack, & Tadin, 2011; Van Wezel & Britten, 2002). Both phenomena are observed in MT even during tasks that do not require WM or during anesthesia (Glasser et al., 2011; Osborne et al., 2004; Priebe et al., 2002; Van Wezel & Britten, 2002). Thus, it is likely that directional signals observed after the sample offset, during the early periods of WM maintenance, are due to such confounding factors (Bisley et al., 2004).

REDUNDANCY OF WM REPRESENTATIONS IN AREAS MST AND LPFC

It is somewhat puzzling that the neural correlates of WM are present in several multiple association areas of the visual dorsal pathway. We have demonstrated memory selectivity for motion direction of similar strength in areas MST and lPFC. At least two alternative hypotheses may explain such a redundancy. First, it could be a

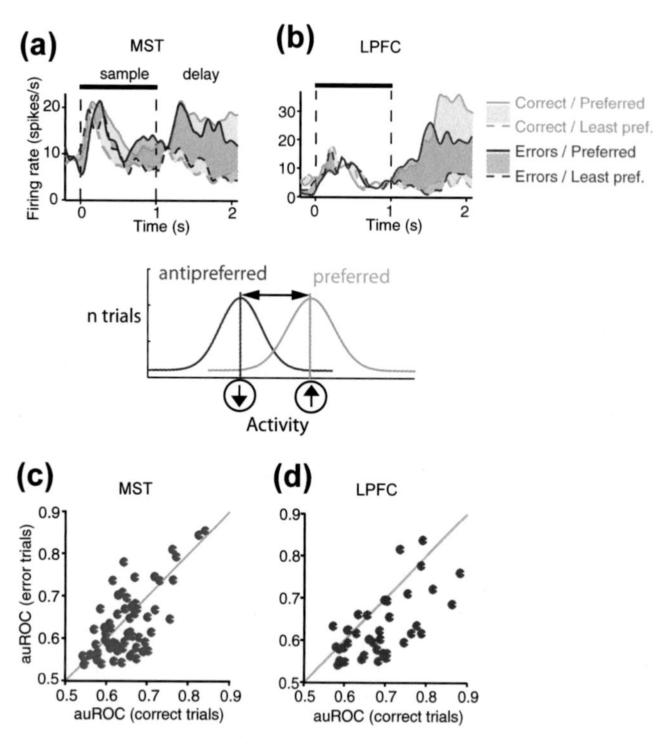

FIGURE 4 Mean firing rate of example MST (a) and lPFC (b) neurons over time in correct (green) and error (red) trials for preferred (solid lines) and least-preferred (dashed lines) sample directions. Colored areas show difference in activity between preferred and least-preferred sample trials. Black horizontal bar and dashed lines delineate sample period. auROC of memory-selective MST neurons ((c), blue dots) and lPFC neurons ((d), red dots) during the delay period in correct trials (horizontal axis) versus error trials (vertical axis). Unity line is shown in gray. The cartoon on top indicates the hypothetical activity of a memory selective neuron during the delay period in preferred and antipreferred sample trials.

consequence of the serial architecture of the visual system. However, in opposition to this hypothesis, MT and MST are also serially connected, but WM representations are only present in the latter area and not in the former. Second, although we have documented redundancy in the WM representations by MST and lPFC, there are essential differences between the signals isolated in each area with respect to other task variables we have not considered yet. In this case, one would anticipate that different areas with redundant representations play a different role in different task components.

One possible scenario is that lPFC neurons receive feedforward inputs from MST encoding the remembered visual motion direction and integrate them with other inputs that encode the cognitive and motor state of the animal to produce a coherent and appropriate behavioral response that is passed on and executed by motor cortices. This hypothesis predicts that trial-to-trial variations in delay activity should be more correlated with the animals' performance in lPFC than in MST neurons. For the example MST and lPFC neurons in Figure 4(a)–(b), delay activity in trials with the sample in the neurons' preferred directions was higher in correct than in error trials. Consequently, the difference in activity between preferred and least-preferred sample trials was larger in correct than in error trials. This result suggests that, in both areas, fluctuations in delay activity are linked to the animals' ability to produce a correct response or an error.

We computed ROC curves for the differences in response between the preferred and antipreferred directions for both correct and error trials. Across neurons and in both areas, the difference in auROC between correct and error trials ($\Delta auROC = auROC_{correct} - auROC_{incorrect}$) was significantly higher than 0, indicating that the ability of neurons to discriminate between the two directions (discriminability) during the delay period was reduced in error relative to correct trials (MST, $p = .02$, lPFC, $p < .001$). Interestingly, this reduction was significantly larger in lPFC than in MST (Figure 4(c) and (d), $p = .038$). Thus, the monkeys' performance in the WM task was correlated with how strongly MST and lPFC neurons encoded the memorized direction, and this effect was greater in lPFC than in MST.

One question derived from this result is whether one can better predict whether the animals would make an error or score a correct response during the WM task from the delay activity in area lPFC relative to MST. We compared the distributions of firing rates during the delay period between correct and error trials using an ROC-based choice

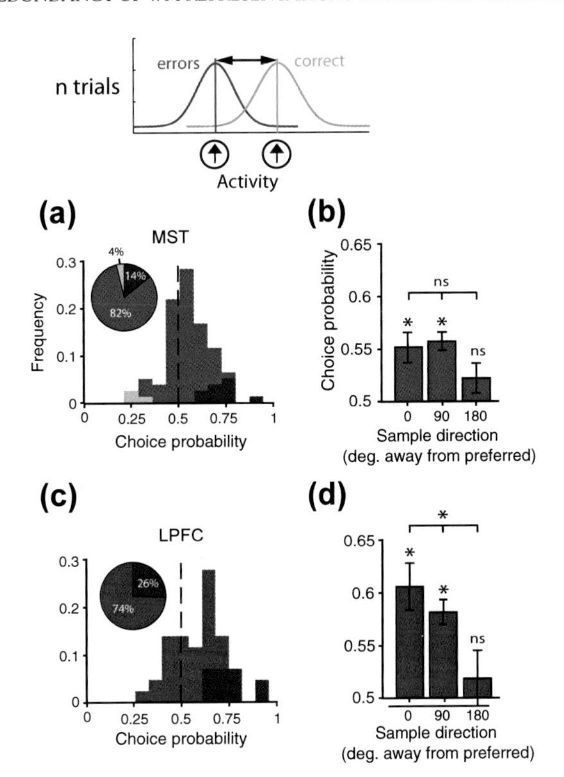

FIGURE 5 Frequency histogram of delay period CP among memory-selective MST (a) and lPFC (c) neurons in preferred-sample trials. Vertical black dashed lines show chance level of CP. In the histograms and pie charts, color tones represent neurons with choice probability significantly above (darker tones), significantly below (lighter tones), and not significantly different (middle tones) from that expected by chance. Mean CP among memory-selective MST (b) and lPFC (d) neurons as a function of the sample direction with respect to each neuron's preferred direction. Error bars denote standard error of the mean; *$p<.05$; ns, nonsignificant. The cartoon on top illustrates a hypothetical distribution of responses for a memory-selective neuron with preferred direction displayed as the sample in correct (green) and error (red) trials during the delay period.

probability analysis (CP) (Figure 5(a); Britten, Newsome, Shadlen, Celebrini, & Movshon, 1996). This was independently done for trials with the sample moving in the neurons' preferred, least-preferred, and intermediate directions. CP was significantly higher than expected by chance in 14% of MST neurons and in 26% of lPFC neurons, whereas it was significantly lower in 4% of MST neurons and in 0% of lPFC neurons (Figure 5(b) and (c); permutation test, $p<.05$). Among all neurons, mean CP across the delay period in preferred-sample trials was significantly higher than chance in both areas ($p<.001$). Moreover, mean CP was significantly higher in lPFC than in MST ($p=.02$). This confirmed that in both areas, the delay activity of neurons correlated with the animals' performance of the WM task, and that it did so more strongly in lPFC than MST.

Interestingly, although the CP for the preferred and orthogonal directions during the delay period was higher than chance in both areas (stars on top of the bars in Figure 5(d) and (e)), mean CP decreased significantly as a function of the difference between the neuron's preferred direction and the memorized direction but only in lPFC (see lines and symbols on top of the bars illustrating the comparisons in Figure 5(d) and (e); repeated-measures analysis of variance; lPFC: $F=4.78$, $p=.01$; MST: $F=2.41$, $p=.09$). Thus, in the lPFC, behavioral decisions are not equally influenced by the activity of the entire neuronal population. This strongly suggests that the contribution of each lPFC neuron to the performance in the WM task depends on the similarity between the neuron's preferred feature (motion direction) and the memorized feature.

Coming back to our observation of redundancy in the coding of WM representations of motion direction in areas MST and lPFC, it is also known that such a redundancy also occurs between parietal and prefrontal cortices during WM for visuospatial locations (Chafee & Goldman-Rakic, 1998) and visual feature categories (Swaminathan & Freedman, 2012) and between inferotemporal and prefrontal cortices during WM for complex objects (Miller et al., 1996). Such a redundancy likely results from the coordinated activity of a distributed and vastly interconnected network of brain areas, each of them playing a different but complementary role. For example, evidence of different roles of the lateral intraparietal area and the lPFC in the filtering of distracting information during WM has been reported (Suzuki & Gottlieb, 2013). Delay activity in lPFC neurons is more robust to distractors than in lateral intraparietal. Similarly, it is possible that activity in MST would be more sensitive to distractor interference than in lPFC.

Despite the similarity between the activity profiles of several MST and lPFC neurons, there were also several differences. WM representations, on average, lasted longer in MST than in lPFC. This suggests that coding is more robust over time in MST, and challenges the view of lPFC as the major contributor to WM maintenance (Miller et al., 1996). Note how this effect differs as compared with other cognitive variables, such as attention, where intensity tends to become stronger as one moves downstream along the visual pathway from sensory to higher-order areas (Buffalo, Fries, Landman, Liang, & Desimone, 2010; Lennert & Martinez-Trujillo, 2011). Interestingly, WM representation strength in lPFC is more predictive of task performance than in MST. This is consistent with the notion that activity becomes more closely linked to behavior further downstream along the chain of sensory-motor processing (Hernandez et al., 2010). Thus, the main role of MST may be to maintain a robust representation of the sample direction, which can be "read out" by lPFC and integrated with behaviorally relevant signals related to reward value and the allocation of attention to produce a meaningful behavioral response. Supporting this hypothesis, lPFC neurons have recently been proposed to encode behavioral choice in a match-to-sample task (e.g., match or nonmatch) (Hussar & Pasternak, 2013).

THE AMPLITUDE OF LOCAL FIELD POTENTIAL OSCILLATIONS IN MT ENCODES THE MEMORIZED DIRECTION

Given the existence of feedback projections from MST and lPFC to MT (Boussaoud et al., 1990; Ninomiya, Sawamura, Inoue, & Takada, 2012), it seems surprising that the memory-selective spiking activity of MST and lPFC neurons do not produce firing rate increases in MT neurons. One possible explanation is that, during WM maintenance, feedback signals from MST and/or lPFC modulate synaptic activity in MT neurons but not strongly enough to cause increases in neuronal firing rates.

From this hypothesis, one may anticipate that the amplitude of local field potentials (LFPs), which reflect changes in local synaptic activity in area MT, encodes the memorized motion directions (Figure 6(a)). Indeed, for the example MT recording site in Figure 6(b), the power of several LFP frequencies was higher while the monkey memorized up-left motion (preferred direction, left panel) relative to when it memorized down-right (least-preferred direction, right panel) motion. For each frequency band, in each MT site, we measured the direction discriminability of the LFPs during the delay period by computing the auROC between the signal power in trials with preferred and least-preferred samples. In all bands, the percentage of sites for which the auROC was significantly higher than expected by chance ranged between 14% and 22% (Figure 6(c); permutation test, $p<.05$). These values were significantly higher than the percentage of false positives (population permutation test, $p<.05$). The mean auROC among significant sites ranged between 0.64 and 0.67 among frequency bands, and all of these values were greater than those expected by chance (permutation test, $p<.05$). These results indicate that the amplitude of LFP oscillations in MT does reflect the memorized directions. In areas MST and lPFC, the amplitude of LFP oscillations also encoded the memorized motion direction in all frequency bands (Figure 6(d) and (e)).

Given that LFPs represent the overall synaptic activity around the recording electrode (Mitzdorf, 1985), which reflects local neuronal interactions as well as feedforward and feedback inputs and that feedforward visual inputs into MT neurons are absent during the delay period of the task, our results suggest that the WM-reflective LFP oscillations may be feedback signals to MT from areas such as MST or lPFC. Such signals would reach MT neurons and be sufficient to modulate the LFP amplitude, but not strong enough to produce increases in firing rate. Similar modulations of LFP without increases in firing rate have been previously documented (Bartolo et al., 2011; Lee et al., 2005).

Importantly, the presence of WM signals in the LFPs recorded in MT may explain the findings of several functional magnetic resonance imaging studies that have reported decoding of visual WM using multivoxel pattern classification analysis from BOLD signals recorded in early human visual cortical areas, including MT (Harrison & Tong, 2009; Riggall & Postle, 2012; Sneve et al., 2012). It has been demonstrated that the amplitude of BOLD signals correlates with LFP amplitude in the absence of changes in neuronal firing rates (Bartolo et al., 2011). Based on this, the most parsimonious explanation for the reported WM-related BOLD signals is that they reflect modulation of LFP signals in early visual areas rather than spiking activity (Harrison & Tong, 2009; Magri, Schridde, Murayama, Panzeri, & Logothetis, 2012). In addition, when drawing conclusions about physiological mechanisms from BOLD measurements, one must take into account that LFP signals may be dissociated from spiking activity (Bartolo et al., 2011) and that spikes and LFPs may have different physiological meanings. Although spikes are the main vehicles of information transmission over long-range connections in the brain, LFPs reflect synaptic activity within a local area, which does not always result in spikes or in information transmission.

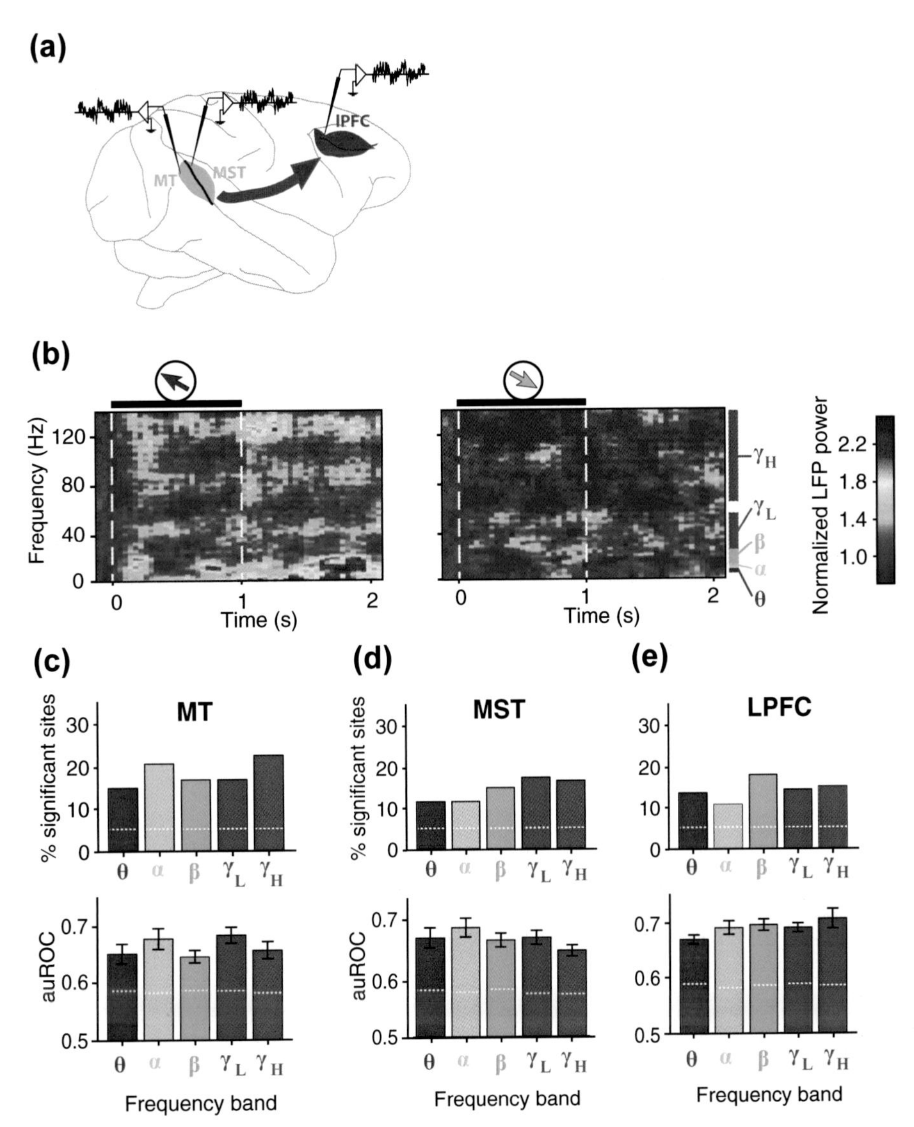

FIGURE 6 (a) Mean normalized LFP spectrogram of an example recording site in MT during trials with preferred (left) and least-preferred (right) sample directions. Black horizontal bars and dashed lines delineate sample period. The range of each frequency band is color-coded: (theta θ=4–8 Hz, alpha α=8–12 Hz, beta β=12–25 Hz, low gamma [γ_L]=25–55 Hz, high gamma [γ_H]=65–135 Hz[26]). (b) Percentage of LFP sites in MT, for each frequency band, during which the LFP power auROC in the delay period was significantly higher than expected by chance (top), and the mean auROC (±standard error) among these selective sites (bottom). (c) and (d) show similar data as (b), but for MST and lPFC, respectively.

It is important to distinguish between feature-based WM signals, described in our study, and the effects of feature-based attention reported in primate visual areas such as MT (McAdams & Maunsell, 1999). Although attention is generally defined as the selection and modulation of sensory inputs, WM is defined as the active maintenance of information in the absence of sensory inputs. These two phenomena may coexist and/or interact in certain circumstances such as visual search, where the searched feature needs to be actively maintained in WM (attentional template). However, feature-based attention can also be fully operational without WM requirements, such as when attending to a target stimulus in the presence of distractors (Treue and Martinez-Trujillo, 1999).

CONCLUSIONS

Current models of WM networks propose that recurrent excitatory connections, characteristic of the cortical architecture of high-order association areas such as lPFC, underlie sustained neuronal firing in the absence of sensory input and therefore support WM maintenance (Ardid, Wang, & Compte, 2007; Goldman-Rakic, 1995). A modeling study has shown that by balancing the strength and offset of recurrent excitatory and inhibitory inputs into a cortical neuron, one can produce delay activity that represents WM (Lim & Goldman, 2013). Our results suggest that such mechanisms may operate not only in high-order association areas but also as early as in the multisensory association area MST, immediately downstream from early visual cortical areas. Rather, the cortical architecture and connectivity of visual areas, such as MT, may primarily act to accurately represent and calibrate the incoming retinal input. The presence of strong feedforward inputs from upstream areas, relatively weak feedback inputs from downstream areas, and possibly stronger lateral inhibitory interactions between neurons (Haider, Hausser, & Carandini, 2013) may explain why sensory areas' spiking activities increase vigorously in response to retinal inputs but rapidly decay in their absence.

One possible scenario is that MST is able to intrinsically maintain WM representations independent of downstream areas, such as lPFC, because it may share a similar cortical architecture. Alternatively, it may be that such cortical architecture is exclusive to such downstream areas, but that sustained activity in MST arises as a consequence of feedback from these areas. All in all, that visual representations in MST neurons arise early during the sample presentation, even before arising in lPFC and persist throughout the delay period, suggests that WM representations in MST are intrinsically built rather than inherited from lPFC. A third possible scenario is that WM representations are actively maintained in MST and are transferred from there to other areas, such as lPFC, through feedforward inputs.

Whichever the case may be, our results indicate that the properties of the cortical architecture in MST—be its local circuitry, its connectivity with downstream areas, or both—that allow it to encode WM representations are likely absent immediately upstream in MT. This implies that a transition in cortical architecture occurs between areas MT and MST. Such a transition may also be present at similar processing stages along the ventral visual pathway (Miller et al., 1996) or other sensory processing streams (Hernandez et al., 2010). Conversely, it may be unique to the dorsal visual pathway network architecture and may not have a clearly defined homolog in the ventral visual pathway. We propose that such a boundary may be an important mechanism for the brain to distinguish representations of current sensory experiences from those imagined or memorized, a function that is impaired in schizophrenia and other hallucinatory mental disorders.

Acknowledgments

This study was supported by grants awarded to J.C.M-T from the Canadian Institutes of Health Research (CIHR), the Canada Research Chairs Program (CRC), and the EJLB Foundation. We thank M. Schneiderman for assistance with electrophysiological recordings. We thank W. Kucharski and S. Nuara for technical assistance.

References

Andersen, R. A., Essick, G. K., & Siegel, R. M. (1987). Neurons of area 7 activated by both visual stimuli and oculomotor behavior. *Experimental Brain Research Experimentelle Hirnforschung Experimentation Cerebrale, 67*, 316–322.

Ardid, S., Wang, X. J., & Compte, A. (2007). An integrated microcircuit model of attentional processing in the neocortex. *Journal of Neuroscience, 27*, 8486–8495.

Baddeley, A. (2012). Working memory: theories, models, and controversies. *Annual Review of Psychology, 63*, 1–29.

Bartolo, M. J., Gieselmann, M. A., Vuksanovic, V., Hunter, D., Sun, L., Chen, X., et al. (2011). Stimulus-induced dissociation of neuronal firing rates and local field potential gamma power and its relationship to the resonance blood oxygen level-dependent signal in macaque primary visual cortex. *European Journal of Neuroscience, 34*, 1857–1870.

Bisley, J. W., Zaksas, D., Droll, J. A., & Pasternak, T. (2004). Activity of neurons in cortical area MT during a memory for motion task. *Journal of Neurophysiology, 91*, 286–300.

Born, R. T., & Bradley, D. C. (2005). Structure and function of visual area MT. *Annual Review of Neuroscience, 28*, 157–189.

Boussaoud, D., Ungerleider, L. G., & Desimone, R. (1990). Pathways for motion analysis: cortical connections of the medial superior temporal and fundus of the superior temporal visual areas in the macaque. *The Journal of Comparative Neurology, 296*, 462–495.

Britten, K. H., Newsome, W. T., Shadlen, M. N., Celebrini, S., & Movshon, J. A. (1996). A relationship between behavioral choice and the visual responses of neurons in macaque MT. *Visual Neuroscience, 13*, 87–100.

Buffalo, E. A., Fries, P., Landman, R., Liang, H., & Desimone, R. (2010). A backward progression of attentional effects in the ventral stream. *Proceedings of the National Academy of Sciences of the United States of America, 107*, 361–365.

Chafee, M. V., & Goldman-Rakic, P. S. (1998). Matching patterns of activity in primate prefrontal area 8a and parietal area 7ip neurons during a spatial working memory task. *Journal of Neurophysiology, 79*, 2919–2940.

Ferrera, V. P., Rudolph, K. K., & Maunsell, J. H. (1994). Responses of neurons in the parietal and temporal visual pathways during a motion task. *Journal of Neuroscience, 14*, 6171–6186.

Funahashi, S., Bruce, C. J., & Goldman-Rakic, P. S. (1989). Mnemonic coding of visual space in the monkey's dorsolateral prefrontal cortex. *Journal of Neurophysiology, 61*, 331–349.

Fuster, J., & Alexander, G. (1971). Neuron activity related to short-term memory. *Science, 173*, 652–654.

Glasser, D. M., Tsui, J. M., Pack, C. C., & Tadin, D. (2011). Perceptual and neural consequences of rapid motion adaptation. *Proceedings of the National Academy of Sciences of the United States of America, 108*, E1080–E1088.

Goldman-Rakic, P. S. (1994). Working memory disfunction in schizophrenia. *Journal of Neuropsychiatry and Clinical Neuroscience, 6*, 348–357.

Goldman-Rakic, P. S. (1995). Cellular basis of working memory. *Neuron, 14*, 477–485.

Haider, B., Hausser, M., & Carandini, M. (2013). Inhibition dominates sensory responses in the awake cortex. *Nature, 493*, 97–100.

Harrison, S. A., & Tong, F. (2009). Decoding reveals the contents of visual working memory in early visual areas. *Nature, 458*, 632–635.

Hernandez, A., Nacher, V., Luna, R., Zainos, A., Lemus, L., Alvarez, M., et al. (2010). Decoding a perceptual decision process across cortex. *Neuron, 66*, 300–314.

Hussar, C. R., & Pasternak, T. (2013). Common rules guide comparisons of speed and direction of motion in the dorsolateral prefrontal cortex. *Journal of Neuroscience, 33*, 972–986.

Kandel, E. R., Schwartz, J. H., Jessell, T. M., Siegelbaum, S. A., & Hudspeth, A. J. (2013). *Principles of neural science* (5th ed.). McGraw Hill Companies.

Lee, H., Simpson, G. V., Logothetis, N. K., & Rainer, G. (2005). Phase locking of single neuron activity to theta oscillations during working memory in monkey extrastriate visual cortex. *Neuron, 45*, 147–156.

Lennert, T., & Martinez-Trujillo, J. (2011). Strength of response suppression to distracter stimuli determines attentional-filtering performance in primate prefrontal neurons. *Neuron, 70*, 141–152.

Lim, S., & Goldman, M. S. (2013). Balanced cortical microcircuitry for maintaining information in working memory. *Nature Neuroscience, 16*, 1306–1314.

Magri, C., Schridde, U., Murayama, Y., Panzeri, S., & Logothetis, N. K. (2012). The amplitude and timing of the BOLD signal reflects the relationship between local field potential power at different frequencies. *Journal of Neuroscience, 32*, 1395–1407.

McAdams, C. J., & Maunsell, J. H. (1999). Effects of attention on the reliability of individual neurons in monkey visual cortex. *Neuron, 23*, 765–773.

Mendoza-Halliday, D., Torres, S., & Martinez-Trujillo, J. C. (2014). Sharp emergence of feature-selective sustained activity along the dorsal visual pathway. *Nature Neuroscience, 17*(9), 1255–1262. http://dx.doi.org/10.1038/nn.3785.

Miller, E., Erickson, C., & Desimone, R. (1996). Neural mechanisms of visual working memory in prefrontal cortex of the macaque. *Journal of Neuroscience, 16*, 5154–5167.

Miller, E. K., Li, L., & Desimone, R. (1991). A neural mechanism for working and recognition memory in inferior temporal cortex. *Science, 254*, 1377–1379.

Mitzdorf, U. (1985). Current source-density method and application in cat cerebral cortex: investigation of evoked potentials and EEG phenomena. *Physiological Reviews, 65*, 37–100.

Ninomiya, T., Sawamura, H., Inoue, K., & Takada, M. (2012). Segregated pathways carrying frontally derived top-down signals to visual areas MT and V4 in macaques. *Journal of Neuroscience, 32*, 6851–6858.

Osborne, L. C., Bialek, W., & Lisberger, S. G. (2004). Time course of information about motion direction in visual area MT of macaque monkeys. *Journal of Neuroscience, 24*, 3210–3222.

Pasternak, T., & Greenlee, M. W. (2005). Working memory in primate sensory systems. *Nature Reviews Neuroscience, 6*, 97–107.

Priebe, N. J., Churchland, M. M., & Lisberger, S. G. (2002). Constraints on the source of short-term motion adaptation in macaque area MT. I. the role of input and intrinsic mechanisms. *Journal of Neurophysiology, 88*, 354–369.

Riggall, A. C., & Postle, B. R. (2012). The relationship between working memory storage and elevated activity as measured with functional magnetic resonance imaging. *Journal of Neuroscience, 32*, 12990–12998.

Sneve, M. H., Alnaes, D., Endestad, T., Greenlee, M. W., & Magnussen, S. (2012). Visual short-term memory: activity supporting encoding and maintenance in retinotopic visual cortex. *Neuroimage, 63*, 166–178.

Super, H., Spekreijse, H., & Lamme, V. A. (2001). A neural correlate of working memory in the monkey primary visual cortex. *Science, 293*, 120–124.

Suzuki, M., & Gottlieb, J. (2013). Distinct neural mechanisms of distractor suppression in the frontal and parietal lobe. *Nature Neuroscience, 16*, 98–104.

Swaminathan, S. K., & Freedman, D. J. (2012). Preferential encoding of visual categories in parietal cortex compared with prefrontal cortex. *Nature Neuroscience, 15*, 315–320.

Treue, S., & Martínez Trujillo, J. C. (1999). Feature-based attention influences motion processing gain in macaque visual cortex. *Nature, 399*(6736), 575–579.

Van Wezel, R. J., & Britten, K. H. (2002). Motion adaptation in area MT. *Journal of Neurophysiology, 88*, 3469–3476.

Zaksas, D., & Pasternak, T. (2006). Directional signals in the prefrontal cortex and in area MT during a working memory for visual motion task. *Journal of Neuroscience, 26*, 11726–11742.

CHAPTER

13

Neurophysiological Mechanisms of Working Memory: Cortical Specialization and Plasticity

Xue-Lian Qi, Xin Zhou, Christos Constantinidis

Department of Neurobiology & Anatomy, School of Medicine, Wake Forest University, Winston-Salem, NC, USA

INTRODUCTION

Single cortical neurons exhibit neural correlates of working memory in the form of sustained discharges, which persist even after the physical stimuli that elicited them are no longer present (Constantinidis, Franowicz, & Goldman-Rakic, 2001b; Funahashi, Bruce, & Goldman-Rakic, 1989; Fuster & Alexander, 1971). Despite long-standing research interest on the neural correlates and mechanisms of working memory generation, several unresolved questions remain, including the specialization of the prefrontal cortex and other cortical areas with respect to working memory; the plasticity of neuronal activation and response properties as a result of working memory training; and the natural changes that neuronal activity related to working memory undergoes during normal development from puberty to adulthood. These questions motivated the series of experiments presented here.

Neurophysiological experiments in nonhuman primates have revealed several cortical and subcortical areas that exhibit sustained discharges during working memory use, including the dorsolateral prefrontal (dlPFC) and posterior parietal cortex (PPC) (Constantinidis & Procyk, 2004). These two areas share many functional properties and exhibit similar patterns of activation, but recent evidence suggests that they may have a subtle specialization with respect to different aspects of working memory (Katsuki & Constantinidis, 2012). If such a specialization does exist, it is important to explore the nature of differences in activation between the two areas and to understand the underlying neural circuit differentiation that allows for this functional specialization.

A second important question concerns the effects of learning, experience, and training on working memory activity. Neurophysiological experiments make it possible to examine neuronal responses during working memory tasks before animals are trained as well as at varying stages of training. This can reveal the nature of information processing in the brain networks responsible for maintaining working memory as well as the effects of task training (Qi & Constantinidis, 2013). Therefore, it is possible to examine how different brain areas process and represent information about stimuli, what neuronal populations are recruited during the execution of the working memory task, and how information about a new task is incorporated in the activity of individual neurons.

A third question that we addressed has to do with the natural development of working memory abilities and concomitant neural changes. Working memory ability appears fairly early in life but is known to undergo a prolonged period of maturation (Fry & Hale, 2000; Gathercole, Pickering, Ambridge, & Wearing, 2004). This increased working memory ability parallels structural changes in the prefrontal cortex (Bunge, Dudukovic, Thomason, Vaidya, & Gabrieli, 2002; Giedd et al., 2006; Nagy, Westerberg, & Klingberg, 2004; Olesen, Nagy, Westerberg, & Klingberg, 2003); however, the underlying neurophysiological changes that mediate improvement in working memory activity have been virtually unexplored. Combined behavioral and neurophysiological experiments in monkeys can reveal the patterns of task performance that characterize the adolescent state as well as sustained activity in the prefrontal cortex. The chapter presents results from a series of experiments in our laboratory that address these issues.

To clarify an issue of terminology, we refer to "working memory" here as synonymous to short-term memory, the ability to maintain any information in mind over a time scale of seconds. In recent years, some authors have reserved

Mechanisms of Sensory Working Memory
http://dx.doi.org/10.1016/B978-0-12-811042-3.00013-X

the term working memory to refer specifically to complex information that needs to be manipulated, as opposed to the memory of simple stimuli (e.g., colored squares) that needs to be maintained without any further transformation, which is referred to as "visual short-term memory" (Todd & Marois, 2004). Some of our experiments explicitly address the passive versus active maintenance of information in memory. However, we describe all of these phenomena as working memory, in the broad sense of the term.

PREFRONTAL AND PARIETAL CONTRIBUTIONS TO WORKING MEMORY

Although first described in the prefrontal cortex, sustained discharges during working memory tasks have since been reported in most areas of the association cortex and several subcortical areas (Constantinidis & Procyk, 2004; Pasternak & Greenlee, 2005). The magnitude and time course of these persistent discharges differs between cortical areas and in different tasks, supporting the idea that different brain areas play distinct roles in the maintenance of working memory. The PPC and dlPFC (Figure 1(a)) have been studied extensively regarding their respective roles in the representation of visual spatial working memory because the two areas are strongly interconnected and share many functional properties (Katsuki & Constantinidis, 2012). Early comparative studies (Chafee & Goldman-Rakic, 1998) investigated prefrontal and parietal activity in a simple spatial working memory task, the oculomotor delayed response task (ODR task; Figure 1(b)). The ODR task involves presentation of a single stimulus, followed by a delay period of a few seconds. When the fixation point is extinguished, the subject is required to generate an eye movement toward the location of the remembered cue. Results showed that both brain areas exhibited persistent activity after the presentation of the cue, with remarkable similarities in the percentage of neurons activated in the two areas and overall level of activity (Chafee & Goldman-Rakic, 1998).

In recent studies, we compared neurophysiological activity from prefrontal and parietal cortex in more complex visuospatial working memory tasks, such as the match/nonmatch task (Figure 1(c)). In this task, monkeys compared two stimuli presented in sequence, with intervening delay periods between them. They then had to decide if the two stimuli appeared in the same spatial location or not. If they did, an eye movement was required to a green choice target; if they did not match, the eye movement had to be directed to a blue choice target. Neurons in both dlPFC and PPC exhibited discharges in the delay periods of the task (Qi et al., 2010). We classified the patterns of delay period activity into two categories. "Sustained" activity was defined by persistent discharges after the presentation of stimulus with a constant or slightly declining rate. Sustained responses are often thought to be the activity that provides a neural correlate of working memory for the preceding stimulus (Goldman-Rakic, 1995). "Anticipatory" responses, on the other hand, accelerated in rate during the delay period. Maximal anticipatory activity often appeared after

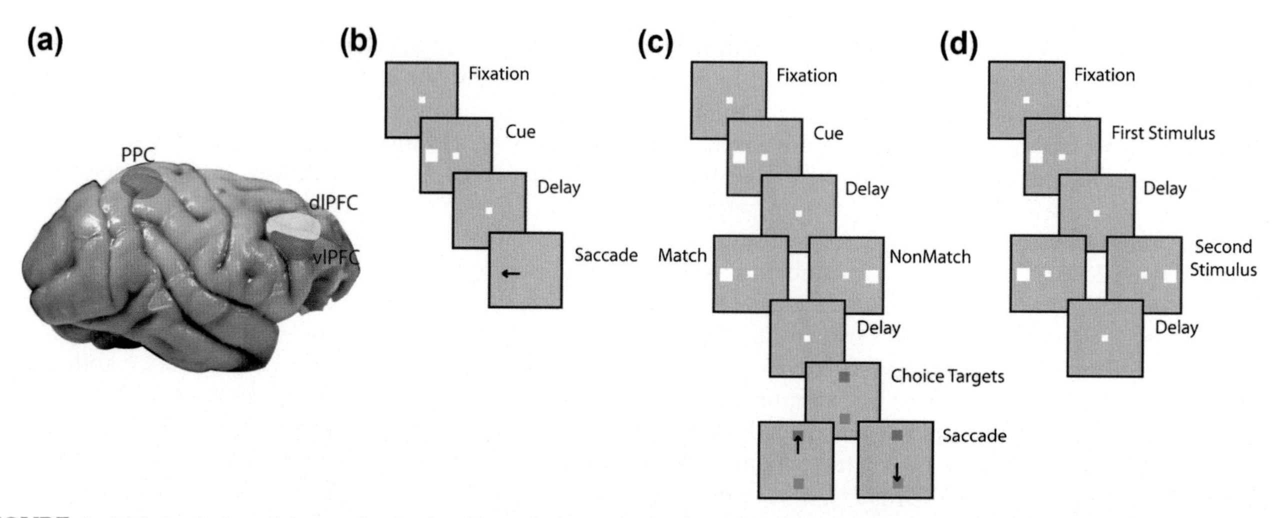

FIGURE 1 (a) Lateral view of the monkey brain with posterior parietal cortex (PPC, red), dorsolateral prefrontal cortex (dlPFC, yellow), and ventrolateral prefrontal cortex (vlPFC, purple) highlighted. (b) Oculomotor delayed response (ODR) task. In this task, monkeys are trained to remember the location of the cue and make an eye movement toward it after the fixation point is turned off. (c) Match/nonmatch task used to test working memory in monkeys. Successive frames illustrate the sequence of presentations on the screen. Monkeys were required to remember the spatial location of two stimuli presented in sequence and make an eye movement to a green target if the two stimuli matched and the blue target if they did not match. (d) Schematic diagram of presentation of the same stimuli, in naive monkeys before working memory training. The sequence of stimuli is identical to that of (c) with the exception of the appearance of choice targets.

presentation of a stimulus out of the receptive field, when the stimulus itself evoked little response. In the match/nonmatch task, both dlPFC and PPC neurons responded robustly with sustained activity above the baseline in both delay periods when stimuli appeared in the receptive field (Figure 2(a)). However, in trials where the cue appeared in the receptive field but was followed by a stimulus outside the receptive field, prefrontal neurons continued to discharge in a robust manner in the second delay period, whereas the activity of parietal neurons returned to baseline (Figure 2(b)). The second stimulus appearing in the receptive field also generated sustained activity (Figure 2(c)). Anticipatory responses with similar time courses were observed in the two areas.

An important point is that the match/nonmatch task requires the animals to remember the initial stimulus and to compare it with a second stimulus presentation. In the match/nonmatch task, after the second stimulus presentation, the animals no longer need to remember either stimulus and simply have to decide whether the two stimuli were the same or not. Nonetheless, we observed responses in the second delay period that continued to represent either the initial or second stimulus (Figure 2(c)). A similar representation of previously presented stimuli has been described in somatosensory working memory studies (Romo, Brody, Hernandez, & Lemus, 1999). In the latter task, monkeys were presented with two vibratory stimuli in sequence separated by delay periods. At the end of the trial, the animals were required to decide if the second frequency was lower or higher than the first. As in our match/nonmatch task, activity of prefrontal neurons continued to represent stimulus attributes in the second delay period.

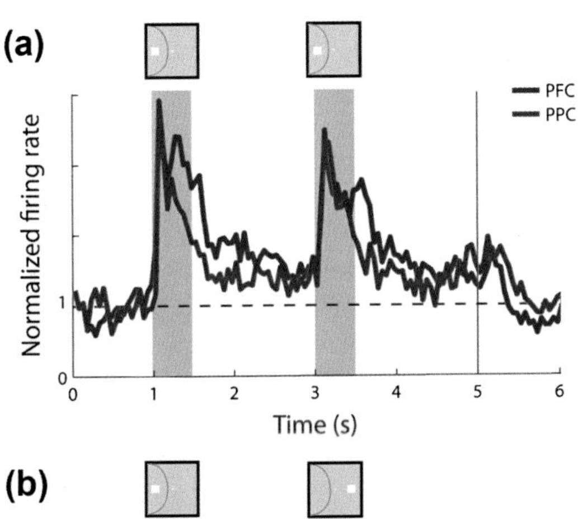

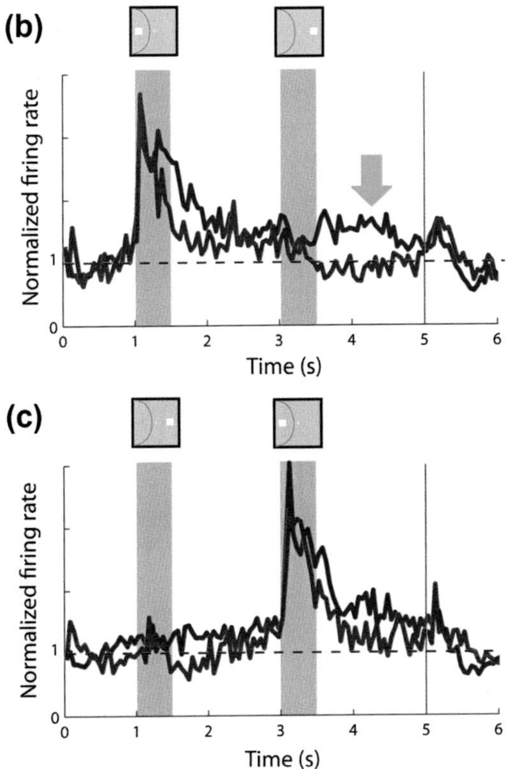

FIGURE 2 Population responses from neurons with sustained activity recorded in the prefrontal cortex (PFC) and posterior parietal cortex (PPC) during performance of the match/nonmatch task. Gray bars represent presentation of stimuli in our out-of-the-receptive-field of each neuron (inset on top). The location of stimulus and position of the receptive field differed for each neuron and is shown here only schematically. Vertical line represents onset of choice targets. (a) Average responses to a cue stimulus presented in the receptive field, followed by match at the same location. (b) Responses to a cue stimulus in the receptive field followed by a nonmatch out of the receptive field. Green arrow represents the delay period following the nonmatch stimulus. (c) Responses to a cue stimulus out of the receptive field followed by a nonmatch in the receptive field. *From Qi et al. (2010).*

The patterns of activity we observed in our combined recordings in the dlPFC and PPC are in agreement with previous studies comparing neuronal activity in the prefrontal cortex and inferior temporal cortex revealing unique properties of the prefrontal cortex. In animals tested with a delayed match-to-sample task requiring monkeys to remember different visual objects, activity of prefrontal neurons activated by the sample stimulus continued to persist and survived the presentation of nonmatch stimuli that the monkeys had to ignore to perform the task (Miller, Erickson, & Desimone, 1996). Other previous studies, examining either the prefrontal or parietal cortex, showed that activity of prefrontal neurons appeared to survive distracting stimulation (di Pellegrino & Wise, 1993), whereas activity of posterior parietal neurons only tracked the most recent stimulus (Constantinidis & Steinmetz, 1996). Recent studies also confirm the unique ability of the prefrontal cortex to filter distractors, compared with the PPC (Suzuki & Gottlieb, 2013). These neurophysiological results parallel the conclusions of human imaging studies (Sakai, Rowe, & Passingham, 2002) and suggest that prefrontal cortex is unique in its ability to maintain information in memory in the face of distraction.

We should emphasize that the difference in activity patterns between areas was subtle and was only evident when we distinguished between the sustained and anticipatory types of persistent activity. We also observed no apparent differences in activity in other tasks such as the spatial version of a delayed match-to-sample task (Qi et al., 2010). The result suggests that the neural apparatus dedicated to the maintenance of information is considerably distributed between the prefrontal and PPC and differences between the two areas are task-dependent. Other studies have also challenged the idea of an absolute dichotomy between the properties of the prefrontal cortex and its cortical afferents. For example, neurons in inferior temporal cortex were shown to represent a stimulus retrieved from long-term memory even when presented between distracting stimuli (Takeda et al., 2005). Another recent study in the inferior temporal cortex suggested that a few inferior temporal neurons continue to represent the sample even after the presentation of a nonmatch stimulus (Woloszyn & Sheinberg, 2009). These results suggest that the properties of prefrontal cortex may be quantitatively rather than qualitatively different from afferents such as the inferior temporal and PPC (Sigala, 2009).

These caveats notwithstanding, the superior ability of the prefrontal cortex to filter distraction, at least in the context of some behavioral tasks, poses the question of how this operation is performed within the prefrontal cortex, and what underlying neural mechanisms endow it with this ability. The prefrontal cortex receives significant dopaminergic innervation, which has been hypothesized to account for its unique physiological properties. Computational models have confirmed that dopamine can serve to stabilize working memory (Durstewitz, Seamans, & Sejnowski, 2000) via an increase of N-methyl-D-aspartate receptor conductance (Chen et al., 2004; Seamans, Durstewitz, Christie, Stevens, & Sejnowski, 2001; Yang & Seamans, 1996; Wang, 2001). However, the effects of dopamine modulation are complex and experimental studies reveal nonmonotonic dosage relationships (Williams & Goldman-Rakic, 1995; Zheng, Zhang, Bunney, & Shi, 1999) and differential physiological effects of dopamine depending on cortical layer, neuron type, and cellular compartment targeted (Gao, Wang, & Goldman-Rakic, 2003; Gonzalez-Islas & Hablitz, 2003; Seamans et al., 2001; Zhou & Hablitz, 1999). Furthermore, the highest concentration of dopaminergic projections targets the medial prefrontal cortex, with only a minor proportion innervating the dlPFC (Lewis, Foote, Goldstein, & Morrison, 1988). For this reason, it is not straightforward to attribute specific functional differences between the dlPFC and PPC to dopaminergic innervation.

Other differences in the intrinsic circuits of the prefrontal and parietal cortex, for example, in terms of the dendritic tree size and numbers of synapses of pyramidal neurons may also be contributing factors in their specialization. Prefrontal pyramidal neurons exhibit the most extensive dendritic trees and highest number of spines of any cortical neurons (Elston, 2000, 2003). The relative composition of interneuron types is also hypothesized to be unique for the prefrontal cortex (Wang, Tegner, Constantinidis, & Goldman-Rakic, 2004). In a series of recent experiments, we addressed the potential physiological consequences of such intrinsic circuit specializations in the dlPFC and PPC.

In one study designed to compare the functional neural circuitry of the two areas, we estimated the strength of effective connectivity between pairs of neurons recorded at various distances of each other, within the dlPFC and within the PPC (Katsuki et al., 2013). We relied on cross-correlation analysis, examining spikes occurring within 5 ms of each other from neurons recorded simultaneously. The strength of this functional interaction is indicative of the strength of intrinsic connections within an area (Constantinidis, Franowicz, & Goldman-Rakic, 2001a). Both PPC and dlPFC exhibited a decrease in effective connectivity as distance between neurons increased, in agreement with prior studies in several cortical areas. Importantly, neurons in the PPC shared a larger percentage of their functional inputs when they were located at short (≤0.3 mm) distances, compared to pairs of neurons recorded at equivalent distances from the dlPFC. We did not record from neurons at distances greater than ~1.5 mm away from each other; however, it is possible that the relative correlation strength between areas reverses at longer distances. In such a case, the difference in correlated firing could be the result of more extended anatomical connections in the dlPFC (Figure 3(b))

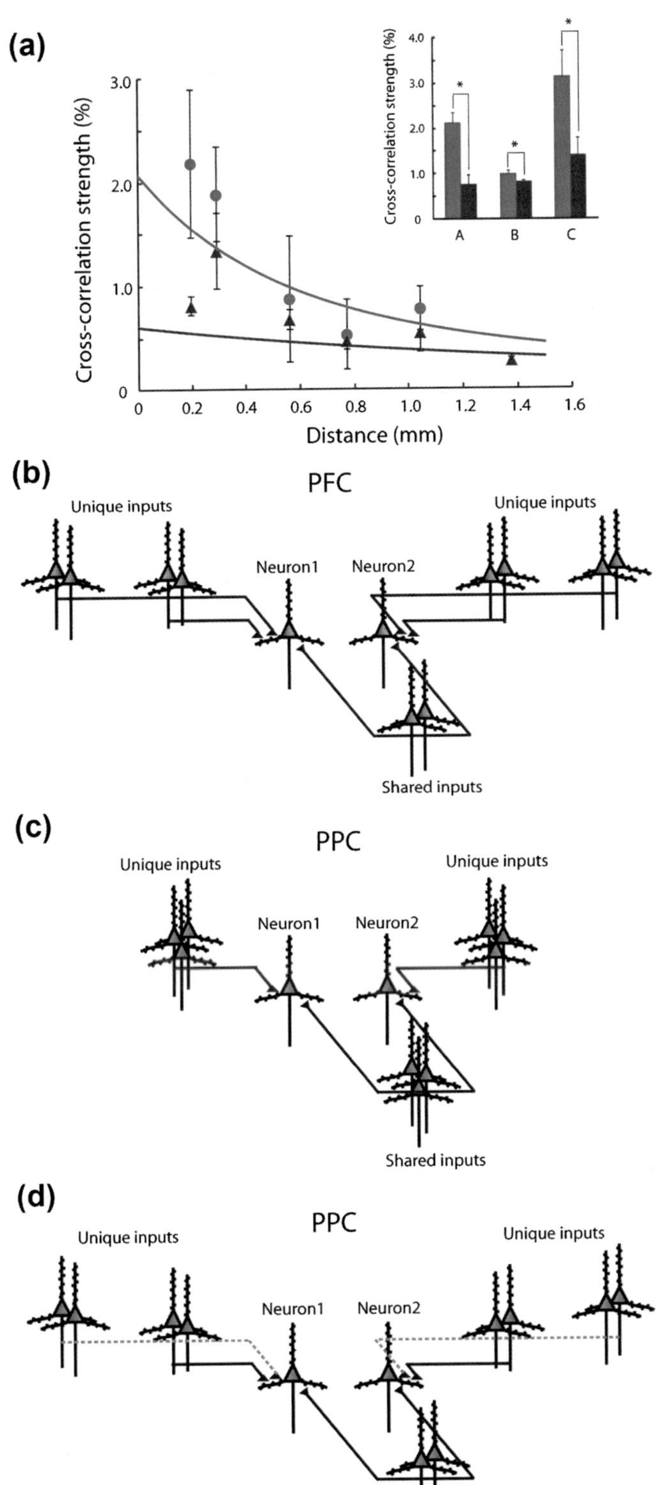

FIGURE 3 (a) Average cross-correlation strength in the dorsolateral prefrontal (red) and posterior parietal cortex (blue). Error bars at each point represent standard error of the mean across monkeys. Inset illustrates average cross-correlation strength computed for neuron pairs with distances of 0.3mm or less for each monkey. Stars indicate statistically significant differences at the 0.05 significance level (t-test). (b) Schematic model of intrinsic functional organization for the prefrontal cortex. Diagram illustrates distributions of synaptic inputs onto a pair of prefrontal neurons. (c) Distributions of posterior parietal inputs. Parietal neurons may integrate inputs over a shorter range of distances compared to prefrontal neurons. (d) An alternative model for posterior parietal organization. Spatial distribution of inputs is identical to prefrontal cortex, but long distance connections exhibit proportionally lower functional strengths (represented with dotted lines). *From Katsuki et al. (2013) with permission.*

compared with the PPC (Figure 3(c)), consistent with anatomical results indicating more extensive dendritic trees of prefrontal neurons, integrating inputs over longer distances (Elston, 2000, 2003). Alternatively, the size of cortical ensembles may be identical in the two areas, but longer distance projections may have greater influence in the dlPFC than in the PPC (Figure 3(d)). This analysis is indicative of direct functional correlates of the anatomical differences discussed previously and point to a unique neural circuitry of the prefrontal cortex.

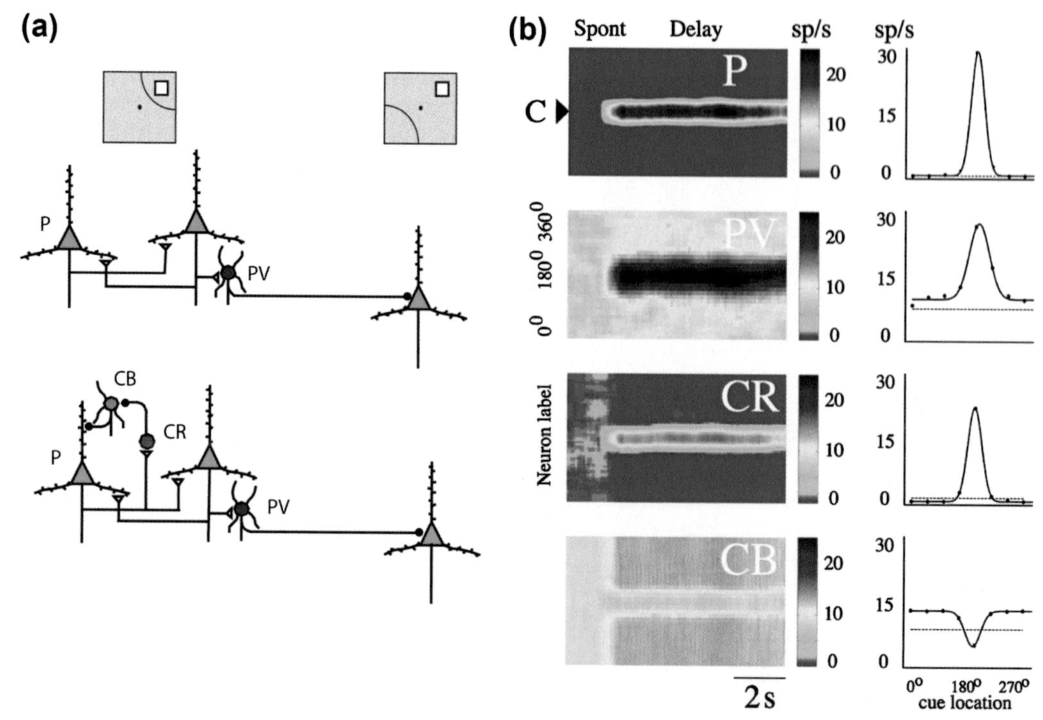

FIGURE 4 (a) Schematic diagram of the prototypical circuit network of posterior parietal cortex and dorsolateral prefrontal cortex. Different types of neurons are indicated as follows: CB, calbindin Interneuron; CR, calretinin Interneuron; P, pyramidal neuron; PV, parvalbumin Interneuron. Open triangles denote excitatory synapses; black circles are inhibitory synapses. Insets on top are meant to illustrate that pyramidal neurons on left side of the figure are driven by a stimulus in the receptive field (shown as an arc), whereas the same stimulus falls out of the receptive field of the neuron on the right side of the figure. (b) Modeling and single neuron examples from prefrontal cortex for P and the three (PV, CB, and CR) inhibitory neuron populations during the cue and delay periods. Instantaneous firing rates are color-coded. Observed neuronal tuning curves (solid lines) during the delay period in the model simulations. Eight different cue positions are used. Dashed lines, spontaneous firing rate during the resting state. In bottom row, three kinds of recorded tuning curves in dorsolateral PFC during an oculomotor delayed response task are shown. An example of inverted tuning curve is shown at the bottom right. *Panel (a) from Zhou et al. (2012); panel (b) from Wang et al. (2004).*

A second intrinsic mechanism that could account for functional differences between the PFC and other cortical areas is the existence of specialized interneuron types, which has been proposed as a unique property of the prefrontal cortex (Wang et al., 2004). Most interneurons in the cortex are parvalbumin-containing neurons (Figure 4), whose physiological properties correspond to the fast spiking category of interneurons, and inhibit neurons with stimulus selectivity different than their own (Krimer et al., 2005; Zaitsev et al., 2005). By some accounts, another cell type, the calbindin-containing interneurons, are more numerous in the prefrontal cortex than in other cortical areas (Elston & Gonzalez-Albo, 2003). Calbindin interneurons (Figure 4) are inhibited by calretinin interneurons and in turn inhibit the dendrites of pyramidal neurons in close vicinity (Figure 4(a)), spatially restricted in vertical columns (Conde, Lund, Jacobowitz, Baimbridge, & Lewis, 1994; Gabbott & Bacon, 1996; Krimer et al., 2005; Zaitsev et al., 2005). The functional consequence of such a circuit in spatial working memory proposed by Wang et al. (2004) is illustrated in Figure 4. After a stimulus has been presented and is maintained in memory, a population of pyramidal cells exhibits persistent activity (Figure 4(b), top row). These neurons activate local parvalbumin interneurons (Figure 4(b), second row), which in turn inhibit pyramidal neurons with different stimulus preference. At the same time, the activated pyramidal neurons excite calretinin interneurons (Figure 4(b), third row), which in turn inhibit calbindin interneurons (Figure 4(b), bottom row). Calbindin interneurons would therefore exhibit inverted tuning (i.e., high baseline firing rate and decrease in firing rate for some stimuli). The precisely localized axons of the calbindin interneurons then further disinhibit the pyramidal neurons that are already activated (Figure 4(a), left). Nonactivated pyramidal neurons (selective for other stimuli) continue to be tonically inhibited by the combined action of parvalbumin interneurons and tonic inhibition of calbindin interneurons in their own microcolumns. As a result, once a stimulus is held in memory in the circuit of Figure 4, a distracting stimulus appearing at a different location is less effective in suppressing the persistent activity of already activated pyramidal neurons that are both mutually excited and receiving less inhibition by calbindin interneurons (Wang et al., 2004). Apart from resistance to the interference of actual distracting stimuli, the tonic action of calbindin interneurons during working memory suppresses the baseline activity of neurons tuned for stimuli away from the remembered stimulus, further sharpening the representation of the remembered stimulus in the population and essentially reducing the noise in the circuit during working memory.

We have previously described such neurons with "inverted tuning" during the delay period of a working memory in the prefrontal cortex (Wang et al., 2004), but no comparative neurophysiological data exist from other cortical areas to support a unique prefrontal role. For this reason, we were motivated to compare the relative incidence and properties of physiologically identified neurons with inverted tuning in the dlPFC and PPC of the same monkeys, performing working memory tasks. The results of these experiments indicated that neurons with inverted tuning during the delay periods of working memory tasks were much more numerous in the dlPFC than the PPC of monkeys (Zhou, Katsuki, Qi, & Constantinidis, 2012). Additionally, only prefrontal neurons with inverted tuning exhibited biophysical properties consistent with narrow-spiking interneurons. Lastly, only PFC neurons possessed spatial tuning to visual stimuli of comparable width to pyramidal neuron tuning, which would be necessary to effectively disinhibit pyramidal neurons during working memory. These results support a mechanism of intrinsic circuit connectivity that can lead to unique neurophysiological properties of the prefrontal cortex during working memory.

EFFECTS OF WORKING MEMORY TRAINING ON PREFRONTAL ACTIVITY

The vast majority of studies investigating the neurophysiology of the prefrontal cortex and its involvement in cognitive functions (such as those reviewed in the previous section) have been performed in animals highly trained to execute stereotypical behavioral tasks. In recent years, the impact of training on brain activation has received considerable attention. Training in working memory tasks has been proved effective as a remediating intervention in cases of brain injury, aging, and mental disorders (Anguera et al., 2013; Jaeggi, Buschkuehl, Jonides, & Shah, 2011; Klingberg, 2010; Klingberg, Forssberg, & Westerberg, 2002; Westerberg et al., 2007; Wexler, Anderson, Fulbright, & Gore, 2000). Tangible changes in blood oxygen level–dependent activation patterns in the prefrontal cortex have been revealed after working memory training in normal adults, suggesting functional differences after training (Dahlin, Neely, Larsson, Backman, & Nyberg, 2008; Fletcher, Buchel, Josephs, Friston, & Dolan, 1999; Hempel et al., 2004; McNab et al., 2009; Moore, Cohen, & Ranganath, 2006; Nyberg et al., 2003; Olesen, Westerberg, & Klingberg, 2004).

The nature of the functional organization of the prefrontal cortex has also been a matter of debate that can be informed by the effects of training. Anatomical projections from the PPC target mostly the dlPFC (areas 8 and 46) in monkeys, whereas projections from the inferior temporal cortex and ventral visual stream, innervate areas 12 and 45 of the ventrolateral prefrontal (vlPFC – Figure 1(a)) cortex (Cavada & Goldman-Rakic, 1989; Constantinidis & Procyk, 2004; Petrides & Pandya, 1984; Selemon & Goldman-Rakic, 1988). This relative segregation of anatomical inputs led researchers to postulate that the prefrontal cortex is organized in a "domain-specific" fashion, with spatial information predominantly represented in the dorsal prefrontal cortex and feature information in the ventral prefrontal cortex (Wilson, Scalaidhe, & Goldman-Rakic, 1993). These conclusions were challenged by some subsequent studies, suggesting that dorsal prefrontal neurons can exhibit significant shape selectivity and ventral neurons spatial selectivity, after monkeys have been trained in tasks that required them to remember both the location and identify of a stimulus (Rainer, Asaad, & Miller, 1998; Rao, Rainer, & Miller, 1997). This "integrative" theory postulates that the prefrontal cortex, because of its intrinsic organization that places parietal and temporal inputs in relative proximity to each other, has the capacity to integrate spatial and feature information for the needs of cognitive tasks, and that functional differentiation is primarily the effect of training.

To address these unresolved questions, a series of neurophysiological recordings were performed in animals that were initially naive to working memory tasks and were subsequently trained in a task. Our strategy was to present the exact same stimuli that would eventually be incorporated in the behavioral task to monkeys that were only required to fixate, using the same stimulus duration and sequence of presentation, except for the appearance of choice targets (Figure 1(d)). The animals were rewarded simply for maintaining fixation. We used two sets of stimuli: a spatial set involving presentation of a white square stimulus at nine spatial locations, and a shape set consisting of eight white geometric shapes. These stimuli were not novel at the onset of neurophysiological experiments because they were presented to the animals repeatedly over a period of weeks before the beginning of experiments. We then recorded throughout the dlPFC and vlPFC, sampling all neurons isolated from our electrodes in an unbiased fashion. The method allowed us to address if there are systematic differences between prefrontal regions before imposing task demands for the recognition and memory of stimuli (Meyer, Qi, Stanford, & Constantinidis, 2011). We went on to train the same animals to perform the match/nonmatch working memory task, requiring the animals to recognize and remember the two stimuli presented in sequence and decide if the stimuli were the same or not (Figure 1(c)).

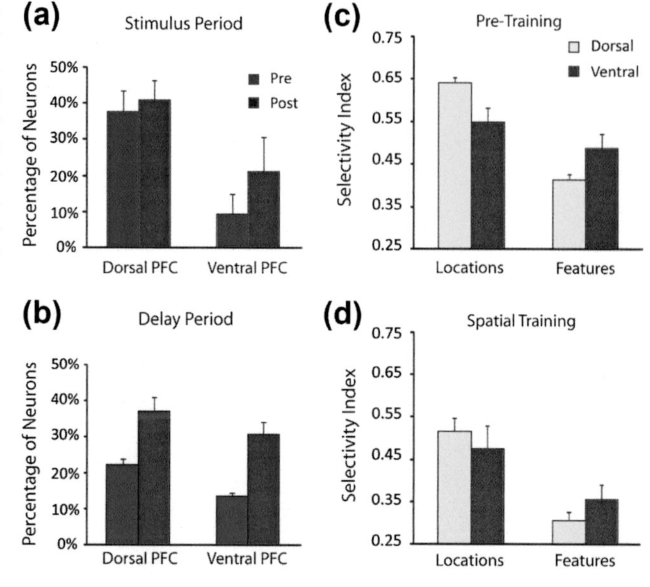

FIGURE 5 Summary of changes in neural activity after training to perform a working memory task. (a) Percentage of neurons activated by a visual stimulus of the spatial working memory set in the dorsolateral and ventrolateral prefrontal cortex, before and after training in a working memory task. Error bars represent standard error of the mean across neurons. (b) Percentage of neurons active during the delay period. (c) Average values of spatial and shape Selectivity Index for each region of the prefrontal cortex before training. Error bars represent standard error of data averaged across monkeys. (d) As in (c), for data from monkeys trained in the spatial working memory task. *Panels (a) and (b) from Qi et al. (2011), with permission; panels (c) and (d) from Meyer et al. (2011).*

We then repeated recordings in the same cortical areas and we were able to determine the nature of changes that follow working memory training (Qi, Meyer, Stanford, & Constantinidis, 2011).

Our recordings before any training revealed significant differences between the dorsal and ventral aspects of the prefrontal cortex in terms of the two regions' responsiveness to stimuli (Figure 5(a)), their selectivity for spatial and nonspatial information (Figure 5(c)), and the effects of training (Figure 5(a) and (b)). A significantly higher percentage of dlPFC than vlPFC neurons responded to any of the visual stimuli we used in our experiment (Figure 5(a), blue bars). We quantified spatial and shape feature selectivity using a selectivity index defined as the difference of the maximum minus the minimum response to the stimuli in each set, divided by their sum. Before training, average selectivity was higher in the dlPFC than in the vlPFC for spatial locations, whereas selectivity was higher in the vlPFC than the dlPFC for shape features (Figure 5(c)). After training, we observed an overall increase in the percentage of neurons that were activated by stimuli (Figure 5(a), compare red and blue bars). Approximately 35% of all neurons sampled were active during execution of either the spatial or feature working memory task compared with 20% before training. The same stimuli also elicited an increased firing rate after training. Importantly, in animals trained only to perform a spatial working memory task, no effect of increased firing rate was observed during the delay period for shape stimuli, which continued to be presented passively (Meyer et al., 2011). The result suggests that the increased activation was specifically related to the working memory task. The increase in activation after training affected the two prefrontal subdivisions disproportionally; a much greater proportion of vlPFC neurons became active after training. Training altered the functional properties of neurons in both the dorsal and ventral subdivisions, but significant functional differences in terms of selectivity for spatial stimuli between the areas remained (Figure 5(d)). The results from these experiments are consistent with the idea that before any training in a working memory task the dlPFC is specialized for spatial information, whereas the vlPFC is for feature information. Neurons in the vlPFC were largely unresponsive to the fairly small stimulus set used in our task, also consistent with precise specialization for the features of stimuli. Behavioral training rendered many of these stimuli relevant to vlPFC neurons and recruited a much larger number of them (Meyer et al., 2011; Qi et al., 2011).

An interesting observation we made in the course of these experiments is that even before training naive monkeys, a population of prefrontal neurons continued to be active after the offset of visual stimuli (Figure 6). Activity did not decay after the offset of the stimulus but persisted throughout the entire delay period (Meyer, Qi, & Constantinidis, 2007). The persistent activity was very similar to the responses recorded in trained animals performing working memory tasks. Discharges exhibited selectivity for the location and features of the preceding stimulus, indicating that persistent activity represented stimulus properties rather than being related to nonspecific factors such as the anticipation of the end of the trial or the expectation of reward (Meyer et al., 2007). Location-specific persistent activity was observed after both the first and second stimulus, when two stimuli were presented in sequence. But the persistent activity generally did not survive a second stimulus presented out of receptive field (Meyer et al., 2007). Working memory is assumed to be an active process requiring conscious effort (Frith & Dolan, 1996; Postle, 2006). A widely used definition of working memory emphasizes its dynamic nature and its role in the integration and

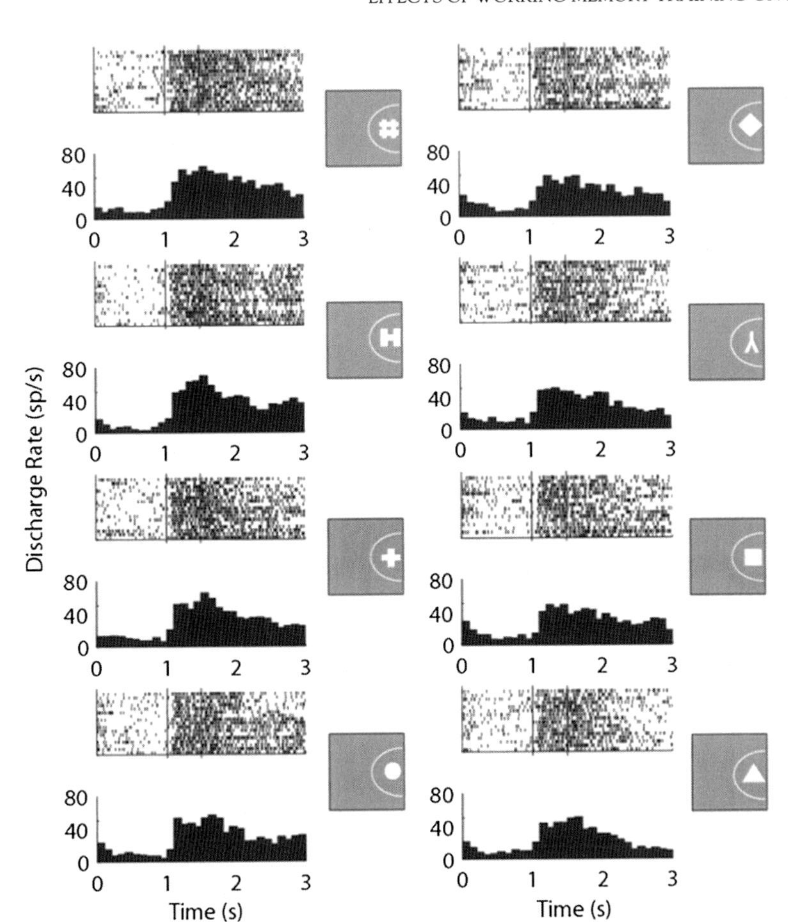

FIGURE 6 Example of a single prefrontal cortex neuron in a monkey naive to training, exhibiting persistent responses after the offset of the cue (two vertical lines). Raster plots and histograms illustrate sustained responses are shown to different stimuli, rank-ordered by stimulus preference.

manipulation of information for the guidance of behavior as opposed to passive storage (Baddeley, 2012). The results from our experiments indicate that the neural systems implicated in working memory are also active during passive stimulus presentation. This implies that persistent activity is not an explicit effect of performing working memory tasks. Such a definition is intuitive because we are able to recall stimuli even when we are not explicitly prompted to remember them. Therefore, it appears that activity present during the active maintenance of working memory is generated in an automatic fashion during passive exposure to sensory stimuli that may have no behavioral relevance. Neither training nor effortful execution of a behavioral task are necessary for its generation. Even task-irrelevant stimuli generate neuronal responses that outlast the physical stimulation and could provide a buffer for working memory available for a number of other possible functions (Meyer et al., 2007). At the same time, willful execution of the task after training does produce changes in neuronal discharges. The increase in the percentage of neurons that exhibited persistent activity (Figure 5(b)) and also the increase in their firing rate are consistent with the results observed in human studies after training (Dahlin et al., 2008; Fletcher et al., 1999; Hempel et al., 2004; McNab et al., 2009; Moore et al., 2006; Nyberg et al., 2003; Olesen et al., 2004).

Another type of firing rate modulation evident in working memory tasks, independent of persistent activity, is that of differential responses to a stimulus when it is preceded by the same stimulus or not, and thus appears as a match or a nonmatch (Romo et al., 1999; Romo & Salinas, 2003). Such modulation of responses can also be viewed as a neural correlate of the remembered stimulus, particularly because the magnitude of response difference to match and nonmatch stimuli differs systematically in correct and error trials in the prefrontal cortex (Zaksas & Pasternak, 2006). Neural activity observed for the repeated (matching) presentation of a stimulus is typically reduced compared with the activity elicited by the identical stimulus appearing as a nonmatch, a phenomenon not only limited to neuronal firing rate but observed also in blood oxygen level–dependent activation and event-related brain potentials, and termed repetition suppression (Grill-Spector, Henson, & Martin, 2006).

The working memory tasks depicted in Figure 1(c) required monkeys to judge whether the two stimuli appeared at matching spatial locations. Because the identical stimulus presentations were used before any training (Figure 1(d)), it was possible to compare responses with match and nonmatch sequences in naive and trained animals.

As was the case for persistent activity, a population of prefrontal neurons was found to signal if the two stimuli matched, even before monkeys were trained to make a comparison (Qi, Meyer, Stanford, & Constantinidis, 2012). Approximately 20% of prefrontal neurons responding to visual stimuli exhibited firing rate modulation based on the match or nonmatch stimulus (an overall preference for the two or an interaction between spatial preference and match/nonmatch status). Equal percentages of neurons (50%) had an overall preference for match or nonmatch stimuli, although across the population of neurons average responses for a match were reduced overall compared with those for a nonmatch. It appears that prefrontal cortex automatically encodes some stimulus relationships even when there is no explicit requirement for comparison, providing relational information that may be drawn upon to guide decision making (Qi & Constantinidis, 2013).

Changes that we observed after training regarding match/nonmatch effects involved an increase in the percentage of neurons exhibiting significant differences in activity between match and nonmatch stimuli, a greater percentage of neurons preferring nonmatch stimuli, and a greater percentage of neurons representing information about the first stimulus during the presentation of the second stimulus (Qi et al., 2012). After training, the magnitude of response difference to match and nonmatch stimuli also differed systematically in correct and error trials, consistent with the idea that this activity guides the subject's decision. Changes after training were even more pronounced for feature stimuli, which were not as discriminable as the spatial stimuli at the outset of training.

Learning to perform the working memory task entails recognition and remembering the visual stimuli as well as association of the match condition with the green choice target and the nonmatch with the blue one (Figure 1(c)). Understanding how the new information about this task element is incorporated into existing neuronal activity is important for understanding how the brain learns to represent new information critical for the working memory task. To address this question, we examined the neuronal response to different shapes before and after training and quantified the information about the identity of the stimulus and the match or nonmatch status of the stimulus by using a neural-population classifier (Meyers, Qi, & Constantinidis, 2012). We found that information about the stimuli itself could be reliably decoded with the classifier both before and after training, but the information about the matching status of shape stimuli was essentially absent before training. After training, robust information about the match/nonmatch status of the shape stimuli could be decoded from the prefrontal cortex (Meyers et al., 2012). This information was encoded in a dynamic population code, with different neurons carrying information at different time intervals during the trial, consistent with recent reports of dynamic coding in the prefrontal cortex (Machens, Romo, & Brody, 2010; Meyers, Freedman, Kreiman, Miller, & Poggio, 2008; Warden & Miller, 2007) and other cortical areas (Crowe, Averbeck, & Chafee, 2010). On the single neuron level, a small population of neurons highly selective for the match/nonmatch status of shape stimuli contained almost all the task-relevant information that was present after training (Meyers et al., 2012). At the same time, a larger population of neurons exhibited task relevant information to a lesser degree. Information about the task was present in the same neurons that were selective for shape information, often manifesting itself in relatively short time windows, in the midst of other large firing rate modulations that occurred throughout the trial. In essence, prefrontal neurons multiplexed different types of information, some of which were present before training and some that only appeared after training in the task.

In a final set of analyses of working memory training effects, we examined differences on factors other than firing rate before and after training. Guided by recent findings demonstrating that variability of discharge rates and correlation between neuronal discharges are indicative of cognitive operations (Churchland et al., 2010; Cohen & Maunsell, 2009), we identified changes in discharge variability, quantified by the Fano factor of spike counts (variance divided by the mean), and the trial-to-trial correlation between neurons (termed spike–count correlation, or noise correlation). Analysis of prefrontal recordings obtained using the same stimuli before and after training revealed that the Fano factor of neuronal responses (Figure 7(a)) exhibited a decrease of approximately 7.5% after training (Qi & Constantinidis, 2012b). This was comparable in magnitude to effects of attention previously described in the context of behavioral tasks (Churchland et al., 2010; Cohen & Maunsell, 2009). The time course of the Fano factor modulation during the second delay period was virtually identical before and after training, even though no judgment was required at the end of the period for the passive fixation condition (Qi & Constantinidis, 2012b).

Changes in noise correlation among prefrontal neurons after training were similar with changes in Fano factor. We relied on spike count correlation between simultaneously recorded neurons in the prefrontal cortex (Figure 7(b)). It is thought that high levels of spike count correlation represent correlated noise that limit the amount of information that can be decoded by neuronal populations (Zohary, Shadlen, & Newsome, 1994). In fact, effects of attention appear to account for behavioral benefits by strongly modulating correlated noise (Cohen & Maunsell, 2009). Spike count correlation demonstrated an approximately 50% overall decrease after training among prefrontal neurons (Qi & Constantinidis, 2012a). These results indicate that representation of stimulus information is improved during the working memory task after training by virtue of a decrease in variability and correlation of neuronal responses.

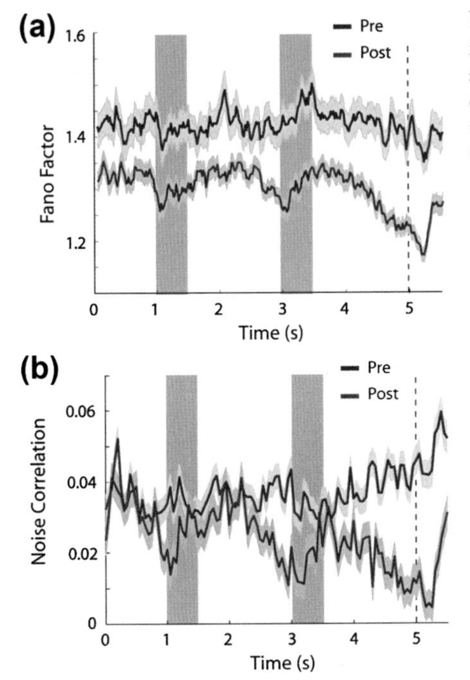

FIGURE 7 Changes in discharge variability and correlation after training. (a) Average Fano factor, a measure of firing rate variability, is plotted as a function of time across all neurons and stimulus presentations, before and after training. (b) Average noise correlation in the firing rate of pairs of neurons recorded simultaneously from separate electrodes, plotted as a function of time, before and after training. *Panel (a) from Qi and Constantinidis (2012a); panel (b) from Qi and Constantinidis (2012b).*

CHANGES IN WORKING MEMORY PERFORMANCE AND ACTIVITY AFTER PUBERTY

Working memory performance increases throughout childhood and adolescence, particularly when measured with working memory tasks that require preservation of information in the face of interference or distraction (Fry & Hale, 2000; Gathercole et al., 2004). Data from human studies provide strong evidence that working memory capacity does not reach a mature level in childhood, but instead increases throughout adolescence; some studies have observed performance increase linearly with age in tasks of visuospatial working memory and executive control (Fry & Hale, 2000; Gathercole et al., 2004). These improvements of performance have been found to parallel long-term structural changes in the prefrontal cortex and other interconnected brain regions, such as myelination and increased axonal thickness in the underlying white matter as well as changes in gray matter (Bunge et al., 2002; Giedd et al., 2006; Nagy et al., 2004; Olesen et al., 2003).

Similar patterns of working memory ability and prefrontal cortical maturation have been observed in nonhuman primates, making that a suitable animal model to study the underlying neurophysiological basis of these changes. To compare the developmental time course of monkeys and humans, it is useful to consider that the male rhesus monkey (*Macaca mulatta*) reaches puberty at approximately 3.5 years of age and full sexual maturity at 5 (Herman, Zehr, & Wallen, 2006; Plant, Ramaswamy, Simorangkir, & Marshall, 2005), equivalent to the human ages of 11 and 16 years, respectively (Crone & Dahl, 2012). Monkeys have a median life span of 25 years and can live up to 40 in captivity, also consistent with a rate of aging roughly three times that of humans (Roth et al., 2004). The monkey prefrontal cortex continues to undergo anatomical maturation during adolescence and early adulthood, similar to the human pattern of development (Bourgeois, Goldman-Rakic, & Rakic, 1994; Fuster, 2002), though some developmental studies suggest the prepubertal, 2- to 3-year-old macaque already begins to undergo some of the biochemical and anatomical changes that characterize the human adolescent prefrontal cortex (Hoftman & Lewis, 2011; Lewis, 1997).

Monkeys begin to develop working memory abilities in infancy. Monkeys as young as 4 months old can already complete delayed response tasks with 2–5 s delay periods at more than 90% accuracy, comparable to 12-month-old human infants (Diamond, 1990). Similarly, 9-month-old monkeys can perform delayed alternation tasks at high levels of performance (Alexander & Goldman, 1978). Cortical inactivation and lesion studies suggest a complex time course of prefrontal cortex involvement in cognitive functions during early development. Cooling of the prefrontal cortex at ages of 19–31 months has negligible effects (<10% decline) on performance of the delayed response task, which monkeys of this age are already able to perform essentially error-free (Alexander & Goldman, 1978). Modest effects on delayed response performance are also observed after complete bilateral lesions of the prefrontal cortex in infant, 4-month-old monkeys (Diamond & Goldman-Rakic, 1989). In contrast, prefrontal cooling when monkeys are older than 33 months has substantial effects (>20% decline) on delayed response and delayed alternation tasks

(Alexander, 1982; Alexander & Goldman, 1978). These results suggest, first, that the prefrontal cortex is not fully developed before 2.5–3 years of age; second, that the execution of working memory tasks at younger ages depends on other brain areas; and, third, that working memory functions become increasingly dependent on prefrontal activation after the age of 2.5–3 years (Alexander & Goldman, 1978). Yet, it is possible that more complex working memory tasks, or tasks requiring other types of executive function are associated with prefrontal activity that emerges later in life.

Despite this evidence, it is still unclear whether the activity of prefrontal cortical neurons in peripubertal monkeys is immature with respect to their ability to implement cognitive capacities such as working memory. To address this question, we recently tested monkeys of around the age of puberty with the oculomotor delayed response task (Figure 1(b)), and investigated the relationship between behavioral performance and neuronal activity (Zhou et al., 2013). We also used a variant of this task that involved presenting a distractor stimulus after the cue. An important caveat for these experiments was that the delay period used in the task was only 1.5 s, rather than 3–3.5 s typically used in previous studies in adult animals, because of the propensity of the animals to make errors during the delay period. Developmental measures were tracked on a quarterly basis over a period before, during, and after the behavioral training and neurophysiological recordings (Zhou et al., 2013).

The adolescent monkeys made several different types of errors in the working memory tasks, including breaking fixation during the (relatively short) delay period, or failing to fixate for the requisite duration at the saccade target location. Such nonspatial memory errors may be related to impulsivity and inability to maintain attention on the task, which may be caused by lack of working memory for the task itself (Zhou et al., 2013). Consistent with this view, once monkeys made an eye movement at the conclusion of the delay period, they could perform at greater than 90% accuracy, comparable to that reported in the ODR task for adult animals (Chafee & Goldman-Rakic, 1998; Funahashi, Bruce, & Goldman-Rakic, 1991). Furthermore, the addition of a distracting stimulus had no significant impact on performance, also consistent with prior findings in adult animals (Powell & Goldberg, 2000). These results suggest that monkeys around the time of puberty already have the ability to perform at least a simple spatial working memory task while resisting the effect of a distracting stimulus.

Neurophysiological recordings obtained from the adolescent monkeys during the execution of the working memory tasks (Figure 8) revealed robust delay period activity, with levels of firing rate, response variability, and spatial tuning similar to those reported in previous studies in adult animals (Zhou et al., 2013). The neurophysiological data further suggested that error trials, both spatial and nonspatial, were characterized by diminished neuronal activation specifically during the delay period. Based on the strong evidence of increasing prefrontal activation from childhood into adulthood, single neuron activity in a working memory task such at the ODR task might be expected to be weaker in immature subjects. When considering the large number of errors observed in the task, which were associated with decreased activation in the delay period, neuronal responses in the adolescent PFC may indeed be viewed as weaker than that of adult animals. However, we also found that the peripubertal prefrontal cortex was capable of generating robust delay period activity in correct ODR trials. It is possible that subtle, further increases in activity are present after adulthood, or that higher overall activation during may only be evident for more complex or challenging tasks.

To gain further insight into the basis of these physiological changes, we compared the strength of intrinsic effective connectivity between pairs of neurons recorded in peripubertal moneys with neurons recorded in adult monkeys. Cross-correlation analysis of neurons recorded at distances of 0.5–1.0 mm from each other demonstrated that the average magnitude of functional connections measured between neurons was lower overall in the prefrontal cortex of peripubertal monkeys (Zhou et al., 2014). This difference was due to negative functional connections (indicative of inhibitory interactions) being stronger and more prevalent in peripubertal compared with adult monkeys, whereas the positive connections showed similar distributions in the two groups. These results suggested that changes in the intrinsic connectivity of prefrontal neurons are responsible for physiological differences observed after adolescence.

CONCLUSIONS

The prefrontal cortex is part of a broader network of cortical and subcortical brain areas activated during working memory. Neurophysiological experiments in nonhuman primates demonstrate that prefrontal cortex exhibits subtle but distinct physiological properties, not observed in other areas, such as the ability to filter distracting stimuli from relevant information. Prefrontal circuits are also characterized by unique patterns of connectivity and neural composition, including a wider spread of intrinsic connections and unique interneuron types that can produce unique functional properties.

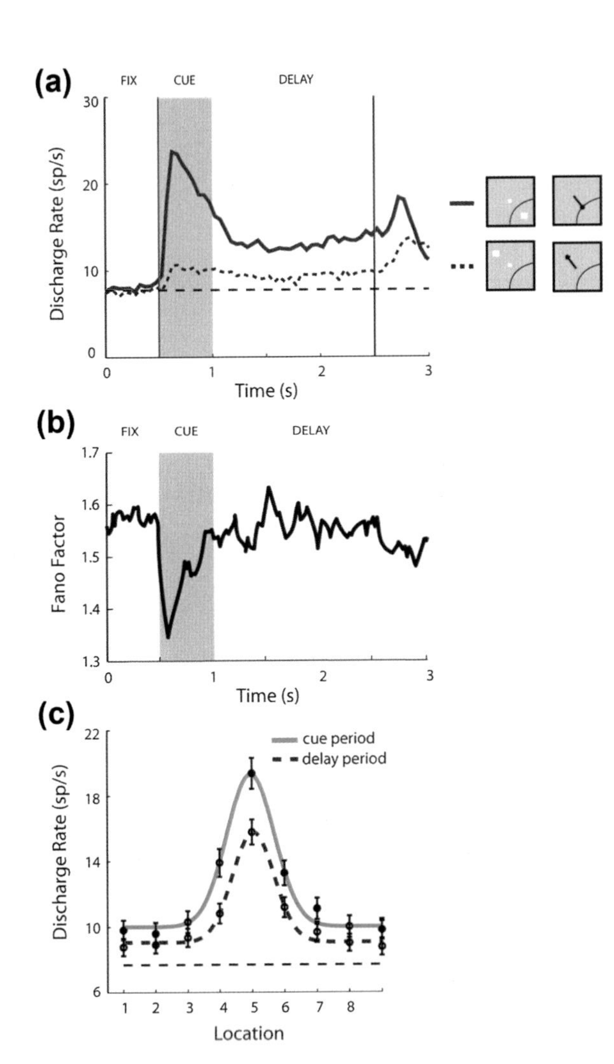

FIGURE 8 (a) Population responses of neurons with significantly elevated firing rate during the delay period of the ODR task. Average peri-stimulus time histogram (PSTH) following the best cue location (solid line) or the diametrically opposed location (dotted line). Insets next to the PSTHs are schematic illustrations of the position of the stimulus relative to the receptive field, which differed for each neuron. Horizontal line represents baseline fixation discharge rate. Labels indicate fixation (FIX), cue presentation (CUE), and delay periods (DELAY). (b) Mean Fano factor values of neuronal discharges for the same population of neuron as in (a). (c) Population tuning curves from the cue period (solid line) and from the delay period (dashed line). *From Zhou et al. (2013).*

Working memory activity in the prefrontal cortex is evident in the form of persistent discharges during the delay intervals of behavioral tasks. Individual neurons exhibit at least some neural correlates of working memory before training. Training and execution of a working memory task does induce changes in neural activity. A higher percentage of neurons is activated after training, and higher discharge rates are present during the task. Novel information is also reflected in the activity of prefrontal neurons after training. This appears to be achieved through the emergence of a small population of neurons that are highly specialized for the new information, while continuing to represent information about stimulus attributes that were present before training.

Plasticity also manifests itself, over the course of development, which for the prefrontal cortex extends after puberty. Adolescent monkeys can execute working memory tasks, around the time of puberty, though a high percentage of nonspatial memory errors are present. Neurophysiological data indicate lower levels of prefrontal activation overall, even though the adolescent prefrontal cortex is capable of generating robust working memory activity in successful trials.

Collectively, these results shed light on the neural mechanisms underlying working memory the brain areas that mediate it and the nature of its plasticity during adolescence and after training in adults. Neurophysiological experiments are likely to continue to be valuable in elucidating the neural correlates and mechanisms of working memory.

Acknowledgments

This work was supported by the National Institutes of Mental Health under award numbers R01 EY16773, R01 EY17077, and R33 MH86946; and by the Tab Williams Family Endowment Fund. We wish to thank Kathini Palaninathan for technical support and Anthony Elworthy for valuable comments on the manuscript.

References

Alexander, G. E. (1982). Functional development of frontal association cortex in monkeys: behavioural and electrophysiological studies. *Neuroscience Research Program Bulletin, 20*, 471–479.

Alexander, G. E., & Goldman, P. S. (1978). Functional development of the dorsolateral prefrontal cortex: an analysis utilizing reversible cryogenic depression. *Brain Research, 143*, 233–249.

Anguera, J. A., Boccanfuso, J., Rintoul, J. L., Al-Hashimi, O., Faraji, F., Janowich, J., et al. (2013). Video game training enhances cognitive control in older adults. *Nature, 501*, 97–101.

Baddeley, A. (2012). Working memory: theories, models, and controversies. *Annual Review of Psychology, 63*, 1–29.

Bourgeois, J. P., Goldman-Rakic, P. S., & Rakic, P. (1994). Synaptogenesis in the prefrontal cortex of rhesus monkeys. *Cerebral Cortex, 4*, 78–96.

Bunge, S. A., Dudukovic, N. M., Thomason, M. E., Vaidya, C. J., & Gabrieli, J. D. (2002). Immature frontal lobe contributions to cognitive control in children: evidence from fMRI. *Neuron, 33*, 301–311.

Cavada, C., & Goldman-Rakic, P. S. (1989). Posterior parietal cortex in rhesus monkey: II. Evidence for segregated corticocortical networks linking sensory and limbic areas with the frontal lobe. *Journal of Comparative Neurology, 287*, 422–445.

Chen, G., Greengard, P., & Yan, Z. (2004). Potentiation of NMDA receptor currents by dopamine D1 receptors in prefrontal cortex. *Proceedings of the National Academies of Science of the United States of America, 101*, 2596–2600.

Chafee, M. V., & Goldman-Rakic, P. S. (1998). Matching patterns of activity in primate prefrontal area 8a and parietal area 7ip neurons during a spatial working memory task. *Journal of Neurophysiology, 79*, 2919–2940.

Churchland, M. M., Yu, B. M., Cunningham, J. P., Sugrue, L. P., Cohen, M. R., Corrado, G. S., et al. (2010). Stimulus onset quenches neural variability: a widespread cortical phenomenon. *Nature Neuroscience, 13*, 369–378.

Cohen, M. R., & Maunsell, J. H. (2009). Attention improves performance primarily by reducing interneuronal correlations. *Nature Neuroscience, 12*, 1594–1600.

Conde, F., Lund, J. S., Jacobowitz, D. M., Baimbridge, K. G., & Lewis, D. A. (1994). Local circuit neurons immunoreactive for calretinin, calbindin D-28k or parvalbumin in monkey prefrontal cortex: distribution and morphology. *Journal of Comparative Neurology, 341*, 95–116.

Constantinidis, C., Franowicz, M. N., & Goldman-Rakic, P. S. (2001a). Coding specificity in cortical microcircuits: a multiple electrode analysis of primate prefrontal cortex. *Journal of Neuroscience, 21*, 3646–3655.

Constantinidis, C., Franowicz, M. N., & Goldman-Rakic, P. S. (2001b). The sensory nature of mnemonic representation in the primate prefrontal cortex. *Nature Neuroscience, 4*, 311–316.

Constantinidis, C., & Procyk, E. (2004). The primate working memory networks. *Cognitive and Affective Behavioral Neuroscience, 4*, 444–465.

Constantinidis, C., & Steinmetz, M. A. (1996). Neuronal activity in posterior parietal area 7a during the delay periods of a spatial memory task. *Journal of Neurophysiology, 76*, 1352–1355.

Crone, E. A., & Dahl, R. E. (2012). Understanding adolescence as a period of social-affective engagement and goal flexibility. *Nature Reviews Neuroscience, 13*, 636–650.

Crowe, D. A., Averbeck, B. B., & Chafee, M. V. (2010). Rapid sequences of population activity patterns dynamically encode task-critical spatial information in parietal cortex. *Journal of Neuroscience, 30*, 11640–11653.

Dahlin, E., Neely, A. S., Larsson, A., Backman, L., & Nyberg, L. (2008). Transfer of learning after updating training mediated by the striatum. *Science, 320*, 1510–1512.

Diamond, A. (1990). Developmental time course in human infants and infant monkeys, and the neural bases of, inhibitory control in reaching. *Annals of New York Academy of Science, 608*, 637–669; discussion 669–676.

Diamond, A., & Goldman-Rakic, P. S. (1989). Comparison of human infants and rhesus monkeys on Piaget's AB task: evidence for dependence on dorsolateral prefrontal cortex. *Experimental Brain Research, 74*, 24–40.

Durstewitz, D., Seamans, J. K., & Sejnowski, T. J. (2000). Dopamine-mediated stabilization of delay-period activity in a network model of prefrontal cortex. *Journal of Neurophysiology, 83*, 1733–1750.

Elston, G. N. (2000). Pyramidal cells of the frontal lobe: all the more spinous to think with. *Journal of Neuroscience, 20*, RC95.

Elston, G. N. (2003). The pyramidal neuron in occipital, temporal and prefrontal cortex of the owl monkey (Aotus trivirgatus): regional specialization in cell structure. *European Journal of Neuroscience, 17*, 1313–1318.

Elston, G. N., & Gonzalez-Albo, M. C. (2003). Parvalbumin-, calbindin-, and calretinin-immunoreactive neurons in the prefrontal cortex of the owl monkey (Aotus trivirgatus): a standardized quantitative comparison with sensory and motor areas. *Brain Behavior and Evolution, 62*, 19–30.

Fletcher, P., Buchel, C., Josephs, O., Friston, K., & Dolan, R. (1999). Learning-related neuronal responses in prefrontal cortex studied with functional neuroimaging. *Cerebral Cortex, 9*, 168–178.

Frith, C., & Dolan, R. (1996). The role of the prefrontal cortex in higher cognitive functions. *Cognitive Brain Research, 5*, 175–181.

Fry, A. F., & Hale, S. (2000). Relationships among processing speed, working memory, and fluid intelligence in children. *Biological Psychology, 54*, 1–34.

Funahashi, S., Bruce, C. J., & Goldman-Rakic, P. S. (1989). Mnemonic coding of visual space in the monkey's dorsolateral prefrontal cortex. *Journal of Neurophysiology, 61*, 331–349.

Funahashi, S., Bruce, C. J., & Goldman-Rakic, P. S. (1991). Neuronal activity related to saccadic eye movements in the monkey's dorsolateral prefrontal cortex. *Journal of Neurophysiology, 65*, 1464–1483.

Fuster, J. M. (2002). Frontal lobe and cognitive development. *Journal of Neurocytology, 31*, 373–385.

Fuster, J. M., & Alexander, G. E. (1971). Neuron activity related to short-term memory. *Science, 173*, 652–654.

Gabbott, P. L., & Bacon, S. J. (1996). Local circuit neurons in the medial prefrontal cortex (areas 24a,b,c, 25 and 32) in the monkey: I. Cell morphology and morphometrics. *Journal of Comparative Neurology, 364*, 567–608.

Gao, W. J., Wang, Y., & Goldman-Rakic, P. S. (2003). Dopamine modulation of perisomatic and peridendritic inhibition in prefrontal cortex. *Journal of Neuroscience, 23*, 1622–1630.

Gathercole, S. E., Pickering, S. J., Ambridge, B., & Wearing, H. (2004). The structure of working memory from 4 to 15 years of age. *Developemental Psychology, 40*, 177–190.

Giedd, J. N., Clasen, L. S., Lenroot, R., Greenstein, D., Wallace, G. L., Ordaz, S., et al. (2006). Puberty-related influences on brain development. *Molecular and Cellular Endocrinology, 254–255*, 154–162.

Goldman-Rakic, P. S. (1995). Cellular basis of working memory. *Neuron, 14*, 477–485.

Gonzalez-Islas, C., & Hablitz, J. J. (2003). Dopamine enhances EPSCs in layer II-III pyramidal neurons in rat prefrontal cortex. *Journal of Neuroscience, 23*, 867–875.

Grill-Spector, K., Henson, R., & Martin, A. (2006). Repetition and the brain: neural models of stimulus-specific effects. *Trends in Cognitive Science, 10*, 14–23.

Hempel, A., Giesel, F. L., Garcia Caraballo, N. M., Amann, M., Meyer, H., Wustenberg, T., et al. (2004). Plasticity of cortical activation related to working memory during training. *American Journal of Psychiatry, 161*, 745–747.

Herman, R. A., Zehr, J. L., & Wallen, K. (2006). Prenatal androgen blockade accelerates pubertal development in male rhesus monkeys. *Psychoneuroendocrinology, 31*, 118–130.

Hoftman, G. D., & Lewis, D. A. (2011). Postnatal developmental trajectories of neural circuits in the primate prefrontal cortex: identifying sensitive periods for vulnerability to schizophrenia. *Schizophrenia Bulletin, 37*, 493–503.

Jaeggi, S. M., Buschkuehl, M., Jonides, J., & Shah, P. (2011). Short- and long-term benefits of cognitive training. *Proceedings of the National Academy of Sciences of the United States of America, 108*, 10081–10086.

Katsuki, F., & Constantinidis, C. (2012). Unique and shared roles of the posterior parietal and dorsolateral prefrontal cortex in cognitive functions. *Frontiers in International Neuroscience, 6*, 17.

Katsuki, F., Qi, X. L., Meyer, T., Kostelic, P. M., Salinas, E., & Constantinidis, C. (2013). Differences in intrinsic functional organization between dorsolateral prefrontal and posterior parietal cortex. *Cerebral Cortex, 24*, 9.

Klingberg, T. (2010). Training and plasticity of working memory. *Trends in Cognitive Science, 14*, 317–324.

Klingberg, T., Forssberg, H., & Westerberg, H. (2002). Training of working memory in children with ADHD. *Journal of Clinical and Experimental Neuropsychology, 24*, 781–791.

Krimer, L. S., Zaitsev, A. V., Czanner, G., Kroner, S., Gonzalez-Burgos, G., Povysheva, N. V., et al. (2005). Cluster analysis-based physiological classification and morphological properties of inhibitory neurons in layers 2–3 of monkey dorsolateral prefrontal cortex. *Journal of Neurophysiology, 94*, 3009–3022.

Lewis, D. A. (1997). Development of the prefrontal cortex during adolescence: insights into vulnerable neural circuits in schizophrenia. *Neuropsychopharmacology, 16*, 385–398.

Lewis, D. A., Foote, S. L., Goldstein, M., & Morrison, J. H. (1988). The dopaminergic innervation of monkey prefrontal cortex: a tyrosine hydroxylase immunohistochemical study. *Brain Research, 449*, 225–243.

Machens, C. K., Romo, R., & Brody, C. D. (2010). Functional, but not anatomical, separation of "what" and "when" in prefrontal cortex. *Journal of Neuroscience, 30*, 350–360.

McNab, F., Varrone, A., Farde, L., Jucaite, A., Bystritsky, P., Forssberg, H., et al. (2009). Changes in cortical dopamine D1 receptor binding associated with cognitive training. *Science, 323*, 800–802.

Meyer, T., Qi, X. L., & Constantinidis, C. (2007). Persistent discharges in the prefrontal cortex of monkeys naive to working memory tasks. *Cerebral Cortex, 17*(Suppl. 1), i70–i76.

Meyer, T., Qi, X. L., Stanford, T. R., & Constantinidis, C. (2011). Stimulus selectivity in dorsal and ventral prefrontal cortex after training in working memory tasks. *Journal of Neuroscience, 31*, 6266–6276.

Meyers, E. M., Freedman, D. J., Kreiman, G., Miller, E. K., & Poggio, T. (2008). Dynamic population coding of category information in inferior temporal and prefrontal cortex. *Journal of Neurophysiology, 100*, 1407–1419.

Meyers, E. M., Qi, X. L., & Constantinidis, C. (2012). Incorporation of new information into prefrontal cortical activity after learning working memory tasks. *Proceedings of the National Academies of Science of the United States of America, 109*, 4651–4656.

Miller, E. K., Erickson, C. A., & Desimone, R. (1996). Neural mechanisms of visual working memory in prefrontal cortex of the macaque. *Journal of Neuroscience, 16*, 5154–5167.

Moore, C. D., Cohen, M. X., & Ranganath, C. (2006). Neural mechanisms of expert skills in visual working memory. *Journal of Neuroscience, 26*, 11187–11196.

Nagy, Z., Westerberg, H., & Klingberg, T. (2004). Maturation of white matter is associated with the development of cognitive functions during childhood. *Journal of Cognitive Neuroscience, 16*, 1227–1233.

Nyberg, L., Sandblom, J., Jones, S., Neely, A. S., Petersson, K. M., Ingvar, M., et al. (2003). Neural correlates of training-related memory improvement in adulthood and aging. *Proceedings of the National Academies of Science of the United States of America, 100*, 13728–13733.

Olesen, P. J., Nagy, Z., Westerberg, H., & Klingberg, T. (2003). Combined analysis of DTI and fMRI data reveals a joint maturation of white and grey matter in a fronto-parietal network. *Brain Research Cognitive Brain Research, 18*, 48–57.

Olesen, P. J., Westerberg, H., & Klingberg, T. (2004). Increased prefrontal and parietal activity after training of working memory. *Nature Neuroscience, 7*, 75–79.

Pasternak, T., & Greenlee, M. W. (2005). Working memory in primate sensory systems. *Nature Reviews Neuroscience, 6*, 97–107.

di Pellegrino, G., & Wise, S. P. (1993). Effects of attention on visuomotor activity in the premotor and prefrontal cortex of a primate. *Somatosensory and Motor Research, 10*, 245–262.

Petrides, M., & Pandya, D. N. (1984). Projections to the frontal cortex from the posterior parietal region in the rhesus monkey. *Journal of Comparative Neurology, 228*, 105–116.

Plant, T. M., Ramaswamy, S., Simorangkir, D., & Marshall, G. R. (2005). Postnatal and pubertal development of the rhesus monkey (Macaca mulatta) testis. *Annals of the New York Academy of Science, 1061*, 149–162.

Postle, B. R. (2006). Working memory as an emergent property of the mind and brain. *Neuroscience, 139*, 23–38.

Powell, K. D., & Goldberg, M. E. (2000). Response of neurons in the lateral intraparietal area to a distractor flashed during the delay period of a memory-guided saccade. *Journal of Neurophysiology, 84*, 301–310.

Qi, X. L., & Constantinidis, C. (2012a). Correlated discharges in the primate prefrontal cortex before and after working memory training. *European Journal of Neuroscience, 36*, 3538–3548.

Qi, X. L., & Constantinidis, C. (2012b). Variability of prefrontal neuronal discharges before and after training in a working memory task. *PLoS ONE, 7*, e41053.

Qi, X. L., & Constantinidis, C. (2013). Neural changes after training to perform cognitive tasks. *Behavioral Brain Research, 241*, 235–243.

Qi, X. L., Katsuki, F., Meyer, T., Rawley, J. B., Zhou, X., Douglas, K. L., et al. (2010). Comparison of neural activity related to working memory in primate dorsolateral prefrontal and posterior parietal cortex. *Frontiers in Systems Neuroscience, 4*, 12.

Qi, X. L., Meyer, T., Stanford, T. R., & Constantinidis, C. (2011). Changes in prefrontal neuronal activity after learning to perform a spatial working memory task. *Cerebral Cortex, 21*, 2722–2732.

Qi, X. L., Meyer, T., Stanford, T. R., & Constantinidis, C. (2012). Neural correlates of a decision variable before learning to perform a match/nonmatch task. *Journal of Neuroscience, 32*, 6161–6169.

Rainer, G., Asaad, W. F., & Miller, E. K. (1998). Memory fields of neurons in the primate prefrontal cortex. *Proceedings of the National Academies of Science of the United States of America, 95*, 15008–15013.

Rao, S. C., Rainer, G., & Miller, E. K. (1997). Integration of what and where in the primate prefrontal cortex. *Science, 276*, 821–824.

Romo, R., Brody, C. D., Hernandez, A., & Lemus, L. (1999). Neuronal correlates of parametric working memory in the prefrontal cortex. *Nature, 399*, 470–473.

Romo, R., & Salinas, E. (2003). Flutter discrimination: neural codes, perception, memory and decision making. *Nature Reviews. Neuroscience, 4*, 203–218.

Roth, G. S., Mattison, J. A., Ottinger, M. A., Chachich, M. E., Lane, M. A., & Ingram, D. K. (2004). Aging in rhesus monkeys: relevance to human health interventions. *Science, 305*, 1423–1426.

Sakai, K., Rowe, J. B., & Passingham, R. E. (2002). Active maintenance in prefrontal area 46 creates distractor-resistant memory *Nature. Neuroscience, 5*, 479–484.

Seamans, J. K., Durstewitz, D., Christie, B. R., Stevens, C. F., & Sejnowski, T. J. (2001). Dopamine D1/D5 receptor modulation of excitatory synaptic inputs to layer V prefrontal cortex neurons. *Proceedings of the National Academies of Science of the United States of America, 98*, 301–306.

Selemon, L. D., & Goldman-Rakic, P. S. (1988). Common cortical and subcortical targets of the dorsolateral prefrontal and posterior parietal cortices in the rhesus monkey: evidence for a distributed neural network subserving spatially guided behavior. *Journal of Neuroscience, 8*, 4049–4068.

Sigala, N. (2009). Visual working memory and delay activity in highly selective neurons in the inferior temporal cortex. *Frontiers in Systems Neuroscience, 3*, 11.

Suzuki, M., & Gottlieb, J. (2013). Distinct neural mechanisms of distractor suppression in the frontal and parietal lobe. *Nature Neuroscience, 16*, 98–104.

Takeda, M., Naya, Y., Fujimichi, R., Takeuchi, D., & Miyashita, Y. (2005). Active maintenance of associative mnemonic signal in monkey inferior temporal cortex. *Neuron, 48*, 839–848.

Todd, J. J., & Marois, R. (2004). Capacity limit of visual short-term memory in human posterior parietal cortex. *Nature, 428*, 751–754.

Wang, X. J., Tegner, J., Constantinidis, C., & Goldman-Rakic, P. S. (2004). Division of labor among distinct subtypes of inhibitory neurons in a cortical microcircuit of working memory. *Proceedings of the National Academies of Science of the United States of America, 101*, 1368–1373.

Wang, X. J. (2001). Synaptic reverberation underlying mnemonic persistent activity. *Trends in Neurosciences, 24*, 455–463.

Warden, M. R., & Miller, E. K. (2007). The representation of multiple objects in prefrontal neuronal delay activity. *Cerebral Cortex, 17*(Suppl. 1), i41–i50.

Westerberg, H., Jacobaeus, H., Hirvikoski, T., Clevberger, P., Ostensson, M. L., Bartfai, A., et al. (2007). Computerized working memory training after stroke–a pilot study. *Brain Injury, 21*, 21–29.

Wexler, B. E., Anderson, M., Fulbright, R. K., & Gore, J. C. (2000). Preliminary evidence of improved verbal working memory performance and normalization of task-related frontal lobe activation in schizophrenia following cognitive exercises. *American Journal of Psychiatry, 157*, 1694–1697.

Williams, G. V., & Goldman-Rakic, P. S. (1995). Modulation of memory fields by dopamine D1 receptors in prefrontal cortex. *Nature, 376*, 572–575.

Wilson, F. A., Scalaidhe, S. P., & Goldman-Rakic, P. S. (1993). Dissociation of object and spatial processing domains in primate prefrontal cortex. *Science, 260*, 1955–1958.

Woloszyn, L., & Sheinberg, D. L. (2009). Neural dynamics in inferior temporal cortex during a visual working memory task. *Journal of Neuroscience, 29*, 5494–5507.

Yang, C. R., & Seamans, J. K. (1996). Dopamine D1 receptor actions in layers V-VI rat prefrontal cortex neurons in vitro: modulation of dendritic-somatic signal integration. *Journal of Neuroscience, 16*, 1922–1935.

Zaitsev, A. V., Gonzalez-Burgos, G., Povysheva, N. V., Kroner, S., Lewis, D. A., & Krimer, L. S. (2005). Localization of calcium-binding proteins in physiologically and morphologically characterized interneurons of monkey dorsolateral prefrontal cortex. *Cerebral Cortex, 15*, 1178–1186.

Zaksas, D., & Pasternak, T. (2006). Directional signals in the prefrontal cortex and in area MT during a working memory for visual motion task. *Journal of Neuroscience, 26*, 11726–11742.

Zheng, P., Zhang, X. X., Bunney, B. S., & Shi, W. X. (1999). Opposite modulation of cortical N-methyl-D-aspartate receptor-mediated responses by low and high concentrations of dopamine. *Neuroscience, 91*, 527–535.

Zhou, F. M., & Hablitz, J. J. (1999). Dopamine modulation of membrane and synaptic properties of interneurons in rat cerebral cortex. *Journal of Neurophysiology, 81*, 967–976.

Zhou, X., Katsuki, F., Qi, X. L., & Constantinidis, C. (2012). Neurons with inverted tuning during the delay periods of working memory tasks in the dorsal prefrontal and posterior parietal cortex. *Journal of Neurophysiology, 108*, 31–38.

Zhou, X., Zhu, D., Katsuki, F., Qi, X. L., Lees, C. J., Bennett, A. J., et al. (2014). Age-dependent changes in prefrontal intrinsic connectivity. *Proceedings of the National Academies of Science of the United States of America, 111*, 3853–3858.

Zhou, X., Zhu, D., Qi, X. L., Lees, C. J., Bennett, A. J., Salinas, E., et al. (2013). Working memory performance and neural activity in the prefrontal cortex of peri-pubertal monkeys. *Journal of Neurophysiology, 110*, 2648–2660.

Zohary, E., Shadlen, M. N., & Newsome, W. T. (1994). Correlated neuronal discharge rate and its implications for psychophysical performance. *Nature, 370*, 140–143.

CHAPTER

14

Neural and Behavioral Correlates of Auditory Short-Term and Recognition Memory

Amy Poremba

Department of Psychology, Behavioral and Cognitive Neuroscience Division, University of Iowa, Iowa City, IA, USA

INTRODUCTION

Language and communication rely heavily on the processing of auditory signals. Two major components of that signal processing are object identification and memory. Remembering each word across a sentence and its meaning or the melody notes in a song, are just two of the many daily operations that require these identification and memory processes. After auditory objects are encoded, several forms of memory play a role when processing sounds (e.g., associative memories where an auditory signal is linked with a particular consequence, response, or meaning, short-term or working memory, and recognition memory, both for previously conditioned stimuli and online for stimulus information to be used during the current context). This chapter will review findings for neural encoding for both short-term memory and recognition memory utilizing auditory cues in nonhuman primates.

Some very complex human behaviors such as communication, language, and aspects of social processing rely on auditory memory among other sensory processing systems. Although left hemisphere specialization for speech and language in humans is well documented (Price et al., 1996), evidence for the neural origins of language development in animals and humans is sparse. During language use, we must remember known signals such as familiar words but must also remember across gaps between words, across sentences, and conversations. Ethological evidence favors the notion that vocal calls of monkeys are precursors of human communication, in part because they provide critical information to other members of the species, which rely on them for survival and social interactions (Cheney & Seyfarth, 1990), whereas competing theories suggest that human communication may have evolved from physical gestures (Corballis, 2009). However, in either case, important vocal communication signals need to be remembered over both short and long periods, depending on the context, and recognition memory is needed in either case.

Although processing, recognition, and memory of complex auditory objects are essential to communication, we are at the beginning of neural investigations to ascertain the neural circuits for auditory recognition and short-term memory or working memory, let alone long-term memory (e.g., Clarke et al., 2002; Diehl, Lotto, & Holt, 2004; Fritz, Mishkin, & Saunders, 2005; Ng, Plakke, & Poremba, 2014; Plakke, Ng, & Poremba, 2013). How is auditory information represented in the cortex, stored in memory, and used for behavior? Even though the original input through the peripheral receptors appears to be very different between sensory systems (e.g., rods, cones, hair cells), the overall outcome of identifying objects, and encoding for cross-modal interactions, may rely on similar processing systems across modalities. Examining parallel strategies employed in another closely related sensory system, such as the visual system, provides a point of reference for the studies reviewed in this chapter. There is a growing body of evidence for a similar organization between visual and auditory sensory systems (Poremba et al., 2003) based on the finding that the cortical visual system is separated into at least two processing streams (Ungerleider & Mishkin, 1982): a dorsal stream specialized for spatial processing and a ventral stream for object processing. There are several studies that suggest the auditory system may be organized in a similar fashion (e.g., Kaas & Hackett, 2000; Kikuchi, Horwitz, & Mishkin, 2010; Poremba et al., 2003; Rämä et al., 2004; Rauschecker & Tian, 2000; Romanski et al., 1999).

Mechanisms of Sensory Working Memory
http://dx.doi.org/10.1016/B978-0-12-811042-3.00014-1

Many studies support the existence of at least two processing pathways in the auditory system, including those studying analogous functions of these pathways, the dorsal pathway for spatial location and the ventral pathway for object identification with possible memory correlates for each pathway's associated function as well (e.g., Kaas & Hackett, 2000; Poremba et al., 2004; Rämä et al., 2004; Rauschecker & Tian, 2000). For example, in humans, a functional dissociation for spatial and nonspatial auditory information during a working memory study for the location versus the identity of human voices was shown in dorsal versus ventral processing pathways respectively, supporting a functional division between the dorsal/ventral pathways (Rämä et al., 2004). In this chapter, we are particularly interested in exploring the ventral auditory pathway and its projection areas in the prefrontal cortex (PFC) during the functional processing of auditory information for short-term and recognition memory.

Exploring Auditory Short-Term Memory and Recognition Memory

Two types of memory that are critical in navigating our daily lives are short-term memory and recognition memory. Because sensory information is often only available for brief amounts of time, adaptive behavior is frequently dependent on selectively remembering relevant information. The process of maintaining this information in the absence of the actual stimulus is considered short-term memory and offers ecological advantages, with levels of short-term memory performance positively correlated with performance on several tasks such as attention, general intelligence, and reading ability (Baddeley, 2003).

Recognition memory is a major component of declarative memory, which also plays a large role in the rich cognitive lives of humans and allows the ability to realize that you have encountered with clarity (i.e., recollection), or a sense of familiarity, the events, objects, or people you have previously encountered. Impairment of recognition memory figures prominently in Alzheimer's disease and maladies in which cognitive dementia is prominent. Visual recognition memory, one of the most studied forms, relies heavily on the ventral processing stream including the temporal lobes and includes processing in the PFC, among other regions.

There are numerous studies of the neural correlates of visual short-term and recognition memory. Overall, the results suggest that short-term memory usually takes the form of sustained increases, with some decreases also noted, in firing rate during the memory delay (i.e., retention interval; e.g., Fuster & Alexander, 1971; Miller, Erickson, & Desimone, 1996; Shafi et al., 2007). Sustained firing activity during the memory delays also reflects retention of information for visual working/recognition memory (e.g., Colombo & Gross, 1994; Curtis & D'Esposito, 2003; Miller, Li, & Desimone, 1993; Miyashita & Chang, 1988; Nakamura & Kubota, 1995, 1996). During visual recognition memory, neural correlates are usually observed as significantly enhanced (match enhancement) or suppressed (match suppression) neuronal activity upon re-presentation of a preceding stimulus (Miller et al., 1996). These types of neural correlates for visual short-term and recognition memory have primarily been observed for nonspatial tasks in the visual areas of the temporal lobe and PFC (Constantinidis & Procyk, 2004; Fuster, Bauer, & Jervey, 1985; Miller & Desimone, 1994; Miller et al., 1996, 1993). There have been few studies of pure auditory short-term and recognition memory; this may result from the difficulty in training nonhuman primates to perform complex auditory tasks requiring memory (Bigelow, Rossi, & Poremba, 2014; Fritz et al., 2005; Munoz-Lopez, Mohedano-Moriano, & Insausti, 2010; Scott, Mishkin, & Yin, 2012). The studies discussed in the following section will suggest that the underlying neural circuits for auditory and visual short-term and recognition memory are similar in many respects on functional and organizational levels (Poremba & Bigelow, 2013).

Delayed Matching-to-Sample Task

To test both auditory short-term and recognition memory, we have been using a commonly employed test, a variation of an earlier delayed matching-to-sample (DMS) task used with visual stimuli (D'Amato, 1973). Monkeys (adult *Macaca mulatta*, males and females), are trained to perform the auditory DMS task through a series of behavioral approximations (Ng, Plakke, & Poremba, 2009). In every recording session eight pre-selected sounds from a large database are used during the memory task (Figure 1). In a given trial, cue 1 (a sample stimulus) is first presented, followed by a retention interval before cue 2 (a test stimulus) is presented. After cue 2, subjects were required to wait 1 s before having an opportunity to make their response (i.e., wait time period). Then, on each trial, a Plexiglas response button is lit from behind to signal the possible response period of 1.5 s. The standard memory delay, inserted between two sound stimuli (i.e., interstimulus interval), is 5 s. On match trials, the two sounds are the same and a correct response is a button press (i.e., a go response), resulting in the delivery of a small food reward. On nonmatch trials, the two sounds are different and a correct response is scored if the monkey avoids touching the button (i.e., a no-go response), with no subsequent food delivery. Sometimes subjects receive a 100- to 500-ms air puff (a mild punishment)

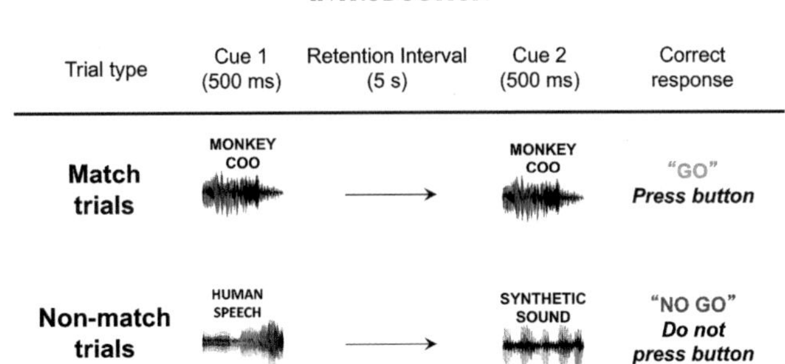

FIGURE 1 Schematic of the auditory DMS short-term memory task. The sound stimuli were naturalistic and artificial sound exemplars. After the cue 2 test stimulus, a preresponse wait period occurred, after which the response button was lit on both match and nonmatch trials to signal the response window.

when they press the button during nonmatch trials (no-go conditions). The mild air puffs are delivered semirandomly during training to discourage erroneous or impulsive responding. This is a go/no-go, asymmetrical reward paradigm, which was adopted because of the difficulty in training nonhuman primates. The auditory DMS task uses: (1) short-term memory across the delay between the two auditory cues, wherein there is no reminder of the physical stimulus last presented, and (2) recognition memory to determine if the second stimulus matches the first stimulus.

Neural Recordings in the Dorsal Temporal Pole

The dorsal temporal pole (dTP) consists of granular and dysgranular areas and has extensive connections with auditory and auditory-related regions (e.g., superior temporal gyrus (STG); parabelt areas; limbic thalamus; amygdala; hippocampus; lateral, orbital, and medial prefrontal cortices) (Barbas, Ghashghaei, Rempel-Clower, & Xiao, 2002; Kondo, Saleem, & Price, 2005; Markowitsch, Emmans, Irle, Streicher, & Preilowski, 1985; Moran, Mufson, & Mesulam, 1987; Romanski et al., 1999; Yeterian & Pandya, 1991). The dTP is at the end of the proposed ventral processing pathway for auditory object identification (Rauschecker & Scott, 2009) and has been implicated in tasks associated with auditory perception such as pattern discrimination (Iversen & Mishkin, 1973; Weiskrantz & Mishkin, 1958), discrimination of monkey vocalizations (Heffner & Heffner, 1984, 1986), voice and vocalization sensitive areas in humans and monkeys (Belin, Zatorre, & Ahad, 2002; Poremba et al., 2004), auditory short-term memory in monkeys (Colombo, D'Amato, Rodman, & Gross, 1990; Fritz et al., 2005), and auditory retrieval and recognition of objects in humans and primates (Fritz et al., 2005; Nakamura et al., 2001; Tranel, 2006).

Lesions of the anterior STG, including the temporal pole, impair performance during delays at 5s and beyond (Fritz et al., 2005). For the lesion study, monkeys were trained on an auditory version of DMS with randomly varying delays of 2–50s and retested after being given different types of bilateral temporal lobe lesions. Monkeys that received complete medial temporal removals or removal of only the rostral third of the STG were severely impaired even at the shortest delays, whereas those given lesions limited to the rhinal (i.e., perirhinal/entorhinal) cortices, which are known to produce severe impairment in both visual and tactile recognition, were unaffected even at the longest delays. These results suggested that the critical neural substrates for recognition in the auditory and visual modalities are anatomically distinct within the temporal lobe. The deficit in auditory recognition after medial temporal lesions may be due to a disconnection between the STG and other critical areas for memory function located outside the temporal lobe. Recent anatomical studies show that the auditory regions in PFC receive strong projections from the STG (Kondo, Saleem, & Price, 2003; Muñoz, Mishkin, & Saunders, 2009; Romanski et al., 1999), strengthening the hypothesis that a disconnection between STG and prefrontal regions may account for the deficit in auditory recognition memory after medial temporal lesions.

Lesions in the human temporal pole and anterior STG are also disruptive to auditory recognition (Clarke et al., 2002). Human imaging studies demonstrate that categorization of environmental sounds and speech comprehension involves the left temporal regions (Engelien et al., 1995; Zatorre, Meyer, Gjedde, & Evans, 1996). Our positron emission tomography imaging studies in monkeys also suggest the temporal polar region, particularly the left hemisphere, in processing of species-specific monkey vocalizations (Poremba et al., 2004). In normal human brains, the anterior temporal pole is activated while listening to sentences but not to scrambled sentences (Vandenberghe, Nobre, & Price, 2002). Even though there is no direct evidence that the human temporal pole is the same structure as the nonhuman primate temporal pole, it may be a functional homolog (Ding, Van Hoesen, Cassell, & Poremba, 2009). These studies, along with the anatomical and functional analogies between visual and auditory nervous systems, suggest that the temporal pole might be important in auditory short-term and recognition memory.

Single-unit activity of dTP neurons was recorded in rhesus monkeys performing a DMS task, as described previously, using simple to complex auditory stimuli (Ng et al., 2014). The findings suggested that dTP is a continuation of hierarchical cortical processing within the primate auditory nervous system, similar to the ventral "what" stream of the visual nervous system, and encodes individual, unique, auditory stimuli. Single-unit activity of dTP neurons was recorded in rhesus monkeys performing a DMS task using simple to complex auditory stimuli (Ng et al., 2014). In the analysis of single units, more than 50% of the neurons recorded in dTP showed significant evoked activity, including increases and decreases in activity, on match trials compared with 35% on nonmatch trials. Neurons of dTP encoded several task-relevant events during the DMS task, and the encoding of auditory cues in this region was associated with accurate recognition performance.

In the population analyses of dTP, most of the trial differences occurred within the first 100 ms of the cue onset. Specifically, population activity in dTP showed a very early developing match suppression mechanism starting within the first 30–60 ms to identical sound stimuli, analogous to that observed in the visual object identification pathway located ventral to dTP in inferior temporal cortex (ITC) and ventral temporal pole (vTP) (Desimone, 1996; Liu, Murray, & Jagadeesh, 2009; Nakamura & Kubota, 1996). This match suppression phenomenon has also been observed in the visual object identification pathway (e.g., ITC and vTP; Baylis & Rolls, 1987; Miller & Desimone, 1994; Miller et al., 1993; Nakamura & Kubota, 1995, 1996). Later in the trial sequence by the cue 2 offset period, match trials had increased activity above nonmatch trials exhibiting match enhancement.

During the retention interval, a little more than 20% of the neurons showed activity changes and these changes were intermittent and not sustained. This intermittent activity is in contrast to the larger percentage of neurons in higher-order visual cortices that can show sustained retention interval activity along areas in the ventral object identification pathway such as ITC and the vTP (Miller & Desimone, 1994; Miyashita & Chang, 1988; Nakamura & Kubota, 1995, 1996; Woloszyn & Sheinberg, 2009).

Neural correlates of visual working and recognition memory are generally thought of respectively as sustained delay activity, and activity that is modulated in response to identical stimulus presentations (match suppression or enhancement) (Desimone, 1996; Miller et al., 1996; Nakamura & Kubota, 1996). The reduced dTP population activity during repeated sound presentation (i.e., cue 2 during match trials) is a robust signature of recognition memory with the auditory system and similar to vTP and ITC along the visual object identification pathway (Baylis & Rolls, 1987; Miller & Desimone, 1994; Miller et al., 1993; Nakamura & Kubota, 1995; Woloszyn & Sheinberg, 2009). Studies in humans also show reduced activity in anterior STG when auditory stimuli were repeated or verbal stimuli were used during a short-term memory task (Buchsbaum & D'Esposito, 2009; Buchsbaum, Padmanabhan, & Berman, 2011; Dehaene-Lambertz et al., 2006).

Mechanisms that underlie match suppression are still under investigation (Liu et al., 2009; McMahon & Olson, 2007; Sawamura, Orban, & Vogels, 2006). During any given trial of the current task, match suppression may indicate that a recent, repeated sound is familiar within the context of a single trial. This type of sound presentation may be more quickly and efficiently processed than a relatively novel or different sound during nonmatch, sometimes referred to as bottom-up processing (Desimone, 1996; Grill-Spector, Henson, & Martin, 2006). Also in the dTP, there was increased neuronal firing to cue 2 on nonmatch trials when a different sound was presented, and this may provide an additional mechanism for differentiating between match and nonmatch trials that is complementary to the suppression mechanism.

The auditory memory correlates in dTP exhibit recognition memory both in their early suppression to the matching stimulus on cue 2 and later development of match enhancement. This is somewhat similar to higher-order visual areas that usually exhibit suppression or enhancement. However, retention interval activity is far less substantial and reliable than in vTP and ITC during visual working memory paradigms (Miller & Desimone, 1994; Miyashita & Chang, 1988; Nakamura & Kubota, 1995, 1996; Woloszyn & Sheinberg, 2009). This lack of consistent spike activity during the delay period may have functional consequences, as behaviorally memory performances are less robust and there are shorter forgetting thresholds for auditory stimuli compared with visual ones in nonhuman primates (Buffalo et al., 1999; D'Amato & Colombo, 1985; Fritz et al., 2005; Scott et al., 2012; Wright & Rivera, 1997). This transient, intermittent retention interval activity may be related to the slower acquisition of, and behavioral performance on, auditory recognition tasks and lowered performance at longer delays (Fritz et al., 2005; Funahashi, 2006; Ng, Malloy, Mishkin, & Poremba, 2006). However the retention interval activity in higher-order visual areas such as ITC can be more susceptible to intervening stimuli compared with regions in the prefrontal cortices (Desimone, 1996). That opens the question to whether sustained activation during the retention interval is the optimal mechanism for maintaining memory traces or rather reflects an attentional influence from a behaviorally engaging task. Another possibility is that other regions may be responsible for auditory memory traces for short-term memory and the PFC was the next possibility given that when using visual cues, this area shows cue-induced maintenance of activity across the delay in addition to recognition memory correlates (e.g., Bodner, Kroger, & Fuster, 1996; Desimone, 1996; Funahashi, 2006; Miller et al., 1996; Plakke et al., 2013).

Neuronal Recordings in the PFC

The lateral PFC (lPFC) was identified early on as an important neural substrate of visual short-term and recognition memory. Visual tasks that rely on short-term memory such as delayed response and DMS are severely impaired after bilateral lesions of the PFC (Goldman-Rakic, 1987; Jacobsen, 1935; Mishkin & Manning, 1978). Neuronal activity in the PFC demonstrates visual short-term memory correlates by exhibiting sustained changes, often elevation, but sometimes suppression, during the retention interval of short-term memory tasks (e.g., Fuster & Alexander, 1971; Miller et al., 1996; Shafi et al., 2007). Visual recognition memory is also apparent in the PFC primarily exhibiting match enhancement while some cells exhibit match suppression (e.g., Miller et al., 1996). The lPFC is hypothesized to be a general working/short-term memory area (Fuster, 2008; Gazzaley et al., 2007; Warden & Miller, 2010), but the difficulty in training monkeys to perform an auditory-only task assessing short-term and recognition memory has made the number of studies addressing the contribution of the PFC across modalities sparse (Arnott, Binns, Grady, & Alain, 2004; Rämä & Courtney, 2005).

The lPFC receives multimodal information from a number of regions and auditory information from the anterolateral, mediolateral, and caudolateral parabelt areas of the superior temporal region and the medial geniculate nucleus of the thalamus (Kaas & Hackett, 1998; Muñoz et al., 2009; Romanski et al., 1999). We have recently recorded from Brodmann's area 46, including dorsal and ventral banks of the principal sulcus (i.e., lPFC) (Plakke et al., 2013). We hypothesized that neurons in area 46 would encode both short-term and recognition memory as they do in visual tasks.

Single-unit activity of lPFC (area 46) was recorded in monkeys performing the same DMS task as the dTP recording experiment described in the previous section. The analysis of single units indicated that the number of cells with activity changes increased on match versus nonmatch trials supported by a clear match enhancement effect starting at 200–300 ms in the population activity and continuing until the animal's response (Figure 2). The increase in cue 2 activity for matching stimuli compared with nonmatching stimuli, and in relation to cue 1, is a robust recognition

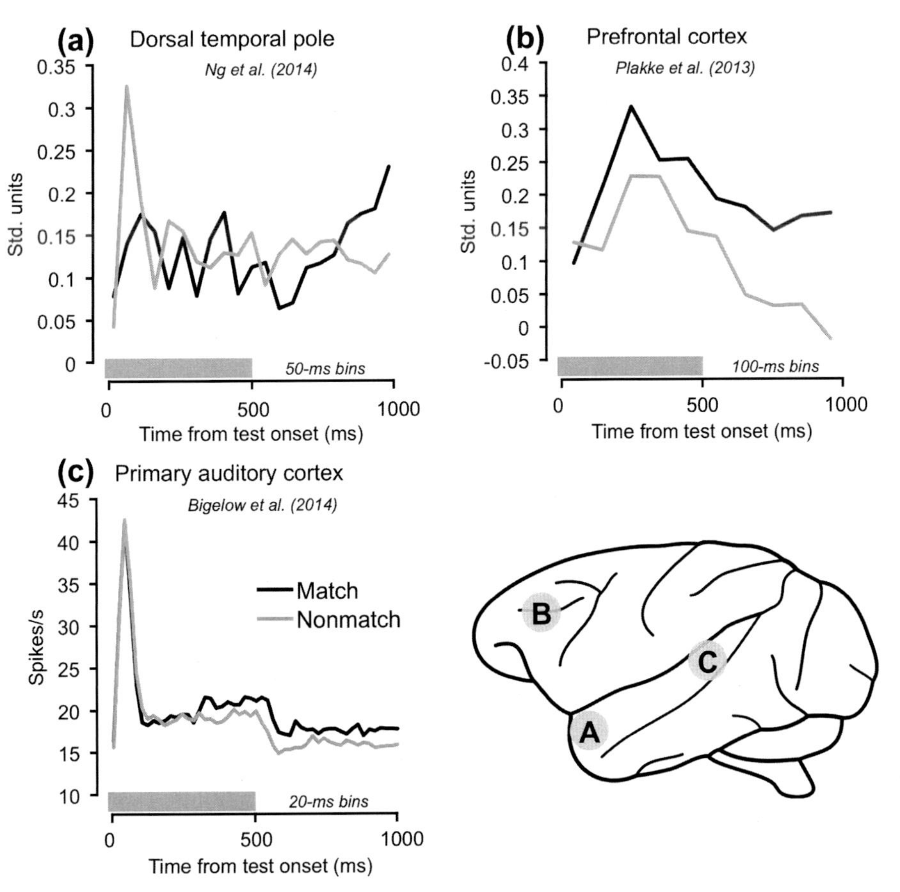

FIGURE 2 Comparison of the population activity in three cortical areas for matching and nonmatching test sounds (cue 2). The recording locations for panels A, B, and C are designated with their respective panel labels on the left hemisphere of the monkey brain representation, lower right. The gray bars above the abscissae indicate the test stimulus presentation period (0–500 ms from cue onset). Each of the summarized experiments was conducted using the same subjects and auditory short-term memory task.

memory correlate. The match enhancement correlate occurs later and is more robust than the early suppression recognition memory correlate seen in dTP. Neurons of the lPFC encoded several other task events such as response and reward conditions as well.

Single-unit activity during the retention interval was evident in only 7–15% of the cells across the delay period. The retention interval activity was typically early or late or intermittent, but rarely sustained to even a small degree. During the "early" portion of the delay 19% of the cells fired significantly above baseline and 21% of cells were active during the "late" portion of the interval. The largest portion of the cells, 41%, was active during the period between cue 2 and the response period on correct match trials.

A small percentage of neurons (10%) demonstrated increasing activity toward the end of the retention interval leading to the cue 2 presentations. This type of activity in visual DMS task is sometimes referred to as "climbing activity" (Asaad, Rainer, & Miller, 1998; Funahashi, Bruce, & Goldman-Rakic, 1989; Wallis, Anderson, & Miller, 2001). Although there was a small increase in retention interval activity in lPFC compared with dTP, it still sharply contrasts with the retention interval activity in visual memory tasks wherein up to 75% of cells demonstrate delay activity, and sustained delay activity is more commonplace (Asaad, Rainer, & Miller, 2000). It may be that there are still other regions that show robust activity during the retention interval because the monkeys can perform the auditory DMS task at 5-s delays. However, this lack of robust activity in lPFC, and perhaps a lack of strong memory encoding at these shorter memory delays, may help to account for the monkeys' inability to perform the DMS task at memory delays longer than 2 min. The memory encoding may be enough to support short-term memory but not enough to support longer term memory. Another possibility is that auditory memory may operate on a different time scale wherein auditory memory fades in seconds to a few minutes, whereas visual memory can last for minutes to hours on standard memory tasks.

The lPFC, despite differences in short-term memory neural correlates, appears to serve a common function for auditory and visual processing when it comes to recognition memory correlates. It is unknown whether this common function within the lPFC is encoded at the level of a single neuron for both modalities or within the same region but encoded with separate neurons.

Similar to visual DMS, neurons of both the dorsal and ventral lPFC demonstrated match enhancement (Figure 2). The latency of this developing enhancement for auditory stimuli occurred later than the match suppression observed in the dTP. However, match enhancement develops earlier in lPFC than in dTP. This match enhancement in lPFC starts during the cue 2 presentation and continues during the cue 2 offset and preresponse period, whereas dTP match enhancement starts near the end of the cue 2 offset period. This suggests that although the dTP neurons encode the match/nonmatch distinction first, the decision to make a response may first occur in the lPFC. One finding, which was a departure from visual DMS tasks, was the lack of robust memory delay activity. Some neurons fired intermittently during the delay, a small percentage ramped up their firing rate before the onset of the second cue 2, and a few neurons had small but sustained activity. Overall, this observed activity pattern was far different than when visual cues are used in a similar task. Also, this pattern of activity in lPFC was only slightly stronger during the memory delay than the dTP activity.

Although recognition was strong in dlPFC during the auditory DMS task, as it was in dTP, the delay memory correlate was weaker than during visual memory tasks. Additional research with visual short-term memory tasks suggests that although the PFC plays a prominent role, other areas earlier in the sensory processing pathway can also be involved (Constantinidis & Procyk, 2004; Pasternak & Greenlee, 2005; Postle, 2006). Visual short-term memory correlates have even been observed in the primary visual cortex and other early visual areas in the occipital lobe (Emrich, Riggall, Larocque, & Postle, 2013; Sligte, Scholte, & Lamme, 2009; Supèr, Spekreijse, & Lamme, 2001). Furthermore, there have been reports of auditory short-term memory correlates for simple tone frequencies in primary auditory cortex as well as auditory thalamus (Gottlieb, Vaadia, & Abeles, 1989; Sakurai, 1990). The current framework suggests that there may be several regions over which sensory inputs are integrated, task-relevant information is encoded, and information flowing in both bottom-up and top-down directions may be necessary to modulate goal-directed behavior (Fuster, 2008; Miller & Cohen, 2001).

Neuronal Recordings in A1: Primary Auditory Cortex

Experiments on auditory tasks that monkeys and other animals can learn, such as classical and operant conditioning, show neural plasticity changes in the auditory cortex (e.g., Blake, Heiser, Caywood, & Merzenich, 2006; Fritz, Elhilali, & Shamma, 2007; Takahashi, Yokota, Funamizu, Kose, & Kanzaki, 2011; Weinberger, 2010). Both neural recordings in auditory cortex and lesion studies suggest that this region is important for some aspects of associative memory (for a review, see Poremba, Bigelow, & Rossi, 2013). But although the elegant work of several laboratories demonstrates that plasticity is a hallmark of A1 with rapid shifting of frequency encoding from simple conditioning (e.g., Weinberger, 2010), questions remain about how complex signal processing and memory processes may be instantiated in the primary auditory cortex.

Together, the studies reviewed herein indicate that temporal pole neurons exhibit match suppression very early in the cue 2 presentation period for matching stimuli and increased activity for nonmatching cue 2 presentations, whereas match enhancement develops quite late after the cue 2 offset. This contrasts with the lPFC, which develops match enhancement earlier, during the cue 2 presentation and exhibits neuronal activity correlated with behavioral outcome before the response period. Both regions exhibit minimal short-term memory correlates during the retention interval compared with those observed in visual DMS tasks. A few lesion and recording studies in animals and studies of humans with brain damage provide partial evidence for a role of primary cortical areas in auditory recognition and short-term memory (Adriani et al., 2003; Fritz et al., 2005; Kusmierek, Malinowska, & Kowalska, 2007). For example, in humans, greater lateralization of damage to the left hemisphere within the primary auditory cortex impairs auditory recognition memory (Adriani et al., 2003). In monkey primary auditory cortex, Benson, Hienz, and Goldstein (1981) showed increased neuronal activity during delay periods in a task that required stimulus memory to perform sound identification or sound localization. It is also possible that primary auditory cortex might encode short-term and/or recognition memory because of the many areas of possible processing that exist in the lower auditory stream before information is projected to the primary auditory cortex. Additionally, correlates of visual short-term memory have been observed in early visual areas within the occipital lobe including primary visual cortex (Emrich et al., 2013; Sligte et al., 2009; Supèr et al., 2001), and the mediodorsal nucleus of the thalamus (Fuster & Alexander, 1973).

Primary auditory cortices in the monkey are defined anatomically as areas of cortex receiving projections from the auditory thalamus (medial geniculate nucleus) and subdivided into the core, belt, and parabelt regions (Kaas & Hackett, 2000). Individual neuronal recordings demonstrate several tonotopic maps and also some regional activity to monkey vocalizations (Morel, Garraghty, & Kaas, 1993; Rauschecker, 1998; Recanzone, 2000) and they project anteriorly along the STG to the temporal pole as well as to other more rostral regions such as the PFC (Galaburda & Pandya, 1983; Kaas & Hackett, 2000; Moran et al., 1987). With a tremendous amount of auditory processing occurring even before stimulus information reaches cortical levels, and no robust delay memory encoding further along the ventral object pathway in the temporal pole nor in the lPFC (Ng, 2011; Plakke et al., 2013), the earlier core regions may contribute to encoding particular aspects of the behavioral DMS task and exhibit correlates of short-term and/or recognition memory. Because dTP and PFC both exhibited recognition memory correlates but sparse short-term memory correlates across the retention interval, we assessed the neural correlates of auditory DMS in primary auditory cortex as well (Bigelow, Rossi, & Poremba, in press).

Several units in the primary auditory cortex exhibited significant changes in firing rate for portions of the retention interval. These changes were also rarely sustained, but were most frequently observed during the early and late portions of the retention interval. This inhibition was observed more frequently than excitation. The retention interval activity in primary auditory cortex was prominent at the population level as a decrease in firing in the roughly 1500ms leading up to the onset of cue 2. Neurons tuned to multiple frequencies tend to show a stronger decrease in activity during the delay period. Sustained changes in firing rate during studies of visual short-term memory tasks are interpreted as a neural representation of the sensory cue (Shafi et al., 2007), despite its absence from the environment. These changes in A1 firing rates across the retention interval were more prominent toward the end of the interval and may reflect the anticipation of the oncoming cue 2 or allow for an increased signal-to-noise ratio for behaviorally relevant sounds by decreasing the baseline firing rate before cue 2. These possibilities will need to be explored in the future.

At the population level, responses elicited on match trials were briefly suppressed early in the sound period relative to nonmatch trials. However, during the latter portion of the sound from approximately 300ms on, firing rates increased significantly for match trials and remained elevated throughout the wait period exhibiting match enhancement. These patterns of activity related to recognition memory correlates were similar to dTP and lPFC in our earlier studies. The overall findings suggest that early match suppression occurs in both A1 and the dTP, whereas later match enhancement occurs first in the PFC, followed by A1 and later in dTP. Because match enhancement occurs first in the PFC, we speculate that enhancement observed in A1 and dTP may reflect top-down feedback. Overall, our findings suggest that A1 forms part of the larger neural system recruited during auditory short-term memory.

Neuronal Recordings in R and RT: Additional Core Auditory Cortex Regions

There are two other core auditory regions in addition to A1, which are more rostral along the STG. Each of these core auditory regions (A1, R, and RT) receives information from the medial geniculate nucleus lemniscal pathway (Hackett, Stepniewska, & Kaas, 1998; Kaas & Hackett, 2000). These regions have not been explored during the auditory DMS task and may offer more clues as to whether neural correlates for recognition memory are occurring and their timing as well as whether there are regions that engender more robust retention interval activity when the monkey is being asked to perform a high-level cognitive task with simple and complex stimuli. With several stages of auditory processing occurring even before the information reaches cortical levels and no observed robust delay memory further along the ventral object pathway, the earlier core regions may contribute to encoding particular aspects of the behavioral DMS task. With the memory delay activity more prevalent in A1 than dTP and recognition encoding occurring in dTP before A1, the possibility exists that other regions such as R and RT within the core auditory area along the ventral pathway and bracketed by A1 and dTP, may also encode important components of the auditory DMS task. We have collected preliminary data from regions R and RT in one animal using the same auditory DMS paradigm (personal observations by author A.P.).

The preliminary evidence suggests that match suppression also occurs in both R and RT in the first 100 ms, similar to our findings in A1. At approximately 300 ms, during cue 2, there is more activity on match compared with nonmatch trials (i.e., match enhancement and this continues throughout the cue 2 offset and preresponse decision period). The timing of the development of match enhancement is very similar, though perhaps slightly slower, to the recording profile of the lPFC. These areas both exhibit strong recognition memory correlates.

Short-term memory correlates in region R appear very similar to A1 in that there is a decrease in the population activity that occurs prior to the cue 2 onset at approximately 2000–2500 ms and this is slightly earlier that A1 activity decreases during the retention interval. One prominent difference in retention interval activity was observed in region RT, which has the most dynamic response observed in any region to date during the retention interval. Population activity decreased during the first 600–800 ms of the retention interval, followed by a small, but steady increase above baseline until a subsequent decrease about 800–1000 ms before the onset of cue 2. Additionally, in the local field potential data (Figure 3), region R has a large significant evoked component during cue 2 offset responses to the nonmatch trials only, whereas RT has a large significant evoked component to the match trials only. This preliminary evidence suggests that there may be separate match and nonmatch processing mechanisms and regions.

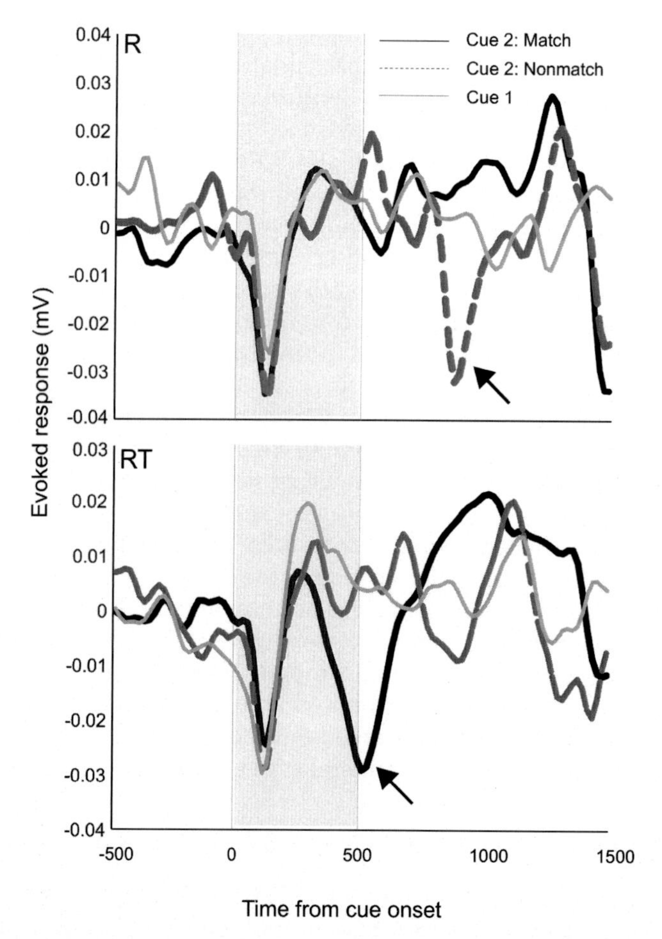

FIGURE 3 Local field potential (LFP) data from regions R and RT collected during the auditory DMS task comparing cue 1 with cue 2 of match and nonmatch trials averaged over 240 trials for individual sites. The gray regions represent the cue period. The arrows indicate the large components evoked differentially to match and nonmatch trials at different latencies. These differential, large component responses were visible in 60% of RT and 68% of R recording sites.

DISCUSSION

Sensory integration plays an important role in object identification and memory (e.g., Kulahci, Drea, Rubenstein, & Ghazanfar, 2013), and delineating where and how these disparate pieces of information overlap or become interpreted in a similar manner is important to ascertain. We have been interested in the similarities because they may be helpful in understanding how pieces of information from different sensory systems can be unified into a cohesive concept and whether the neural processing and organizing mechanisms are conserved across systems. The differences are also of great interest because they may illuminate the divergence in processing and help us to understand what limitations may exist in the system when combining or synthesizing a richer interpretation of the available sensory information.

In terms of behavioral performance, there are several similarities with some major differences evident as well. First of all, the monkeys can learn a version of DMS in both the auditory and visual modalities but with a large variance in the speed of acquisition and lower levels of asymptotic performance. Typically, when using visual stimuli during a noncomputerized DMS task, the average number of trials to acquire asymptotic behavioral performance with trial unique stimuli and a retention interval of 10s is 360 trials, which at 20 trials per day is a total of 18 training days (Mishkin & Delacour, 1975). This is in sharp contrast to acquisition for auditory versions of this task that can take roughly 6 months to a year to learn at 120 trials per day, or approximately 15,000 trials, with only a 5-s memory delay (Fritz et al., 2005; Ng et al., 2006). In fact, several laboratories have tried training animals in auditory DMS with limited success because, overall, it is quite difficult and requires some differences from training in the visual task (e.g., a lot of shaping, serial presentation of the stimuli, a go/no-go version of the task) (Fritz et al., 2005; Ng et al., 2006).

Once the monkeys do learn the auditory DMS task though, there are several similarities with visual memory performance. Proactive interference wherein the memory of a previous stimulus affects performance in the present occurs for both visual and auditory stimuli in analogous ways (Bigelow & Poremba, 2013a, 2013b). Generally, auditory short-term memory is highly susceptible to proactive interference after stimulus repetition, suggesting that a memory trace must first be established for any later presentations to decrease behavior (Bigelow & Poremba, 2013a). Also, as the stimulus set size grows smaller, proactive interference becomes worse and as intertrial intervals increase the effects of proactive interference lessen (Bigelow & Poremba, 2013b). This suggests that some mechanisms for memory across modalities must be similar, particularly across short retention intervals and basic recognition memory.

One caveat concerning basic memory mechanisms is a difference in auditory primacy memory as assessed by Wright and colleagues (Wright & Rivera, 1997). Whereas in visual processing, an increase in memory performance for both the early stimuli presented from a list (primacy) and the stimuli toward the end of the list (recency) are evident, auditory stimuli only exhibit a recency effect with no evidence of a primacy effect. This suggests that as we move toward long-term memory capabilities, the visual system is capable of longer retention spans for objects. This parallels the finding that auditory long-term memory, if it exists beyond associatively conditioned stimuli (for review, Poremba et al., 2013) and has a retention span far shorter than the visual and somatosensory systems, both of which can maintain object representations for several hours (Mishkin, 1979), whereas, short-term or working memory behavioral capabilities are similar with performance up to 2 min possible on the auditory task (Fritz et al., 2005; Ng et al., 2006).

Despite the similarities in the neuronal encoding of recognition memory as reviewed in this chapter, retention interval encoding in the analogous areas investigated so far is not as robust for auditory stimuli as for visual stimuli. Overall, the strongest results were found in the primary auditory cortex, and these differed from the visual system neuronal activity in other regions during the retention interval. Normally, with visual stimuli, many neurons exhibit robust firing increases during the retention interval and some of these are sustained across the entire interval. Firing rate changes during the retention interval occur sporadically to auditory stimuli in the dTP and dlPFC. Decreases in firing toward the end of the retention interval were more common in the primary auditory cortex, with the a mixture of increases and decreases in region RT. Nonhuman primate auditory and visual system processing pathways are similar in their anatomical arrangement, functional processing streams, and recognition memory encoding, but they differ in their behavioral memory performance, task acquisition, and memory retention neural correlates. Overall, the neural encoding during the retention interval is less robust than during visual delays. This difference in retention interval encoding may be a matter of exploring more regions or it may underlie the differences we see between auditory and visual memory. The less robust auditory memory may help to explain the disparities that are seen between human and nonhuman communication levels.

One possibility for modality difference in memory, as mentioned in the introduction, is the amount of subcortical processing that occurs in the auditory system versus the visual system. This may allow for quick and more thorough processing of the auditory stimuli before this information reaches the cortical processing stages. Another possibility accounting for memory capability differences between the auditory and visual systems are their cortical projections. The rhinal cortices facilitate memory processing and storage but the auditory projections are sparse to these regions (Munoz-Lopez et al., 2010), and lesions of the rhinal cortices significantly impair visual but not auditory DMS task performance.

There is also a large difference between auditory and visual stimuli in that the majority of natural auditory sounds have a temporal quality to them. Sounds can change frequently and quickly across time for the duration of the stimulus. Although moving visual images can have the same type of changes across time as auditory ones during the duration of the stimulus, many visual images are static and the viewer frequently has the opportunity to return to the visual image. However, many auditory stimuli, unless recorded, are unique in their presentations and cannot be revisited. Even the same speaker can produce the same word with different qualities when asked to repeat the sound. A recent paper about nonhuman primates reported very little dependence on temporal cues (Scott et al., 2012).

It may be that the best memory for auditory stimuli is engendered by auditory associative learning, the pairing of a consequence, such as aversive or appetitive reinforcement, with an auditory stimulus (for review, Poremba et al., 2013). These types of associative conditioning, whether classical or operant, tend to use simple stimuli, such as pure tones, and can be remembered over hours to days; much like visual stimuli and unlike the trial unique stimuli presented during DMS tasks. Other species, such as dogs, include some exceptional individual examples of animals that can learn over a 100 auditory cues when associated with particular visual stimuli or actions, and rewarded with food or attention (Kaminski, Call, & Fischer, 2004). This type of long-term auditory memory, where meaning is attached to the stimulus, is in sharp contrast to the maximum 2 min that has been shown with trial-unique, but familiar, stimuli or auditory memory (Fritz et al., 2005; Ng et al., 2006) and list length memory (Wright & Rivera, 1997).

Categorization may be another difference between auditory and visual system processing that leads to differences in memory capabilities. Many visual items are categorized on their physical properties such as shape, texture, color, placement in space in addition to their perceived state, liquid, solid, or gas, and their assigned or learned function. Studies in nonhuman primates show the ability to categorize visual objects (Freedman & Miller, 2008) and there are a large number of studies in other species as well (e.g., Castro & Wasserman, 2014; Wasserman, Castro, & Freeman, 2012). There are some studies in nonhuman primates showing perceptual categorization of auditory stimuli (e.g., Neider, 2012; Russ, Lee, & Cohen, 2007) and yet other studies with nonhuman primates suggesting more abstract categorization of auditory stimuli such as the number of auditory stimuli and the categorization of monkey vocalizations for specific food types (for review please see Tsunada & Cohen, 2014). The ability to categorize may be due to memory capabilities, or categorization may encourage long-term memory.

Other possibilities include physical differences between the input systems or molecular differences during memory storage that underlie the difference between visual and auditory short- and long-term memory systems. We have also considered the possibility that it may be a disadvantage to remember sound or unassociated sounds over the long-term. Perhaps by necessity, the system is set up to work properly without long-term memories. It may be advantageous to remember where visual items are because you might want to retrieve them later or be able to navigate in low light through your house at night. Nature and/or humans do not tend to completely rearrange objects in the three-dimensional landscape on a constant basis. This contrasts with the auditory system in which important stimuli are constantly rearranging themselves and rearranging to some extent the content of the auditory stream. At any point in a conversation in a coffee shop, the table will look the same, the person across the table will look the same, but the conversational sounds of others, and the music in the background has continuously been changing. In fact, auditory sounds in the background that are not important are necessary to tune out (i.e., not remember) because they may interfere with actively listening for and hearing the important stimuli when they do occur. For example, foreign languages are extremely easy to tune out and sound like gibberish unless we know some of the associated word meanings or a word sounds similar to our known languages. We may only want to remember sounds when they have a real-world consequence or are attached to a specific meaning.

In spite of the difference in memory performance for auditory and visual short-term memory, the similarities are numerous in terms of recognition memory, general form of retention interval encoding, overall layout of the anatomical processing pathways, involvement of multiple regions such as PFC and temporal lobe, and encoding of task-relevant events. When comparing across the regions, we have recorded from dTP, lPFC, and core auditory regions, which are interesting differences in the progression of match suppression to match enhancement occur (Figure 3). Match suppression happens quite early in the core auditory cortical regions and the dTP, within the first 100 ms (Figure 2). This suppression is the first indication in any area we have recorded from that a matching test stimulus has been detected. This type of match signal may project to other cortical regions such as the PFC to engender match enhancement and eventually lead to a decision and behavioral response. Match enhancement then develops about 300 ms after the test stimulus onset in the lPFC, and this pattern of activity to the matching stimulus is retained throughout the cue offset and preresponse periods (Figure 2). These enhancement effects were very early in the lPFC and of a large magnitude. This is in contrast to the late developing match enhancement in dTP and A1. We suggest that the early match suppression reflects bottom-up processing that detects change or no-change in the acoustic environment, whereas the late developing match enhancement in these regions may be the result of top-down feedback from the PFC in which task-relevant information is processed and the appropriate behavioral response is selected. Overall, these studies, and those of others, have highlighted numerous similarities and differences in the neural processing of short-term memory for auditory and visual stimuli. Continued studies will be necessary to form a unified theory of recognition memory both within the auditory system and across modalities.

Acknowledgments

Thank you to Drs Bethany Plakke and Chi-Wing Ng, and James Bigelow for their dedication to their graduate studies and the many projects contained within this chapter. I'd also like to thank Ryan Opheim, Breein Rossi, Iva Zdilar and many undergraduates for their assistance with data collection and to Drs Mortimer Mishkin and Richard Saunders for their support and assistance as well over the years. This research was supported by funding awarded to Amy Poremba from the University of Iowa and a grant from the National Institute on Deafness and Other Communication Disorders (DC0007156).

References

Adriani, M., Maeder, P., Meuli, R., Thiran, A. B., Frischknecht, R., Villemure, J. G., et al. (2003). Sound recognition and localization in man: specialized cortical networks and effects of acute circumscribed lesions. *Experimental Brain Research, 153*, 591–604.

Arnott, S. R., Binns, M. A., Grady, C. L., & Alain, C. (2004). Assessing the auditory dual-pathway in humans. *NeuroImage, 1*, 401–408.

Asaad, W. F., Rainer, G., & Miller, E. K. (1998). Neural activity in the primate prefrontal cortex during associative learning. *Neuron, 21*, 1399–1407.

Asaad, W. F., Rainer, G., & Miller, E. K. (2000). Task-specific neural activity in the primate prefrontal cortex. *Journal of Neurophysiology, 84*, 451–459.

Baddeley, A. (2003). Working memory: looking back and looking forward. *Nature Reviews Neuroscience, 4*, 829–839.

Barbas, H., Ghashghaei, H. T., Rempel-Clower, N. L., & Xiao, D. (2002). Anatomic basis of functional specialization in prefrontal cortices in primates. In J. Grafman (Ed.), *Handbook of neuropsychology* (pp. 1–27). Amsterdam: Elsevier.

Baylis, G. C., & Rolls, E. T. (1987). Responses of neurons in the inferior temporal cortex in short term and serial recognition memory tasks. *Experimental Brain Research, 65*, 614–622.

Belin, P., Zatorre, R. J., & Ahad, P. (2002). Human temporal-lobe response to vocal sounds. *Cognitive Brain Research, 13*, 17–26.

Benson, D. A., Hienz, R. D., & Goldstein, M. H., Jr. (1981). Single-unit activity in the auditory cortex of monkeys actively localizing sound sources: spatial tuning and behavioral dependency. *Brain Research, 219*, 249–267.

Bigelow, J., & Poremba, A. (2013a). Auditory memory in monkeys: costs and benefits of proactive interference. *American Journal of Primatology, 75*, 425–434.

Bigelow, J., & Poremba, A. (2013b). Auditory proactive interference in monkeys: the roles of stimulus set size and intertrial interval. *Learning and Behavior, 41*, 319–332.

Bigelow, J., Rossi, B. N., & Poremba, A. (2014). Neural encoding of recognition memory in primary auditory cortex. *Frontiers in Neuroscience, 8*, 250.

Blake, D. T., Heiser, M. A., Caywood, M., & Merzenich, M. M. (2006). Experience-dependent adult cortical plasticity requires cognitive association between sensation and reward. *Neuron, 52*, 371–381.

Bodner, M., Kroger, J., & Fuster, J. M. (1996). Auditory memory cells in dorsolateral prefrontal cortex. *Neuroreport, 7*, 1905–1908.

Buchsbaum, B. R., & D'Esposito, M. (2009). Repetition suppression and reactivation in auditory-verbal short-term recognition memory. *Cerebral Cortex, 19*, 1474–1485.

Buchsbaum, B. R., Padmanabhan, A., & Berman, K. F. (2011). The neural substrates of recognition memory for verbal information: spanning the divide between short- and long-term memory. *Journal of Cognitive Neuroscience, 23*, 978–991.

Buffalo, E. A., Ramus, S. J., Clark, R. E., Teng, E., Squire, L. R., & Zola, S. M. (1999). Dissociation between the effects of damage to perirhinal cortex and area TE. *Learning and Memory, 6*, 572–599.

Castro, L., & Wasserman, E. A. (2014). Pigeons' tracking of relevant attributes in categorization learning. *Journal of Experimental Psychology: Animal Learning and Cognition, 40*, 195–211.

Cheney, D. L., & Seyfarth, R. M. (1990). *How monkeys see the world*. Chicago, IL: University of Chicago Press.

Clarke, S., Bellmann-Thiran, A., Maeder, P., Adriani, M., Vernet, O., Regli, L., et al. (2002). What and where in human audition: selective deficits following focal hemispheric lesions. *Experimental Brain Research, 147*, 8–15.

Colombo, M., D'Amato, M. R., Rodman, H. R., & Gross, C. G. (1990). Auditory association cortex lesions impair auditory short-term memory in monkeys. *Science, 247*, 336–338.

Colombo, M., & Gross, C. G. (1994). Responses of inferior temporal cortex and hippocampal neurons during delayed matching to sample in monkeys (*Macaca fascicularis*). *Behavioral Neuroscience, 108*, 443–455.

Constantinidis, C., & Procyk, E. (2004). The primate working memory networks. *Cognitive, Affective, and Behavioral Neuroscience, 4*, 444–465.

Corballis, M. C. (2009). Language as gesture. *Human Movement Science, 28*, 556–565.

Curtis, C. E., & D'Esposito, M. (2003). Persistent activity in the prefrontal cortex during working memory. *Trends in Cognitive Science, 7*, 415–423.

Dehaene-Lambertz, G., Dehaene, S., Anton, J. L., Campagne, A., Ciuciu, P., Dehaene, G. P., et al. (2006). Functional segregation of cortical language areas by sentence repetition. *Human Brain Mapping, 27*, 360–371.

Desimone, R. (1996). Neural mechanisms for visual memory and their role in attention. *Proceedings of the National Academy of Sciences of the United States of America, 93*, 13494–13499.

Diehl, R. L., Lotto, A. J., & Holt, L. L. (2004). Speech perception. *Annual Review of Psychology, 55*, 149–179.

Ding, S. L., Van Hoesen, G. W., Cassell, M. D., & Poremba, A. (2009). Parcellation of human temporal polar cortex: a combined analysis of multiple cytoarchitectonic, chemoarchitectonic, and pathological markers. *The Journal of Comparative Neurology, 514*, 595–623.

D'Amato, M. R. (1973). Delayed matching and short-term memory in monkeys. In G. H. Bower (Ed.), *The psychology of learning and motivation: Advances in research and theory* (pp. 227–269). New York: Academic Press.

D'Amato, M. R., & Colombo, M. (1985). Extent and limits of the matching concept in monkeys (*Cebus paella*). *Journal of Experimental Psychology Animal Behavior Processes, 11*, 25–51.

Emrich, S. M., Riggall, A. C., Larocque, J. J., & Postle, B. R. (2013). Distributed patterns of activity in sensory cortex reflect the precision of multiple items maintained in visual short-term memory. *The Journal of Neuroscience, 33*, 6516–6523.

Engelien, A., Silbersweig, D., Stern, E., Huber, W., Doring, W., Frith, C., et al. (1995). The functional anatomy of recovery from auditory agnosia. A PET study of sound categorization in a neurological patient and normal controls. *Brain, 118*, 1395–1409.

Freedman, D. J., & Miller, E. K. (2008). Neural mechanisms of visual categorization: insights from neurophysiology. *Neuroscience and Biobehavioral Reviews, 32*, 311–329.

Fritz, J. B., Elhilali, M., & Shamma, S. A. (2007). Adaptive changes in cortical receptive fields induced by attention to complex sounds. *Journal of Neurophysiology, 98*, 2337–2346.

Fritz, J., Mishkin, M., & Saunders, R. C. (2005). In search of an auditory engram. *Proceedings of the National Academy of Sciences of the United States of America, 102*, 9359–9364.

Funahashi, S. (2006). Prefrontal cortex and working memory processes. *Neuroscience, 139*, 251–261.

Funahashi, S., Bruce, C. J., & Goldman-Rakic, P. S. (1989). Mnemonic coding of visual space in the monkey's dorsolateral prefrontal cortex. *Journal of Neurophysiology, 61*, 331–349.

Fuster, J. M. (2008). Overview of prefrontal functions: the temporal organization of behavior. In J. M. Fuster (Ed.), *The prefrontal cortex* (pp. 333–385). Boston, MA: Academic Press.

Fuster, J. M., & Alexander, G. E. (1971). Neuron activity related to short-term memory. *Science, 173*, 652–654.

Fuster, J. M., & Alexander, G. E. (1973). Firing changes in cells of the nucleus medialis dorsalis associated with delayed response behavior. *Brain Research, 61*, 79–91.

Fuster, J. M., Bauer, R. H., & Jervey, J. P. (1985). Functional interactions between inferotemporal and prefrontal cortex in a cognitive task. *Brain Research, 330*, 299–307.

Galaburda, A. M., & Pandya, D. N. (1983). The intrinsic architectonic and connectional organization of the superior temporal region of the rhesus monkey. *Journal of Comparative Neurology, 221*, 169–184.

Gazzaley, A., Rissman, J., Cooney, J., Rutman, A., Seibert, T., Clapp, W., et al. (2007). Functional interactions between prefrontal and visual association cortex contribute to top-down modulation of visual processing. *Cerebral Cortex* Suppl. 1, i125–i135.

Goldman-Rakic, P. S. (1987). Circuitry of primate prefrontal cortex and the regulation of behavior by representational memory. In F. Plum (Ed.), *Handbook of physiology* (Vol. 5) (pp. 373–417). Bethesda, MD: American Physiological Society.

Gottlieb, Y., Vaadia, E., & Abeles, M. (1989). Single unit activity in the auditory cortex of a monkey performing a short term memory task. *Experimental Brain Research, 74*, 139–148.

Grill-Spector, K., Henson, R., & Martin, A. (2006). Repetition and the brain: neural models of stimulus-specific effects. *Trends in Cognitive Science, 10*, 14–23.

Hackett, T. A. (2008). Anatomical organization of the auditory cortex. *Journal of the American Academy of Audiology, 19*, 774–779.

Heffner, H. E., & Heffner, R. S. (1984). Temporal lobe lesions and perception of species-specific vocalizations by macaques. *Science, 226*, 75–76.

Heffner, H. E., & Heffner, R. S. (1986). Effect of unilateral and bilateral auditory cortex lesions on the discrimination of vocalizations by Japanese macaques. *Journal of Neurophysiology, 56*, 683–701.

Iversen, S. D., & Mishkin, M. (1973). Comparison of superior temporal and inferior prefrontal lesions on auditory and non-auditory tasks in rhesus monkeys. *Brain Research, 55*, 355–367.

Jacobsen, C. F. (1935). Functions of the frontal association area in primates. *Archives of Neurologic Psychiatry, 33*, 558–569.

Kaas, J. H., & Hackett, T. A. (1998). Subdivisions of auditory cortex and levels of processing in primates. *Audiology and Neurootology, 3*, 73–85.

Kaas, J. H., & Hackett, T. A. (2000). Subdivisions of auditory cortex and processing streams in primates. *Proceedings of the National Academy of Sciences of the United States of America, 97*, 11793–11799.

Kaminski, J., Call, J., & Fischer, J. (2004). Word learning in a domestic dog: evidence for "fast mapping." *Science, 304*, 1682–1683.

Kikuchi, Y., Horwitz, B., & Mishkin, M. (2010). Hierarchical auditory processing directed rostrally along the monkey's supratemporal plane. *Journal of Neuroscience, 30*, 13021–13030.

Kondo, H., Saleem, K. S., & Price, J. L. (2003). Differential connections of the temporal pole with the orbital and medial prefrontal networks in macaque monkeys. *Journal of Comparative Neurology, 465*, 499–523.

Kondo, H., Saleem, K. S., & Price, J. L. (2005). Differential connections of the perirhinal and parahippocampal cortex with the orbital and medial prefrontal networks in macaque monkeys. *Journal of Comparative Neurology, 493*, 479–509.

Kulahci, I. G., Drea, C. M., Rubenstein, D. I., & Ghazanfar, A. A. (2013). Individual recognition through olfactory-auditory matching in lemurs. *Proceedings of Biological Science, 281*, 20140071.

Kusmierek, P., Malinowska, M., & Kowalska, D. M. (2007). Different effects of lesions to auditory core and belt cortex on auditory recognition in dogs. *Experimental Brain Research, 180*, 491–508.

Liu, Y., Murray, S. O., & Jagadeesh, B. (2009). Time course and stimulus dependence of repetition-induced response suppression in inferotemporal cortex. *Journal of Neurophysiology, 101*, 418–436.

Markowitsch, H. J., Emmans, D., Irle, E., Streicher, M., & Preilowski, B. (1985). Cortical and subcortical afferent connections of the primate's temporal pole: a study of rhesus monkeys, squirrel monkeys, and marmosets. *Journal of Comparative Neurology, 242*, 425–458.

McMahon, D. B., & Olson, C. R. (2007). Repetition suppression in monkey inferotemporal cortex: relation to behavioral priming. *Journal of Neurophysiology, 97*, 3532–3543.

Miller, E. K., & Cohen, J. D. (2001). An integrative theory of prefrontal cortex function. *Annual Review of Neuroscience, 24*, 167–202.

Miller, E. K., & Desimone, R. (1994). Parallel neuronal mechanisms for short-term memory. *Science, 263*, 520–522.

Miller, E. K., Erickson, C. A., & Desimone, R. (1996). Neural mechanisms of visual working memory in prefrontal cortex of the macaque. *Journal of Neuroscience, 16*, 5154–5167.
Miller, E. K., Li, L., & Desimone, R. (1993). Activity of neurons in anterior inferior temporal cortex during a short-term memory task. *Journal of Neuroscience, 13*, 1460–1478.
Mishkin, M. (1979). Analogous neural models for tactual and visual learning. *Neuropsychologia, 17*, 139–151.
Mishkin, M., & Delacour, J. (1975). An analysis of short-term visual memory in the monkey. *Journal of Experimental Psychology: Animal Behavior Processes, 4*, 326–334.
Mishkin, M., & Manning, F. J. (1978). Non-spatial memory after selective prefrontal lesions in monkeys. *Brain Research, 143*, 313–323.
Miyashita, Y., & Chang, H. S. (1988). Neuronal correlate of pictorial short-term memory in the primate temporal cortex. *Nature, 331*, 68–70.
Moran, M. A., Mufson, E. J., & Mesulam, M. M. (1987). Neural inputs into the temporopolar cortex of the rhesus monkey. *Journal of Comparative Neurology, 256*, 88–103.
Morel, A., Garraghty, P. E., & Kaas, J. H. (1993). Tonotopic organization, architectonic fields, and connections of auditory cortex in macaque monkeys. *Journal of Comparative Neurology, 335*, 437–459.
Munoz-Lopez, M. M., Mohedano-Moriano, A., & Insausti, R. (2010). Anatomical pathways for auditory memory in primates. *Frontiers in Neuroanatomy, 4*, 129.
Muñoz, M., Mishkin, M., & Saunders, R. C. (2009). Resection of the medial temporal lobe disconnects the rostral superior temporal gyrus from some of its projection targets in the frontal lobe and thalamus. *Cerebral Cortex, 19*, 2114–2130.
Nakamura, K., Kawashima, R., Sugiura, M., Kato, T., Nakamura, A., Hatano, K., et al. (2001). Neural substrates for recognition of familiar voices: a PET study. *Neuropsychologia, 39*, 1047–1054.
Nakamura, K., & Kubota, K. (1995). Mnemonic firing of neurons in the monkey temporal pole during a visual recognition memory task. *Journal of Neurophysiology, 74*, 162–178.
Nakamura, K., & Kubota, K. (1996). The primate temporal pole: its putative role in object recognition and memory. *Behavioural Brain Research, 77*, 53–77.
Neider, A. (2012). Supramodal numerosity selectivity of neurons in primate prefrontal and posterior parietal cortices. *Proceedings of the National Academy of Sciences of the United States of America, 109*, 1860–1865.
Ng, C. W. (2011). Behavioral and neural correlates of auditory encoding and memory functions in Rhesus Macaques. *Iowa Research Online (Thesis), 1041*, 1–173.
Ng, C., Malloy, M., Mishkin, M., & Poremba, A. (2006). Auditory delayed matching-to-sample in monkeys: comparison of training methods. *Society of Neuroscience* Abstract, Program No. 573 10, 547.
Ng, C. W., Plakke, B., & Poremba, A. (2009). Primate auditory recognition memory performance varies with sound type. *Hearing Research, 256*, 64–74.
Ng, C. W., Plakke, B., & Poremba, A. (2014). Neural correlates of auditory recognition memory in the primate dorsal temporal pole. *Journal of Neurophysiology, 111*, 455–469.
Pasternak, T., & Greenlee, M. W. (2005). Working memory in primate sensory systems. *Nature Reviews Neuroscience, 6*, 97–107.
Plakke, B., Ng, C. W., & Poremba, A. (2013). Neural correlates of auditory recognition memory in primate lateral prefrontal cortex. *Neuroscience, 244*, 62–76.
Poremba, A., & Bigelow, J. (2013). Neurophysiology of attention and memory processing. In Y. E. Cohen, A. N. Popper, & R. R. Fay (Eds.), *Neural correlates of auditory cognition* (pp. 215–250). New York, NY: Springer.
Poremba, A., Bigelow, J., & Rossi, B. (2013). Processing of communication sounds: contributions of learning, memory, and experience. *Hearing Research, 305*, 31–44.
Poremba, A., Malloy, M., Saunders, R. C., Carson, R. E., Herscovitch, P., & Mishkin, M. (2004). Species-specific calls evoke asymmetric activity in the monkey's temporal poles. *Nature, 427*, 448–451.
Poremba, A., Saunders, R. C., Crane, A. M., Cook, M., Sokoloff, L., & Mishkin, M. (2003). Functional mapping of the primate auditory system. *Science, 299*, 568–572.
Postle, B. R. (2006). Working memory as an emergent property of the mind and brain. *Neuroscience, 139*, 23–38.
Price, C. J., Wise, R. J., Warburton, E. A., Moore, C. J., Howard, D., Patterson, K., et al. (1996). Hearing and saying. The functional neuro-anatomy of auditory word processing. *Brain, 119*, 919–931.
Rämä, P., & Courtney, S. M. (2005). Functional topography of working memory for face or voice identity. *NeuroImage, 24*, 224–234.
Rämä, P., Poremba, A., Sala, J. B., Yee, L., Malloy, M., Mishkin, M., et al. (2004). Dissociable functional cortical topographies for working memory maintenance of voice identity and location. *Cerebral Cortex, 14*, 768–780.
Rauschecker, J. P. (1998). Cortical processing of complex sounds. *Current Opinion in Neurobiology, 8*, 516–521.
Rauschecker, J. P., & Scott, S. K. (2009). Maps and streams in the auditory cortex: nonhuman primates illuminate human speech processing. *Nature Neuroscience, 12*, 718–724.
Rauschecker, J. P., & Tian, B. (2000). Mechanisms and streams for processing of "what" and "where" in auditory cortex. *Proceedings of the National Academy of Sciences of the United States of America, 97*, 11800–11806.
Recanzone, G. H. (2000). Response profiles of auditory cortical neurons to tones and noise in behaving macaque monkeys. *Hearing Research, 150*, 104–118.
Romanski, L. M., Tian, B., Fritz, J., Mishkin, M., Goldman-Rakic, P. S., & Rauschecker, J. P. (1999). Dual streams of auditory afferents target multiple domains in the primate prefrontal cortex. *Nature Neuroscience, 2*, 1131–1136.
Russ, B. E., Lee, Y. S., & Cohen, Y. E. (2007). Neural and behavioral correlates of auditory categorization. *Hearing Research, 229*, 204–212.
Sakurai, Y. (1990). Cells in the rat auditory system have sensory-delay correlates during the performance of an auditory working memory task. *Behavioral Neuroscience, 104*, 856–868.
Sawamura, H., Orban, G. A., & Vogels, R. (2006). Selectivity of neuronal adaptation does not match response selectivity: a single-cell study of the FMRI adaptation paradigm. *Neuron, 49*, 307–318.
Scott, B. H., Mishkin, M., & Yin, P. (2012). Monkeys have a limited form of short-term memory in audition. *Proceedings of the National Academy of Sciences of the United States of America, 109*, 12237–12241.
Shafi, M., Zhou, Y., Quintana, J., Chow, C., Fuster, J., & Bodner, M. (2007). Variability in neuronal activity in primate cortex during working memory tasks. *Neuroscience, 146*, 1082–1108.

Sligte, I. G., Scholte, H. S., & Lamme, V. A. (2009). V4 activity predicts the strength of visual short-term memory representations. *Journal of Neuroscience, 29*, 7432–7438.

Supèr, H., Spekreijse, H., & Lamme, V. A. (2001). A neural correlate of working memory in the monkey primary visual cortex. *Science, 293*, 120–124.

Takahashi, H., Yokota, R., Funamizu, A., Kose, H., & Kanzaki, R. (2011). Learning-stage-dependent, field-speicific, map plasticity in the rat auditory cortex during appetitive operant conditioning. *Neuroscience, 199*, 243–258.

Tranel, D. (2006). Impaired naming of unique landmarks is associated with left temporalpolar damage. *Neuropsychology, 20*, 1–10.

Tsunada, J., & Cohen, Y. E. (2014). Neural mechanisms of auditory categorization: from across brain areas to within local microcircuits. *Frontiers in Neuroscience, 8*, 161.

Ungerleider, L. G., & Mishkin, M. (1982). In D. J. Ingle, M. A. Goodale, & R. J. W. Masfield (Eds.), *Analysis of visual behavior* (pp. 549–586). Cambridge, MA: MIT Press.

Vandenberghe, R., Nobre, A. C., & Price, C. J. (2002). The response of left temporal cortex to sentences. *Journal of Cognitive Neuroscience, 14*, 550–560.

Wallis, J. D., Anderson, K. C., & Miller, E. K. (2001). Single neurons in prefrontal cortex encode abstract rules. *Nature, 411*, 953–956.

Warden, M. R., & Miller, E. K. (2010). Task-dependent changes in short-term memory in the prefrontal cortex. *Journal of Neuroscience, 30*, 15801–15810.

Wasserman, E. A., Castro, L., & Freeman, J. H. (2012). Same-different categorization in rats. *Learning & Memory, 19*, 142–145.

Weinberger, N. M. (2010). The cognitive auditory cortex. In A. Rees, & A. Palmer (Eds.), *The Oxford handbook of auditory science: The auditory brain* (pp. 439–475). Oxford: Oxford University Press.

Weiskrantz, L., & Mishkin, M. (1958). Effects of temporal and frontal cortical lesions on auditory discrimination in monkeys. *Brain, 81*, 406–414.

Woloszyn, L., & Sheinberg, D. L. (2009). Neural dynamics in inferior temporal cortex during a visual working memory task. *Journal of Neuroscience, 29*, 5494–5507.

Wright, A. A., & Rivera, J. J. (1997). Memory of auditory lists by rhesus monkeys (*Macaca mulatta*). *Journal of Experimental Psychology: Animal Behavior Processes, 23*, 441–449.

Yeterian, E. H., & Pandya, D. N. (1991). Corticothalamic connections of the superior temporal sulcus in rhesus monkeys. *Experimental Brain Research*, 83, 168–284.

Zatorre, R. J., Meyer, E., Gjedde, A., & Evans, A. C. (1996). PET studies of phonetic processing of speech: review, replication and reanalysis. *Cerebral Cortex, 6*, 21–30.

CHAPTER

15

Brain Activity Related to the Retention of Tones in Auditory Short-Term Memory

Sophie Nolden

CERNEC, BRAMS, CRBLM, Département de Psychologie, Université de Montréal, Montreal, QC, Canada

This chapter reviews brain activity that contributes to auditory short-term memory (ASTM). Thereby, the focus is on the retention of low-level sensory features for a short period of time. Short-term memory is an important capacity that is involved in many cognitive processes. It allows us to retain sensory stimuli after they are gone from the environment, and thereby contributes importantly to, for example, learning, social behavior, and verbal communication. In the auditory domain, short-term memory is important for auditory scene analysis because many acoustic stimuli do not gain their meaning until they are interpreted in their temporal context. To participate successfully in a spoken conversation or to enjoy an artful piece of music, acoustic events such as chords or phonemes must be related to the preceding stimuli, which are no longer present in the environment and—if not self-produced—are often not immediately reproducible.

To guarantee the availability of stimulus representations after the stimulus is gone from the environment, the brain must retain activity patterns that code for these representations. Sensory representations must be held in an active state, or the brain must at least be in a state that allows easily reproducing those representations. As a result, previous experiences might importantly influence which information is retained and how it is retained. When we hear the beginning of the second movement of Beethoven's Symphony No. 9 and we want to retain what we have just heard, we might probably not only retain the representations of the sound stimuli themselves. Instead, we might recognize the music, possibly also retrieve the name of the piece, or remember an attended concert and the emotions related to it, and hold these additional codes in memory as well. It is even thinkable that we do not need to retain any sound representations at all in memory because we can reproduce the activity patterns coding for those sound representations at any time based on the long-term memory content that we associated with it (for example, the name of the piece). The processes taking place during the retention of auditory stimuli are thus manifold and complex. In this chapter, the focus is mainly on the retention of low-level sensory features and targets fundamental mechanisms of ASTM.

Retention of sensory stimuli in short-term memory has been associated with perceptual processes, especially when it comes to the retention of simple sensory stimuli that are not easily to be associated with implicit or explicit contents from long-term memory. For example, D'Esposito (2007) has suggested that activity patterns arising from perception (in posterior sensory areas and unimodal association cortices) are held in active state during retention in concert with multimodal association cortices, such as the prefrontal cortex, the parietal cortex, and the hippocampus (see also Goldman-Rakic, 1987; Petrides, 1991, 2005; Postle, 2006; Ruchkin, Grafman, Cameron, & Berndt, 2003). Consequentially, there should be some overlap between brain activity related to the perception and the retention of the same stimulus. The implication of sensory brain areas in short-term memory, namely the idea that stimulus representations preserve sensory features, has also been found for visual mental imagery (for example, Kosslyn, 1980, 2005), based on the observation that mental imagery of visual stimuli recruits similar brain areas as perception. During mental imagery, activity patterns that code for representations of sensory stimuli must be (re-) created, whereas they must be retained after the stimulus is gone in a short-term memory task. The common idea is that stimulus representations preserve sensory features and are supported by sensory brain areas.

Mechanisms of Sensory Working Memory
http://dx.doi.org/10.1016/B978-0-12-811042-3.00015-3

Retention-related brain activity has not only been revealed in sensory areas. Activity in the prefrontal cortex has been repeatedly observed during retention, and it has been suggested that this activity supports the retention of stimulus representations (Postle, 2006; see also D'Esposito, Postle, & Rypma, 2000; Fuster, 1973; Fuster & Alexander, 1971; Postle, Berger, & D'Esposito, 1999). The prefrontal cortex has been argued to be related to a number of functions that support and facilitate the retention of sensory representations, including top-down control over posterior sensory areas (Knight, Staines, Swick, & Chao, 1999) and attentional control including inhibition of interference (Passingham & Rowe, 2002). Some areas in the parietal cortex might also contribute to retention without necessarily compromising sensory representations. They have been found to be involved in retention of visual (Harrison, Jolicœur, & Marois, 2010; Robitaille, Grimault, & Jolicœur, 2009, 2010; Todd & Marois, 2004, 2005) and auditory representations (Gaab, Gaser, Zaehle, Jancke, & Schlaug, 2003; Grimault et al., 2014; Koelsch et al., 2009; Nolden, Grimault, et al., 2013; Zatorre, Evans, & Meyer, 1994). One reason for the observation of this parietal activity could be the attentional demands that come along with short-term memory tasks because parietal structures, in particular the right superior parietal lobe, have been discussed as part of a modality-independent network for attention (Belin et al., 1998; Farah, Wong, Monheit, & Morrow, 1989; Pardo, Fox, & Raichle, 1991; Paus et al., 1997; Zatorre, Mondor, & Evans, 1999).

Given that a close link has been claimed between perceptual and mnemonic processes, one would expect differences in retention-related brain activity between different sensory modalities, although there might be a partial overlap in related brain activity. Different sensory modalities process specific types of environmental information and rely on distinct neural systems. This should consequently also be reflected in brain activity related to retention. In the literature, there is a preponderance of studies on visual short-term memory, especially on verbal visual short-term memory, but expected differences between the retention of visual and auditory representations justify a focus on distinct sensory modalities (Lefebvre et al., 2013). Stimulus representations in different sensory modalities not only contain information about different physical stimulus properties, but their retention might also differ regarding the applied memory strategies, as for example internal rehearsal. Different sensory modalities deserve specific consideration, as well as the kind of material that is to be retained. In the auditory domain, especially language-related and music-related stimuli may be retained using specific memory strategies. The kind of to-be-memorized stimuli should therefore always be considered when interpreting brain activity observed during an ASTM task.

Although much research has been conducted on short-term memory in the past few decades, there are still many open questions that concern brain activity during retention. In the following sections, research that has isolated an electrophysiological correlate of retention that is specific for the auditory modality will be reviewed. Thereby, the role of capacity limits of ASTM will be discussed, as well as a tentative attempt to clarify how tone representations are retained, especially the role of internal rehearsal. Finally, studies that target the brain structures related to retention of tones will be presented.

THE SUSTAINED ANTERIOR NEGATIVITY—AN ELECTROPHYSIOLOGICAL CORRELATE OF THE RETENTION OF TONES

A series of recent studies has aimed to isolate an electrophysiological correlate of the retention of low-level sensory features in ASTM (Alunni-Menichini et al., 2014; Guimond et al., 2011; Lefebvre et al., 2013; Nolden, Bermudez et al., 2013). To isolate brain activity related to simple retention of tones in ASTM, three useful strategies were applied. The first strategy concerns the isolation of brain activity related to mnemonic activity from non-mnemonic activity that takes place at the same time. When participants are engaged in a short-term memory task, we expect that brain activity measured during this task does not only reflect memory-related processes. We would instead expect that many different processes take place during the task including non-mnemonic processes, as for example the perception of a visually presented fixation cross. One way to isolate memory-related activity is to vary memory-related demands parametrically. A significant part of mnemonic brain activity should then be influenced by this parametric manipulation, whereas non-mnemonic brain activity should be independent from it (Sternberg, 1969). A parametric manipulation of memory demands can be accomplished by varying the number of stimuli that have to be retained (memory set size). Brain activity related to retention should increase when more items are retained, thus reflecting the increased implication of the mnemonic system. Besides memory set size, performance in the memory task should also be taken into account when interpreting the measured brain activity. The memory set size indicates how many items are presented and should be retained by the participants, but does not tell us how many items the participants actually retain. Individuals vary in memory capacity and the number of to-be-retained items might sometimes exceed memory span. Therefore, indexes of memory capacity that can be related to specific memory loads have been proposed. One of these indexes is the memory capacity index K (Pashler, 1988; Cowan, 2001). It is calculated for every participant and for every load condition based on the participant's behavioral responses and represents the average number of retained stimuli in each memory load condition. It provides an estimate of the number of retained items, thus taking individual differences into account, whereas memory set size reflects the task demands in form of the number of to-be-retained items. Brain activity related to short-term memory should vary with the demands of the memory task; therefore, the number to-be-retained tones were varied parametrically.

The second strategy concerns the isolation of brain activity related to retention from brain activity related to other memory-related processes, such as encoding or retrieval. This can be achieved by using a design that allows independent control of distinct processing phases, as for example Sternberg's memory task (Sternberg, 1966). Sternberg's memory task can be divided into three distinct phases: presentation of memory tones, delay, and probe presentation. First, memory tones are presented and presumably encoded. Then, representations of these tones are presumably retained until probe presentation, when they have to be compared with the probe. Brain activity related to retention can thus be measured during the delay period (retention interval). In the very beginning of the delay period, there might still be some brain activity related to sensory persistence or encoding of memory tones, so that brain activity in the latter part of the delay period might more exclusively reflect retention, when excluding the last 100 ms or so before probe presentation, which might be confounded with brain activity related to the anticipation of the probe (Cheyne, Bakhtazad, & Gaetz, 2006). Hence, the time window that is most suitable for the study of brain activity related to retention is quite precise. Methods with fine temporal resolution, such as electroencephalography (EEG) and magnetoencephalography (MEG), are useful here because they directly capture electric potentials and magnetic fields caused by ongoing brain activity with submillisecond temporal precision. In addition to an excellent temporal resolution, MEG also provides a relatively good spatial resolution and can therefore be used to localize brain areas that contribute to the observed brain activity, including that related to the retention of tones.

The third strategy concerns the careful choice of acoustic stimuli in order to isolate brain activity related to the retention of simple sensory features. Material requiring more sophisticated memory strategies and the retention of additional information (for example, semantic codes) was avoided in order to isolate brain activity related to the retention of low-level sensory features. In particular, material that is associated with implicit or explicit long-term memory content should be controlled because participants might develop specific memory strategies (consider for example studies on the memory performance of chess experts, Chi, 1978; Schneider, Gruber, Gold, & Opwis, 1993). In the auditory domain, language-related and music-related stimuli are likely to induce specific processes during retention (see, for example, early electrophysiological studies demonstrating distinct brain activity for the retention of verbal acoustic material and musical notes, Lang, Starr, Lang, Lindinger, & Deecke, 1992; Pelosi, Hayward, & Blumhardt, 1998; Pratt, Michalewski, Barrett, & Starr, 1989; Pratt, Michalewski, Patterson, & Starr, 1989a, 1989b). The occurrence of complex language-related or music-related processes can be minimized by choosing tones that cannot be easily linked to musical scales, chords, instruments, or language. Brain imaging studies have provided evidence for the special role occupied by language-related material (Penney, 1989; Schumacher et al., 1996). Recently, brain imaging and electrophysiological studies have also isolated brain activity related to some aspects of the processing of music (for example, related to the processing of musical scale structure and harmonic relations, Besson, Faïta, Peretz, Bonnel, & Requin, 1998; Brattico, Tervaniemi, Näätänen, & Peretz, 2006; Janata et al., 2002; Koelsch, Gunter, & Friederici, 2000; Leino, Brattico, Tervaniemi, & Vuust, 2007; for expertise and learning see also Koelsch & Jentschke, 2008; Koelsch, Jentschke, Sammler, & Mietchen, 2007), which could possibly influence brain activity during retention. The strategy to use simple nonverbal and "non-musical" tones maximizes the possibility that activity patterns arising from perception are held in active state and that brain activity observed during retention is related to those stimulus representations, in addition to brain activity that helps to maintain the representations.

Based on these research strategies, an event-related potential (ERP) has been isolated and has been argued to constitute an electrophysiological correlate of the retention of tones measured with EEG. This ERP has been named sustained anterior negativity (SAN), because it is a sustained component that becomes more negative with increasing memory load, and is observed at frontocentral electrode sites. For example, Lefebvre et al. (2013) asked participants to retain sequences of two, four, or six pure tones varying in pitch. The sequences were preceded by five, three, or one white noise stimuli, respectively, to create sequences of equal duration. After a retention interval of 2000 ms, these tones were compared with a test sequence and participants indicated if the two sequences were the same (which was true in half of the trials) or different. The tones differing in pitch were so-called "non-musical" tones. The relation of the frequencies of the tone stimuli did not correspond to the well-tempered scale as the frequencies embracing an octave have been divided into seven equidistant parts (and not 12, as for the well-tempered scale, Trehub, Schellenberg, & Kamenetsky, 1999). The authors observed a sustained ERP in the latter part of the retention interval that became more negative with increasing memory load. This component was observed at frontocentral electrode sites, mainly at AFz, at which some other ERPs as the auditory N1 were also observed (see Figure 1, see also Guimond et al., 2011). Importantly, this component has been considered to be due to an active memory process exceeding early sensory register (Atkinson & Shiffrin, 1968; Neisser, 1967; see Nolden, Bermudez, et al., 2013 for further discussion) because it was not observed in a control condition with the same stimulation but with a task that was unrelated to short-term memory (the task was to indicate if the last two tones of the second sequence were ascending or descending in pitch).

The SAN also showed some other interesting characteristics related to memory capacity. Individual memory capacity was estimated with the memory capacity index K (Pashler, 1988; Cowan, 2001), which was calculated by multiplying memory load by the difference of the proportions of correct rejections and false alarms. This was done for every participant and every load condition, resulting in an estimator for the average number of items retained in each load condition per participant. The mean K can be predicted by the change in amplitude of the SAN (Alunni-Menichini et al., 2014; Lefebvre et al., 2013). Figure 2 displays the correlation of the mean K and the amplitude difference of the SAN between load conditions two and six. Participants who show a greater change in amplitude with increasing memory load also have a higher mean memory capacity, whereas those with a smaller change in amplitude have a lower memory capacity (Lefebvre et al., 2013). Note that the SAN has also been observed when participants retained simultaneously presented tones, so it most likely reflects the retention of individual tones rather than the complexity of sequences (Guimond et al., 2011).

The SAN is thus closely related to memory span. Memory span has been extensively investigated, and many studies have shown that short-term memory capacity is limited (Alvarez & Cavanagh, 2004; Brady, Konkle, & Alvarez, 2011; Cowan, 2005; Luck & Vogel, 1997; Wilken & Ma, 2004; Zhang & Luck, 2008). Memory span in ASTM has been investigated in an EEG experiment, using the SAN as an electrophysiological correlate of memory capacity (Alunni-Menichini et al., 2014). Sequences of two, four, six, and eight pure tones differing in pitch were presented and had to be retained for 2000 ms. The high memory loads of this experiment systematically

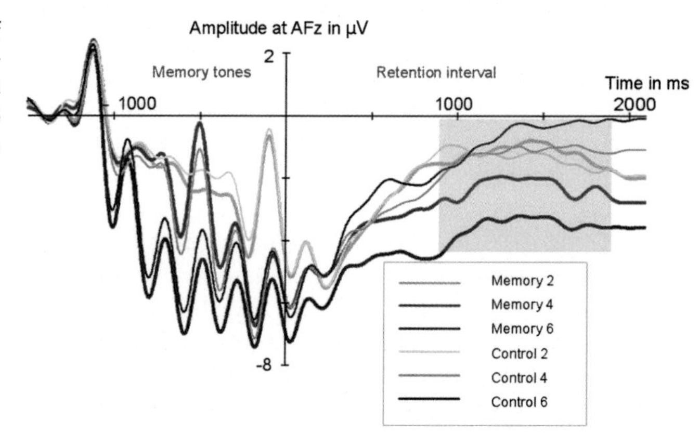

FIGURE 1 Event-related potentials of the auditory experiment of Lefebvre et al. (2013) at electrode site AFz. In the latter part of the retention interval (900–1900 ms after the onset of the retention interval, green window), a sustained component that varied with memory load was observed in memory blocks. *Adapted from Lefebvre et al. (2013, Figure 3, p. 2944) with permission.*

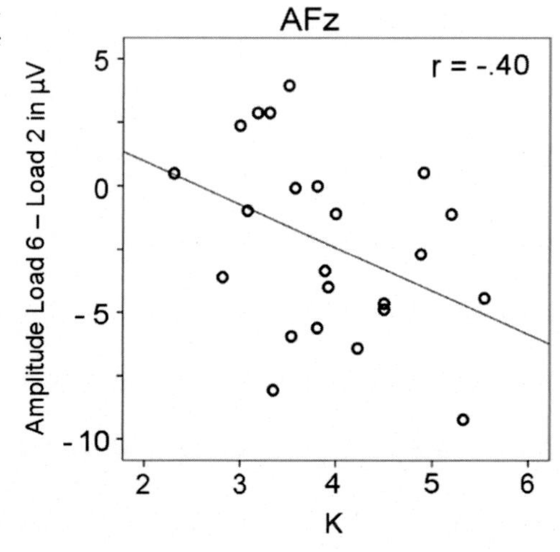

FIGURE 2 Correlation between electrophysiological data and mean memory capacity of the auditory experiment of Lefebvre et al. (2013). The difference between the SAN of condition load 3 and load 1 at AFz was correlated with the mean memory capacity index K. *Adapted from Lefebvre et al. (2013, Figure 5, p. 2945) with permission.*

exceeded the memory span of most participants, because the corresponding K (based on performance in the memory task) did not increase for higher memory load conditions, and reached a plateau at a relatively small load. The SAN showed a similar pattern as the memory capacity index K and became monotonically more negative for low memory loads that did not yet exceed memory span, but reached a plateau for higher memory loads that exceeded memory span. These electrophysiological results reflect capacity limits for ASTM, and a correlation with individual differences between electrophysiological data and behavioral responses provided strong support for the validity of the SAN as a correlate of brain activity related to the retention of tones.

Some further observations concerning the SAN are in line with the idea that brain activity related to the retention of sensory stimuli overlaps with brain activity related to the perception of these stimuli. First, the SAN has been isolated at frontocentral electrode sites where other auditory ERPs, such as the auditory N1, have also been observed (Guimond et al., 2011). The SAN appears to reflect the active retention of auditory representations, rather than automatic sensory responses to the stimuli. Second, Lefebvre et al. (2013) conducted a visual short-term memory experiment that was very similar to their ASTM experiment. Instead of retaining sequences of tones, participants retained sequences of simple visual stimuli. In this visual short-term memory experiment, they did not observe an SAN. Instead, they observed a sustained posterior contralateral negativity (SPCN), which was in turn specific for the visual modality. As the SAN, the SPCN is sensitive to memory load, reaches a plateau for higher memory loads that exceed memory span, and presumably reflects that sensory properties of the stimuli are preserved, for instance as information about the hemifield of the stimulus presentation is still preserved seconds after the stimuli are gone from the environment (Gratton, Shin, Fabiani, Chapter 7, in this volume; Klaver, Talsma, Wijers, Heinze, & Mulder, 1999; Vogel & Machizawa, 2004). An important difference to the SAN is the lateralized posterior scalp distribution, thereby establishing a link to visual perceptual processes. Third, the SAN has been found for different auditory stimuli and presentation modes; for example, for sequences of tones differing in pitch (Alunni-Menichini et al., 2014; Lefebvre et al., 2013), simultaneously presented tones with a single probe tone (Guimond et al., 2011), and sequences of tones differing in timbre (Nolden, Bermudez, et al., 2013).

In sum, the SAN can be considered as a modality-specific electrophysiological correlate of the retention of fundamental characteristics of auditory stimulation. It is a sustained ERP in the latter part of the retention interval that becomes more negative with an increasing number of retained tones and is correlated with individual differences in memory capacity.

THE ROLE OF INTERNAL REHEARSAL IN AUDITORY SHORT-TERM MEMORY

How are tones retained in ASTM? An influential model of short-term memory (Baddeley & Hitch, 1974) claimed that visual and auditory items are retained in different memory components, the so-called visuospatial sketch pad and the phonological loop. In a more recent version of the model, a multimodal episodic buffer has been added as a new component (Baddeley, 2000). Acoustic stimuli, verbal and nonverbal, are claimed to be held in a phonological store and to be constantly refreshed in a phonological loop. The idea of a loop in which representations of acoustic stimuli are repeated over and over again was taken on by many studies, even though some authors argued that verbal and nonverbal acoustic items are treated in different subsystems (Bertz, 1995; Pechmann & Mohr, 1992; see also Williamson, Baddeley, & Hitch, 2010). In case of a loop-like mechanism, brain activity that codes for representations of tones would be constantly re-created, possibly following a certain temporal pattern. In a store-like mechanism, on the other hand, representations of tones would be retained without constant refreshment. In that case, brain activity coding for representations of tones would be held in active state by a mechanism other than the phonological loop. Both possibilities can explain how tones are held in ASTM and are not exclusive. They could also interact. There might be a mechanism that can maintain, in an active state, a number of representations of auditory stimuli, and this mechanism could be "read out" into a loop if the content is compatible (e.g., speakable, singable), and this loop could probably provide a way to extend the life, and possibly expand the capacity, of the representations in the auditory store.

Based on these notions, a series of studies aimed to find out whether internal rehearsal is the predominant mechanism for the retention of sensory features in ASTM. EEG and MEG experiments were conducted to find out if rehearsal in a loop, that it is to say the constant refreshment of tone representations, is necessary for retention (Guimond et al., 2011; Nolden, Grimault, et al., 2013). Participants performed ASTM tasks that minimized the potential use of rehearsal, namely experiments on the retention of simultaneously presented tones and with articulatory suppression of the phonological loop.

Although internal rehearsal in a loop may help retention, it might be more useful as a mnemonic strategy in certain stimulus constellations than in others. Sequences of tones might be relatively easy to be rehearsed in a loop.

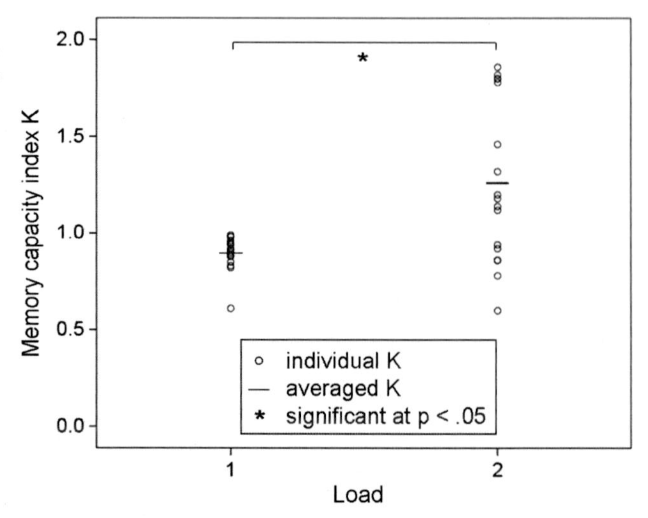

FIGURE 3 Memory capacity results of Nolden, Grimault, et al. (2013). Memory capacity index K for each participant, and the group average, for memory load 1 and memory load 2. *From Nolden, Grimault, et al. (2013, Figure 2, p. 388) with permission.*

We might, for example, hear a sequence of tones, and during retention, we loop through the sequence again and again by refreshing the representations of the different tones in the same order in which we have heard them. This would be especially useful when the order of the individual tones of the sequence matters for the task. If, on the other hand, simultaneously presented tones have to be retained, looping through what we have just heard might be more difficult, especially when the task requires the segregation of the percept into distinct tone representation. When simultaneously presented tones have to be retained as individual items, there is no temporal order in the stimulus presentation, which might make rehearsal in a loop more difficult to use as a memory strategy. In addition, most humans cannot produce more than one tone at a time.

According to these considerations, Nolden, Grimault, et al. (2013) used simultaneously presented pure tones differing in pitch in a short-term memory task to create a design that minimizes the probability of the use of rehearsal in a loop. Three load conditions were used: (1) a white noise stimulus was presented to both ears at the same time and no retention was required (load 0), (2) a pure tone was presented to both ears at the same time (load 1), (3) or two different pure tones were presented simultaneously, one to each ear (load 2). These tones were retained during a retention interval of 2000 ms and then compared with a single probe tone presented to both ears. The task was to indicate if the probe tone had been presented at the beginning of the trial, or not. The use of a single probe encouraged the participants to build individual tone representations in the load 2 condition, leading to the retention of two distinct items in memory rather than a single complex sound (e.g., a chord). Brain activity was measured with MEG. The behavioral data revealed that accuracy decreased with increasing memory load (load 0: 98%, load 1: 95%, load 2: 82%). Accuracy was still above the chance level in condition load 2, and the capacity index K suggested that participants retained more tones in condition load 2 than condition load 1, even though the variance between participants also increased (see Figure 3). During the retention interval, event-related magnetic fields varied with memory load. Figure 4 shows the signal on all sensors overlaid. Figure 4(a) shows the difference between the conditions load 1 and load 0, and Figure 4(b) shows the difference between the conditions load 2 and load 0. In the latter part of the retention interval, the signal increased monotonically with increasing memory load, as can be seen when comparing the amplitude of the signal during retention with the amplitude of the signal during the baseline period before stimulus onset (−400 to −200 ms). In agreement with the three research principles presented previously, the experiment revealed load-sensitive brain activity. Participants showed good memory performance when rehearsal in a loop was unlikely to occur; thus, the results point to the possible explanation that holding the tonal representations in active state might importantly contribute to retention.

A similar experiment also used simultaneously presented tones (Guimond et al., 2011). In addition, the hypothesized phonological loop was suppressed with concurrent articulation (Murray, 1968: Schendel & Palmer, 2007). One pure tone was presented to both ears at the same time (load 1), or two different pure tones differing in pitch were presented simultaneously (load 2, one tone to each ear). These tones were retained for 2000 ms and then compared with a single probe tone. In addition, in half of the blocks, participants counted overtly from one to 10 during the retention interval in order to suppress the use of a phonological loop. Brain activity was measured with EEG. Accuracy was lower for condition load 2 than condition load 1, and also lower in suppression blocks than in the no-suppression blocks. Although participants performed generally less well in suppression blocks, the typical load

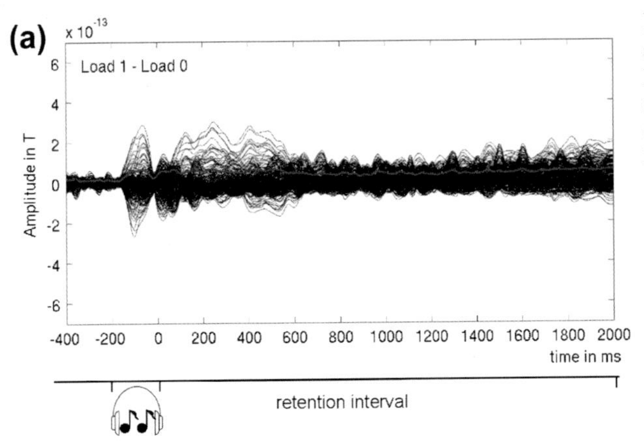

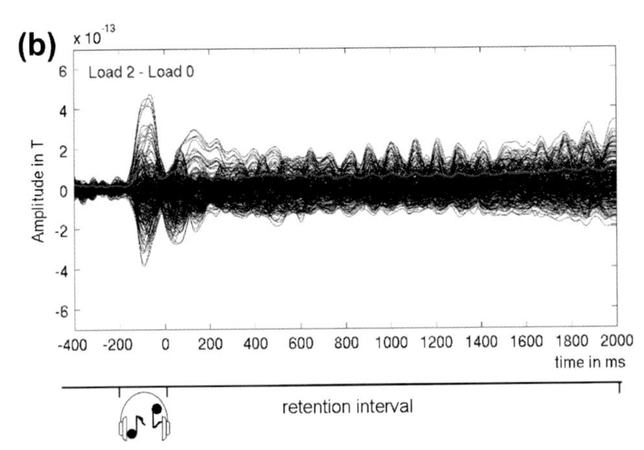

FIGURE 4 Differences of grand average event-related magnetic fields on all sensors of Nolden, Grimault, et al. (2013). Each black line represents one MEG sensor; the red line represents global field power. The sounds were presented from −200 to 0 ms, as indicated by the symbols underneath the graph. (a) Difference between load 1 and load 0. (b) Difference between load 2 and load 0. *From Nolden, Grimault, et al. (2013, Figure 4, p. 389) with permission.*

effect in the behavioral data suggest that there was a similar pattern in blocks with and without articulatory suppression. This was also reflected in the electrophysiological data. There was a main effect of suppression and of memory load. In blocks with articulatory suppression, the amplitude was in general more positive than in blocks without articulatory suppression. In blocks with and without articulatory suppression, an SAN was revealed and the SAN became more negative with increasing memory load. The load effect was thus present in both cases. Given the presence of a clear load-dependent SAN even in the presence of concurrent articulation, the authors suggested that the memory-related load effect was more likely to originate in a mechanism of active maintenance that did not rely on a loop-like internal rehearsal than via such a loop.

It is also remarkable that the results obtained at the sensor level for the retention of simultaneously presented memory tones are very similar to results obtained with sequences of tones, both in EEG (Alunni-Menichini et al., 2014; Lefebvre et al., 2013) and MEG experiments (Grimault et al., 2014). When participants retain sequences of tones, rehearsal in a loop might be easier than with simultaneously presented tones. In both cases—simultaneous and sequential tones—brain activity varied with memory load, thus suggesting that participants successfully retained the tones no matter which mnemonic processes would be more convenient. Although results with simultaneous and sequential tones are very similar on the surface level, source analyses on MEG data suggested some interesting differences that will be discussed in the last part of this chapter.

Based on these experiments, it is thus possible to assume that rehearsal in a loop is not always the predominant mechanism for the retention of tones and that a more fundamental storage mechanism likely plays a substantial role (see also Nolden, Bermudez, et al., 2013 for further discussion). Note that experiments with simultaneously presented tones and articulatory suppression only minimized the possibility of rehearsal in a loop. It is still thinkable that some rehearsal-based processes took place during retention, although they do not appear likely. But, because experiments with simultaneous and sequential tones have revealed similar load-sensitive brain activity (and not a degradation form sequential tones to simultaneous tones, as would have been expected if rehearsal in a loop is the predominant mechanism of retention), it is more plausible to suppose that the brain uses rehearsal and storage during retention based on the task demands, and dynamically applies the mnemonic activity that is most suitable for the given stimulus constellation and context of the memory task. Retention may thus not always predominantly rely on rehearsal in a loop, although internal rehearsal can certainly support retention and might constitute a suitable mnemonic strategy in many situations.

BRAIN AREAS RELATED TO THE RETENTION OF TONES

In the past few decades, imaging methods have been used to reveal brain areas contributing to the retention of tones, among them positron emission tomography (PET), functional magnetic resonance imaging (fMRI), and MEG. Thereby, several experimental approaches have been used, including contrasts between memory and control conditions, parametric designs with memory load manipulations, and correlations between brain activity and individual performance.

Some of these studies have used musical stimuli in their designs; it is therefore possible that the revealed brain areas contribute to the retention of low-level sensory features as well as to the retention of music-specific features. One approach was to use sequences of tones whereby participants indicated if two tones that occurred at certain positions in the sequence were the same or different. In other words, a single tone had to be retained while other tones were presented, one of them being the test tone. In a PET study, participants listened to melodies of eight tones and indicated either if the first tone had a higher or lower pitch than the second ("second tone comparison"), or if the first tone had a higher or lower pitch than the last ("eighth tone comparison"). Thus, in the eighth tone comparison condition, participants had to maintain one tone in ASTM for about 3s and to ignore intervening tones while they could compare the tones immediately in the second tone comparison condition. In the second tone comparison condition, the right inferior frontal lobe was more activated than in a passive listening condition. In the eighth tone comparison condition, the right inferior frontal lobe, the right temporal lobe, the parietal, and the insular cortex were more activated than in a passive listening condition. The authors concluded that frontal and temporal cortices in the right hemisphere are part of a network that underpins the retention of pitch in short-term memory (Zatorre et al., 1994).

Comparison of two tones within a sequence has also been used in a very similar, more recent study (Gaab et al., 2003). Brain activity was measured with fMRI and sparse temporal sampling was used to avoid contamination by scanner noise. Participants listened to a sequence that contained six or seven tones differing in pitch. They compared the first tone of the sequence with the last or second-to-last tone, which was indicated by instruction after the sequence. They thus had to retain the first tone of the sequence and did not need to retain some of the following tones. They also had to retain the last two tones of the sequence to do the task, which occurred at different positions depending on whether a six- or a seven-tone sequence had been presented. The authors observed that the superior temporal gyrus, the supramarginal gyrus, frontal regions, parietal regions, and the cerebellum were more activated in the memory task than in a motor control task. Interestingly, they found that the left inferior frontal gyrus was activated, whereas Zatorre et al. (1994) found that the right inferior frontal gyrus was more activated in their eighth tone comparison condition (which is closest to the memory condition of Gaab and collaborators). Gaab and collaborators also found that activity in the supramarginal gyrus (especially in the left hemisphere) and the dorsolateral cerebellum were significantly correlated with individual performance in the memory task. These two experiments in which participants compared tones that occurred at different positions in a sequence both revealed the implication of temporal as well as frontal and parietal areas in the retention of tones (Gaab et al., 2003; Zatorre et al., 1994). They were less conclusive about the lateralization of the implicated brain areas. Zatorre and collaborators found a preponderance of areas in the right hemisphere, whereas Gaab and collaborators found more areas in the left hemisphere to be related to the retention of tones. It thus seems that there is no clear lateralization when it comes to the retention of individual tones and that both hemispheres contribute to the retention of tones.

Other authors have focused more on the retention of sequences of tones (during a silent delay phase), whereby the retention of music and the role of musical expertise also played a role. In an fMRI study by Koelsch et al. (2009), participants were asked to remember either the pitch or the vowel of a sequence of sung syllables, with or without simultaneous singing of a well-known song (suppression conditions). Compared with a passive listening condition, a network of activity in premotor cortices, the planum temporale, the inferior parietal lobe, the anterior insula, subcortical structures, and the cerebellum were involved in the retention of pitch and vowel information. The authors concluded that similar structures were involved in the retention of pitch and speech information. However, it is not clear whether participants encoded and retained either pitch or vowel identity singly or always encoded and retained both, treating the presented sounds as a complex unit incorporating both dimensions. The design might thus not have been sensitive for differences between the retention of pitch or speech. Overall, the results reveal the implication of similar structures in ASTM as found in previous studies, including temporal, frontal, and parietal areas and the cerebellum (Gaab et al., 2003; Zatorre et al., 1994). It thus appears that a relatively broad network of brain areas is recruited for the retention of acoustic material. The specific demands of each task may presumably determine how these implicated structures work together.

Another fMRI study demonstrated how the retention of tones is influenced by musical expertise, which is presumably accompanied by the predominant use of music-specific retention mechanisms (Schulze, Mueller, & Koelsch, 2011). The study targeted brain activity related to mnemonic strategies that could be used when retaining sequences of tones. Two kinds of stimuli were used in a memory sequence: (1) tones of the same musical scale or (2) tones without a tonal relation to each other. When tones belonged to the same musical scale, they might be more easily chunked and might also be more easily related to implicit or explicit knowledge of music. The task was completed by a group of musicians and a group of non-musicians. Contrasting these two populations is especially interesting because musicians might use different or additional mnemonic strategies compared with musically untrained participants. In particular, associating parts of the tonal sequences to long-term memory content or to chunk tones of the tonal sequences might have been easier for musicians than for non-musicians, as suggested by the behavioral results. Musicians performed better when retaining tones of the same musical scale compared to retaining tones that did not belong to the same musical scale. When retaining tones of the same musical scale, musicians also performed better than non-musicians, who did not show performance differences between the two kinds of sequences. Only musicians showed increased brain activity in the right superior frontal gyrus, the right inferior precentral gyrus, the right premotor cortex, and the left intraparietal sulcus for the tonal sequences compared with the atonal sequences and compared with the non-musicians. The authors argue that these brain areas might be especially important for strategies that make use of the melodic structure of sound sequences. These results are very interesting because they show that the recruitment of brain areas related to the retention of tones depends on expertise and applied mnemonic strategy (see also Schulze, Zysset, Mueller, Friederici, & Koelsch, 2011). The cortical networks that underpin the retention of tones can thus be very complex and dynamic and depend highly on mnemonic strategies and stimulus properties.

Two recent MEG studies used designs with "non-musical" tones to reveal brain areas underpinning the retention of low-level stimulus features in ASTM (Grimault et al., 2014; Nolden, Grimault, et al., 2013). One important goal was to minimize the impact of music-specific encoding and retention mechanisms; therefore, the authors used similar strategies to those already used for the isolation of the SAN: a parametric variation of memory load, isolation of brain activity during the delay period, and the use of simple tones that cannot easily be associated to language-related or music-related long-term memory contents. MEG is a useful method here, because it is a time-sensitive method that allows the isolation of brain activity related to retention from brain activity related to other phases of the ASTM task. Its spatial resolution is adequate to perform reliable source localization models that reveal the implicated brain structures.

In their MEG study, Nolden, Grimault, et al. (2013) varied memory load to isolate brain areas supporting the retention of tones in ASTM (see previous section for methodological details). Source localizations were performed with the maximum entropy of the mean approach (Amblard, Lapalme, & Lina, 2004; Grova et al., 2006), interpolated into individual anatomic scans of the participants obtained with magnetic resonance imaging, and normalized in a common template. Multiple linear regression analyses were performed on the sources obtained during the retention interval using a general linear model (Cox, 1996) with the localized sources as a function of individual memory capacity. Individual memory capacity was estimated with the memory capacity index K that was calculated for each load condition and each participant. The model thus targeted brain activity that increased when the number of retained tones in short-term memory increased. K is a functionally useful regressor, because it represents how many tones were actually retained (and not how many tones should be retained) and because it takes individual differences in memory capacity into account. The model revealed significant clusters in the right superior temporal gyrus, right inferior temporal gyrus, left superior temporal gyrus, right superior parietal lobule, right precuneus, and the right inferior frontal gyrus (see Figure 5). As in previous imaging studies based on contrasts between memory and control conditions (Gaab et al., 2003; Koelsch et al., 2009; Zatorre et al., 1994), a network of temporal, frontal, and parietal areas was revealed to be related to the retention of tones. As in some previous work, there was a certain dominance of right hemispheric structures.

Grimault et al. (2014) conducted a very similar MEG experiment using the same tone stimuli. Instead of simultaneously presented tones (Nolden, Grimault, et al., 2013), they presented sequences of seven stimuli. The first stimulus was always a white noise, and within the six following stimuli, there were two, four, or six tones differing in pitch (the remaining sounds consisted of white noise that had the same duration as the tones). These sequences were retained for 2000 ms and then compared with a test sequence, which was the same or different from the memory sequence. The authors used the same source localization method and the same general linear model based on the memory capacity index K as Nolden, Grimault, et al. (2013). The model revealed that activity in the left and right superior parietal

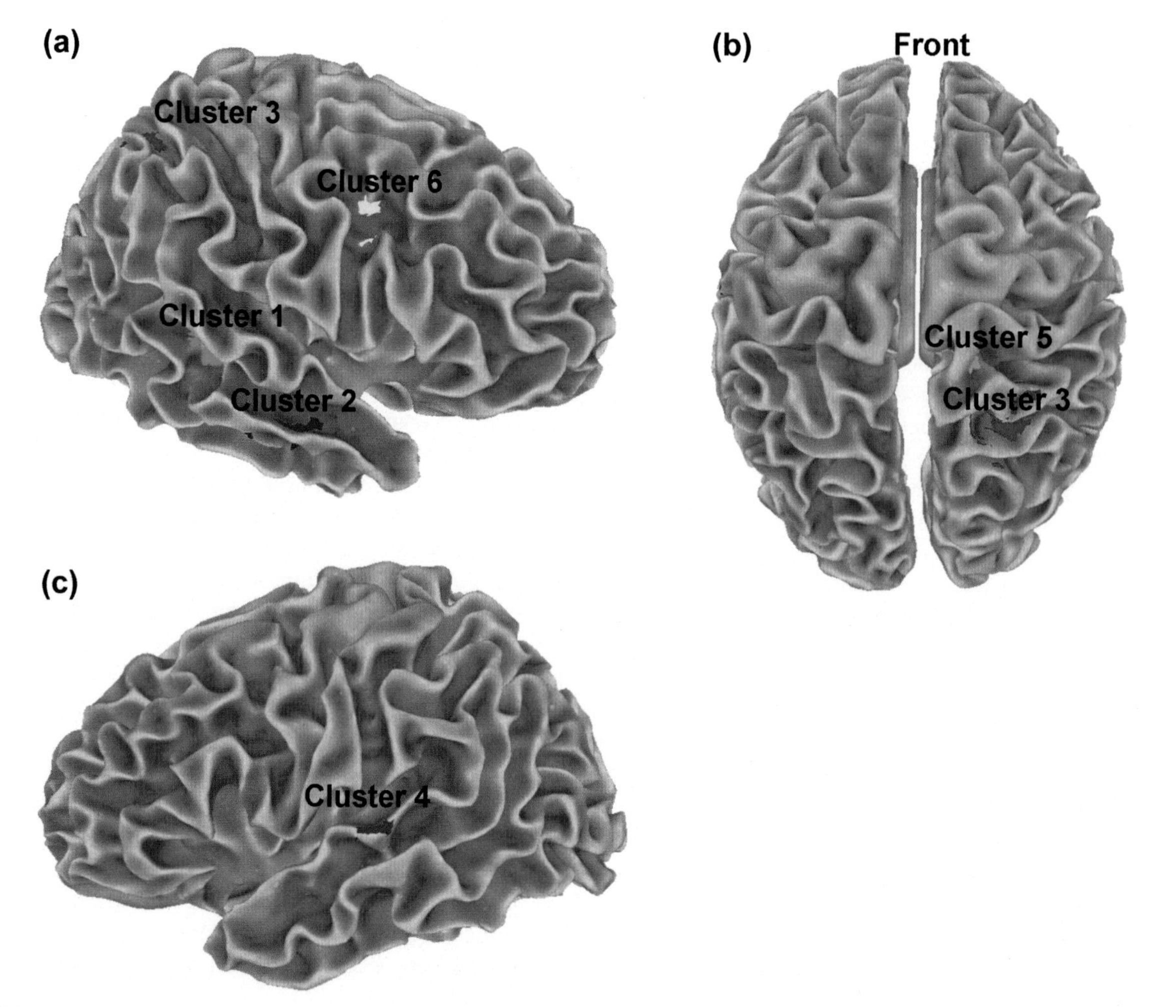

FIGURE 5 Results of the covariation analyses on the source localization of Nolden, Grimault, et al. (2013). Results are shown as *t*-statistical maps superimposed on a three-dimensional brain anatomical image of the white matter–gray matter interface. (a) Right brain view, with clusters 1, 2, 3, and 6. (b) Top brain view, with clusters 3, and 5. (c) Left brain view, with cluster 4. *From Nolden, Grimault, et al. (2013, Figure 6, p. 390) with permission.*

lobe, left inferior frontal gyrus/precentral gyrus and right middle/inferior frontal gyrus/precentral gyrus, left middle temporal gyrus and right middle/inferior temporal gyrus, left and right superior/middle temporal gyrus, left superior/middle frontal gyrus, and right parahippocampal gyrus increased when more tones were retained. As in previous studies, temporal and frontal structures were recruited for the retention of tones differing in pitch.

It is quite interesting to compare the MEG experiments from Nolden, Grimault, et al. (2013) and Grimault et al. (2014), because they used the same tone frequencies, the same method to measure brain activity, and the same model to localize brain areas based on behavioral data. The main difference was the presentation mode (simultaneously vs sequentially presented tones) and the probe (single probe tone vs tone sequence), which may induce the use of distinct mnemonic strategies. Hence, the differences between these two studies might reflect how different memory strategies change the involved networks of brain activity. In the following sections, the role of the brain areas revealed by these two MEG studies will be discussed.

Both studies revealed that activity in superior temporal gyri increased with an increasing number of retained tones, consistent with previous brain-imaging studies on the retention of pitch (Gaab et al., 2003; Koelsch et al., 2009; Zatorre et al., 1994). For both sequentially and simultaneously presented tones, the implication of this structure was almost symmetric (Grimault et al., 2014; Nolden, Grimault, et al., 2013). Secondary auditory cortices are located in these areas. Their implication in the retention of tones supports recent models of short-term memory that propose that retention of sensory stimuli takes place in brain areas involved in their perception, perhaps as a reactivation of patterns that arose during perception (D'Esposito, 2007; D'Esposito et al., 2000; Postle, 2006; Postle et al., 1999; Ruchkin et al., 2003). The close relation between perception and retention could also account for the differences concerning the lateralization between the two studies. Whereas the involved clusters were almost symmetric for sequentially presented sounds (Grimault et al., 2014), there was a preponderance for right hemispheric structures supporting the retention of simultaneously presented tones (Nolden, Grimault, et al., 2013). There are some indications that the right auditory cortex has a relative specialization for fine spectral processing and that the left auditory cortex has a relative specialization for fine temporal processing (for a review see Zatorre, Belin, & Penhune, 2002). This could be one possible explanation for the inconsistent results regarding hemispheric specialization revealed by previous studies on the retention of tones. We conjecture that simultaneously presented tones required a more elaborate spectral analysis than sequentially presented tones because distinct tone representations had to be isolated from a simultaneous presentation. Sequentially presented tones, on the other hand, might have promoted temporal processing because the order of the stimuli was task-relevant, which was not the case for simultaneously presented tones. These differences might have led to a stronger implication of the right auditory cortex for simultaneously presented tones than for sequentially presented tones. Consequentially, the stronger activation of the auditory cortex of one hemisphere may also have supported the recruitment of a network of other brain areas in the same hemisphere. If the hemispheric differences between the two MEG studies are indeed from relative specializations in perceptual processing, it is interesting to observe them long after encoding, during the retention interval. Hence, tone representations would retain various sensory features of the original presentation, as in visual short-term memory (VSTM, Dell'Acqua, Sessa, Toffanin, Luria, & Jolicœur, 2010; Fortier-Gauthier, Moffat, Dell'Acqua, McDonald, & Jolicœur, 2012; see also Gratton, 1998; Gratton, Corballis, & Jain, 1997; Gratton et al., Chapter 7 in this volume). Even though relative specialization of the auditory cortices could explain variations in results regarding the implication of the right and left hemisphere in ASTM (e.g., Gaab et al., 2003; Grimault et al., 2014; Nolden, Grimault, et al., 2013; Zatorre et al., 1994), more research is needed to clarify why group studies sometimes point to a stronger implication of the right than the left hemisphere in ASTM and vice versa, possibly with a stronger focus on individual differences.

For sequentially presented tones, more frontal structures were revealed than for simultaneously presented tones, for instance, middle frontal gyri and precentral gyrus in the left hemisphere. Those frontal structures that were not present for simultaneously presented tones might possibly be important for the retention of tone sequences, perhaps regarding the encoding of a tone contour, the use of chunking, or perhaps rehearsal in a loop, given that all of these possibilities are plausible for sequentially presented tones but minimized for simultaneously presented tones. The retention of simultaneous tones was supported by areas in the right parietal cortex that showed increased activity when the number of retained tones increased, namely the right superior parietal lobe and the right precuneus (Nolden, Grimault, et al., 2013). Parietal areas supporting the retention of sequentially presented tones were less expanded and not lateralized (Grimault et al., 2014).

Interestingly, parietal structures have also been argued to be involved in the retention of visual representations (Grimault et al., 2009; Harrison et al., 2010; Robitaille et al., 2009, 2010; Todd & Marois, 2004, 2005). A common finding is that brain activity in the intraparietal sulcus varies with the number of retained visual items; therefore, the intraparietal sulcus has been suggested as an important node in the representations of visual stimuli (Todd & Marois, 2004) or to code for the spatial location of the items (Harrison et al., 2010). MEG studies on the retention of tones did typically not emphasize the retention of the spatial location of tones (Nolden, Grimault, et al., 2013; Grimault et al., 2014) and revealed activity in parietal structures (e.g., superior parietal cortex) varying with memory load as well, even when the presentation of the tones was identical for both ears (Grimault et al., 2014). In addition, musicians showed greater activity in the intraparietal sulcus when they retained tonal sequences compared with atonal sequences in an fMRI study. This effect was not present for nonmusicians (Schulze, Mueller, et al., 2011). Although it is possible that the parietal lobe assumes different functions in ASTM and in VSTM, areas in the parietal lobe might also play a more general, modality-independent role in short-term memory. One could speculate that the parietal lobe, especially the intraparietal sulcus, is recruited when sensory representations are referred to an internal or external reference, or whenever relationships between stimuli are important, as might be the case for visual items in a certain spatial configuration or tones in a specific musical scale. The current state of knowledge does not allow drawing strong conclusions about the role of the parietal cortex in ASTM or in short-term memory in general. Nonetheless, it is remarkable to find an overlapping structure in ASTM and in VSTM with fMRI and MEG, especially given that ERPs related to the retention of tones or visual stimuli, as the SAN or the SPCN, typically do not show spatial overlap in measured voltage fields. Future research is needed to elucidate the role of the parietal cortex in ASTM, including its functional relation to activity in other brain areas.

In conclusion, temporal, frontal, and parietal areas seem to underpin the retention of tones in ASTM. These structures have been found in PET (Zatorre et al., 1994) and fMRI studies (Gaab et al., 2003; Koelsch et al., 2009) that contrasted memory and control conditions. The cerebellum also seems to be important because activity in the cerebellum has been found in fMRI studies (Gaab et al., 2003; Koelsch et al., 2009). Activity in the cerebellum is further correlated with memory capacity (Gaab et al., 2003). MEG experiments; (Grimault et al., 2014; Nolden, Grimault, et al., 2013) that correlated individual memory capacity to brain activity also pointed to temporal, frontal, and parietal structures (activity in the cerebellum was not analyzed in these MEG studies).

Despite these similarities between different studies, the obtained results were never identical and revealed different cerebral networks that depended on specific task properties and memory strategies. Brain areas related to the retention of tones are not preponderantly located in one hemisphere either, even though many studies revealed the dominance of either the right or the left hemisphere. We might therefore conclude that both hemispheres contribute to the retention of tones, whereas lateralized brain networks are recruited depending on specific task demands. The cortical network that is implicated in the retention of tones is thus dynamic and adapts to specific task demands.

CONCLUSIONS

In this chapter, research on brain activity related to the retention of tones in ASTM was reviewed. EEG studies have isolated the SAN, an electrophysiological correlate of the retention of tones. The SAN is a sustained ERP that occurs during the retention of tones, it is observed at frontocentral electrode sites, varies with memory load, and is correlated with individual differences in memory capacity. The SAN is also specific for the auditory modality, thus supporting the notion that short-term memory is at least partly modality specific. Studies on the brain structures based on PET, fMRI, and MEG, underpinning retention pointed to the implication of temporal, frontal, parietal structures, and the cerebellum. These studies further revealed that the brain network implicated in the retention of tones adapts dynamically to specific task requirements.

References

Alunni-Menichini, K., Guimond, S., Bermudez, P., Nolden, S., Lefebvre, C., & Jolicœur, P. (2014). Saturation of auditory short-term memory causes a plateau in the sustained anterior negativity event-related potential. *Brain Research, 1592*, 55–64.

Alvarez, G. A., & Cavanagh, P. (2004). The capacity of visual short-term memory is set both by visual information load and by number of objects. *Psychological Science, 15*, 106–111.

Amblard, C., Lapalme, E., & Lina, J.-M. (2004). Biomagnetic source detection by maximum entropy and graphical models. *IEEE Transactions on Biomedical Engineering, 51*(3), 427–442.

Atkinson, R. C., & Shiffrin, R. M. (1968). Human memory: a proposed system and its control processes. In K. W. Spence, & J. T. Spence (Eds.), *The psychology of learning and motivation: Advances in research and theory* (Vol. 2). New York: Academic Press.

Baddeley, A. (2000). The episodic buffer: a new component of working memory? *Trends in Cognitive Science, 4*, 417–423.

Baddeley, A. D., & Hitch, G. J. (1974). Working memory. In G. A. Bower (Ed.), *The psychology of learning and motivation* (pp. 47–89). New York, NY: Academic Press.

Belin, P., McAdams, S., Smith, B., Savel, S., Thivard, L., Samson, S., et al. (1998). The functional anatomy of sound intensity discrimination. *The Journal of Neuroscience: The Official Journal of the Society for Neuroscience, 18*(16), 6388–6394.

Berz, W. L. (1995). Working memory in music: a theoretical model. *Music Perception, 12*, 353–364.

Besson, M., Faïta, F., Peretz, I., Bonnel, A.-M., & Requin, J. (1998). Singing in the brain: independence of lyrics and tunes. *Psychological Science, 9*(6), 494–498.

Brady, T. F., Konkle, T., & Alvarez, G. A. (2011). A review of visual memory capacity: beyond individual items and towards structured representations. *Journal of Vision, 11(5)* 4, 1–34.

Brattico, E., Tervaniemi, M., Näätänen, R., & Peretz, I. (2006). Musical scale properties are automatically processed in the human auditory cortex. *Brain Research, 1117*, 162–174.

Cheyne, D., Bakhtazad, L., & Gaetz, W. (2006). Spatiotemporal mapping of cortical activity accompanying voluntary movements using an event-related beamforming approach. *Human Brain Mapping, 27*(3), 213–229.

Chi, M. T. H. (1978). Knowledge structures and memory development. In R. S. Siegler (Ed.), *Children's thinking: What develops?* (pp. 73–96). Hillsdale, New York: Erlbaum.

Cowan, N. (2001). The magical number 4 in short-term memory: a reconsideration of mental storage capacity. *Behavioral and Brain Sciences, 24*, 87–185.

Cowan, N. (2005). *Working memory capacity*. Hove, East Sussex, UK: Psychology Press.

Cox, R. W. (1996). AFNI: software for analysis and visualization of functional magnetic resonance neuroimages. *Computers and Biomedical Research, 29*, 162–173.

Dell'Acqua, R., Sessa, P., Toffanin, P., Luria, R., & Jolicœur, P. (2010). Orienting attention to objects in visual short-term memory. *Neuropsychologia, 48*, 419–428.

D'Esposito, M. (2007). From cognitive to neural models of working memory. *Philosophical Transactions of the Royal Society B: Biological Sciences, 362*(1481), 761–772.

D'Esposito, M., Postle, B. R., & Rypma, B. (2000). Prefrontal cortical contributions to working memory: evidence from event-related fMRI studies. *Experimental Brain Research, 133*, 3–11.

Farah, M. J., Wong, A. B., Monheit, M. A., & Morrow, L. A. (1989). Parietal lobe mechanisms of spatial attention: modality-specific or supramodal? *Neuropsychologia, 27*(4), 461–470.

Fortier-Gauthier, U., Moffat, N., Dell'Acqua, R., McDonald, J. J., & Jolicœur, P. (2012). Contralateral cortical organization of information in visual short-term memory: evidence from lateralized brain activity during retrieval. *Neuropsychologia, 59*, 1748–1758.

Fuster, J. M. (1973). Unit activity in prefrontal cortex during delayed-response performance: neural correlates of transient memory. *Journal of Neurophysiology*, *36*(1), 61–78.

Fuster, J. M., & Alexander, G. E. (1971). Neuron activity related to short-term memory. *Science*, *173*, 652–654.

Gaab, N., Gaser, C., Zaehle, T., Jancke, L., & Schlaug, G. (2003). Functional anatomy of pitch memory–an fMRI study with sparse temporal sampling. *NeuroImage*, *19*, 1417–1426.

Goldman-Rakic, P. S. (1987). Circuitry of primate prefrontal cortex and regulation of behavior by representational memory. In V. B. Mountcastle, & F. Plum (Eds.), *Handbook of physiology, Vol. 5: The nervous system, higher functions of the brain* (pp. 373–417). Bethesda, MD: American Physiological Society.

Gratton, G. (1998). The contralateral organization of visual memory: a theoretical concept and a research tool. *Psychophysiology*, *35*, 638–647.

Gratton, G., Corballis, P. M., & Jain, S. (1997). Hemispheric organization of visual memories. *Journal of Cognitive Neuroscience*, *9*, 92–104.

Gratton, G., Shin, E., & Fabiani, M. (2015). *Hemispheric organization of visual memory: Analyzing working memory with brain measures* (Chapter 7, this volume).

Grimault, S., Nolden, S., Lefebvre, C., Vachon, F., Hyde, K., Peretz, I., et al. (2014). Brain activity is related to individual differences in the number of items stored in auditory short-term memory for pitch: evidence from magnetoencephalography. *NeuroImage*, *94*, 96–106.

Grimault, S., Robitaille, N., Grova, C., Lina, J.-M., Dubarry, A.-S., & Jolicœur, P. (2009). Oscillatory activity in parietal and dorsolateral prefrontal cortex during retention in visual short-term memory: additive effects of spatial attention and memory load. *Human Brain Mapping*, *30*, 3378–3392.

Grova, C., Daunizeau, J., Lina, J.-M., Bénar, C. G., Benali, H., & Gotman, J. (2006). Evaluation of EEG localization methods using realistic simulations of interictal spikes. *NeuroImage*, *29*, 734–753.

Guimond, S., Vachon, F., Nolden, S., Lefebvre, C., Grimault, S., & Jolicœur, P. (2011). Electrophysiological correlates of the maintenance of the representation of pitch objects in acoustic short-term memory. *Psychophysiology*, *48*, 1499–1508.

Harrison, A., Jolicœur, P., & Marois, R. (2010). "What" and "Where" in the intraparietal sulcus: an fMRI study of object identity and location in visual short-term memory. *Cerebral Cortex*, *20*(10), 2478–2485.

Janata, P., Birk, J. L., Van Horn, J. D., Leman, M., Tillmann, B., & Bharucha, J. J. (2002). The cortical topography of tonal structures underlying western music. *Science*, *298*, 2167–2170.

Klaver, P., Talsma, D., Wijers, A. A., Heinze, H.-J., & Mulder, G. (1999). An event-related brain potential correlate of visual short-term memory. *Neuroreport*, *10*, 2001–2005.

Knight, R. T., Staines, W. R., Swick, D., & Chao, L. L. (1999). Prefrontal cortex regulates inhibition and excitation in distributed neural networks. *Acta Psycholica*, *101*, 159–178.

Koelsch, S., Gunter, T., & Friederici, A. D. (2000). Brain indices of music processing: "Nonmusicians" are musical. *Journal of Cognitive Neuroscience*, *12*(3), 520–541.

Koelsch, S., & Jentschke, S. (2008). Short-term effects of processing musical syntax: an ERP study. *Brain Research*, *1212*, 55–62.

Koelsch, S., Jentschke, S., Sammler, D., & Mietchen, D. (2007). Untangling syntactic and sensory processing: an ERP study of music perception. *Psychophysiology*, *44*, 476–490.

Koelsch, S., Schulze, K., Sammler, D., Fritz, T., Müller, K., & Gruber, O. (2009). Functional architecture of verbal and tonal working memory: an FMRI study. *Human Brain Mapping*, *30*, 859–873.

Kosslyn, S. M. (1980). *Image and mind*. Cambridge, MA: Harvard University Press.

Kosslyn, S. M. (2005). Mental images and the brain. *Cognitive Neuropsychology*, *22*(3/4), 333–347.

Lang, W., Starr, A., Lang, V., Lindinger, G., & Deecke, L. (1992). Cortical DC potential shifts accompanying auditory and visual short-term memory. *Electroencephalography and Clinical Neurophysiology*, *82*(4), 285–295.

Lefebvre, C., Vachon, F., Grimault, S., Thibault, J., Guimond, S., Peretz, I., et al. (2013). Distinct electrophysiological indexes for the maintenance of acoustic and visual stimuli in short-term memory. *Neuropsychologia*, *51*(13), 2939–2952.

Leino, S., Brattico, E., Tervaniemi, M., & Vuust, P. (2007). Representations of harmony rules in the human brain: further evidence from event-related potentials. *Brain Research*, *1142*, 169–177.

Luck, S. J., & Vogel, E. K. (1997). The capacity of visual working memory for features and conjunctions. *Nature*, *390*, 279–281.

Murray, D. J. (1968). Articulation and acoustic confusability in short-term memory. *Journal of Experimental Psychology*, *78*, 679–684.

Neisser, U. (1967). *Cognitive psychology*. New York: Appleton-Century-Crofts.

Nolden, S., Bermudez, P., Alunni-Menichini, K., Lefebvre, C., Grimault, S., & Jolicœur, P. (2013). Electrophysiological correlates of the retention of tones differing in timbre in auditory short-term memory. *Neuropsychologia*, *51*(13), 2740–2746.

Nolden, S., Grimault, S., Guimond, S., Lefebvre, C., Bermudez, P., & Jolicœur, P. (2013). The retention of simultaneous tones in auditory short-term memory: a magnetoencephalography study. *NeuroImage*, *82*, 384–392.

Pardo, J. V., Fox, P. T., & Raichle, M. E. (1991). Localization of a human system for sustained attention by positron emission tomography. *Nature*, *349*(6304), 61–64.

Pashler, H. (1988). Familiarity and visual change detection. *Perception and Psychophysics*, *44*(4), 369–378.

Passingham, R. E., & Rowe, J. B. (2002). Dorsal prefrontal cortex: maintenance in memory or attentional selection? In D. T. Stuss, & R. T. Knight (Eds.), *Principles of frontal lobe function* (pp. 221–232). Oxford: Oxford University Press.

Paus, T., Zatorre, R. J., Hofle, N., Caramanos, Z., Gotman, J., Petrides, M., et al. (1997). Time-related changes in neural systems underlying attention and arousal during the performance of an auditory vigilance task. *Journal of Cognitive Neuroscience*, *9*(3), 392–408.

Pechmann, T., & Mohr, G. (1992). Interference in memory for tonal pitch: implications for a working-memory model. *Memory & Cognition*, *20*, 314–320.

Pelosi, L., Hayward, M., & Blumhardt, L. D. (1998). Which event-related potentials reflect memory processing in a digit-probe identification task? *Cognitive Brain Research*, *6*, 205–218.

Penney, C. G. (1989). Modality effects and the structure of short-term verbal memory. *Memory and Cognition*, *17*(4), 398–422.

Petrides, M. (1991). Monitoring of selections of visual stimuli and the primate frontal cortex. *Proceedings of the Royal Society of London. Series B: Biological Sciences*, *246*(1317), 293–298.

Petrides, M. (2005). Lateral prefrontal cortex: architectonic and functional organization. *Philosophical Transactions of the Royal Society B: Biological Sciences, 360*(1456), 781–795.

Postle, B. R. (2006). Working memory as an emergent property of the mind and brain. *Neuroscience, 139*(1), 23–38.

Postle, B. R., Berger, J. S., & D'Esposito, M. (1999). Functional neuroanatomical double dissociation of mnemonic and executive control processes contributing to working memory performance. *Proceedings of the National Academy of Sciences, 96*(22), 12959–12964.

Pratt, H., Michalewski, H. J., Barrett, G., & Starr, A. (1989). Brain potentials in a memory-scanning task. I. Modality and task effects on potentials to the probes. *Electroencephalography and Clinical Neurophysiology, 72*(5), 407–421.

Pratt, H., Michalewski, H. J., Patterson, J. V., & Starr, A. (1989a). Brain potentials in a memory-scanning task. II. Effects of aging on potentials to the probes. *Electroencephalography and Clinical Neurophysiology, 72*(6), 507–517.

Pratt, H., Michalewski, H. J., Patterson, J. V., & Starr, A. (1989b). Brain potentials in a memory-scanning task. III. Potentials to the items being memorized. *Electroencephalography and Clinical Neurophysiology, 73*(1), 41–51.

Robitaille, N., Grimault, S., & Jolicœur, P. (2009). Bilateral parietal and contralateral responses during maintenance of unilaterally encoded objects in visual short-term memory: evidence from magnetoencephalography. *Psychophysiology, 46*(5), 1090–1099.

Robitaille, N., Marois, R., Todd, J., Grimault, S., Cheyne, D., & Jolicœur, P. (2010). Distinguishing between lateralized and nonlateralized brain activity associated with visual short-term memory: fMRI, MEG, and EEG evidence from the same observers. *NeuroImage, 53*(4), 1334–1345.

Ruchkin, D. S., Grafman, J., Cameron, K., & Berndt, R. S. (2003). Working memory retention systems: a state of activated long-term memory. *Behavioral and Brain Sciences, 26*, 709–777.

Schendel, Z. A., & Palmer, C. (2007). Suppression effects on musical and verbal memory. *Memory & Cognition, 35*, 640–650.

Schneider, W., Gruber, H., Gold, A., & Opwis, K. (1993). Chess expertise and memory for chess positions in children and adults. *Journal of Experimental Child Psychology, 56*, 328–349.

Schulze, K., Mueller, K., & Koelsch, S. (2011). Neural correlates of strategy use during auditory working memory in musicians and non-musicians. *European Journal of Neuroscience, 33*, 189–196.

Schulze, K., Zysset, S., Mueller, K., Friederici, A. D., & Koelsch, S. (2011). Neuroarchitecture of verbal and tonal working memory in nonmusicians and musicians. *Human Brain Mapping, 32*(5), 771–783.

Schumacher, E. H., Lauber, E., Awh, E., Jonides, J., Smith, E. E., & Koeppe, R. A. (1996). PET evidence for an amodal verbal working memory system. *NeuroImage, 3*(2), 79–88.

Sternberg, S. (1966). High-speed scanning in human memory. *Science, 153*, 652–654.

Sternberg, S. (1969). The discovery of processing stages: extensions of Donders' method. *Acta Psychologica, 30*, 276–315.

Todd, J. J., & Marois, R. (2004). Capacity limit of visual short-term memory in human posterior parietal cortex. *Nature, 428*(6984), 751–754.

Todd, J. J., & Marois, R. (2005). Posterior parietal cortex activity predicts individual differences in visual short-term memory capacity. *Cognitive, Affective, & Behavioral Neuroscience, 5*(2), 144–155.

Trehub, S. E., Schellenberg, E. G., & Kamenetsky, S. B. (1999). Infants' and adults' perception of scale structure. *Journal of Experimental Psychology: Human Perception and Performance, 25*(4), 965–975.

Vogel, E. K., & Machizawa, M. G. (2004). Neural activity predicts individual differences in visual working memory capacity. *Nature, 428*, 748–751.

Wilken, P., & Ma, W. J. (2004). A detection theory account of change detection. *Journal of Vision, 4(12)* 11, 1120–1135.

Williamson, V. J., Baddeley, A. D., & Hitch, G. J. (2010). Musicians' and nonmusicians' short-term memory for verbal and musical sequences: comparing phonological similarity and pitch proximity. *Memory and Cognition, 38*(2), 163–175.

Zatorre, R. J., Belin, P., & Penhune, V. B. (2002). Structure and function of auditory cortex: music and speech. *Trends in Cognitive Sciences, 6*(1), 37–46.

Zatorre, R. J., Evans, A. C., & Meyer, E. (1994). Neural mechanisms underlying melodic perception and memory for pitch. *The Journal of Neuroscience, 14*(4), 1908–1919.

Zatorre, R. J., Mondor, T. A., & Evans, A. C. (1999). Auditory attention to space and frequency activates similar cerebral systems. *NeuroImage, 10*(5), 544–554.

Zhang, W., & Luck, S. J. (2008). Discrete fixed-resolution representations in visual working memory. *Nature, 453*, 233–235.

CHAPTER

16

The Interplay Between Auditory Attention and Working Memory

Claude Alain[1,2,3], *Stephen R. Arnott*[1], *Susan Gillingham*[1,2], *Ada W.S. Leung*[1,4,5], *Jeffrey Wong*[1]

[1]Rotman Research Institute, Baycrest Centre for Geriatric Care, Toronto, ON, Canada; [2]Department of Psychology, University of Toronto, ON, Canada; [3]Institute of Medical Sciences, University of Toronto, ON, Canada; [4]Department of Occupational Therapy, University of Alberta, Edmonton, AB, Canada; [5]Centre for Neuroscience, University of Alberta, Edmonton, AB, Canada

Current theories of auditory attention aim to understand the neural networks underlying attentional selection and how they relate to working memory processes that allow us to manipulate dynamically and prioritize coexisting information in short-term memory. Historically, attention research was primarily focused on how sustained, selective, and/or divided attention modulated the processing of incoming sensory information. In auditory attention experiments, participants are usually instructed to listen for infrequent target stimuli embedded in a predefined stimulus stream. The task-relevant stimulus stream may be defined by a salient feature such as location (e.g., right versus left ear), pitch (e.g., low- versus high-pitched sounds) or both (e.g., low tone/left ear versus high tone/right ear). Participants have to keep in mind the task instructions and a *perceptual representation* of the incoming task-relevant stimuli to detect less salient target stimuli (e.g., slightly louder or longer stimuli). In this context, working memory is critical for maintaining both task goals and predefined target representations, such that an incoming stimulus may be compared with the perceptual representation of the target stimulus. The duration of this comparison process is likely determined by how similar the perceptual representation is to its working memory counterpart. In other studies, attention is biased on a trial-by-trial basis by presenting a cue before the presentation of stimuli. Here, the task set (goals and priorities) changes from trial to trial. When the cue correctly orients attention to the target, participants' responses are quicker and more accurate than when there is no informative cue or when the cue is invalid (e.g., cueing a wrong location) (e.g., Bedard, el Massioui, Pillon, & Nandrino, 1993; McDonald, Teder-Salejarvi, Di Russo, & Hillyard, 2005; Mondor & Zatorre, 1995; Posner, Snyder, & Davidson, 1980). Such a cueing paradigm requires participants to update the perceptual representation in working memory. Functional magnetic resonance imaging (fMRI) studies have revealed an enhanced blood oxygen level–dependent (BOLD) response in a frontoparietal network during selective attention as well as during the deployment of attention to a cued target within both visual (Corbetta & Shulman, 2002; Giesbrecht, Woldorff, Song, & Mangun, 2003; Hopfinger, Buonocore, & Mangun, 2000; Shulman et al., 1999) and auditory contexts (Hill & Miller, 2010; Shomstein & Yantis, 2004; Shomstein & Yantis, 2006).

The notion that attention can be focused on incoming external stimuli as well as internal trains of thought can be traced back to William James' nineteenth-century book, *Principles of Psychology.* This idea was reintroduced and restudied using modern advanced technology (Chun & Johnson, 2011; Johnson & Hirst, 1993). In particular, behavioral and neuroimaging experiments examined the role of attention during visual short-term (Chun & Johnson, 2011; Johnson & Hirst, 1993; Nobre et al., 2004; Nobre & Stokes, 2011) and long-term memory tasks that engaged retrieval and recognition processes (Burianova, Ciaramelli, Grady, & Moscovitch, 2012; Cabeza et al., 2011; Ciaramelli, Grady, Levine, Ween, & Moscovitch, 2010; Ciaramelli, Grady, & Moscovitch, 2008; Grillon, Johnson, Krebs, & Huron, 2008; Patai, Doallo, & Nobre, 2012). This research supports the concept of reflective attention (Chun & Johnson, 2011; Johnson & Hirst, 1993), in

Mechanisms of Sensory Working Memory
http://dx.doi.org/10.1016/B978-0-12-811042-3.00016-5

which an individual is focused on internal representations in short-term memory or information retrieved from long-term memory.

Reflective attention is best illustrated experimentally using a variant of the delayed-match-to-sample task. In these paradigms, a memory array composed of one or several items is first presented and participants must encode the information it contains. Then, the array is removed and participants must maintain its contents in memory. During this retention interval, a cue (dubbed the retro-cue) may be presented to orient attention to a target representation in their memory. After the delay, participants are presented with a change detection task and must determine whether the new item matches one of the previously encoded items. In such paradigms, the retro-cue reflectively directs the attention of participants to a particular object, thus relying on top-down control processes to refresh (i.e., bring to the foreground) (Johnson, Mitchell, Raye, D'Esposito, & Johnson, 2007; Miller, Verstynen, Johnson, & D'Esposito, 2008; Raye, Johnson, Mitchell, Greene, & Johnson, 2007; Yi, Turk-Browne, Chun, & Johnson, 2008) or rehearse (Baddeley, 1992) the retro-cued object. In this way, reflective attention appears analogous to working memory. It involves dynamic maintenance, manipulation, and prioritization of coexisting and currently important object representations in short-term memory. In the current chapter, we refer to short-term memory as an active store in which information remains available for a short period of time. The duration of this store varies substantially, spanning from a few seconds, if active maintenance or rehearsal is prevented, to minutes, depending on the material presented and the perceptual context (Cowan, 1984; Cowan, 1993).

As we delve further into understanding how the brain processes information, we find ourselves in the position of trying to consolidate new evidence of interaction and possibly even overlap between attention and memory using our traditional conceptualizations of these processes as separable entities deserving of individual labels. On the one hand, we can easily conceptualize the behavioral difference between attending to something and creating a memory of it, if only for a short time. Yet even the plethora of literature using tasks that explicitly try to capture either working memory or selective attention shows that the neural networks supporting these cognitive domains overlap substantially and vary as a function of task demands (Chun & Johnson, 2011; Nobre et al., 2004). Further, we rely on non-memory operations to help explain that working memory itself is not a unitary function, but can instead be divided into at least storage (maintenance) and processing (manipulation) components (Baddeley, 1992). Most of these operations tend to fall under the umbrella of frontal attentional processes, but they range from a mass "central executive" that organizes and integrates the two components (Baddeley, 1992) or "works with memory" (Moscovitch, 1992); to non-centralized, separable executive processes of goal establishment/task setting (D'Esposito & Badre, 2012; Stuss et al., 2005; Stuss, Binns, Murphy, & Alexander, 2002) and monitoring (the "epoptic process") (Petrides, 1991; Petrides, 1995; Petrides, 2005; Petrides, 2012; Petrides & Milner, 1982; Petrides, Tomaiuolo, Yeterian, & Pandya, 2012), to reflectively attending to the contents of working memory (Chun & Johnson, 2011; Nobre et al., 2004). Hence, it is not always clear whether the processes and neural underpinnings of attention and working memory can be distinguished from each other.

Despite a growing acceptance of the interplay between attention and memory, there has been a paucity of work exploring the commonality and differences between auditory attention and working memory (Fougnie, 2008; Huang, Seidman, Rossi, & Ahveninen, 2013). In the next sections, we review studies that examine this relationship in an effort to determine whether they can be distinguished from each other. First, we will examine the current views on how auditory information from the external world is received and processed in the early stages, followed by a comparison of the domain-specific nature of auditory working memory and attention. For example, studies have shown that auditory selective attention generates a reliable difference in patterns of neural activity when attention is focused on pitch as opposed to sound location (e.g., Alain, Arnott, Hevenor, Graham, & Grady, 2001; Altmann, Henning, Doring, & Kaiser, 2008; Hill & Miller, 2010; Krumbholz, Eickhoff, & Fink, 2007; Maeder et al., 2001). It is thus possible that working memory may also maintain and manipulate particular features of a sound object (e.g., pitch, spatial location, intensity) independently of other features. After that discussion, we will review studies that focused on cognitive processes that are a hallmark of working memory (e.g., maintenance, manipulation), detailing their similarities and differences from attention. Finally, we review studies that focused on directly manipulating the cognitive processes underlying attention or working memory and assessed how such manipulation affects performance and/or brain activity.

CURRENT VIEWS OF AUDITORY ATTENTION

Early work on auditory attention sought to understand how one can selectively attend to a conversation in the midst of other, less relevant sounds in the environment. This is illustrated by the cocktail party scenario (Cherry, 1953), in which a listener can readily focus his or her attention on one conversation while filtering out less relevant co-occurring conversations. Typically, speech comprehension in noisy environments is constrained by our capacity to hold critical information online in working memory while keeping irrelevant information (external stimuli or internal thought) from intruding. For effective speech reception to take place, listeners must be able to group sound elements coming from one source together (i.e., one speaker) and segregate those from other sources (e.g., another speaker). Current theories of auditory selective attention posit facilitation as well as suppression mechanisms that possibly have an important role in parsing concurrent sound sources as well as enhancing figure-ground segregation when focusing attention to someone amidst a conversational background (Alain, Arnott, & Dyson, 2013). Studies using fMRI have revealed attention-related changes in modality-specific and supra-modal areas including the prefrontal and parietal cortices (e.g., Hill & Miller, 2010; Petkov et al., 2004; Salmi, Rinne, Koistinen, Salonen, & Alho, 2009; Tzourio et al., 1997). Although modality-specific areas are important for sound segregation to occur (e.g., Alain et al., 2005; Gutschalk et al., 2005), the supra-modal brain areas are likely important for controlling the flow and organization of sensory information in modality-specific areas and deploying attention to task-relevant information (Alain et al., 2013).

Auditory attention has often been characterized in spatial or frequency terms, and likened to a spotlight or filter that moves around, applying processing resources to whatever falls within a selected spatial or frequency region (e.g., Broadbent, 1962; LaBerge, 1983). Other models postulate resource allocation on the basis of perceptual objects, in which attending to a particular object enhances processing of all features of that object (e.g., Chen & Cave, 2008; Duncan, 1984). Recent models of auditory attention have been consistent with the latter conception of attention, in which underlying units of selection are discrete objects or streams. Critically, attending to one component of an auditory object facilitates the processing of other properties of that same object (Alain & Arnott, 2000; Shinn-Cunningham, 2008).

According to the object-based account of auditory attention (Alain & Arnott, 2000), low-level properties of incoming concurrent sounds including spectro-temporal structure, location, and onset are analyzed and perceptually grouped into distinct sound object representations. These sound objects or streams, derived from a pre-attentive segmentation of the auditory scene, form the basic units for attentional selection. Evidence for pre-attentive segmentation of incoming acoustic data into sound objects has been obtained from behavioral and event-related brain potential (ERP) studies demonstrating that sequential (Cusack, Carlyon, & Robertson, 2000; Snyder, Alain, & Picton, 2006) and concurrent sound segregation (Alain & Izenberg, 2003; Dyson, Alain, & He, 2005) can occur independent of a listener's attention. After the auditory scene has been partitioned into sound object representations, a selection process allows an individual to focus on a particular object and switch attention from one object representation to another. Importantly, both bottom-up (i.e., stimulus salience) and top-down demands (i.e., internal goals) determine which sound(s) in our surroundings are selectively attended.

FEATURE-SPECIFIC PROCESSING: A COMPARISON BETWEEN AUDITORY ATTENTION AND WORKING MEMORY

Feature-specific attention effects refer to neural modulation when attending to particular stimulus features (Alain et al., 2013). Here, the term feature refers to perceptual characteristics of acoustic events such as duration, pitch, timbre, location, and/or loudness. Feature-specific attention effects are often investigated using a rapid serial auditory presentation of stimuli that vary randomly in pitch (e.g., low- and high-pitched sound) and location (e.g., left and right ears). Participants are usually asked to attend to a subset of stimuli that are defined by a clearly identifiable feature(s) (e.g., pitch and/or location) to detect target stimuli. The target sounds are constructed in such a way that participants must focus attention on the task-relevant stream to detect the more subtle difference between the standard and target stimuli within the attended stream. Woods and colleagues used such a paradigm in conjunction with ERP recordings to investigate feature-specific attention effects (e.g., attend to high-pitch tones) as well as object-based attention effects that rely on the conjunction of sound features (e.g., attend to high-pitch tones in the left ear) (Woods & Alain, 1993, Woods & Alain, 2001; Woods, Alho, & Algazi, 1994). The effects of selective attention were isolated by subtracting ERPs to tones with no target features from those elicited by tones sharing one, two, or three target features. Results revealed distinct attention-related ERP signatures over fronto-central scalp regions for processing task-relevant pitch and location features. These feature-specific attention effects are long lasting and overlap in time, which has been taken as evidence to suggest that pitch and location features are processed in parallel and are represented in separate cortical fields (Degerman, Rinne, Sarkka, Salmi, & Alho, 2008; Woods & Alain, 2001; Woods et al., 1994). These paradigms have also shown that the neural correlates of auditory feature conjunction differed from the sums of single-feature attention effects. The feature conjunction effects began about 60 ms after

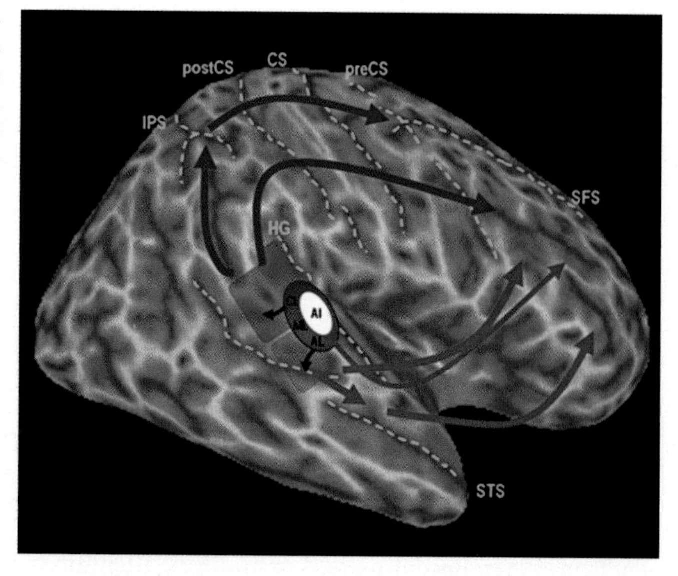

FIGURE 1 Schematic diagram of the dual-pathway model of auditory cortical processing. The dorsal (spatial) and ventral (what) pathways are denoted by blue and red arrows, respectively. AL, anterolateral belt; IPS, intraparietal sulcus; postCS, postcentral sulcus; CL, caudolateral belt; CS, central sulcus; ML, middle lateral belt; preCS, precentral sulcus; SFS, superior frontal sulcus; HG, Heschl's gyrus; STS, superior temporal sulcus. *Adapted from Rauschecker and Tian (2000); Romanski, Tian, et al. (1999).*

the feature-specific effects and lasted for several hundred milliseconds (Woods & Alain, 1993; Woods & Alain, 2001; Woods et al., 1994). Woods and Alain (2001) proposed the facilitatory interactive feature analysis model to account for these temporally overlapping ERP feature-specific and feature conjunction effects. In this model, the neuronal responses to single features undergo active amplification during the conjunction process, which accounts for the nonlinearity of the ERP attention effects.

Attention, Working Memory and Dual-Pathway Model. Anatomically, these feature-specific attention effects follow the dual-pathway model of auditory scene analysis in which sound identity and location are preferentially processed along the ventral (*what*) and dorsal (*where*) pathways (Arnott & Alain, 2011; Arnott, Binns, Grady, & Alain, 2004; Rauschecker & Tian, 2000). Similar to the visual system, the frontoparietal network, in conjunction with primary auditory cortices, has been shown to differentially activate when comparing brain activation between attending to sound locations versus pitch (e.g., Alho, Rinne, Herron, & Woods, 2014; Degerman, Rinne, Salmi, Salonen, & Alho, 2006). Figure 1 shows a schematic of the dual pathway model for audition, highlighting two primary pathways that originate from the caudal and rostral portion of the auditory cortex. Generally, the areas involved include the inferior and superior frontal gyri, dorsal precentral sulcus, inferior parietal sulcus, superior parietal lobule, and auditory cortex. The parietal cortex is recruited to a greater extent when attending to sound location whereas the anterior and superior temporal gyri are recruited more strongly when attending to pitch (e.g., Alho et al., 2014; Degerman et al., 2006; Krumbholz et al., 2007).

Like auditory attention, auditory working memory involves the coordinated effort of widely distributed regions that include, but are not limited to, the auditory cortices, superior temporal gyrus including the Heschl's gyrus and planum temporale, parietal cortices, and inferior prefrontal gyrus (e.g., Alain, He, & Grady, 2008; Buchsbaum, Olsen, Koch, & Berman, 2005; Rama et al., 2004). The auditory traces are thought to be represented in auditory cortices (Brechmann et al., 2007; Buchsbaum et al., 2005), whereas the prefrontal and parietal activity may index the deployment of attention to sound object representation and likely have an important role in keeping these representations in working memory for comparison with incoming stimuli (Brechmann et al., 2007; Buchsbaum et al., 2005). Although different auditory features can be retained and manipulated in working memory, few studies have directly compared memory-related activity for simple acoustic features such as pitch and location. In one fMRI study, Alain et al. (2001) used a delayed comparison discrimination task in which an initial sound (S1) was followed, after a delay, by a second sound (S2) that was either identical to S1 or differed in pitch and/or location (Figure 2(a)). Depending on the task instructions, participants were required to indicate whether S2 had the same, lower, or higher pitch than S1, or whether S2 was presented at the same, leftward, or rightward position to S1. The performance of participants on the pitch and location tasks was monitored closely, and the discriminability of the stimuli was adjusted to ensure that attentional demands were similar and the difference in neural activity could not be due to task difficulty. Results showed that the ventrolateral prefrontal cortex (i.e., inferior frontal gyrus) was more strongly activated during the pitch than during the sound location task (Figure 2(b)). Conversely, the superior frontal sulcus and parietal cortex were more strongly activated during the location than during the pitch judgment task. Using a different paradigm and stimuli, Maeder et al. (2001) also found similar dissociations between ventral and dorsal brain areas, with greater

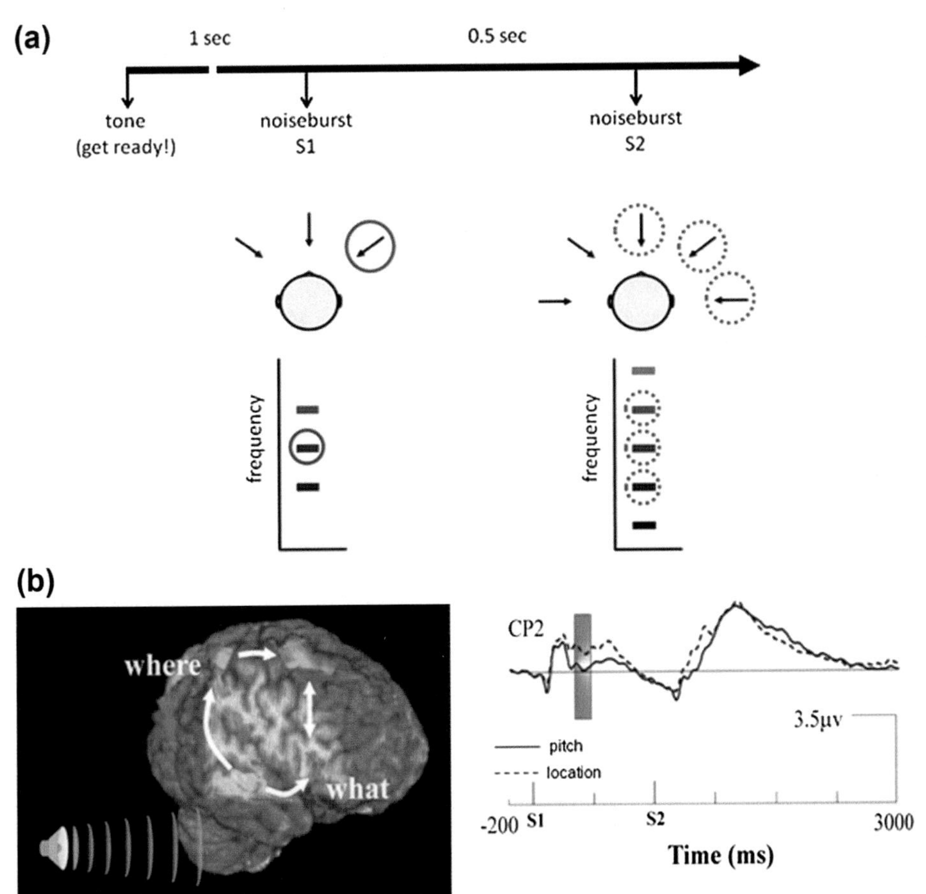

FIGURE 2 (a) Schematic of a trial. There are five possible pitches and locations. The first stimulus (S1) is presented in one of three locations and one of three possible pitches: for example, the one circled. S2 can then appear in one of three locations (the original location or one on either side) at one of three pitches (the original, one higher, or one lower). Depending on task instructions, the subject compares S1 and S2 on the basis of the location or the pitch. (b) Brain areas showing differential activity during auditory identification and localization tasks *(from Alain et al., 2001)*. Greater activity during location, compared with pitch, was seen in superior frontal sulcus and superior parietal cortex. Greater activity during the pitch task, compared with location, was seen in inferior frontal gyrus and superior temporal gyrus. Other regions where location was greater than pitch resulted from decreased activity during pitch rather than increased activity during the location task.

fMRI activation for sound object recognition and sound source localization, respectively. In both of those studies, the enhanced activity in the inferior frontal gyrus may reflect articulatory processes and semantic encoding associated with comparing the identity of the two consecutive stimuli. The frontal activation may also index a sense of familiarity (Ley, Vroomen, & Formisano, 2014), which would be greater for namable sound objects than for sound location. In contrast, the enhanced activity in parietal cortex may reflect attention to memory such as a shifting attention from one representation to another, which may be prominent when participants are making decisions regarding the relative spatial location of two stimuli. These findings were confirmed in subsequent meta-analyses of neuroimaging studies (Alho et al., 2014; Arnott et al., 2004) and are in accordance with those of the visual working memory system (e.g., Ungerleider & Haxby, 1994; Yantis & Serences, 2003).

In the study by Alain et al. (2001), the same participants took part in an ERP experiment to assess the timing of neural activation in response to specific task instructions. Interestingly, between 300 and 500 ms after S1 presentation, greater positivity was observed over the inferior frontotemporal scalp regions during the pitch task whereas greater positivity was observed over the centro-parietal scalp regions during the location task (Figure 2(B), right panel). This effect of task instruction on ERPs elicited during the retention interval suggests that different neuronal population processes the maintenance of pitch and location information in working memory for an eventual comparison with the second stimulus (S2). The S2 probe generated sensory-evoked responses as well as negative–positive potentials around 350 and 800 ms after sound onset. The ERP modulation during the 300- to 400-ms interval preceded the late positive complex at parietal sites, peaking at about 800 ms after sound onset, and was thought to index a comparison process between the probe and the representation of S1 held in working memory. The effects of task instruction on ERPs provide converging evidence that biasing attention toward processing the pitch (what) or location (where) of a sound object engages distinct ventral and dorsal processing streams, respectively.

Subsequent studies using magnetoencephalography have shown enhanced gamma–band activity localized over posterior parietal cortex and left inferior frontal/anterior temporal regions during the delay and response period to sound location and auditory pattern, respectively (Kaiser, Ripper, Birbaumer, & Lutzenberger, 2003; Lutzenberger, Ripper, Busse, Birbaumer, & Kaiser, 2002). Similarly, these feature-specific effects for auditory working memory have been observed during the n-back task (Alain, McDonald, Kovacevic, & McIntosh, 2009; Anourova et al., 1999; Anurova et al., 2003).

Together, these findings provide converging evidence for feature-specific effects in auditory working memory and are consistent with the notion that our representation of auditory events is closely related to our perceptual experience (Buchsbaum et al., 2005). The overlap of neural activation, regardless of the task-relevant features, has been taken as evidence of domain-related anatomical areas between tasks of selective attention and tasks of working memory. Likewise, evidence from visual studies has shown that similar brain regions are engaged for both perceptual and reflective attention (e.g., Nobre et al., 2004; Roth, Johnson, Raye, & Constable, 2009). Specifically, reflective attention, like perceptual attention, can be feature-based. Furthermore, these findings reveal the interplay between attention and working memory, with participants able to guide attention to specific sound features, which in turn modulates neural activity during the maintenance and manipulation of items in memory.

Despite strong dissociations in many studies, such as the ones reviewed above, other studies have failed to find differences in ERPs during working memory for pitch or location. For example, using a delayed-match-to-sample and an n-back task, Rama et al. (2000) found enhanced ERP amplitude over the parietal and occipital scalp region during the retention interval with increased working memory load. However, they found no difference in amplitude distribution as a function of sound feature or interaction between sound feature and working memory load. Increases in ERP amplitude with increased working memory load could indicate that an area is involved in the maintenance of task-relevant features. Nonspecific increases in ERP amplitude with increasing working memory load would be expected from areas more involved in higher-order attention demands than with feature-specific processing. In the study by Rama et al., the small difference or lack of difference in ERPs as a function of a sound feature to be remembered may be partly related to the ERP analysis, which focused on a particular component and subset of scalp electrodes. Hence, it remains possible that feature-specific effects might have been undetected because of the choice of scalp electrodes and ERP time windows entered into the analysis.

Notwithstanding similarities in neural activation patterns in different tasks (selective attention versus working memory) and modalities (auditory versus visual), some studies have shown differences in intentional behavior to attend to something as opposed to remembering something. Rinne, Koistinen, Salonen, and Alho (2009) compared fMRI activity when participants were engaged in a pitch discrimination task against an n-back task using the same stimuli. Results showed specific memory-related effects in the inferior parietal lobule. Moreover, they manipulated the task difficulty of both tasks and found that increasing the difficulty of the pitch discrimination task enhanced BOLD activity in anterior superior temporal gyrus areas to primary auditory cortex but showed no significant changes in BOLD responses in the inferior parietal lobule. Conversely, increasing working memory demands enhanced BOLD response in the inferior parietal lobule but had little impact on brain activation measured in areas rostral to primary auditory cortex. Similar findings were observed in a neuropsychological study in which patients with focal brain lesions in inferior parietal cortex showed impaired short-term sound representation compared with age-matched controls or patients with lesions in inferior frontal cortex (Baldo & Dronkers, 2006). Together, these findings suggest that the inferior parietal lobule has a unique role in maintaining and/or updating information in working memory.

Working Memory and Auditory Distraction. A few studies have attempted to investigate the interactive relationship between attention and working memory. For instance, a study using a dichotic listening paradigm found that healthy adults with lower working memory capacity were more likely to notice their own name in the unattended stream than participants with high capacity (Conway, Cowan, & Bunting, 2001). Furthermore, adults with low capacity also showed larger amplitude of an ERP component (i.e., P3a) thought to index attentional capture (Yurgil & Golob, 2013). This suggests that individuals with lower working memory capacity are less able to filter out unwanted distracting task-irrelevant information than those with higher working memory capacity. However, because these findings were correlational in nature, they cannot demonstrate a causal role for working memory in auditory selective attention. In addition, evidence suggests that these correlations might only hold in situations where participants are required to ignore semantically meaningful information (i.e., their own names) and not in situations where the possible interference is non-semantic (Beaman, 2004).

Dalton Santangelo and Spence (2009) devised an experiment to investigate more directly the relationship between attention and working memory. They examined whether a listener's ability to localize a brief target sound would be affected by varying verbal working memory load (i.e., remembering one or six digits). The target sound (i.e., a short, continuous white noise burst) was presented from one of four possible speakers (upper left, lower left, upper right, or lower right) while a distractor sound consisting of amplitude-modulated white noise was presented in the opposite hemispace either at the same (congruent) or different (incongruent) elevation. Overall, participants' performance was worse when the distractor was incongruent compared with when it was congruent. The magnitude of the distractor interference was greater with increasing working memory load. Furthermore, participants who had lower working memory capacity were more susceptible to interference effects. These findings are in line with those from studies showing that participants with lower working memory capacity are more sensitive to auditory distraction (Conway et al., 2001; Yurgil & Golob, 2013), which is consistent with a difficulty in filtering out task-irrelevant stimuli.

Using a different experimental paradigm in which targets and distractors may share an acoustic feature (e.g., duration), Berti and Schröger (2003) found the opposite pattern and showed that increasing working memory demands reduced distractor interference, which was accompanied by a decrease in an ERP component indexing attentional capture (i.e., P3a) (see also Yurgil & Golob, 2013) and reorientation of attention to distractors (reorienting negativity). These results provide compelling evidence demonstrating a causal role for working memory in selective attention. Specifically, increasing working memory demand appears to attenuate the processing allocated to task-irrelevant stimuli. This conclusion would fit with the proposal that the specific role of working memory in selective attention is to keep the current task priorities in mind (e.g., Lavie, Hirst, de Fockert, & Viding, 2004). This type of function would seem likely to be modality independent because similar findings have been reported for visual (de Fockert, Rees, Frith, & Lavie, 2001; Lavie et al., 2004), auditory (Dalton, Santangelo, et al., 2009), and tactile information (Dalton, Lavie, et al, 2009).

One difficulty in assessing the similarities and differences between attention and working memory is the lack of separation between underlying cognitive operations that are reflected in brain activity changes. For instance, it is unclear whether the difference between selective attention and working memory tasks reflects maintenance and/or manipulation processes, or results from the executive/attention component biasing processing toward the task-relevant sound feature. One reason for this is the design of task paradigms; the earlier fMRI studies (e.g., Alain et al., 2001; Maeder et al., 2001) used block design paradigms in which it is difficult to dissociate stimulus- from response-related processes. That is, some of the differences in early fMRI studies could partly reflect differences in response selection, preparation, and execution. In the following paragraphs, we review studies that focus on specific components as well as those that used mixed block event–related designs that allow the separation of stimulus- and response-related activity.

PROCESS-SPECIFIC ACTIVATION DURING ATTENTION AND WORKING MEMORY

Working memory includes subsystems that are responsible for maintaining and manipulating information in sensory memory. However, it was difficult to identify neural correlates of these processes in the earlier studies because the experimental design was such that encoding-, maintenance- and response-related activities were often embedded within the fMRI responses. To address this, Rama et al. (2004) used a longer retention interval to investigate whether domain specificity occurred during the maintenance of auditory information. In this delayed recognition paradigm, participants were presented with human voices from different locations (S1) for about 1.5 s, followed by a 4.5-s retention interval and then a test stimulus. During this time, participants indicated whether the two stimuli were identical. The experiment also included a control condition in which participants were told that they did not need to remember the stimuli and only needed to press a button when the test stimulus (i.e., S2) was presented. The stimuli from the control task were scrambled and presented with the same timing as in the memory task. Using different regressors time-locked on the S1, delay, and S2 onsets, they were able to capture differential neural activity during the delay period, such that the dorsal prefrontal (superior frontal sulcus/anterior cingulate gyrus) and parietal cortices were significantly more activated during location trials. Conversely, ventral prefrontal cortex and the anterior portion of the insula showed greater activation during the voice trials. Interestingly, this preferential response to the voice identity in ventral prefrontal cortex continued into the recognition test period. However, the double dissociation was observed only during maintenance and not during encoding or recognition. These findings suggest that, during auditory working memory, maintenance of spatial and non-spatial information modulates activity preferentially in a dorsal and ventral auditory pathway, respectively.

Arnott et al. (Arnott, Grady, Hevenor, Graham, & Alain, 2005) used a similar approach to investigate whether the functional segregation in what and where processing would differ during the retention and the manipulation of information in working memory. They used an event-related fMRI design with a variant of the delayed-match-to-sample task that included a cue before S1 instructing participants to either maintain the location of S1 or its identity, or do nothing. In addition, participants were presented with a verbal question after the delay that instructed how S1 and S2 were to be compared. This design allowed the examination of domain-specific effects separately between the maintenance and comparison of complex sounds in working memory. The first analysis replicated results of earlier studies and showed a ventral and dorsal dissociation for sound identity and sound location, respectively. Subsequent analyses used different regressors for the maintenance and manipulation periods. During the maintenance period, results showed greater activity in the superior temporal gyrus and insula when participants maintained sound object identity than when they maintained its location in working memory. In contrast, only a small area near the frontal pole showed greater activity when participants maintained spatial location rather than sound identity. During the manipulation period, Arnott et al. (2005) found greater activity in dorsal regions including precuneus, and posterior temporal regions after participants were asked to compare the spatial location of S1 and S2 (e.g., "Which sound was most left/right/central/lateral?"). They suggested that holding spatial and non-spatial information in mind activates dorsal and ventral brain areas, respectively, whereas manipulating both types of information activated the dorsolateral prefrontal cortex. This study not only shows that a domain-specific dissociation exists during the retention period, but also suggests that the division of labor into what and where information processing is a fundamental principle of functional organization in the brain across sensory domains for both perception and working memory. The results from Arnott et al. (2005) and Rama et al. (2004) also suggest that the dorsolateral prefrontal cortex has a general domain-independent role in auditory working memory necessary for top-down control of attention. The differences in prefrontal cortex activation for maintenance and manipulation components of working memory may reflect differences in attentional processing, and provide further evidence that maintenance and manipulation of sensory representations in working memory are fundamentally unique and independent processes.

In the studies reviewed above, the activation in the dorsal stream was interpreted as an index of auditory spatial working memory. However, it remains possible that the activity in the dorsal region reflects response-related processes rather than memory-related activity. For instance, the dorsal activity could partly reflect response compatibility, which would show greater activity during the location tasks because participants' responses are relative to the sound position. In an effort to tease apart these two possibilities, Alain et al. (2008) used a variant of the one-back task that included occasional targets embedded in a block of standard stimuli. Participants were presented with natural everyday sounds from three different semantic categories: animal (e.g., dog bark, bird chirping), human (e.g., cough, laugh), and musical instrument (e.g., flute, clarinet). Stimuli were presented from three possible azimuth locations relative to straight ahead (−90°, 0°, and +90°) using head-related transfer functions to mimic the acoustic effects of the head and ears of an average listener (Wenzel, Arruda, Kistler, & Wightman, 1993). Occasionally, the sound identity or location was repeated (i.e., target). Participants were told to press a button as quickly as possible whenever they heard repetition of either sound category or location irrespective of changes in the other task-irrelevant dimension. The stimulus set was identical in both the category and location tasks, and only task instruction differed (i.e., attend category versus attend location). Task difficulty was also adjusted such that the attentional demand for both tasks was comparable. Importantly, the stimuli were presented with a fixed stimulus onset asynchrony of 2 s with the inter-target interval varying between 4 and 12 s.

This n-back task variant was used during a mixed block/event-related fMRI design, which allowed the identification of a sustained signal that persisted across an entire block of trials (i.e., task set) while separating trial-related activity (i.e., response to the occasional target). The fMRI data were analyzed with the general linear model using separate regressors for sustained and transient activity in both listening conditions. The sustained (i.e., task-related) and event-related activities were modeled simultaneously using boxcar function with a width equal to the duration of the block and gamma function time-locked to the stimulus repetition (i.e., target), respectively. This analysis revealed more sustained activity in the right dorsal brain regions, including the inferior parietal lobule and superior frontal sulcus, during the location task after accounting for transient activity related to target detection and the motor response. Conversely, there was greater sustained activity in the left superior temporal gyrus and left inferior frontal gyrus during the category task compared with the location task. Transient target-related activity in both tasks was associated with enhanced signal in the left pre– and post–central gyrus, prefrontal cortex, and bilateral inferior parietal lobule. These results suggest dual roles for the right inferior parietal lobule in auditory working memory—one involved in monitoring and updating sound location independent of motor responding, and another that underlies the integration of sensory and motor functions.

Using another n-back task, Leung and Alain (2011) sought to examine the pattern of neural activity induced by working memory load for sound identity or location. They reasoned that specific memory-related activity would emerge as an interaction between tasks and load. Specifically, different brain areas would be more relevant for maintaining sound identification and location depending on the demand placed on the ventral (what) and dorsal (where) pathways. As in prior studies, working memory for sound identity generated greater activity in ventral brain areas, whereas working memory for sound location was associated with greater activity in dorsal brain regions. More importantly, Leung and Alain found an interaction between task and working memory load. The increase in working memory load during the category task was associated with increased activity in the right inferior parietal lobule, whereas no such load-related changes in BOLD signal were observed during the location task. Conversely, increasing working memory load for the location task was associated with increased BOLD signal in a different area of the inferior parietal lobule bilaterally, as well as the left supramarginal gyrus, the left middle frontal gyrus, the left transverse gyrus, and the left superior temporal gyrus. The right inferior parietal lobule regions for the location task were ventral and anterior to those observed in the category task. For both tasks, the peak activation in the right inferior parietal lobule correlated with accuracy. These findings indicate that specific regions in the inferior parietal lobule are selectively recruited as a function of working memory demands (Figure 3). A similar functional segregation in parietal cortex was recently observed in meta-analysis of neuroimaging studies of working memory tasks (Rottschy et al., 2013).

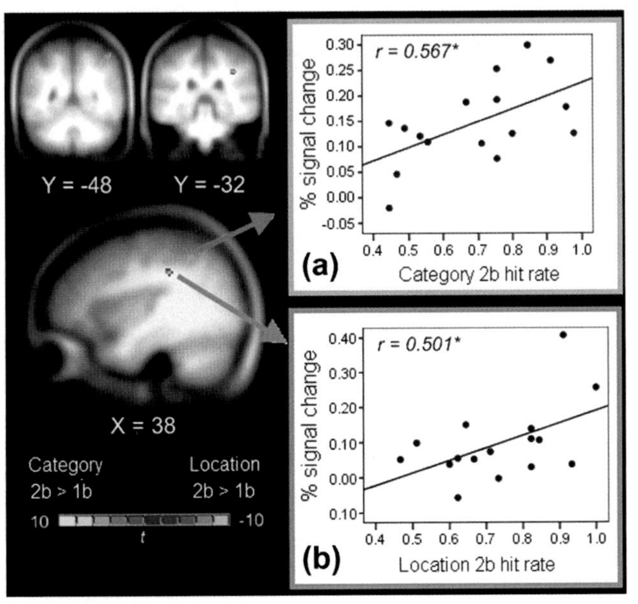

FIGURE 3 Brain images showing the neural activation of identity load and location load on working memory one-back and two-back tasks. The orange region represents stronger neural activation in the two-back than the one-back category working memory task. The blue region represents stronger neural activation in the two-back than the one-back location working memory task. (A) Significant positive correlation between the hit rate of two-back category working memory task and percent signal change in the orange region, a dorsolateral region of the inferior parietal lobe. (B) Significant positive correlation between the hit rate of one-back location working memory task and percent signal change in the blue region, a ventromedial region of the inferior parietal lobe. 1b = one-back. 2b = two-back *(reference to Leung & Alain, 2011).*

Other findings from Leung and Alain (2011) differ from those of prior studies that revealed domain-specific processing within the prefrontal cortex (Poremba et al., 2003; Rama & Courtney, 2005). Goldman-Rakic (1996) proposed a model that postulates a modular organization of working memory based on the type of information processing. In that model, domain specificity refers to the registration of any sensory modality into an information processing system for feature analysis. For auditory stimuli, studies have reported pathways from the posterior and anterior auditory association cortex to the dorsolateral prefrontal regions for spatial processing and ventrolateral prefrontal regions for object processing (Bushara et al., 1999; Rama et al., 2004; Romanski, Bates, & Goldman-Rakic, 1999). In the study by Leung and Alain (2011), domain specificity was observed in the inferior parietal lobule but not the prefrontal cortex. One possibility is that the functional dissociation of the inferior parietal lobule assists in diverting domain-specific information to specific regions of the prefrontal cortex. Another possibility is that the prefrontal regions index other task-related processes rather than working memory.

So far, evidence from fMRI studies suggests that auditory working memory may be supported by ventral (what) and dorsal (where) pathways. However, whether and to what extent this dissociation exists when participants listen passively to auditory stimuli remains equivocal. Alain et al. (Alain, Shen, Yu, & Grady, 2010) used an adaptation paradigm to distinguish memory-related function from goal-directed action. Participants passively listened (no response required) to sequences of four sounds presented either at the same location or at four different locations while the BOLD response was measured using fMRI. Participants were not required to maintain or manipulate auditory stimuli. Consequently, the difference in fMRI adaption reflected the encoding of sound location invariance, which could be important for orienting attention to changes in sound source location. Alain et al. found greater inferior parietal lobule activation for changes in sound location than for sounds presented at the same location (Figure 4). The inferior parietal lobule activation overlapped with areas that were observed during an auditory spatial working memory task. Similar findings were reported by Altmann et al., who used an event-related fMRI design to examine the neural activity associated with processing stimulus invariance (Altmann, Bledowski, Wibral, & Kaiser, 2007). In their study, participants listened passively to a series of repetitive spatial animal vocalizations with regular transition into a different vocalization (what), location (where), or both vocalization and location. Greater activity along the bilateral anterior superior temporal gyrus and superior temporal sulcus, the planum polare, lateral Heschl's gyrus, and anterior planum temporale was present for identifying vocalization changes. For location changes, significant brain activation was observed in bilateral posterior superior temporal gyrus, planum temporal, and inferior parietal lobule. This finding supports the role of parietal cortex in auditory spatial working memory, which can be dissociated from response selection and execution. Together, findings from these fMRI adaptation studies are consistent with the proposal that object and spatial auditory working memory, like object and auditory spatial attention, is supported by separate neural systems.

The studies reviewed above provide strong support that the inferior parietal lobule is essential in monitoring and updating sound location in working memory independent of motor responding (Alain et al., 2008; Alain et al., 2010). This memory-related activity was little affected by variation in response mode and was present for variations in

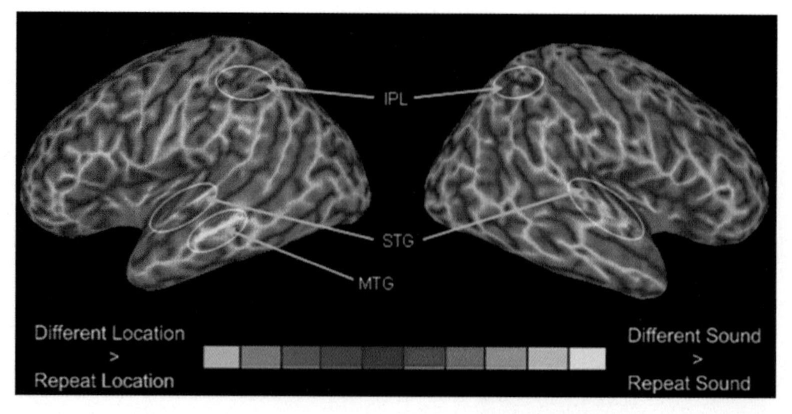

FIGURE 4 Three-dimensional view revealing enhanced activation for changes in sound category (difference sound, repeat location) versus changes in sound location (same sound, different location). The yellow color indicates greater activity for changes in sound identity, whereas areas in blue correspond to changes in sound location. IPL, inferior parietal lobule; STG, superior temporal gyrus; MTG, media temporal gyrus. *From Alain et al. (2010).*

sound location during passive listening with no stimulus–response mapping (Alain et al., 2010). However, the role of inferior parietal lobule is not limited to spatial working memory; it may also contribute to monitoring and updating sound identity in working memory, as shown in Leung and Alain (2011).

An approach to exploring the relationship between attention and working memory uses factorial tasks in which the load of maintaining and updating information is varied independently from attentional demands by changing the number of distractors. Using fMRI, Huang et al. (2013) measured changes in BOLD response while participants performed an auditory continuous performance task, which involved listening to a sequence of spoken letters and pressing a button whenever a pre-specified target letter was presented in the sequence. The levels of auditory–verbal working memory load and amount of attentional interference were manipulated to allow separation of activity specific to memory from that associated with filtering task-irrelevant information. Huang et al. found a double dissociation between prefrontal cortical sub-areas associated selectively with either auditory attention or working memory. Specifically, anterior dorsolateral prefrontal cortex and the right anterior insula were selectively activated by increasing working memory load, whereas sub-regions of middle lateral prefrontal cortex and inferior frontal cortex were associated with interference only. Furthermore, they observed a super-additive interaction between interference and load in left medial superior frontal cortex. Huang et al. suggested that in this area, activations not only overlap but reflect a common resource pool recruited by increased attentional and working memory demands. Indices of working memory-specific suppression of anterolateral non-primary auditory cortices and attention-specific suppression of primary auditory cortex were also found, possibly reflecting suppression/interruption of sound-object processing of irrelevant stimuli during continuous task performance. These results show that auditory attention and working memory can be dissociated within the prefrontal cortex.

CONCLUDING REMARKS

Although we can ascribe different perceptual and cognitive operations when we are asked to pay attention to something versus when we are asked to keep something in mind over a period of time, the brain's approach to effecting those behaviors is by no means clear or straightforward, or at least may not follow within the boundaries that our operational definitions provide for what it means to attend or to memorize. Traditionally, selective attention and working memory functions have been studied separately. Yet, increasing direct and indirect evidence suggests that there is considerable overlap in the neural network supporting these two core functions. Even when there is anatomical dissociation based on how the brain processes sensory features of incoming information, overlap remains.

Study of the direct interaction between attention and working memory in the auditory domain is only just beginning. This is in part because of our rigorous commitment to defining cognitive functions as either an attention or a memory process. As we begin to try to elucidate what is occurring inside the brain after the perception of external information, it may be necessary for us to break down component processes that lead to the attending or memorizing behavior. In the context of evidence of their similarity, we may have to debate whether reflective attention and some aspects of working memory are indeed the same function simply because we have labeled them differently. Eventually, we may instead have to approach our study of the inner nature of the brain by asking more specific questions related to understanding of the processes involved when we perceive, attend to, and remember something.

The difference in this approach may seem subtle and unimportant to some. However, we argue that there is both empirical and clinical value in thinking about cognition in terms of component processes rather than cognitive domains. From a research perspective, it may necessarily change the variables that we choose as measurements of a process. Instead of labeling separate tasks as attention or working memory, and then attempting to make post hoc comparisons of their similarities and differences, can we instead design tasks of increasing cognitive demands and assess the location and timing of activation as different processes are required to come online (or at least are required in sufficiently greater amounts that they are noticeably measurable)? We will also have a framework with which to approach the question of similarities and differences across sensory modalities. The term "component process" suggests that there is an underlying procedure for managing information regardless of the type of information. Therefore, are there component processes that are shared between modalities, and at what point does modality-specific processing occur? Given the amount of evidence with regard to the overlap between attention and working memory, we can begin the initial examination of these component processes.

From a clinical perspective, if a person exhibits a problem with working memory or attention, we are already in a position to understand that there are at least several potential places within our established networks where problems may occur, simply because we now believe that constructs such as working memory have multiple components. However, we are not yet ready to have a measurement handy that will help us focus on the functioning of each component process. The implications of this are potentially enormous, especially as we come to understand that many disorders that were once well-described based on specific symptomatic outward behaviors may actually have multiple variants believed to be dependent on the brain area(s) and processes affected. For examples, the various dementias, amyotrophic lateral sclerosis, Parkinson disease, chemotherapy-related cognitive impairment, and even age-related hearing loss each now have several if not tens of separable neurological and neuropsychological profiles, including non-motor dysfunction. If we continue to approach their assessment from the perspective of gross cognitive domains, we may miss out on understanding key processes that contribute to their overlap and differentiation. Furthermore, the sharing of knowledge for potential treatments between disorders (i.e., we may be led unnecessarily astray by patients reporting problems focusing versus problems remembering things) will be lost. Considering that attention and memory, and specifically working memory, tend to be the first noticeably relevant sources of cognitive failure that affect daily functioning, it is imperative for us to find a deeper understanding of their connection. Future research using a combination of structural and functional imaging will likely help isolate the processes that are specific to auditory attention from those related to working memory.

References

Alain, C., & Arnott, S. R. (2000). Selectively attending to auditory objects. *Frontiers in Biosciences, 5*, D202–D212.

Alain, C., Arnott, S. A., & Dyson, B. (2013). Varieties of auditory attention. In K. N. Ochsner, & S. M. Kossline (Eds.), *The Oxford handbook of cognitive neuroscience: Volume I core topics* (pp. 215–236). Oxford: Oxford University Press.

Alain, C., Arnott, S. R., Hevenor, S., Graham, S., & Grady, C. L. (2001). What" and "where" in the human auditory system. *Proceedings of the National Academy of Sciences of the United States of America, 98*(21), 12301–12306.

Alain, C., He, Y., & Grady, C. (2008). The contribution of the inferior parietal lobe to auditory spatial working memory. *Journal of Cognitive Neuroscience, 20*(2), 285–295.

Alain, C., & Izenberg, A. (2003). Effects of attentional load on auditory scene analysis. *Journal of Cognitive Neuroscience, 15*(7), 1063–1073.

Alain, C., McDonald, K. L., Kovacevic, N., & McIntosh, A. R. (2009). Spatiotemporal analysis of auditory "what" and "where" working memory. *Cerebral Cortex, 19*(2), 305–314.

Alain, C., Reinke, K., McDonald, K. L., Chau, W., Tam, F., Pacurar, A., et al. (2005). Left thalamo-cortical network implicated in successful speech separation and identification. *Neuroimage, 26*(2), 592–599.

Alain, C., Shen, D., Yu, H., & Grady, C. (2010). Dissociable memory- and response-related activity in parietal cortex during auditory spatial working memory. *Frontiers in Psychology, 1*, 202.

Alho, K., Rinne, T., Herron, T. J., & Woods, D. L. (2014). Stimulus-dependent activations and attention-related modulations in the auditory cortex: a meta-analysis of fMRI studies. *Hearing Research, 307*, 29–41.

Altmann, C. F., Bledowski, C., Wibral, M., & Kaiser, J. (2007). Processing of location and pattern changes of natural sounds in the human auditory cortex. *Neuroimage, 35*(3), 1192–1200.

Altmann, C. F., Henning, M., Doring, M. K., & Kaiser, J. (2008). Effects of feature-selective attention on auditory pattern and location processing. *Neuroimage, 41*(1), 69–79.

Anourova, I., Rama, P., Alho, K., Koivusalo, S., Kahnari, J., & Carlson, S. (1999). Selective interference reveals dissociation between auditory memory for location and pitch. *Neuroreport, 10*(17), 3543–3547.

Anurova, I., Artchakov, D., Korvenoja, A., Ilmoniemi, R. J., Aronen, H. J., & Carlson, S. (2003). Differences between auditory evoked responses recorded during spatial and nonspatial working memory tasks. *Neuroimage, 20*(2), 1181–1192.

Arnott, S. R., & Alain, C. (2011). The auditory dorsal pathway: orienting vision. *Neuroscience & Biobehavioral Reviews, 35*(10), 2162–2173.

Arnott, S. R., Binns, M. A., Grady, C. L., & Alain, C. (2004). Assessing the auditory dual-pathway model in humans. *Neuroimage, 22*(1), 401–408.

Arnott, S. R., Grady, C. L., Hevenor, S. J., Graham, S., & Alain, C. (2005). The functional organization of auditory working memory as revealed by fMRI. *Journal of Cognitive Neuroscience, 17*(5), 819–831.

Baddeley, A. (1992). Working memory. [Research support, non-U.S. Gov't review]. *Science, 255*(5044), 556–559.

Baldo, J. V., & Dronkers, N. F. (2006). The role of inferior parietal and inferior frontal cortex in working memory. *Neuropsychology, 20*(5), 529–538.

Beaman, C. P. (2004). The irrelevant sound phenomenon revisited: what role for working memory capacity? *Journal of Experimental Psychology Learning Memory and Cognition, 30*(5), 1106–1118.

Bedard, M. A., el Massioui, F., Pillon, B., & Nandrino, J. L. (1993). Time for reorienting of attention: a premotor hypothesis of the underlying mechanism. *Neuropsychologia, 31*(3), 241–249.

Berti, S., & Schröger, E. (2003). Working memory controls involuntary attention switching: evidence from an auditory distraction paradigm. *European Journal of Neuroscience, 17*(5), 1119–1122.

Brechmann, A., Gaschler-Markefski, B., Sohr, M., Yoneda, K., Kaulisch, T., & Scheich, H. (2007). Working memory specific activity in auditory cortex: potential correlates of sequential processing and maintenance. *Cerebral Cortex, 17*(11), 2544–2552.

Broadbent, D. E. (1962). Attention and the perception of speech. *Sci Am, 206*, 143–151.

Buchsbaum, B. R., Olsen, R. K., Koch, P., & Berman, K. F. (2005). Human dorsal and ventral auditory streams subserve rehearsal-based and echoic processes during verbal working memory. [Research Support, N.I.H., intramural]. *Neuron, 48*(4), 687–697.

Burianova, H., Ciaramelli, E., Grady, C. L., & Moscovitch, M. (2012). Top-down and bottom-up attention-to-memory: mapping functional connectivity in two distinct networks that underlie cued and uncued recognition memory. *Neuroimage, 63*(3), 1343–1352.

Bushara, K. O., Weeks, R. A., Ishii, K., Catalan, M. J., Tian, B., Rauschecker, J. P., et al. (1999). Modality-specific frontal and parietal areas for auditory and visual spatial localization in humans. *Nature Neuroscience, 2*(8), 759–766.

Cabeza, R., Mazuz, Y. S., Stokes, J., Kragel, J. E., Woldorff, M. G., Ciaramelli, E., et al. (2011). Overlapping parietal activity in memory and perception: evidence for the attention to memory model. *Journal of Cognitive Neuroscience, 23*(11), 3209–3217.

Chen, Z., & Cave, K. R. (2008). Object-based attention with endogenous cuing and positional certainty. *Perception & Psychophysics, 70*(8), 1435–1443.

Cherry, E. C. (1953). Some experiments on the recognition of speech with one and with two ears. *Journal of Acousitcal Society of America, 25*, 975–979.

Chun, M. M., & Johnson, M. K. (2011). Memory: enduring traces of perceptual and reflective attention. [Research support, N.I.H., extramural review]. *Neuron, 72*(4), 520–535.

Ciaramelli, E., Grady, C., Levine, B., Ween, J., & Moscovitch, M. (2010). Top-down and bottom-up attention to memory are dissociated in posterior parietal cortex: neuroimagingand and neuropsychological evidence. *Journal of Neuroscience, 30*(14), 4943–4956.

Ciaramelli, E., Grady, C. L., & Moscovitch, M. (2008). Top-down and bottom-up attention to memory: a hypothesis (AtoM) on the role of the posterior parietal cortex in memory retrieval. *Neuropsychologia, 46*(7), 1828–1851.

Conway, A. R., Cowan, N., & Bunting, M. F. (2001). The cocktail party phenomenon revisited: the importance of working memory capacity. *Psychonomic Bulletin & Review, 8*(2), 331–335.

Corbetta, M., & Shulman, G. L. (2002). Control of goal-directed and stimulus-driven attention in the brain. [Research support, non-U.S. Gov't research support, U.S. Gov't, P.H.S. review]. Nature reviews. *Neuroscience, 3*(3), 201–215.

Cowan, N. (1984). On short and long auditory stores. *Psychological Bulletin, 96*(2), 341–370.

Cowan, N. (1993). Activation, attention, and short-term memory. *Memory & Cognition, 21*(2), 162–167.

Cusack, R., Carlyon, R. P., & Robertson, I. H. (2000). Neglect between but not within auditory objects. *Journal of Cognitive Neuroscience, 12*(6), 1056–1065.

D'Esposito, M., & Badre, D. (2012). Combining the insights derived from lesion and fMRI studies to understand the function of the prefrontal cortex. In B. Levine, & F. I. M. Craik (Eds.), *Mind and the frontal lobes: Cognition, behavior, and brain imaging* (pp. 93–108). New York: Oxford University Press.

Dalton, P., Lavie, N., & Spence, C. (2009). The role of working memory in tactile selective attention. [Research support, non-U.S. Gov't]. *Quarterly Journal of Experimental Psychology, 62*(4), 635–644.

Dalton, P., Santangelo, V., & Spence, C. (2009). The role of working memory in auditory selective attention. *Quarterly Journal of Experimental Psychology (Colchester), 62*(11), 2126–2132.

de Fockert, J. W., Rees, G., Frith, C. D., & Lavie, N. (2001). The role of working memory in visual selective attention. [Research support, non-U.S. Gov't]. *Science, 291*(5509), 1803–1806.

Degerman, A., Rinne, T., Salmi, J., Salonen, O., & Alho, K. (2006). Selective attention to sound location or pitch studied with fMRI. *Brain Research, 1077*(1), 123–134.

Degerman, A., Rinne, T., Sarkka, A. K., Salmi, J., & Alho, K. (2008). Selective attention to sound location or pitch studied with event-related brain potentials and magnetic fields. *European Journal of Neuroscience, 27*(12), 3329–3341.

Duncan, J. (1984). Selective attention and the organization of visual information. *Journal of Experimental Psychology: General, 113*(4), 501–517.

Dyson, B. J., Alain, C., & He, Y. (2005). Effects of visual attentional load on low-level auditory scene analysis. *Cognitive, Affective, & Behavioral Neuroscience, 5*(3), 319–338.

Fougnie, D. (2008). The relationship between attention and working memory. In N. B. Johnson (Ed.), *New research on short-term memory* (pp. 1–45). Nova Science Publishers.

Giesbrecht, B., Woldorff, M. G., Song, A. W., & Mangun, G. R. (2003). Neural mechanisms of top-down control during spatial and feature attention. [Clinical trial research support, non-U.S. Gov't research support, U.S. Gov't, non-P.H.S. research support, U.S. Gov't, P.H.S.]. *Neuroimage, 19*(3), 496–512.

Goldman-Rakic, P. S. (1996). Regional and cellular fractionation of working memory. [Review]. *Proceedings of the National Academy of Sciences of the United States of America, 93*(24), 13473–13480.

Grillon, M. L., Johnson, M. K., Krebs, M. O., & Huron, C. (2008). Comparing effects of perceptual and reflective repetition on subjective experience during later recognition memory. [Research support, N.I.H., extramural research support, non-U.S. Gov't]. *Consciousness and Cognition, 17*(3), 753–764.

Gutschalk, A., Micheyl, C., Melcher, J. R., Rupp, A., Scherg, M., & Oxenham, A. J. (2005). Neuromagnetic correlates of streaming in human auditory cortex. *Journal of Neuroscience, 25*(22), 5382–5388.

Hill, K. T., & Miller, L. M. (2010). Auditory attentional control and selection during cocktail party listening. [Research support, N.I.H., extramural]. *Cerebral Cortex, 20*(3), 583–590.

Hopfinger, J. B., Buonocore, M. H., & Mangun, G. R. (2000). The neural mechanisms of top-down attentional control. [Research support, non-U.S. Gov't research support, U.S. Gov't, non-P.H.S. Research support, U.S. Gov't, P.H.S.]. *Nature Neuroscience, 3*(3), 284–291.

Huang, S., Seidman, L. J., Rossi, S., & Ahveninen, J. (2013). Distinct cortical networks activated by auditory attention and working memory load. *Neuroimage, 83*, 1098–1108.

Johnson, M. K., & Hirst, W. (1993). MEM: memory subsystems as processes. In A. A. Collins, S. S. Gathercole, M. M. Conway, & P. E. Morris (Eds.), *Theories of memory*. East Sussex, England: Erlbaum.

Johnson, M. R., Mitchell, K. J., Raye, C. L., D'Esposito, M., & Johnson, M. K. (2007). A brief thought can modulate activity in extrastriate visual areas: top-down effects of refreshing just-seen visual stimuli. *Neuroimage, 37*(1), 290–299.

Kaiser, J., Ripper, B., Birbaumer, N., & Lutzenberger, W. (2003). Dynamics of gamma-band activity in human magnetoencephalogram during auditory pattern working memory. *Neuroimage, 20*(2), 816–827.

Krumbholz, K., Eickhoff, S. B., & Fink, G. R. (2007). Feature- and object-based attentional modulation in the human auditory "where" pathway. *Journal of Cognitive Neuroscience, 19*(10), 1721–1733.

LaBerge, D. (1983). Spatial extent of attention to letters and words. *Journal of Experimental Psychology: Human Perception and Performance, 9*(3), 371–379.

Lavie, N., Hirst, A., de Fockert, J. W., & Viding, E. (2004). Load theory of selective attention and cognitive control. [Clinical trial randomized controlled trial research support, non-U.S. Gov't]. *The Journal of Experimental Psychology: General, 133*(3), 339–354.

Leung, A. W., & Alain, C. (2011). Working memory load modulates the auditory "what" and "where" neural networks. *Neuroimage, 55*(3), 1260–1269.

Ley, A., Vroomen, J., & Formisano, E. (2014). How learning to abstract shapes neural sound representations. *Frontiers in Neuroscience, 8*, 132.

Lutzenberger, W., Ripper, B., Busse, L., Birbaumer, N., & Kaiser, J. (2002). Dynamics of gamma-band activity during an audiospatial working memory task in humans. *Journal of Neuroscience, 22*(13), 5630–5638.

Maeder, P. P., Meuli, R. A., Adriani, M., Bellmann, A., Fornari, E., Thiran, J. P., et al. (2001). Distinct pathways involved in sound recognition and localization: a human fMRI study. *Neuroimage, 14*(4), 802–816.

McDonald, J. J., Teder-Salejarvi, W. A., Di Russo, F., & Hillyard, S. A. (2005). Neural basis of auditory-induced shifts in visual time-order perception. *Nature Neuroscience, 8*(9), 1197–1202.

Miller, B. T., Verstynen, T., Johnson, M. K., & D'Esposito, M. (2008). Prefrontal and parietal contributions to refreshing: an rTMS study. *Neuroimage, 39*(1), 436–440.

Mondor, T. A., & Zatorre, R. J. (1995). Shifting and focusing auditory spatial attention. *Journal of Experimental Psychology: Human Perception and Performance, 21*(2), 387–409.

Moscovitch, M. (1992). Memory and working-with-memory: a component process model based on modules and Central systems. *Journal of Cognitive Neuroscience, 4*(3), 257–267.

Nobre, A. C., Coull, J. T., Maquet, P., Frith, C. D., Vandenberghe, R., & Mesulam, M. M. (2004). Orienting attention to locations in perceptual versus mental representations. *Journal of Cognitive Neuroscience, 16*(3), 363–373.

Nobre, A. C., & Stokes, M. G. (2011). Attention and short-term memory: crossroads. *Neuropsychologia, 49*(6), 1391–1392.

Patai, E. Z., Doallo, S., & Nobre, A. C. (2012). Long-term memories bias sensitivity and target selection in complex scenes. *Journal of Cognitive Neuroscience, 24*(12), 2281–2291.

Petkov, C. I., Kang, X., Alho, K., Bertrand, O., Yund, E. W., & Woods, D. L. (2004). Attentional modulation of human auditory cortex. *Nature Neuroscience, 7*(6), 658–663.

Petrides, M. (1991). Monitoring of selections of visual stimuli and the primate frontal cortex. *Proceedings of the Royal Society B: Biological Sciences, 246*(1317), 293–298.

Petrides, M. (1995). Functional organization of the human frontal cortex for mnemonic processing. Evidence from neuroimaging studies. *Annals of the New York Academy of Sciences, 769*, 85–96.

Petrides, M. (2005). Lateral prefrontal cortex: architectonic and functional organization. *Philosophical Transactions of the Royal Society, B: Biological Sciences, 360*(1456), 781–795.

Petrides, M. (2012). The mid-dorsolateral prefronto-parietal network and the epoptic process. In D. T. Stuss & R. T. Knight (Eds.), *Principles of Frontal Lobe Function*, (2nd ed.). (pp. 79–89). New York, NY: Oxford University Press.

Petrides, M., & Milner, B. (1982). Deficits on subject-ordered tasks after frontal- and temporal-lobe lesions in man. *Neuropsychologia, 20*(3), 249–262.

Petrides, M., Tomaiuolo, F., Yeterian, E. H., & Pandya, D. N. (2012). The prefrontal cortex: comparative architectonic organization in the human and the macaque monkey brains. *Cortex, 48*(1), 46–57.

Poremba, A., Saunders, R. C., Crane, A. M., Cook, M., Sokoloff, L., & Mishkin, M. (2003). Functional mapping of the primate auditory system. [Research support, U.S. Gov't, P.H.S.]. *Science, 299*(5606), 568–572.

Posner, M. I., Snyder, C. R., & Davidson, B. J. (1980). Attention and the detection of signals. [Research support, U.S. Gov't, non-P.H.S.]. *Journal of Experimental Psychology, 109*(2), 160–174.

Rama, P., & Courtney, S. M. (2005). Functional topography of working memory for face or voice identity. [Comparative study research support, non-U.S. Gov't research support, U.S. Gov't, P.H.S.]. *Neuroimage, 24*(1), 224–234.

Rama, P., Paavilainen, L., Anourova, I., Alho, K., Reinikainen, K., Sipila, S., et al. (2000). Modulation of slow brain potentials by working memory load in spatial and nonspatial auditory tasks. *Neuropsychologia, 38*(7), 913–922.

Rama, P., Poremba, A., Sala, J. B., Yee, L., Malloy, M., Mishkin, M., et al. (2004). Dissociable functional cortical topographies for working memory maintenance of voice identity and location. *Cerebral Cortex, 14*(7), 768–780.

Rauschecker, J. P., & Tian, B. (2000). Mechanisms and streams for processing of "what" and "where" in auditory cortex. *Proceedings of the National Academy of Sciences of the United States of America, 97*(22), 11800–11806.

Raye, C. L., Johnson, M. K., Mitchell, K. J., Greene, E. J., & Johnson, M. R. (2007). Refreshing: a minimal executive function. *Cortex, 43*(1), 135–145.

Rinne, T., Koistinen, S., Salonen, O., & Alho, K. (2009). Task-dependent activations of human auditory cortex during pitch discrimination and pitch memory tasks. *Journal of Neuroscience, 29*(42), 13338–13343.

Romanski, L. M., Bates, J. F., & Goldman-Rakic, P. S. (1999). Auditory belt and parabelt projections to the prefrontal cortex in the rhesus monkey. *Journal of Comparative Neurology, 403*(2), 141–157.

Romanski, L. M., Tian, B., Fritz, J., Mishkin, M., Goldman-Rakic, P. S., & Rauschecker, J. P. (1999). Dual streams of auditory afferents target multiple domains in the primate prefrontal cortex. *Nature Neuroscience*, *2*(12), 1131–1136.

Roth, J. K., Johnson, M. K., Raye, C. L., & Constable, R. T. (2009). Similar and dissociable mechanisms for attention to internal versus external information. *Neuroimage*, *48*(3), 601–608.

Rottschy, C., Caspers, S., Roski, C., Reetz, K., Dogan, I., Schulz, J. B., et al. (2013). Differentiated parietal connectivity of frontal regions for "what" and "where" memory. [Research support, N.I.H., extramural research support, non-U.S. Gov't]. *Brain Structure & Function*, *218*(6), 1551–1567.

Salmi, J., Rinne, T., Koistinen, S., Salonen, O., & Alho, K. (2009). Brain networks of bottom-up triggered and top-down controlled shifting of auditory attention. *Brain Research*, *1286*, 155–164.

Shinn-Cunningham, B. G. (2008). Object-based auditory and visual attention. *Trends in Cognitive Sciences*, *12*(5), 182–186.

Shomstein, S., & Yantis, S. (2004). Control of attention shifts between vision and audition in human cortex. *Journal of Neuroscience*, *24*(47), 10702–10706.

Shomstein, S., & Yantis, S. (2006). Parietal cortex mediates voluntary control of spatial and nonspatial auditory attention. *Journal of Neuroscience*, *26*(2), 435–439.

Shulman, G. L., Ollinger, J. M., Akbudak, E., Conturo, T. E., Snyder, A. Z., Petersen, S. E., et al. (1999). Areas involved in encoding and applying directional expectations to moving objects. [Research support, non-U.S. Gov't research support, U.S. Gov't, P.H.S.]. *The Journal of Neuroscience: The Official Journal of The Society for Neuroscience*, *19*(21), 9480–9496.

Snyder, J. S., Alain, C., & Picton, T. W. (2006). Effects of attention on neuroelectric correlates of auditory stream segregation. *Journal of Cognitive Neuroscience*, *18*(1), 1–13.

Stuss, D. T., Alexander, M. P., Shallice, T., Picton, T. W., Binns, M. A., Macdonald, R., et al. (2005). Multiple frontal systems controlling response speed. *Neuropsychologia*, *43*(3), 396–417.

Stuss, D. T., Binns, M. A., Murphy, K. J., & Alexander, M. P. (2002). Dissociations within the anterior attentional system: effects of task complexity and irrelevant information on reaction time speed and accuracy. *Neuropsychology*, *16*(4), 500–513.

Tzourio, N., Massioui, F. E., Crivello, F., Joliot, M., Renault, B., & Mazoyer, B. (1997). Functional anatomy of human auditory attention studied with PET. *Neuroimage*, *5*(1), 63–77.

Ungerleider, L. G., & Haxby, J. V. (1994). 'What' and 'where' in the human brain. *Current Opinion in Neurobiology*, *4*(2), 157–165.

Wenzel, E. M., Arruda, M., Kistler, D. J., & Wightman, F. L. (1993). Localization using nonindividualized head-related transfer functions. *Journal of the Acoustical Society of America*, *94*(1), 111–123.

Woods, D. L., & Alain, C. (1993). Feature processing during high-rate auditory selective attention. *Perception & Psychophysics*, *53*(4), 391–402.

Woods, D. L., & Alain, C. (2001). Conjoining three auditory features: an event-related brain potential study. *Journal of Cognitive Neuroscience*, *13*(4), 492–509.

Woods, D. L., Alho, K., & Algazi, A. (1994). Stages of auditory feature conjunction: an event-related brain potential study. *The Journal of Experimental Psychology: Human Perception and Performance*, *20*(1), 81–94.

Yantis, S., & Serences, J. T. (2003). Cortical mechanisms of space-based and object-based attentional control. *Current Opinion in Neurobiology*, *13*(2), 187–193.

Yi, D. J., Turk-Browne, N. B., Chun, M. M., & Johnson, M. K. (2008). When a thought equals a look: refreshing enhances perceptual memory. *Journal of Cognitive Neuroscience*, *20*(8), 1371–1380.

Yurgil, K. A., & Golob, E. J. (2013). Cortical potentials in an auditory oddball task reflect individual differences in working memory capacity. *Psychophysiology*, *50*, 1263–1274.

CHAPTER

17

Neuroimaging of the Mind's Ear Using Representational Similarity Analysis

Rhodri Cusack, Annika C. Linke

Brain and Mind Institute, Western University, London, ON, Canada

To make sense of the dynamic and fleeting soundscape that reaches the ear, it is necessary to hold auditory information in short-term memory or bring it back to mind using imagery. Here we show how functional magnetic resonance imaging and representational similarity analysis (RSA) can be used to unravel how the brain processes complex sound information during these different mental processes. Compared with standard neuroimaging methods, RSA allows many features of complex sounds to be probed simultaneously and as such makes it possible to investigate neural representations in much greater detail. Our results show that even in sensory cortex, representations of the same complex sounds differ during auditory short-term memory and imagery, and provide initial evidence that performance on these tasks is related to changes in which information about a stimulus becomes encoded.

Picture a scene in a suspense movie—a dark night, an unexpected sound—followed by an alarmed "What was that?" It is time for the protagonist and the viewer to examine their auditory memory. Many sounds are fleeting (in the real world, as well as at the movies) and auditory memory is essential if we are to analyze, learn, or compare sounds. Sometimes we need to preserve something we have just heard, for which we use *auditory short-term memory* or *auditory working memory*. At other times, a sound from long ago is brought back to life in the mind, a kind of mental rejuvenation called *auditory imagery*. We have used human neuroimaging with functional magnetic resonance imaging (fMRI) to understand the neurocognitive basis of these processes of the mind's ear.

The first section probes the neural basis of auditory memory. Whereas neuroimaging and electrophysiology have given us a guide to what brain regions are involved in auditory memory, they have placed few constraints on the type of mental memory representations contained within these regions. We have addressed this using the method of *representational similarity analysis* to probe the nature of these mental codes.

The second section examines the flexibility of these representations across tasks. Some flexibility is already well established, with memory representations affected by the timescale of maintenance, the type of memorandum, and the attentional state of the viewer at encoding. We will ask whether subtler distinctions are important, by testing whether differences in task, within a timescale, and for the same stimuli, affect the nature of auditory mental representations.

The final section investigates the effect of individual differences on representations of the mind's ear. We relate behavioral performance to neuroimaging measures of activity and the fidelity of representation. For visual short-term memory, we have provided evidence that individual differences with important associations to cognitive aptitude and educational performance are a result of strategic differences in encoding, rather than fundamental differences in memory capacity. In audition, individual differences also suggest that encoding strategy is important.

THE REPRESENTATION OF AUDITORY MEMORIES: CONNECTING THE MIND AND BRAIN

Limitations of Conventional Methods

Our goal here is to understand what kind of mental code is used to represent auditory memories. Although informative in other ways, many existing experiments tell us little about this. For example, consider a neuroimaging or

Mechanisms of Sensory Working Memory
http://dx.doi.org/10.1016/B978-0-12-811042-3.00017-7

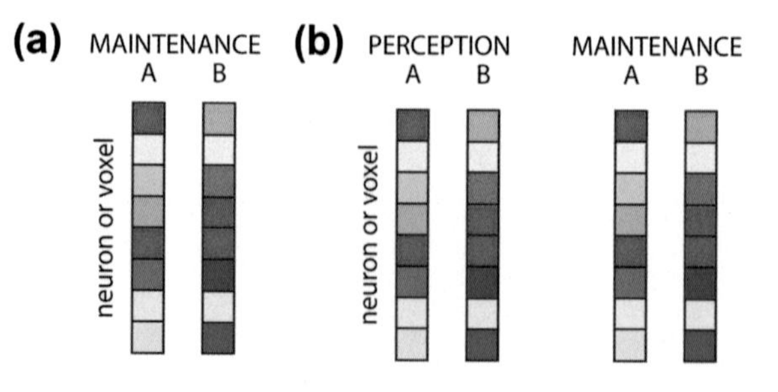

FIGURE 1 (a) A reliable difference in the pattern of activity during the memory maintenance of two items (A and B) shows that a brain region is involved in working memory but does not reveal what features of the items are being encoded. (b) Similarity of codes across perception and maintenance also places few constraints on the mental code for memories.

electrophysiological recording experiment in which for each trial, one of two different items had to be remembered (e.g., Fuster, 1973). During the memory maintenance period, some brain regions or sets of neurons have reliably different patterns of activity for the two items (Figure 1(a)). This tells us that the brain region is recruited during memory maintenance, and that there is some form of mental representation of the items during the memory delay. But it does not reveal what aspect of the two items is being represented. As two arbitrary examples, it could be a sensory-like code, the brain's equivalent of a photographic snapshot; or it could be an abstract verbal label.

Next, consider experiments in which activity is compared between perception and memory maintenance (e.g., Harrison & Tong, 2009). Imagine it is possible to use the pattern of activity evoked by the two objects during perception to decode which of the two objects is being remembered from the maintenance period activity (Figure 1(b)). This tells us that there is some common representation of the objects across perception and memory maintenance. But it does not inform us of the nature of this representation. During perception, we know cortex represents characteristics of the physical stimulus. Similarity between perception and memory is therefore consistent with the perseverance of a code that is like the original stimulus. However, it is also consistent with other codes. Many other codes are rapidly derived during perception: semantic, aesthetic, emotional, and so on, and the similarity between perception and memory might reflect shared representation of one of these more abstract codes.

Another potential window on the problem is to use the location of the brain regions recruited by maintenance to infer the code being used (Arnott, Grady, Hevenor, Graham, & Alain, 2005). For example, in vision, recruitment of the brain region middle temporal (MT) during maintenance could be taken as a fingerprint of the representation of motion information. There are important limitations to this approach. It requires consistent mapping between the type of information stored and the brain region. Put another way, if MT encodes motion in a perceptual experiment, can we be certain this is what it is doing in a maintenance period? As we shall see later, this assumption of a stable mapping can be violated. Furthermore, even this guarded inference will be impossible if there is little consensus among researchers regarding mapping between features and brain regions, as is the case for audition.

Representational Similarity Analysis

A strategy that addresses these problems is RSA (Kriegeskorte, Mur, Ruff, & Kiani, 2008). The idea is to use a richer stimulus set containing a minimum of three, and preferably more stimuli. From a model of some aspect of the stimuli (e.g., the frequency of a sound) a representational dissimilarity matrix (RDM) is calculated, which describes the (dis) similarity of each stimulus to every other stimulus on this feature (Figure 2, left). Where the stimulus feature is scalar (as is the case for frequency), the absolute or squared difference might be used to calculate each entry in the RDM, and when it is a vector (as might be the case for a quantification of timbre) the similarity of pairs of vectors might be measured using correlation (strictly, 1-correlation, because we need a measure that is larger for more dissimilar things), Mahalanobis distance, or Euclidean distance.

A parallel analysis is applied to the neural data (Figure 2, right). The activation pattern evoked by a stimulus across voxels (in neuroimaging) or neurons (in electrophysiology) is recorded. An RDM is then calculated that describes the (dis)similarity of the activation pattern evoked by each stimulus to that evoked by every other stimulus. Again, the correlation, Mahalanobis, or Euclidean distances or another measure of similarity might be used.

The similarity of the feature and neural RDMs are then compared. If they have shared structure, it supports the view that this stimulus feature is represented in that brain region. This method has been effective in teasing apart visual representations during perception (Cusack, Veldsman, Naci, Mitchell, & Linke, 2011; Kriegeskorte et al., 2008; Vicente-Grabovetsky, Carlin, & Cusack, 2014). With a sufficiently rich stimulus set, it allows the disentangling of multiple codes within a single brain region (Cusack et al., 2011; Kravitz, Kriegeskorte, & Baker, 2010). It has also been used to decode visual memories (Xue et al., 2010). Its strength in providing a strong test of psychological models, thus providing a stronger connection between models of the mind and brain (Vicente-Grabovetsky et al., 2014), is increasingly coming into focus.

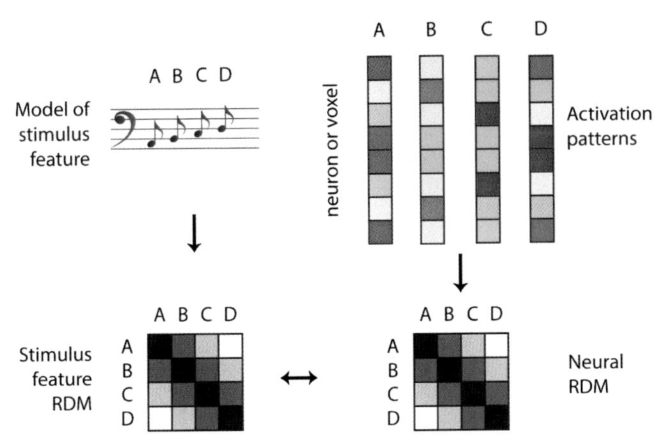

FIGURE 2 (Left) From a model of a feature of the stimulus (e.g., sound frequency) an representational dissimilarity matrix (RDM) is calculated (e.g., using the absolute difference) that describes how dissimilar each stimulus is to every other stimulus on the feature in the model. (Right) The activation patterns to four stimuli (A–D) are compared to create the neural RDM (e.g., using 1-correlation). Similarity of the resulting stimulus feature and neural RDMs then provides evidence that the brain is representing the stimulus feature.

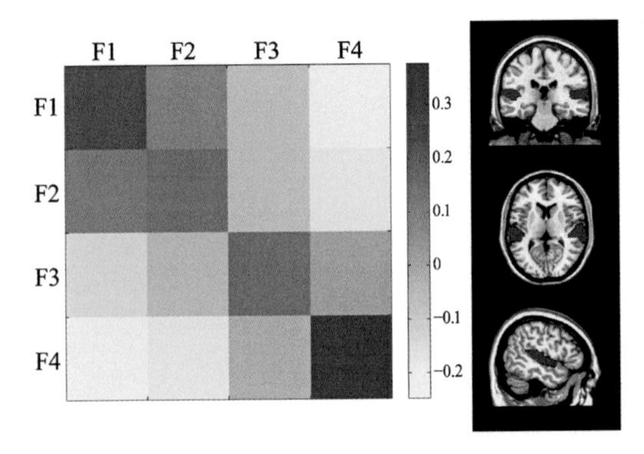

FIGURE 3 Mean correlations of activity patterns within and across frequencies (F1–F4) in auditory cortex (AC) during perception; darker colors indicate higher correlations. The diagonal reflects consistency of activation patterns within the same condition, with the matrix calculated for repetitions of the same condition across but not within the different blocks of the experiment to avoid confounding resulting from temporal proximity of events within a block. These results show that sounds that are more similar in frequency evoke a more similar pattern of activation.

Memory for Pure Tones

In a first evaluation of RSA applied to auditory memory, we presented pure tones chosen from four distinct frequency regions in a change detection task (Linke, Vicente-Grabovetsky, & Cusack, 2011). An event-related design was used to help separate the brain activity from encoding and memory maintenance, with delays of 1 and 10s. Higher-resolution fMRI was used, with isotropic voxels of size approximately 2.4mm.

RSA confirmed that during perception, frequency was represented both in primary auditory cortex (AC) (around Heschl's gyrus [HG]) and in broader AC (Figure 3). This was confirmed with a summary statistic that compared the similarity of activation evoked in trials with the same frequency (i.e., the leading diagonal) with trials of different frequency (i.e., the off-diagonal elements) by fitting a contrast matrix to the RDM ($t[24]=10.363$; $p<.001$).

During memory maintenance, there was general suppression of auditory regions around HG and beyond (Figure 4(a)). This suppression of sensory regions during short-term memory maintenance has since been reported by others (Buchsbaum, Lemire-Rodger, Fang, & Abdi, 2012; Konstantinou, Bahrami, Rees, & Lavie, 2012; Riggall & Postle, 2012). We discuss its functional implications later.

Importantly, RSA showed that frequency was encoded during memory maintenance in HG, with more similar activation patterns in pairs of trials of the same frequency than those with different frequency (pattern distinctiveness $=0.05$; $p<.05$). However, strikingly, when the activity patterns were compared across perception and memory maintenance, we found that in both HG and broader AC, the voxels that were most activated by a particular frequency during perception *were most suppressed* during memory maintenance (Figure 4(b)). This pattern has since been found by others (Buchsbaum et al., 2012).

Memory for Complex Sounds

In a second experiment, we investigated memory for complex sounds (Linke et al., 2015). Twelve sounds of approximately 1.5s were chosen, three each from animate (human and animal vocalizations) and inanimate (object sounds and musical instruments) categories. They were chosen from a larger set to de-correlate their physical differences from category or animacy class. Again, a change detection task was used, with a memory delay chosen from the interval 2–11s. For half of the trials, the stimuli changed slightly in one of three ways

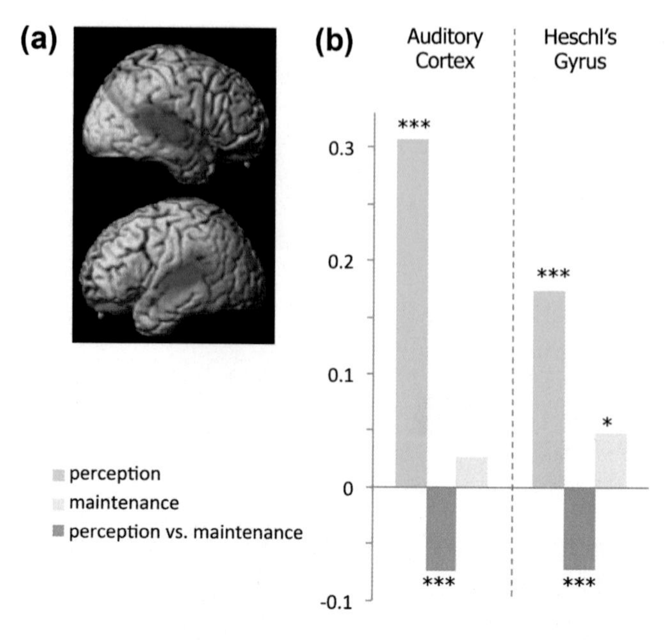

FIGURE 4 (a) Auditory regions were in general suppressed during memory maintenance. (b) During perception and maintenance there was evidence of frequency-specific coding. Asterisks illustrate the results of fitting a contrast matrix (assuming higher values in the leading diagonal) to the perception and maintenance neural RDMs (***$p<.001$; *$p<.05$). However, the voxels that were most activated during perception of a given frequency were most suppressed during maintenance of that frequency, as illustrated by a negative correlation of activity patterns during perception with activity patterns with the same frequencies during maintenance (blue bars; $t[24]=-4.265$, $p<.001$ for auditory cortex (AC) and $t[24]=-4193$, $p<.001$ for HG). The height of the bar graphs quantifies frequency specificity and was calculated as the difference in similarity (correlations in RDMs) of within versus across frequency pattern comparisons.

(frequency, rate, or loudness). A quiet fMRI sequence (Peelle, Eason, Schmitter, Schwarzbauer, & Davis, 2010; Schmitter et al., 2008) was used to reduce acoustic interference.

During perception and memory maintenance, which of the 12 sounds was being presented or remembered could be decoded in three auditory regions (Figure 5). Interestingly, when we looked at the relationship between perception and maintenance, there was again a negative relationship; voxels that were more activated by a given sound during perception were more deactivated during maintenance.

These analyses show that these brain regions are involved in the memory of complex sounds, but they do not reveal the nature of the code. To investigate this, we performed a broader RSA in which we compared three codes: coding for animacy (two levels of animate versus inanimate), category (four levels), and individual objects as above. We found that during this change detection task, only coding at the level of individual objects was observed. This is consistent with a representation of the basic acoustic properties of the stimuli, as would be expected given the subtle nature of the acoustic changes that had to be detected.

Why Suppression During Maintenance?

We suggest four hypotheses for why there is an intriguing reversal of the patterns of activity between perception and maintenance. One hypothesis is that the negative pattern during maintenance in AC reflects suppression of incoming auditory input that might interfere with a memory representation elsewhere in the brain (Sawaki & Luck, 2011; Sylvester, Jack, Corbetta, & Shulman, 2008; Zanto & Gazzaley, 2009). One aspect of our experiments incidentally manipulated this. In the first experiment with pure tones, we used conventional fMRI, which generates substantial acoustic noise. However, in the second experiment, we used a quiet fMRI sequence that is barely audible through the ear defenders worn by the participant. That the negative pattern was similar for different levels of interfering noise suggests that this might not be the cause of the observed suppression.

A second hypothesis is that it is a result of the comparison requirement of a change detection task. If the memorandum is maintained as a negative image, when the second (probe) stimulus arrives, the difference between the two could provide an error signal that cues the answer. However, counter to this idea is evidence that suppression can occur for memory tasks such as recall that do not involve change detection (Buchsbaum et al., 2012; Zvyagintsev et al., 2013).

A third intriguing hypothesis proposed by Buchsbaum and colleagues (Buchsbaum et al., 2012) is that the suppression is a neural marker of the non-perceptual origin of the representations, which could be disrupted in those who experience auditory hallucinations.

A fourth hypothesis is that the apparent stimulus-specific suppression is a result of the limited resolution of fMRI, and at the neural level there is a center-surround pattern of excitation and inhibition, so that neurons with best frequencies at the remembered frequency are activated but those with nearby best frequencies are suppressed. On the resolution of fMRI this appears as a net suppression at the remembered frequency.

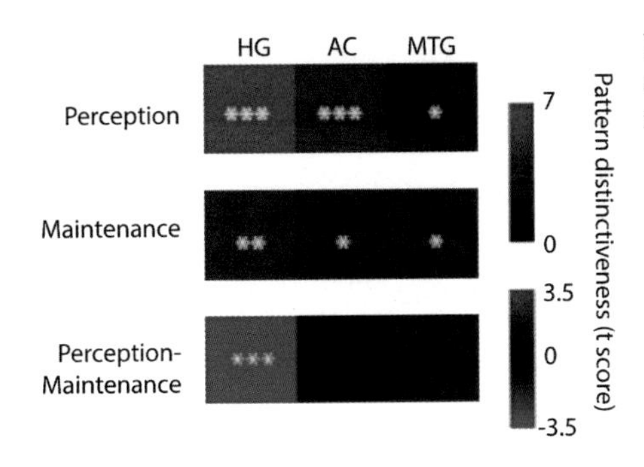

FIGURE 5 Distinctiveness of neural patterns for individual sounds, in the site of primary auditory cortex (AC), Heschl's Gyrus (HG), broader AC, or a higher-level auditory region in the middle temporal gyrus (MTG).

We cannot yet distinguish among these hypotheses for suppression during memory. A future model might also account for evidence of occasional suppression of the AC even during perception (Otazu, Tai, Yang, & Zador, 2009), and for the apparent dissociation between the overall degree of activation or suppression, and the presence of pattern information (LaRocque, Lewis-Peacock, Drysdale, Oberauer, & Postle, 2013; Lewis-Peacock, Drysdale, Oberauer, & Postle, 2012), and ideally provide an explanation at the neural level, perhaps focusing on the role of inhibitory interneurons in controlling cortical circuits (Shamma, 2013).

FLEXIBILITY OF REPRESENTATIONS ACROSS TASKS

A number of factors have been established that influence the form of representations held in mind, including the timescale for maintenance, the type of memorandum, and the attentional state of the viewer at encoding.

First, representations across different timescales are thought to use different mechanisms, with distinctions drawn between sensory memory, short-term memory, and long-term memory (Atkinson & Shiffrin, 1968; Sperling, 1960). Many experiments support such a division in vision, and extensions have been proposed (e.g., fragile memory) (Sligte, Scholte, & Lamme, 2008). How far memory systems in audition parallel those in vision, in both their characteristics and number, is still a matter of controversy (Demany, Semal, Cazalets, & Pressnitzer, 2010). Second, there is substantial evidence that the neural code is affected by the type of stimulus to be held in mind—for example, whether it lends itself to phonological or visuospatial representation (Baddeley & Hitch, 1975). Third, we know that selective attention at encoding can affect what is held in mind (Sperling, 1960). There are quantitative models of the role of selective attention during encoding in short-term visual memory tasks (Bundesen, 1990) and evidence that attention can even select within working memory (LaRocque et al., 2013).

It is unclear whether beyond these broad aspects—timescale, stimulus type, and selective attention—the neural code is still sensitive to task. We consider three qualitatively different possible patterns of variation in the neural code, which are illustrated in Figure 6, for two hypothetical visual memory tasks requiring memory for different aspects of a visual stimulus (orientation or motion). In this illustration, the strength of coding of these two visual features in two visual regions specialized for them (MT and primary visual cortex [V1]) is shown.

One model is that the representations may be invariant to task (Figure 6(a)). That is, V1 represents orientation irrespective of task, and MT represents motion irrespective of task. A second qualitatively different model is that the task might change the balance of recruitment of different brain regions (Figure 6(b)). In vision there are a plethora of representations along the ventral visual stream. Perhaps the representation that is most appropriate is used to remember what is relevant for the task. A third possibility is that type of task might affect representations within brain systems (Figure 6(c)). As discussed in the introduction, it is not clear whether there is consistent mapping between brain region and representation. In some regions of the brain, such as the dorsolateral prefrontal cortex, strong representational flexibility is well accepted (Duncan, 2001; Freedman, Riesenhuber, Poggio, & Miller, 2001). In sensory systems, a consistent mapping is often assumed but without rigorous proof.

Compared with vision, there is less consensus on how AC should be parceled, or what features are represented by different regions during perception. However, we have found evidence that AC is just as modular as visual cortex (Cusack et al., 2013). An effect of memory task on the balance of recruitment between regions, or the code within them, is therefore equally possible in audition.

We conducted a neuroimaging experiment to assess whether there is a single form of representation in working memory for audition, or whether it changes with task, even when the stimulus, maintenance duration, and attentional requirements were controlled. We investigated three brain regions that are considered part of a hierarchy of auditory processing: primary sensory regions including HG, AC association regions on the superior temporal plane,

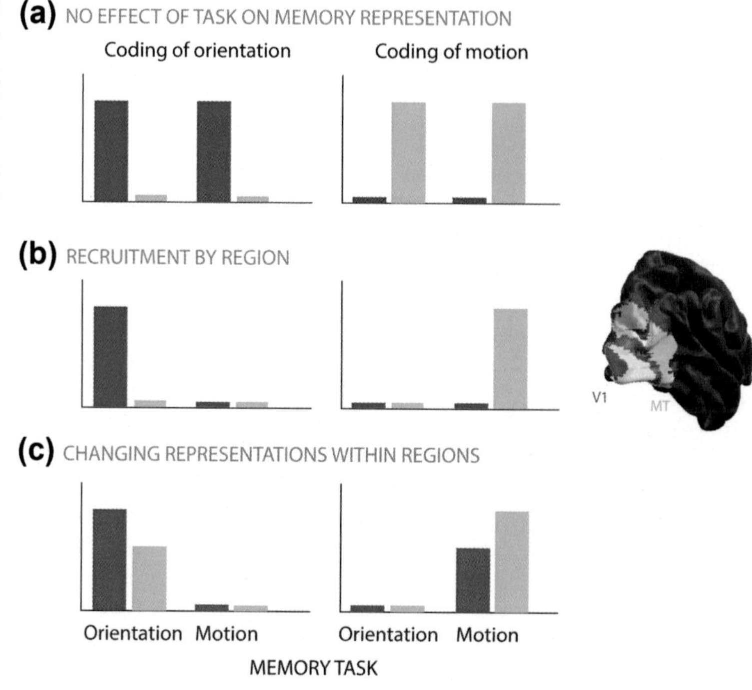

FIGURE 6 Three hypothetical models of the effect of memory task on coding of orientation and motion in two brain regions. a) is an example of memory representation in V1 (purple) and MT (green) being unchanged irrespective of task (orientation vs. motion) performed, b) shows selectivity of a region for a certain task and stimulus, with e.g. V1 coding for orientation in an orientation task only, and c) shows changing representations in the two regions based on the task performed.

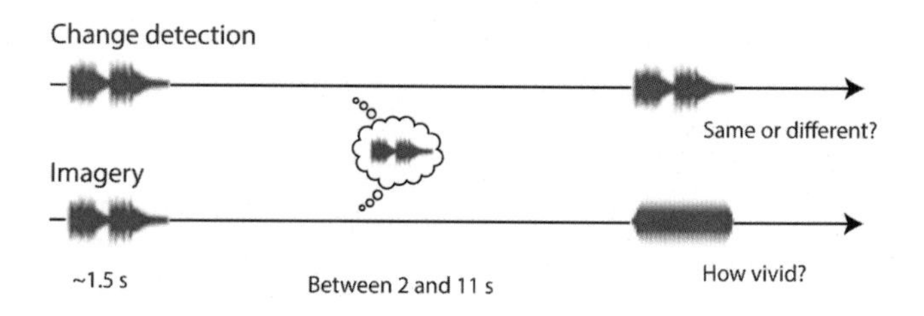

FIGURE 7 Paradigms for the change detection and imagery blocks.

and a more inferior and anterior region on the middle temporal gyrus (MTG) that is sensitive to more abstract stimulus categories (Davis & Johnsrude, 2003). We measured the extent to which task affects the balance of recruitment of these three regions and whether task affects representations within them.

Influence of Task on the Representation of Complex Sounds

Neural representations of sounds were measured for two tasks that differed in their instructions, but not in the physical stimuli presented, the timescale over which the task was performed, or which stimuli had to be attended. One task was change detection, described in the section "Memory for Complex Sounds." The second task required participants to hold the stimuli in mind (i.e., mental imagery) and rate the vividness of their mental representation. The order of the two tasks was counterbalanced. In the imagery task, the initial sounds and the subsequent gap were identical to the change detection task (Figure 7). The second sound was replaced by a sinusoidal tone, and the response of "same or different" was replaced with a rating of the vividness of imagery. The neuroimaging parameters were identical to the change detection condition.

Figure 8 shows the univariate (bulk) brain activity. As expected given the identical stimuli, during perception the brain responses (top row) are similar across tasks. During the delay period (bottom row), there is activity around AC for both change detection and imagery. However, there is substantially more activity in the delay period of the imagery condition, and the pattern of this activity is similar to the perception phase. From these results alone, it might be concluded that imagery causes something like a re-instantiation of the original pattern of activity (Kosslyn, 2003) and a mental image that is like the perception of the original stimulus.

We now turn to RSA, which illustrates the dangers of relying on univariate analyses or assuming consistent mapping between brain region and information. As for change detection, we investigated whether animacy, category, or the individual sounds were represented. The two left matrices in Figure 9 review the results reported earlier for change detection: only identity is coded in all three brain regions, in both perception and memory maintenance. For the imagery task, as for change detection, during perception the individual sounds are encoded in all three regions.

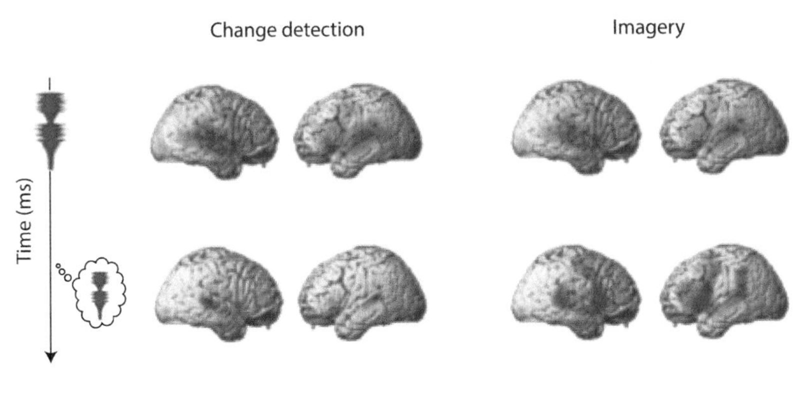

FIGURE 8 Univariate brain activity for perception (top row) and the maintenance period (bottom row) for the change detection and imagery tasks.

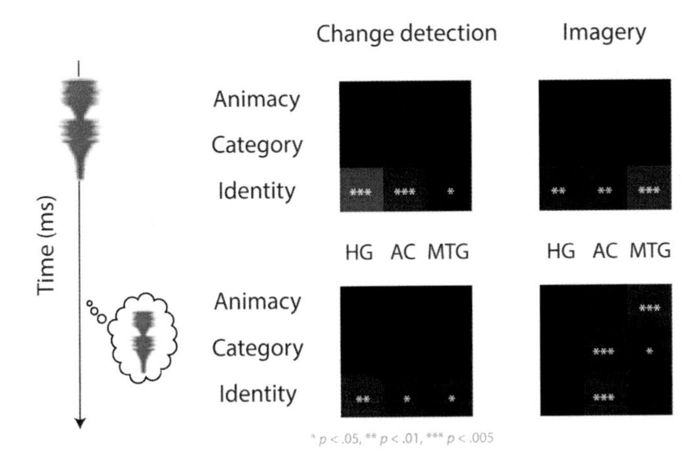

FIGURE 9 Representational similarity analyses (RSA) of coding of three levels of representation (identity, category, or animacy) in three regions (HG, AC, or MTG) for the two tasks (change detection and imagery). Matrices show results (t values) of a general linear model fitting different contrast matrices (assuming greater pattern similarity for sounds that share identity, category, or animacy) to the neural RDMs in the three regions during perception (matrices in first row) and delay (matrices in second row). Higher t values are shown in lighter colors and indicate more distinct information coding (***$p<.001$; **$p<.005$; *$p<.05$).

During the imagery period, however, more abstract levels of coding were seen: category in AC and MTG, and animacy in MTG. Despite the similarity in the univariate patterns of activity for perception and imagery maintenance, the RSA reveals different information being encoded.

The effect of task appears to be both a modulation of the balance of activity across regions (Figure 8) and a change in the information represented within those regions (Figure 9). These strong differences were seen even though the tasks used the same stimuli, had the same temporal requirements, and required attention to all stimuli. The results show within-region changes in representation even in sensory regions, where a consistent mapping between region and representation is often assumed.

FLEXIBILITY OF REPRESENTATIONS ACROSS INDIVIDUALS

Individual differences in visual short-term memory are important indicators of cognitive aptitude and scholastic success (Cowan et al., 2005). We have shown that the correlation between intelligence tests and visual short-term memory depends on the form of the task used to measure memory (Cusack, Lehmann, Veldsman, & Mitchell, 2009). This is the result of differences in encoding strategy used by different individuals when faced with particular tasks (Linke, Vicente-Grabovetsky, Mitchell, & Cusack, 2011). Put another way, in the visual short-term memory tasks we have studied, how people choose to remember something determines their memory, not their memory capacity in itself.

Mirroring this conclusion, our neuroimaging studies have shown that individual differences in memory performance are related to individual differences in brain activity during encoding and not maintenance (Linke, Vicente-Grabovetsky, Mitchell, et al., 2011). This supports the argument that some process happening during encoding—likely the remapping of perceptual information into a form more suitable for memory—is the primary determinant of memory performance in visual short-term memory. Studies of neural representations within brain regions support this model. Using RSA, in two experiments we observed a relationship between memory capacity and the fidelity of representations during encoding, but not maintenance (Vicente-Grabovetsky et al., 2014). In vision, there is a coherent body of work indicating that strategically driven remapping during encoding is a strong modulator of performance on working memory tasks.

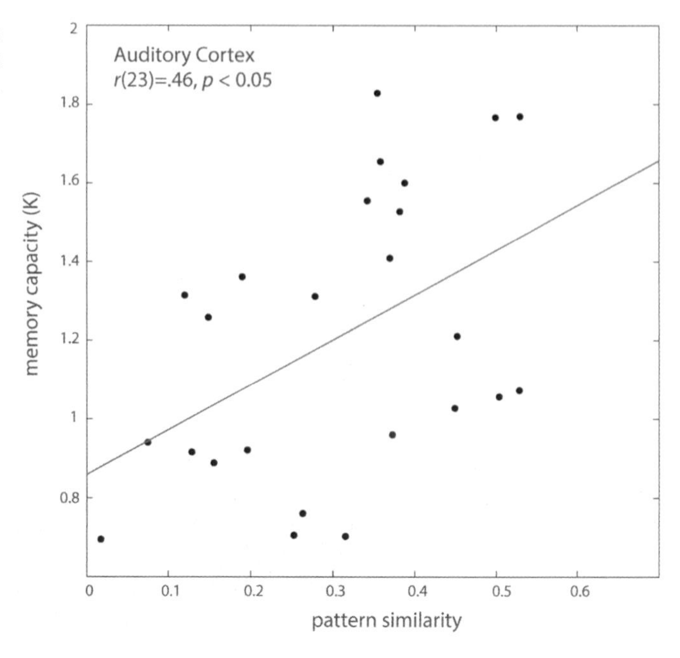

FIGURE 10 Individual differences in memory capacity for tones were predicted by the specificity of the neural code for frequency during the encoding phase of the task (Linke, Vicente-Grabovetsky, & Cusack, 2011).

There is evidence that remapping may also have important influence on audition. In the tone memory task described in the first section, we used RSA to quantify the strength of neural coding for tone frequency during encoding and maintenance strength (Linke, Vicente-Grabovetsky, & Cusack, 2011). We found that individual differences in the strength of coding of tone frequency during encoding, but not maintenance, predicted memory performance (Figure 10). This evidence is far weaker than for vision, where behavioral manipulations, univariate fMRI, and RSA have all converged, but it is nevertheless consistent with an important role for remapping processes in audition.

SUMMARY

To understand the mind's ear, we need to understand its neural and mental representations. The method of RSA has great potential as a way to connect neural codes to the features that they represent: in other words, to connect the mind and brain. Representational similarity analysis has shown that holding simple or complex sounds in memory evokes stimulus-specific suppression in AC. It has also shown that we need to be cautious before assuming that each brain region can only hold one type of code, because even in sensory cortex, tasks that differ subtly can evoke different neural codes. Finally, RSA has provided initial evidence that individual differences in auditory memory may, like vision, be strongly influenced by remapping processes during encoding.

References

Arnott, S. R., Grady, C. L., Hevenor, S. J., Graham, S., & Alain, C. (2005). The functional organization of auditory working memory as revealed by fMRI. *Journal of Cognitive Neuroscience*, *17*(5), 819–831. http://dx.doi.org/10.1162/0898929053747612.

Atkinson, R., & Shiffrin, R. (1968). Human memory: a proposed system and its control processes. In *The psychology of learning and* …Retrieved from http://books.google.ca/books?hl=en&lr=&id=SVxyXuG73wwC&oi=fnd&pg=PA89&dq=atkinson+shiffrin&ots=CxN3uuNV5y&sig=4m3NCOcgaSrfHYbCnp_Dzb1JC94.

Baddeley, A., & Hitch, G. (1975). Working memory. In *The psychology of learning and* …Retrieved from http://books.google.ca/books?hl=en&lr=&id=o5LScJ9ecGUC&oi=fnd&pg=PA47&dq=baddeley+hitch&ots=8ybE4R5kX1&sig=uVsMotL8iJtN0P4sk_k8CmudZMw.

Buchsbaum, B. R., Lemire-Rodger, S., Fang, C., & Abdi, H. (2012). The neural basis of vivid memory is patterned on perception. *Journal of Cognitive Neuroscience*, *24*(9), 1867–1883. http://dx.doi.org/10.1162/jocn_a_00253.

Bundesen, C. (1990). A theory of visual attention. *Psychological Review*. Retrieved from http://psycnet.apa.org/journals/rev/97/4/523/.

Cowan, N., Elliott, E. M., Scott Saults, J., Morey, C. C., Mattox, S., Hismjatullina, A., et al. (2005). On the capacity of attention: its estimation and its role in working memory and cognitive aptitudes. *Cognitive Psychology*, *51*(1), 42–100.

Cusack, R., Daley, M., Herzmann, C., Linke, A. C., Peelle, J. E., Wild, C. J., et al. (2013). No sound consensus. In *Human brain mapping hackathon*. Retrieved from http://ohbm-seattle.github.io/.

Cusack, R., Lehmann, M., Veldsman, M., & Mitchell, D. J. (2009). Encoding strategy and not visual working memory capacity correlates with intelligence. *Psychonomic Bulletin Review*, *16*(4), 641–647. http://dx.doi.org/10.3758/PBR.16.4.641.

Cusack, R., Veldsman, M., Naci, L., Mitchell, D. J., & Linke, A. C. (November 2011). Seeing different objects in different ways: measuring ventral visual tuning to sensory and semantic features with dynamically adaptive imaging. *Human Brain Mapping, 397*, 387–397. http://dx.doi.org/10.1002/hbm.21219.

Davis, M., & Johnsrude, I. (2003). Hierarchical processing in spoken language comprehension. *The Journal of Neuroscience*. Retrieved from http://www.jneurosci.org/content/23/8/3423.short.

Demany, L., Semal, C., Cazalets, J.-R., & Pressnitzer, D. (2010). Fundamental differences in change detection between vision and audition. *Experimental Brain Research, 203*(2), 261–270. Retrieved from http://www.ncbi.nlm.nih.gov/pubmed/20369233.

Duncan, J. (2001). An adaptive coding model of neural function in prefrontal cortex. *Nature Reviews Neuroscience*. Retrieved from http://www.nature.com/nrn/journal/v2/n11/abs/nrn1101-820a.html.

Freedman, D., Riesenhuber, M., Poggio, T., & Miller, E. (2001). Categorical representation of visual stimuli in the primate prefrontal cortex. *Science*. Retrieved from http://www.sciencemag.org/content/291/5502/312.short.

Fuster, J. M. (1973). Unit activity in prefrontal cortex during delayed-response performance: neuronal correlates of transient memory. *Journal of Neurophysiology, 36*(1), 61–78. Retrieved from http://www.ncbi.nlm.nih.gov/pubmed/4196203.

Harrison, S. A., & Tong, F. (2009). Decoding reveals the contents of visual working memory in early visual areas. *Annales de la Societe Royale des Sciences Medicales et Naturelles de Bruxelles, 458*(7238), 632–635. http://dx.doi.org/10.1038/nature07832.

Konstantinou, N., Bahrami, B., Rees, G., & Lavie, N. (2012). Visual short-term memory load reduces retinotopic cortex response to contrast. *Journal of Cognitive Neuroscience, 24*(11), 2199–2210. http://dx.doi.org/10.1162/jocn_a_00279.

Kosslyn, S. M. (2003). Understanding the mind's eye... and nose. *Nature Neuroscience, 6*(11), 1124–1125. http://dx.doi.org/10.1038/nn1103-1124.

Kravitz, D. J., Kriegeskorte, N., & Baker, C. I. (2010). High-level visual object representations are constrained by position. *Cerebral Cortex, 20*(12), 2916–2925. http://dx.doi.org/10.1093/cercor/bhq042.

Kriegeskorte, N., Mur, M., Ruff, D., & Kiani, R. (2008). Matching categorical object representations in inferior temporal cortex of man and monkey. *Neuron, 60*(6), 1126–1141. Retrieved from http://www.sciencedirect.com/science/article/pii/s0896-6273(08)00943-4.

LaRocque, J. J., Lewis-Peacock, J. A., Drysdale, A. T., Oberauer, K., & Postle, B. R. (2013). Decoding attended information in short-term memory: an EEG study. *Journal of Cognitive Neuroscience, 25*(1), 127–142. http://dx.doi.org/10.1162/jocn_a_00305.

Lewis-Peacock, J. A., Drysdale, A. T., Oberauer, K., & Postle, B. R. (2012). Neural evidence for a distinction between short-term memory and the focus of attention. *Journal of Cognitive Neuroscience, 24*(1), 61–79. http://dx.doi.org/10.1162/jocn_a_00140.

Linke, A. C., & Cusack, R. (2015). Flexible information coding in human auditory cortex during perception, imagery and STM of complex sounds. *Journal of Cognitive Neuroscience*. http://dx.doi.org/10.1162/jocn_a_00780.

Linke, A. C., Vicente-Grabovetsky, A., & Cusack, R. (2011). Stimulus-specific suppression preserves information in auditory short-term memory. *Proceedings of the National Academy of Sciences of the United States of America, 108*(31), 12961–12966. http://dx.doi.org/10.1073/pnas.1102118108.

Linke, A. C., Vicente-Grabovetsky, A., Mitchell, D. J., & Cusack, R. (2011). Encoding strategy accounts for individual differences in change detection measures of VSTM. *Neuropsychologia, 49*(6), 1476–1486. Retrieved from http://www.ncbi.nlm.nih.gov/pubmed/21130789.

Otazu, G. H., Tai, L.-H., Yang, Y., & Zador, A. M. (2009). Engaging in an auditory task suppresses responses in auditory cortex. *Nature Neuroscience, 12*(5), 646–654. http://dx.doi.org/10.1038/nn.2306.

Peelle, J. E., Eason, R. J., Schmitter, S., Schwarzbauer, C., & Davis, M. H. (2010). Evaluating an acoustically quiet EPI sequence for use in fMRI studies of speech and auditory processing. *NeuroImage, 52*(4), 1410–1419. http://dx.doi.org/10.1016/j.neuroimage.2010.05.015.

Riggall, A. C., & Postle, B. R. (2012). The relationship between working memory storage and elevated activity as measured with functional magnetic resonance imaging. *The Journal of Neuroscience: The Official Journal of the Society for Neuroscience, 32*(38), 12990–12998. http://dx.doi.org/10.1523/JNEUROSCI.1892-12.2012.

Sawaki, R., & Luck, S. J. (2011). Active suppression of distractors that match the contents of visual working memory. *Visual Cognition, 19*(7), 956–972. http://dx.doi.org/10.1080/13506285.2011.603709.

Schmitter, S., Diesch, E., Amann, M., Kroll, A., Moayer, M., & Schad, L. R. (2008). Silent echo-planar imaging for auditory FMRI. *Magma (New York, N.Y.), 21*(5), 317–325. http://dx.doi.org/10.1007/s10334-008-0132-4.

Shamma, S. A. (2013). Inhibition mediates top-down control of sensory processing. *Neuron, 80*(4), 838–840. http://dx.doi.org/10.1016/j.neuron.2013.11.007.

Sligte, I., Scholte, H., & Lamme, V. (2008). Are there multiple visual short-term memory stores? *PLOS One*. Retrieved from http://dx.plos.org/10.1371/journal.pone.0001699.

Sperling, G. (1960). The information available in brief visual presentations. *Psychological Monographs: General and Applied*. Retrieved from http://psycnet.apa.org/journals/mon/74/11/1/.

Sylvester, C. M., Jack, A. I., Corbetta, M., & Shulman, G. L. (2008). Anticipatory suppression of nonattended locations in visual cortex marks target location and predicts perception. *The Journal of Neuroscience: The Official Journal of the Society for Neuroscience, 28*(26), 6549–6556. http://dx.doi.org/10.1523/JNEUROSCI.0275-08.2008.

Vicente-Grabovetsky, A., Carlin, J. D., & Cusack, R. (2014). Strength of retinotopic representation of visual memories is modulated by strategy. *Cerebral Cortex (New York, N.Y.: 1991), 24*(2), 281–292. http://dx.doi.org/10.1093/cercor/bhs313.

Xue, G., Dong, Q., Chen, C., Lu, Z., Mumford, J. A., & Poldrack, R. A. (2010). Greater neural pattern similarity across repetitions is associated with better memory. *Science, 330*(6000), 97–101. http://dx.doi.org/10.1126/science.1193125.

Zanto, T. P., & Gazzaley, A. (2009). Neural suppression of irrelevant information underlies optimal working memory performance. *The Journal of Neuroscience: The Official Journal of the Society for Neuroscience, 29*(10), 3059–3066. http://dx.doi.org/10.1523/JNEUROSCI.4621-08.2009.

Zvyagintsev, M., Clemens, B., Chechko, N., Mathiak, K. A., Sack, A. T., & Mathiak, K. (2013). Brain networks underlying mental imagery of auditory and visual information. *The European Journal of Neuroscience, 37*(9), 1421–1434. http://dx.doi.org/10.1111/ejn.12140.

C H A P T E R

18

Remembering Touch: Using Interference Tasks to Study Tactile and Haptic Memory

Rebecca Lawson[1], *Alexandra M. Fernandes*[2,3], *Pedro B. Albuquerque*[3], *Simon Lacey*[4]

[1]Department of Experimental Psychology, University of Liverpool, UK; [2]Institute for Energy Technology, Halden, Norway; [3]School of Psychology, University of Minho, Braga, Portugal; [4]Department of Neurology, Emory University, Atlanta, GA, USA

INVESTIGATING OUR SENSE OF TOUCH

We are rarely aware of using our memory for touch and yet research suggests that we are efficient—perhaps surprisingly so—at storing and subsequently accessing memories of how something felt to us. In some cases we may do this explicitly, for example, when deciding whether a melon is ripe or a cake is cooked, but in many more cases we are probably not aware that we are reactivating memories of things that we have touched. Relatively little research has been conducted on memory for touch compared with visual and auditory memory. The proportion of papers on memory for each modality has been stable, with around 70% of search results for visual plus memory, 20% for auditory plus memory, 5% for olfactory plus memory, and less than 5% in total for tactile, haptic, and gustatory plus memory.[1] Gallace and Spence (2009) provided a review of the literature on haptic and tactile memory. More recently, Wang, Bodner, and Zhou (2013; see also Wang et al., 2012) reviewed the neural basis of tactile working memory. Here, we focus on the empirical findings from previous interference studies investigating short-term and working memory for touch and also on evaluating this methodology. We then present two studies investigating interference effects on haptic memory for three-dimensional (3D) objects. In the literature there is some inconsistency in the use of the terms *tactile* and *haptic* perception. In the current chapter, *tactile* refers to passive touch (such as an experimenter pressing a shape against the palm of a participant's hand) whereas *haptics* refers to active touch, in which participants move their body to feel a stimulus (for example, by grasping a mug).

COMPARING MEMORY FOR VISION AND FOR TOUCH

As a likely consequence of the overwhelming concentration on visual and auditory memory, research on tactile and haptic memory has largely focused on comparing performance across modalities. There are difficulties in making direct comparisons across the modalities but it is clear that our sense of touch can efficiently perceive a number of important physical features, such as shape, size, and texture, which can also be extracted by vision. It is therefore possible to devise studies (e.g., Lawson, 2009) for which at least the physical stimuli presented and the perceptual information available from them are better matched than is possible for any other pair of modalities. However, even if this is done, the nature of exploration and information extraction remains fundamentally different across vision and touch. Specifically, vision can detect information at a distance and across a wide area in parallel, whereas touch

[1] Searches were conducted for both the title only and for the abstract using the PsycINFO database and using the University of Liverpool's Web of Science database for the twentieth century and its database for 2000 to 2013.

Mechanisms of Sensory Working Memory
http://dx.doi.org/10.1016/B978-0-12-811042-3.00018-9

is typically restricted to the exploration of near space and to the sequential accumulation of information. Researchers have sometimes tried to reduce these differences, for example, by slowing down the extraction of visual information by forcing participants to view a scene using a small, movable aperture, to try to equate the load on working memory across vision and touch (Craddock, Martinovic, & Lawson, 2011; Loomis, Klatzky, & Lederman, 1991; Martinovic, Lawson, & Craddock, 2012; Monnier & Picard, 2010). However, such measures may artificially handicap processing in one or both senses, reducing the ecological validity of these studies. For example, the only visual stimuli that typically require multiple saccades to be encoded are large-scale scenes.

Differences in working memory load could explain much of the variation in performance, both when comparing across vision and touch and when contrasting different tasks involving touch. As an extreme example, the haptic exploration of two-dimensional (2D) shapes such as raised line drawings is typically extremely slow, effortful, and sequential (Lawson & Bracken, 2011); exploration times exceeding 1 min are commonplace. Building up representations of such stimuli would be expected to involve a substantial load on working memory. Overvliet, Wagemans, and Krampe (2013) found that older adults were worse at recognizing 2D raised line drawings haptically, whereas several studies have indicated that the haptic recognition and matching of 3D objects is similar for older and younger adults (Ballesteros & Reales, 2004; Norman et al., 2006, 2011). According to the image mediation model, sequentially acquired 2D haptic inputs must be translated into visual images that can then be identified using the normal visual processing system (Klatzky & Lederman, 1987; Lederman & Klatzky, 1990; Lederman, Klatzky, Chataway, & Summers, 1990). Haptic recognition of 2D shapes is therefore likely to put a much greater demand on working memory than when recognition is easier, for example for everyday 3D objects, where processing may be mediated directly by the haptic system.

As mentioned, it is possible to try to reduce the differences between vision and touch with respect to the spatiotemporal accumulation of information. For example, visual performance can be matched to be similarly slow and inaccurate as touch. Martinovic et al. (2012) attempted to determine whether haptic object identification was relatively slow and inaccurate compared with vision simply because haptics acquires information serially and more slowly than vision. This difference could in turn result in a greater working memory load for haptics than for vision. The alternative hypothesis is that there are more fundamental processing differences between haptics and vision. To address this issue, we slowed visual processing using a restricted viewing technique in which people moved one finger around on a touch screen to control an aperture that enabled them to see only a small area of the object at a time. We determined the aperture size required to achieve similar recognition performance across vision and touch (Craddock et al., 2011). We then conducted an electroencephalographic experiment to compare the time course of visual and haptic object recognition (Martinovic et al., 2012). Participants had to discriminate familiar (nameable) from unfamiliar (unnamable) objects either visually via aperture viewing or haptically. Mean response times were around 5 s for familiar objects and 9 s for unfamiliar objects in both modalities. We analyzed the evoked and total fronto-central theta-band (5–7 Hz; a marker of working memory) and the occipital upper alpha-band (10–12 Hz; a marker of perceptual processing) locked to stimulus onset. Long-latency modulations of both theta-band and alpha-band activities differed between familiar and unfamiliar objects in haptics. Decreases in total upper alpha-band activity for haptic identification of familiar relative to unfamiliar objects suggested that multisensory extrastriate areas were involved in object processing (see also Deshpande, Hu, Lacey, Stilla, & Sathian, 2010; Lacey, Flueckiger, Stilla, Lava, & Sathian, 2010 for complementary results from functional magnetic resonance imaging studies). In contrast, theta-band activity showed a general increase over time for the slowed-down visual recognition task. Our results suggested that visual and haptic object recognition share some common representations, but, importantly, there are fundamental differences between the senses that are not simply a result of differences in their usual speed of information extraction.

Notwithstanding the technical difficulties in comparing vision and touch, several studies have found efficient cross-modal recognition across these two modalities (Bushnell & Baxt, 1999; Easton, Srinivas, & Greene, 1997; Lacey & Campbell, 2006; Lawson, 2009; Newell, Ernst, Tjan, & Bülthoff, 2001; Norman, Norman, Clayton, Lianekhammy, & Zielke, 2004; Ballesteros & Reales, 1999). Furthermore, this can be done with novel, unfamiliar objects that are difficult to distinguish verbally (see Lawson, 2009), and so cross-modal performance does not have to rely on a verbal strategy involving naming the objects. Cross-modal recognition is likely achieved by integrating visual and haptic representations into a single, multisensory representation accessible to both vision and touch (Lacey, Pappas, Kreps, Lee, & Sathian, 2009).

In a further step, this evidence showing excellent cross-modal visuo-haptic processing, together with the similarity in the processing goals of vision and touch (Lawson, 2009), the importance of spatial encoding in vision and touch (Millar, 1999), and the general superiority of visual processing, has led some to assume that inputs from touch must be recoded into visual representations to be interpreted and stored (Klatzky & Lederman, 1987). However, there is now clear evidence against strong versions of such image mediation accounts (Lederman et al., 1990). For example, studies have shown that congenitally blind individuals can still recognize objects efficiently by touch despite having no visual processes available (e.g., Pietrini et al., 2004). Furthermore a number of studies have concluded that congenitally blind people as well as blindfolded sighted people can use mental imagery and working memory to perform haptic tasks requiring spatial transformation of the input (e.g., Aleman, van Lee, Mantione, Verkoijen, & de Haan, 2001; Marmor & Zaback, 1976). Nevertheless, there is good evidence that haptic shape perception is often mediated by visual imagery, as reviewed by Lacey and Sathian (2013, 2014). In a series of articles, Sathian and colleagues (Deshpande et al., 2010; Lacey et al., 2010, 2014) showed that haptic perception of unfamiliar shape and visual-spatial imagery share a common network whereas haptic perception of familiar shape and visual-object imagery share a different common network (see Kozhevnikov & Blazhenkova, 2013; for a discussion of visual-object and visual-spatial imagery differences).

MEMORY FOR TOUCH

The visual and auditory memory literature has traditionally divided memory into separate component systems such as sensory memory and short-term, long-term, and working memory (Cowan, 1995, 2001, 2008; Linden, 2007). Sensory memory refers to the registers for veridical, modality-specific representations that provide the first memories of inputs to each of our senses. These representations have limited capacity and degrade rapidly over a few seconds (Sinclair & Burton, 1996). They cannot be maintained through rehearsal. Any information to be retained must be transferred from sensory memory to short-term memory. This is another limited capacity store that can retain information for at least around 30 s without rehearsal and for longer with rehearsal (Hill & Bliss, 1968; Sinclair & Burton, 1996). Working memory refers to one component of the short-term memory system that can maintain, rehearse, and actively manipulate sensory information (Baddeley, 2000, 2007; Baddeley & Hitch, 1974). Sensory memory, short-term memory, and working memory are all limited-capacity, temporary stores. Any information that is not transferred from them into long-term memory will soon be lost. Long-term memory, in contrast, has an extremely large capacity and allows information to be stored for many years. We will use these broad distinctions to provide a convenient framework for organizing our review of research on haptic and tactile memory. However, there is no consensus as to whether these distinctions reflect meaningful theoretical divisions (Cowan, 2008; Engle, Tuholski, Laughlin, & Conway, 1999).

In the following sections we will start by briefly reviewing evidence about the most transient and then about the most durable memories for touch (sensory memory, then long-term memory). We will then focus in more detail on the use of interference studies to investigate short-term memory and working memory for touch. Finally, we will describe the results of two new studies investigating the effects of different interference tasks on haptic object recognition.

SENSORY MEMORY IN TOUCH

A number of studies have investigated the capacity and duration of our initial memories for inputs from touch, inspired by Sperling's (1960) studies of visual sensory memory. Bliss, Crane, Mansfield, and Townsend (1966; Hill & Bliss, 1968) investigated sensory memory in touch using a partial report paradigm with air-jet stimulators to three locations of each finger on both hands. Participants could recall around 3.5 positions of the air-jets for full report and around one extra position for partial report. This advantage for partial over full report indicated that there was sensory memory for tactile inputs and that this memory degraded after about 1 s. The benefit for tactile partial report seemed to be much smaller than that for visual partial report. However, participants had to learn to map each of 24 locations to a different letter, so deciding on a response was demanding. This task may thus have measured more complex processing than just immediate tactile memory.

More recently, and using a simpler task, Gallace, Tan, Haggard, and Spence (2008) assessed memory for the location of vibrations over the body using whole report (numerosity judgments) and partial report (spatially cued). Participants recalled up to three stimuli in the whole report condition and up to five using partial report. There was also a trade-off between the number of stimuli presented and the duration of the representation, with faster decay as the number of stimuli increased. Similarly, Auvray, Gallace, and Spence (2011) found that only three item locations could be reported in a full report condition of a tactile short-term memory task whereas up to six items could be recalled using partial report. They also found an overall cost to performance of concurrent articulatory suppression but no interaction with stimulus onset asynchrony, which suggests that verbal mediation did not have an important role in this task.

The whole report capacity of tactile memory for around three spatial locations and its greater capacity for partial report (Auvray et al., 2011; Bliss et al., 1966; Gallace, Tan, & Spence, 2006; Gallace et al., 2008) are similar to the limits of accurate judgments of numerosity for vision (i.e., the limits of subitization) (Ester, Drew, Klee, Vogel, & Awh, 2012; Piazza, Fumarola, Chinello, & Melcher, 2011) which may in turn reflect the capacity of visual working memory, which is also around three (e.g., Anderson, Vogel, & Awh, 2011). However, it remains possible that performance on these spatial location tasks relies on verbal recoding and as such, their results may not reflect people's ability to store purely tactile information (Bancroft, Hockley, & Servos, 2012).

Other approaches have been used to study the capacity and durability of memory for touch. Heller (1987) traced numbers onto people's hands and estimated digit span as up to seven at a slow presentation rate. A number of studies have exploited the suffix effect to compare sensory memory across different modalities. The suffix effect involves presenting an extra item immediately after a to-be-remembered list of items. In a typical, no-suffix control condition for auditory presentation there is a strong recency effect (i.e., superior recall for the final items relative to items in the middle of the list) whereas the recency effect only occurs for the last item for visual presentation. In suffix conditions, however, auditory responses are like visual responses, with no recency or recency only for the last item (e.g., Conrad & Hull, 1968). Watkins and Watkins (1974) touched participants' fingers with a pen and asked them to recall the order of presentation. Recall in control conditions showed a recency effect that disappeared after a tactile suffix, similar to the auditory suffix effect (see also Mahrer & Miles, 1999, 2002). This tactile suffix effect was confirmed by Manning (1980) using letters and nonverbal stimuli. These results suggest that a short-term, tactile sensory memory is available after a stimulus has been presented but that it is eliminated by the presentation of a subsequent tactile stimulus. This conclusion is consistent with results of Nairne and McNabb (1985), who found greater recency with recall of tactile compared with visual sequences.

Together, these studies confirm the existence of a limited-capacity, rapidly decaying, and easily overwritten sensory store for touch similar to the iconic and echoic memory stores proposed for visual and auditory inputs, respectively. The advantage of partial report over whole report appears to be smaller for tactile than for visual presentation and more similar to that for auditory presentation. In addition, the recency effect may be greater for tactile and auditory presentations than for visual sequences. These consistent differences across the modalities may reflect their relative reliance on serial versus parallel processing (e.g., Loomis et al., 1991). However, no firm conclusions can be made without conducting more direct comparisons across the modalities while controlling for variables such as stimulus familiarity and complexity. In addition, the results need to be interpreted with caution because performance in some of these tasks may rely on stored verbal or spatial information instead of, or as well as, exclusively tactile or haptic information (see Gallace & Spence, 2008, 2009; Mahrer & Miles, 2002).

LONG-TERM MEMORY IN TOUCH

One means of assessing long-term memory for touch is to measure our ability to name everyday objects because this requires people to access representations of those objects stored before the start of an experiment. Several studies indicate that we are good at recognizing 3D familiar objects using touch alone (Craddock & Lawson, 2008, 2009a, 2009b; Klatzky, Lederman, & Metzger, 1985; Lawson, 2014). We remain fast and accurate even when non-shape cues such as size, texture, and temperature are removed by presenting plastic scale models of objects and when fine shape discriminations are required: for example, between a typical chair and a bench-chair morph (Lawson, 2009, 2014). Across a range of studies it has been found that haptic object naming is usually possible in 2–6 s with 4–12% error rates. In contrast, response times for naming raised line drawings of familiar objects are often over 60 s and error rates are typically 50–90%, indicating that depth information is crucial for efficient haptic recognition (Lawson & Bracken, 2011). Overall these results indicate that we can readily encode haptic inputs from 3D objects and match these to stored object representations (although we may struggle to encode and interpret 2D stimuli). Furthermore, our sensitivity to perceptual manipulations such as rotations in depth and size changes (Craddock & Lawson, 2008, 2009b; Ernst, Lange, & Newell, 2007; Lawson, 2009, 2011; Newell et al., 2001) and differences between visual versus haptic presentation (Lawson, 2009) argue against haptic object recognition simply being mediated by stored visual or verbal rather than haptic representations.

Few studies have directly investigated the longer-term memory representations used in haptic perception. In one study Nabeta and Kusumi (2008) compared memory across touch and vision. They presented either 100 or 500 everyday objects at study. A subset of these old items together with new objects with different names was then presented in an old/new recognition test. Performance in both modalities was excellent and suggested that memory capacity easily exceeds 500 objects in both haptics and vision. Surprisingly, haptic memory was superior to visual memory, which provides further evidence that long-term haptic memory is not mediated solely by visual or verbal representations.

Pensky, Johnson, Haag, and Homa (2008) provided the strongest evidence about long-term haptic memory. They addressed the issue of verbal recoding directly and tested retention across intervals of up to 1 week. Their results showed that people could encode stable representations of both visually and haptically presented objects. In the study, participants manipulated a set of everyday objects. Then, after 1 h or 1 week, participants performed an old/new recognition test with the study objects and distracters from the same basic-level category (e.g., two different shaped and labeled food cans). Both haptic and cross-modal recognition was well above chance even after 1 week although unimodal haptic recognition decayed faster than unimodal visual recognition. Because participants could not rely on remembering the basic-level names of the study items to succeed at the task, this finding of above-chance recognition after a week shows that long-term haptic representations of objects can be acquired. Haptic recognition was worse than visual recognition, but this may have been because the availability of features of objects was not matched across the modalities. For example, differences between the food can labels could only be perceived visually, not haptically. Thus, overall performance levels across the modalities may simply have reflected differences in the information available to vision versus to touch.

SHORT-TERM MEMORY AND WORKING MEMORY IN TOUCH

A distinction is often drawn between short-term and working memory. However, performance on tasks that supposedly reflect one or the other type of processing typically show high intercorrelations, which suggests that to a large extent, they tap a common construct (Engle et al., 1999; Swanson & Luxenberg, 2009; Withagen, Kappers, Vervloed, Knoors, & Verhoeven, 2013). In the current chapter it is not necessary to distinguish between more passive short-term memory and more active working memory. Because little work has attempted to develop distinct short-term and working memory tasks for touch, we will not separate our review of research on temporary memory for touch based on this distinction.

Temporary memory for touch over delays of several seconds is assumed to require more stable representations than those available using the sensory store. However, tactile working memory is limited in capacity, shows a fair degree of interindividual variability (Bliss & Hämäläinen, 2005), and decays between around 15 and 30 s post-stimulus (Kiphart, Hughes, Simmons, & Cross, 1992). There is also evidence that different modality-specific areas are involved in the short-term maintenance as well as the acquisition of information through vision and touch (Gallace & Spence, 2009; Seemüller, Müller, & Rösler, 2012; Woods, O'Modhrain, & Newell, 2004). It is also important to acknowledge that much research on working memory for touch used stimuli that could be visually or verbally encoded and rehearsed. This can make it difficult to determine whether performance on a specific task is underpinned by modality-specific, tactile, or haptic representations rather than visual or verbal representations. Inputs from touch may be recoded into visual or verbal representations, and indeed, they may also be encoding using motor, kinesthetic, or spatial information that is acquired when touch is used to actively explore stimuli (e.g., see Loomis, Klatzky, & Giudice, 2013; for discussion of the use of amodal spatial images in working memory). The degree to which people rely on alternative representations will likely depend on the individual, the task, and the stimuli.

Several studies have provided evidence that working memory processes occur for inputs from touch (Cornoldi & Vecchi, 2003; Mahrer & Miles, 2002; Miles & Borthwick, 1996) and that modality-specific brain areas support the retention of sensory information from touch (Harris, Miniussi, Harris, & Diamond, 2002; Ricciardi et al., 2006; but see Bancroft, Hogeveen, Hockley, & Servos, 2014). In addition, studies of congenitally blind people have shown that visual experience is not essential to support effective short-term and working memory processes such as forward and backward span tasks, the generation and transformation of spatial images, and object recognition; blind people show superior memory compared with sighted individuals in many cases (Hull & Mason, 1995; Pring, 2008; Rokem & Ahissar, 2009; Swanson & Luxenberg, 2009; Pietrini et al., 2004; Vecchi, 1998; Vecchi, Tinti, & Cornoldi, 2004; Withagen et al., 2013).

In the next sections we will focus on research on temporary memory for touch that has used an interference task methodology to determine the nature of the representations underlying performance in haptic memory tasks. We will first consider general issues arising from the use of this method before reviewing the results of research on memory for touch using interference tasks. Finally, we present two new studies that tested a range of interference tasks to investigate the type of representations used in haptic object recognition.

ISSUES ARISING FROM THE USE OF INTERFERENCE TECHNIQUES

General Principles

Interference (or dual-task) studies require participants to perform a primary task that is of interest and a concurrent secondary task that may rely on the same cognitive processes that are hypothesized to support the primary task. The extent to which performance on the primary task is impaired by the concurrent secondary task is assumed to reflect the extent to which both tasks share common processes. Interference tasks have long been used to investigate structural aspects of working memory: for example, to probe the existence and limits of its visual, spatial, and verbal subcomponents (e.g., Baddeley, Grant, Wight, & Thomson, 1975). Interference techniques have an effect by inducing competition for cognitive resources and not by reducing perceptual sensitivity. Requiring participants to monitor speech while wearing earplugs reduces performance but only by reducing perceptual sensitivity; presenting the speech to be monitored to one ear and different distracting speech to the other ear induces competition and constitutes interference.

Competition for cognitive resources between the primary and secondary tasks results in poorer performance on the primary task compared with a control, single-task condition in which there is no secondary task and also with alternative concurrent secondary tasks that do not compete for the same resources. Using multiple secondary tasks has the benefit of helping to control for the deleterious effects of divided attention and competition for general processing resources in dual-task conditions. A further advantage is that performance on each secondary task alone, without the primary task, can be measured. This allows the secondary tasks to be adjusted to try to equate their difficulty, and then this single-task performance can be compared with dual-task performance to check whether secondary task performance is disrupted in the dual-task condition.

Experimental Design in Haptic Interference Tasks

Following on from the general principles described above, the design of interference experiments needs to take into consideration a number of factors to avoid some obvious pitfalls (Cowan, 1995; Green & Vaid, 1983; Oberauer & Göthe, 2006; Pashler, 1994). Several studies of memory for touch have compared just one dual-task condition (an interference task performed simultaneously with the primary task) to a control condition (the primary task performed alone with no interference) (e.g., Gilson & Baddeley, 1969; Sinclair & Burton, 1996; Sullivan & Turvey, 1972). Here, any interference may not reflect shared task-specific processes but instead may simply reflect domain-general competition for processing capacity such as a common attentional resource. This problem can be ameliorated by using multiple different interference tasks of similar general difficulty and comparing performance across them (e.g., Lacey & Campbell, 2006). A more complex alternative, but one that avoids the problem of variation in the overall difficulty of the interference tasks, is to use multiple primary tasks (e.g., Chan & Newell, 2008); the effect of the same secondary interference tasks can then be compared across the different primary tasks.

An interesting recent approach that sidesteps these issues used transcranial magnetic stimulation (TMS) to improve performance by reducing interference in a working memory task requiring delayed tactile temporal discrimination. Hannula and colleagues demonstrated that using TMS early during the retention period to suppress irrelevant sensory processing of a distracting tactile stimulus (an electrical pulse identical to the experimental stimuli) actually improved response times in the primary working memory task (Hannula et al., 2010). This was probably because TMS reduced activity in the primary somatosensory cortex responsible for representing the distracting tactile stimuli during maintenance of the tactile temporal information for the primary task. This improvement was likely modality-specific in that TMS improved performance during tactile, but not visual, interference (Savolainen et al., 2011). However, one caveat of this latter finding is that tactile interference was greater than visual interference, so it is possible that the difference in the observed TMS effects reflected the amount of interference caused by the distracter rather than its modality.

A second factor to consider with interference studies is that the secondary task should disrupt the process assumed to be engaged by the primary task. For example, articulatory suppression (repeating the same word aloud) is assumed to prevent verbal rehearsal (Baddeley, 2000; Peterson & Peterson, 1959). However, relatively few studies have tried to assess the effectiveness of the secondary interference tasks in achieving this aim, and it is easy to employ secondary tasks uncritically. In a study of visuo-haptic cross-modal scene recognition, Newell, Woods, Mernagh, and Bülthoff (2005) concluded that the spatial relationships between the objects in the scene were not verbally recoded into a description of the scene because there was no effect of verbal interference during encoding. However, in their task, producing a scene description was a semantically and syntactically meaningful process whereas the verbal interference task merely required participants to repeat the word *the*. This would disrupt subvocal rehearsal but it would not be expected to engage semantic or syntactic processes. Thus, the interference task did not tap all of the verbal processes assumed to underlie the primary task. By contrast, Lacey and Campbell (2006) required participants to listen to task-irrelevant but semantically meaningful speech as a secondary task during encoding. The use of verbal labels for stimuli encoded by touch need not be restricted to the names of familiar items. Instead, perceptual qualities (*cold* or *bumpy*) or names reflecting similarities to known objects may support performance. Lacey and Campbell's proposal that subvocal description aided the encoding of unfamiliar objects was supported by their finding that listening to irrelevant speech selectively interfered with the haptic encoding of unfamiliar but not familiar objects. Similarly, care needs to be taken in the use of visual interference tasks. Dynamic visual noise (DVN) is often used in studies of working memory (see Quinn & McConnell, 1996, 1999, 2006; McConnell & Quinn, 2000; Andrade, Kemps, Werniers, May, & Szmalec, 2002). Such studies typically deal with the maintenance of representations that are retrieved from long-term memory and change rates in the DVN of 5–10% are enough to disrupt these stored representations. However, for perceptually derived representations, a change rate of 50% is required (Dean, Dewhurst, Morris, & Whittaker, 2005; Dean, Dewhurst, & Whittaker, 2008) and this DVN rate has been shown to disrupt both visual and haptic encoding of unfamiliar objects (Lacey & Campbell, 2006).

An important and related factor in matching the secondary interference task to the putative processes underlying the primary task is that many, if not most, interference tasks appear to tap multiple cognitive processes. This raises challenges in interpreting the cause of any interference effects found. For example, Millar (1974) required participants to trace standard shapes on the floor in one interference task and to stack 3D objects inside each other in another, whereas Ittyerah and Marks (2007) asked people to place a series of paper clips at regular intervals. All three of these secondary interference tasks included motor, haptic, and spatial components. It may be especially difficult to distinguish haptic interference tasks from motor interference tasks. One approach to resolving this issue was taken in Experiment 1 here, in which the same action (stroking strips of paper) was used in two secondary interference tasks. When no response was required, the task was labeled as motor interference, but when the paper texture had to be discriminated and responded to, it was labeled as haptic interference.

Having selected appropriate interference tasks in accordance with these points, it is important to ensure that the tasks chosen cover all of the competing hypotheses. This has not always been done. For example, Millar (1972) found no significant effect on children's visual, haptic, or cross-modal shape recognition owing to visual interference (viewing cartoon figures) or to verbal interference (digit repetition). However, haptic, motor, and spatial interference conditions were not tested and no positive evidence was revealed about the nature of the representations involved in this task. Similarly, Holtby and D'Angiulli (2012) tested verbal and visual but not haptic interference, whereas Garvill and Molander (1977) used visual and haptic matching tasks as interference but did not include verbal or spatial interference tasks. Using visual, verbal, and haptic interference tasks, Lacey and Campbell (2006) were able to distinguish, at least in part, among visual imagery, dual-code (Paivio, 1986, 2007), and amodal accounts of cross-modal recognition; however, they did not include a spatial interference condition, so this possibility was left open.

Finally, there is the question of when in time to place the interference task relative to the encoding, maintenance, and retrieval stages of the primary haptic or tactile memory task. Interference has been applied during study (e.g., Newell et al., 2005; Lacey & Campbell, 2006; Experiment 1 here), at test, when a response is required (e.g., Lacey & Campbell, 2006), or most commonly, during the study–test retention interval (e.g., Bancroft & Servos, 2011; Chan & Newell, 2008; Garvill & Molander, 1977; Mahrer & Miles, 2002; Millar, 1972; Paz, Mayas & Ballesteros, 2007; Experiment 2 here). Differences in the effects of interference at each of these three stages may simply reflect variation in the difficulty of combining two tasks. Employing interference during the retention interval is generally easier because participants are not simultaneously acquiring information or responding in the primary task, and the study may be simpler to set up because both hands are free. In addition, it is easy to vary interval duration during the retention period. However, there are theoretically well-motivated reasons to believe that different processes are influenced by interference at each stage. During encoding (study) and retrieval (test), interference can be used to investigate whether different formats of representations are involved (e.g., Lacey & Campbell, 2006) whereas interference during the retention interval can be used to examine maintenance and rehearsal mechanisms or decay functions. Even finer temporal distinctions may be important: Some studies have shown that interference effects are influenced by whether interference is applied early or late during the retention period (Bancroft & Servos, 2011; Hannula et al., 2010).

COMPARING ACROSS INTERFERENCE STUDIES INVESTIGATING TOUCH

It should already be clear that there is great variety in the secondary interference tasks that researchers have used to test short-term memory for touch. Unfortunately, tasks sometimes appear to have been chosen idiosyncratically or with no clear theoretical basis. Also many (and probably most) tasks are complex and fail to tap a single cognitive process cleanly. This makes it difficult to interpret any disruptive effects of the task and also makes it likely that different secondary tasks used in the study will both target the same process. This diversity and complexity of secondary tasks could explain apparent inconsistencies in the results found across studies that appear to use similar methods, as we discuss below.

For example, some verbal interference tasks have used simple articulatory suppression (e.g., repeating a single syllable) (Mahrer & Miles, 2002; Miles & Borthwick, 1996; Newell et al., 2005), counting aloud (Ittyerah & Marks, 2007), or passively listening to speech (Lacey & Campbell, 2006). However, many have used much harder verbal tasks: counting backward in threes (Miles & Borthwick, 1996; Millar, 1974), reciting the alphabet backward (Gentaz & Hatwell, 1999), repeating digit strings (Millar, 1972), mental arithmetic (Cohen, Voss, Lepore, & Scherzer, 2010), generating words beginning with a specific letter in Experiment 2 here, and doing a difficult non-word discrimination task in Experiment 1 here.

Visual interference tasks are similarly diverse and it is difficult to ensure that they do not also require spatial processing. Some have not required a response (e.g., looking at DVN (Lacey & Campbell, 2006)) but often they do (such as matching the hue or location of color patches (Chan & Newell, 2008)).

Spatial interference tasks have ranged from simple visual dot-tracking (Lacey & Campbell, 2006) to difficult imagery-based tasks (specifying the direction of turns around a capital letter in Experiment 2 here, or a mental rotation matching task as in Experiment 1 here). In addition, spatial tasks can be presented visually (e.g., Lacey & Campbell, 2006) (Experiment 1 here), verbally (Experiment 2 here), or haptically (feeling a matrix of blocks, imagining moving the blocks, then reporting the final block configuration) (Cohen et al., 2010). Few studies have directly compared visuo-spatial with haptic-spatial interference, and even when this has been done the tasks have been poorly matched across modality (e.g., Paz et al., 2007; Sebastián, Mayas, Manso, & Ballesteros, 2008).

Finally, haptic and tactile interference tasks can be passive (such as an experimenter moving a pen against the participant's fingers) (Mahrer & Miles, 2002), active but requiring no response (manipulating an object in the non-dominant hand) (Lacey & Campbell, 2006), although this might better be described as a motor task; or finding and rotating a block (Paz et al., 2007), or active with a response required (matching the orientation or texture of sandpaper (Chan & Newell, 2008), stacking objects inside each other (Millar, 1974), using both hands to move books between two stacks (Ittyerah & Marks, 2007), moving jigsaw pieces between two containers (Experiment 2 here), or discriminating between samples of paper (Experiment 1 here).

This diversity of interference tasks makes it difficult to compare across studies and the occasional lack of a clear theoretical motivation for task selection makes it hard to draw clear conclusions as to why a given pattern of interference effects occurred. Regrettably, Lacey and Campbell (2006) would confess that they were not especially clear about the rationale for choosing a verbal interference task involving continuous prose text over simple word repetition. However, lack of standardization of interference tasks is by no means restricted to dual-task research on memory for touch (e.g., see Al-Yahya et al., 2011) and creating standard tasks would take considerable work. Furthermore, despite these problems with cross-study comparisons, consistencies in results appear, as we will show later, if one contrasts studies on a like-for-like basis and takes into account when interference task occur.

USING INTERFERENCE TASKS TO INVESTIGATE SHORT-TERM AND WORKING MEMORY IN TOUCH

We have discussed a number of studies that used interference (dual) task methodology to investigate short-term and working memory for touch across a range of different primary tasks. We now describe the results of these studies to evaluate what we have learned from them. We begin with a series of studies that measured memory for the location of a tactile stimulus as the primary task.

Gilson and Baddeley (1969) asked participants to point with their finger to the location of a touch to the forearm by a pen. They responded either immediately or after delays of up to a minute. A control group performed the task alone while a second group counted backward in threes during the retention interval to prevent verbal rehearsal. The accuracy of tactile location for this second, articulatory suppression group deteriorated rapidly over time, with large errors made by delays of 45 s. In contrast, accuracy of the first, control group was relatively good up to 15 s retention and then slowly decayed (broadly consistent with Kiphart et al., 1992), and it remained superior to that of the articulatory suppression group. Presumably the dual-task interference group had to rely on transient sensory memory of the tactile stimuli whereas the single-task group benefitted from being able to transfer information from sensory memory to a more durable short-term memory store. These results were interpreted as indicating that rehearsal was an important component of this information transfer. However, the nature of this rehearsal could not be elucidated from this study; importantly, an alternative account of a general cost of divided attention cannot be ruled out because only one interference task was tested.

Miles and Borthwick (1996) failed to replicate the interaction between task and retention interval reported by Gilson and Baddeley (1969). In their first experiment, although their no-suppression control group outperformed their articulatory suppression group, accuracy in reporting the location of a tactile stimulus to the arm decayed smoothly for both groups as the retention interval increased and there was no difference between groups in this decay rate. This result is consistent with that reported by Sullivan and Turvey (1972), who also had participants report the location of a discrete tactile stimulus on the arm in a control, single-task condition or in a dual-task condition (doing simple arithmetic during the retention interval of up to 60 s). Similar to Miles and Borthwick's findings, accuracy decayed over time for both the single and dual tasks, although performance reached asymptote much sooner in Sullivan and Turvey's study, after just 5 s. Miles and Borthwick argued that if accurate recall of tactile locations relied on rehearsing verbal information, an interaction would be predicted between costs resulting from articulatory suppression and those from increasing the retention interval. Recall with articulatory suppression should decay rapidly over time as the tactile sensory trace decayed, whereas recall with no suppression should be relatively impervious to retention interval because location information could be maintained through rehearsal, as reported by Gilson and Baddeley (1969). However, they did not find this interaction; neither did Sullivan and Turvey (1972). In their third experiment, Miles and Borthwick (1996) reported that both verbal (articulatory suppression) interference and tactile interference (the experimenter moving a pen over the eight tactile test locations during the retention interval) impaired tactile memory, but that doing both of these interference tasks together caused no further drop in performance. They suggested that verbal and tactile interference impaired different processes involved in tactile short-term memory, with verbal interference reducing central processing resources and tactile interference disrupting a sensory tactile memory of the stimuli. However, they may not have had sufficient power in their study to detect differences among the three interference conditions, so this conclusion remains preliminary.

One important issue to note about the interference studies conducted by Gilson and Baddeley (1969), Miles and Borthwick (1996), and Sullivan and Turvey (1972) is that participants were asked only to recall where a tactile stimulus occurred rather than to report anything about its identity. These results, which suggest that both verbal and tactile interference disrupts memory for tactile location, therefore may not generalize to a wider range of tasks that require objects to be recognized or matched. Furthermore, there were important and unexplained differences in the findings across these short-term memory interference studies. These could result from relatively subtle methodological variation. Also, visual imagery of the tactile skin locations might have mediated performance in these studies such that, for example, effects of retention interval reflected decay of visual images rather than decay of tactile representations (Gallace & Spence, 2009).

We now move to review studies that used interference tasks to investigate haptic span as the primary task. Paz et al. (2007) contrasted visual and haptic spatial memory in young adults, older adults, and older adults with mild cognitive impairment. Working memory spans for spatial location information were smaller for haptic presentation (three to four items for young adults) than for visual presentation (five to seven items) in all three groups. However, stimulus presentation was not matched across the two modalities. Visual matrices showed all of the to-be-remembered locations simultaneously, whereas one haptic block was presented at a time in different locations and an interference task was done in between presenting each location. Also, as is common in tasks comparing vision with haptics, the haptic stimuli were presented for much longer to try to produce similar levels of performance. Participants had 5 s to encode the entire visual matrix independent of the number of items, whereas they had 5 s to explore each item in the haptic task. Thus, the haptic task produced a much greater load on working memory than the visual task. More interesting was the comparison between the effects of visual versus haptic interference on visual versus haptic span. The retention interval between encoding and retrieval of the spatial information could be filled by an interference task that was either haptic (rotating a block) or visual (deciding whether two arrows pointed to the same place). Performance was worse when the secondary interference task had the same modality as the primary spatial span task (e.g., when haptic interference occurred during the haptic span task), which suggests that distinct systems were responsible for maintaining visual and haptic spatial information.

In a similar spatial memory task, Sebastián et al. (2008) presented visual and haptic 3 × 3 matrices. There were two targets and one distracter in each matrix. Participants had to remember the target positions and ignore the distracter. A 6-s retention interval was filled with one of four interference tasks: haptic spatial (using the nondominant hand to explore an empty matrix), visual spatial (tracking a continuously moving dot on the screen), verbal (articulatory suppression), or a visual static control (looking at a fixation cross). Given the spatial nature of the primary task, it was perhaps unsurprising that only the two spatial tasks impaired performance. More interestingly, the spatial interference was modality specific, replicating the finding that Paz et al. (2007) reported. When the primary task was haptic spatial recall, only the haptic spatial interference task significantly impaired performance. In contrast, when the primary task was visual spatial recall, only the visual spatial interference task significantly impaired performance. Sebastián et al. suggested that their results indicated a modality-specific component to spatial encoding across vision and haptics. However, the trend was for both visual and haptic spatial interference tasks to impair recall on both the visual and the haptic primary matrix tasks, so they did not produce wholly distinct patterns of interference. A further problem in interpreting their results is that the two spatial interference tasks were both different from each other and did not exclusively tap spatial processing. For example, tracking a moving dot and exploring an empty matrix by touch involve different attentional and motor control processes.

Finally, Cohen et al. (2010) tested working memory by measuring people's visual and haptic span for immediate serial recall of letters. They reported that congenitally blind participants' haptic span for Braille letters was as good as sighted participants' visual span for visually presented letters. This was the case both with and without articulatory suppression (repeatedly saying "bla"). Thus, the capacity of blind people's short-term tactile memory was equal to sighted people's visual memory, at least for these stimuli. In a second experiment, participants who were blind or had only residual vision performed a secondary interference task in addition to the primary haptic span task with Braille letters. Interference was either verbal (mental arithmetic) or haptic-spatial (feeling a matrix of blocks and then being instructed to imagine moving the blocks, and then reporting the final block configuration). Performing these tasks concurrently with the primary task reduced span for both interference conditions, relative to a no-interference control condition. Greatest disruption (floor performance) followed spatial interference. However, the spatial interference task was extremely demanding and it drew on a wide range of different processes (imagery, motor, haptic, verbal, and spatial). It is thus not clear what aspect of this task resulted in the poor span performance.

In summary, these studies by Cohen et al. (2010), Paz et al. (2007), and Sebastián et al. (2008) suggest that haptic interference and spatial interference reduce haptic span relative to no interference. In addition, spatial interference may be modality-specific, with haptic span being disrupted more by haptic-spatial interference than by visuospatial interference.

We next consider the effects of interference on haptic perception tasks. Ittyerah and Marks (2007) investigated unimodal and cross-modal haptic (H) and visual (V) discrimination of curvature. They used five different interference tasks during a 30-s retention interval. Visuospatial-motor interference (placing a set of paper clips at regular, experimenter-specified intervals) and motor interference (using both hands to move books between two piles without vision) disrupted HH and HV discrimination relative to both a no-interference control condition and to verbal interference (counting aloud). Surprisingly, continued exploration of the first stimulus during the retention period did not improve performance relative to the no-interference control although it eliminated the need for rehearsal. No significant differences were found among the five interference conditions for VV and VH discrimination, and there was only a marginal effect of interference, with visuospatial-motor and motor interference generally reducing accuracy relative to the no-interference control, consistent with the pattern for the HH and HV tasks. Fewer participants were tested on the VV and VH tasks than the HH and HV tasks and different (though overlapping) sets of curvature discrimination stimuli were used in these tasks. This means that there may simply have been less power to detect effects in the VV and VH tasks, particularly because performance was the highest possible in a number of conditions, which makes cross-condition comparisons difficult.

Gentaz and Hatwell (1999) tested the influence of interference during a 30-s retention interval for an orientation reproduction task with a rod. A no-interference control condition was compared with a challenging verbal interference task (naming the letters of the alphabet in reverse order) and a haptic interference task (moving a finger between two points along a raised path). Performance in the two interference conditions was similar and was worse than the no-interference condition.

Finally, Seemüller, Fiehler, and Rösler (2011) used a delayed match to sample test with simple two-line angle stimuli that were presented either visually or for passive kinesthetic exploration (the right hand held a stylus that was programmed to move along a trajectory). The interference task involved determining the orientation of an ellipse that was presented visually or by passive kinesthetic exploration as for the primary task. Interference during the retention interval reduced accuracy on the primary matching task relative to a no-interference control condition. Seemüller et al. further argued that this interference was modulated by modality. However, this interaction was not significant in their main analysis. It was only significant in a difference analysis, and there it amounted to only a 2% extra decrease in accuracy for visuospatial relative to kinesthetic interference when matching visually encoded stimuli and a similar extra cost for kinesthetic compared with visuospatial interference when matching kinesthetically encoded stimuli.

These three studies, which tested haptic perception of curvature (Ittyerah & Marks, 2007), orientation (Gentaz & Hatwell, 1999), and kinesthetic perception of angle (Seemüller et al., 2011) as the primary tasks, found discrepant results. Relative to no interference, both visuospatial and motor interference disrupted curvature perception but verbal interference had no effect; verbal and haptic interference both disrupted orientation reproduction, and visual and passive kinesthetic interference both disrupted angle perception. Furthermore, the pattern of results in all three studies was more complex than the summary presented here. It seems that no clear generalizations can be drawn from these findings.

Finally, we consider the effect of interference tasks on haptic shape recognition tasks. Using unfamiliar plastic 3D shapes, Millar (1972) reported that neither verbal interference (repeating digit strings) nor visual interference (viewing colored pictures) during a 9-s retention interval affected haptic shape recognition by children. Subsequently, Millar (1974) compared old/new recognition of unfamiliar 3D shapes by blind and sighted children with interference tasks used during a retention interval of up to 30 s. Both blind and sighted children responded more slowly after verbal interference (counting backward in threes) or haptic/motor interference (stacking 3D shapes) relative to either a no-interference control condition or to a rehearsal task involving tracing with a finger the to-be-remembered shapes on a surface. However, no interference effects were obtained in a second experiment that tested only sighted children. This largely replicated the first experiment but there were fewer repetitions (eight rather than 32) of the shapes used in the primary task. There, there was only an effect of the retention period with faster and more accurate recognition for shorter durations.

In more recent work, Holtby and D'Angiulli (2012) reported that haptic perception of raised line drawings was impaired by both verbal interference (articulatory suppression) and visual interference (DVN) at encoding. They suggested that haptic information was recoded into verbal and visuospatial representations. However, people's ability to recognize raised line drawings is much worse than their recognition of 3D objects (Lawson & Bracken, 2011; Wijntjes, van Lienen, Verstijnen, & Kappers, 2008), probably in large part because of the lack of depth information for such stimuli (Lawson & Bracken, 2011). As such, the recognition of raised line drawings may be unusually reliant on visual recoding (for example, by imagining how the stimulus would look, as proposed by the image mediation account) (see Lederman & Klatzky, 1990; Lederman et al., 1990). Moreover Holtby and D'Angiulli's (2012) task involved the experimenter guiding the participants' finger along the raised line drawings at a constant pace. Such passive encoding is different from our everyday active perception of shape by our hands, which may have further encouraged the use of unusual recoding strategies in this task.

More generally, few previous studies on memory for touch adapted working memory paradigms to take account of the specific strengths and weaknesses of our sense of touch and spatiotemporal differences in stimulus exploration by touch. In particular, studies of haptic shape and object perception rarely permitted the hand to freely explore real everyday 3D objects. An exception is Lacey and Campbell (2006), who investigated the accuracy of cross-modal (haptic-visual and visual-haptic) old/new recognition of familiar and unfamiliar objects in two interference experiments. Participants either touched with their dominant hand or saw a set of objects and then they saw or touched the same objects plus a set of distracters and tried to decide which objects were new. They did a simultaneous interference task at either the encoding stage or the retrieval (recognition) stage. The interference tasks were described as haptic (manipulating an object with their nondominant hand), visual (looking at DVN), or verbal (hearing irrelevant speech through headphones). Interference at encoding did not affect familiar object recognition, although accuracy was over 95%, so ceiling effects may have made it difficult to distinguish between the conditions. Relative to a no-interference control condition, unfamiliar object recognition was disrupted by both verbal and visual interference at encoding but not by haptic interference. There was no main effect of interference at retrieval but performance was accurate and over 95% for familiar objects, so again ceiling effects may have limited the sensitivity to detect interference effects. However, notwithstanding this issue, the disruptive effect of verbal interference for the haptic recognition of unfamiliar objects suggests that verbal strategies for encoding haptically presented stimuli are not restricted to nameable objects and might include subvocal description for unfamiliar objects.

These final four studies used interference tasks to probe shape and object recognition by touch. They employed a wide variety of stimuli and of primary and interference tasks. Nevertheless, their results were consistent in showing widespread interference effects. Specifically, haptic/motor, verbal, and visual interference all disrupted performance relative to a no-interference control condition. In some cases, though, no interference effects were found: for example, when stimuli were repeated fewer times (Millar, 1974).

DRAWING CONCLUSIONS FROM STUDIES OF MEMORY FOR TOUCH USING INTERFERENCE TASKS

We have reviewed a series of 13 studies on memory for touch that used one or more interference tasks. To summarize their findings, studies of memory for tactile location suggested that tactile and verbal interference disrupted performance relative to no interference. Studies of haptic span suggested that both haptic-spatial and visuospatial interference reduced span relative to no interference. Furthermore, this interference was modality-specific, with haptic spans reduced more by haptic-spatial than by visuospatial interference, and vice versa for visual spans. Visuospatial and motor interference both disrupted haptic curvature perception but verbal interference had no effect, whereas verbal and haptic interference both disrupted orientation reproduction, and visuospatial and kinesthetic interference both disrupted kinesthetic angle perception. Finally, studies of haptic shape and object recognition indicated that haptic/motor, verbal, and visual interference all could disrupt performance.

Overall, these results indicate that a range of interference tasks can disrupt memory for touch but that tasks with a spatial element may be particularly effective. However, when interference is found, this may sometimes simply reflect the added difficulty of doing two tasks together (i.e., the interference is not a consequence of targeting a specific component process in the primary task). Similarly, when interference is absent, it could reflect the use of an insufficiently demanding interference task. Studies testing multiple secondary interference tasks have not compared directly the ease of performing each task and the tasks used often seem poorly matched for difficulty. This issue alone could account for the varied effects of the tasks and the problem of drawing clear-cut conclusions from the results of these studies. We now turn to our own data and present two interference studies that investigated haptic object memory. Both studies allowed free manual exploration of complex 3D objects that were moderately or highly familiar. Thus, we used both stimuli and a task well-suited to this modality.

EXPERIMENT 1

In our first study we investigated the effects of four interference conditions (motor, haptic, spatial, and verbal) plus a control no-interference condition on an old/new object recognition task with objects of high and low familiarity. The interference tasks occurred during object encoding to try to elucidate the nature of the encoding strategies used in haptic object recognition. For example, verbal interference would be expected to disrupt verbal encoding strategies such as giving basic-level labels to familiar objects or sub-vocally describing unfamiliar objects (see Lacey & Campbell, 2006).

Method

Participants There were 150 participants. Half were presented with more familiar objects and half with less familiar objects. Within each of these two groups, 15 participants were allocated to each of five interference conditions (motor, haptic, spatial, verbal, and no interference). Participants were students and former students at the University of Minho; they participated for course credit or inclusion in a prize draw with 30 prizes of 10-Euro vouchers.

Materials and Procedure Objects were small enough to be explored with one hand and were presented inside a wooden box to prevent them from being seen. The box was divided in half with the participant's right and left hands going into each half. To prevent participants from hearing noises from the objects as they touched them, participants wore headphones that produced white noise. A computer screen was placed on top of the box and was used to present the stimuli for the verbal and spatial interference tasks. Haptic encoding for the primary task consisted of 3 s of free exploration of each of 50 objects using the nondominant hand. The nondominant hand was used because during piloting, participants stated a preference for using the dominant hand to do the haptic interference task, and Craddock and Lawson (2009a) reported no difference in performance between the dominant and nondominant hand for haptic object recognition. Sound cues were provided through the headphones to start and end object exploration. In the test, each participant felt 25 of the old items that had been presented at encoding and a further 25 new items in random order. They were told to make an old/new decision as quickly and accurately as possible.

All 75 objects presented to a given participant were randomly selected from either a pool of 92 highly familiar, real objects (such as a spoon and a pair of scissors) or a pool of 83 less familiar objects (such as a toy tractor or a plastic chili). The objects were rated for familiarity using a five-point Likert scale (1 = used less than once a year; 5 = used almost every day or more). The set of 92 highly familiar objects had mean ratings of 3.5 (standard deviation (SD) = 0.5) and 83% (SD = 3%) were correctly identified. The set of 83 less familiar objects had mean familiarity ratings of 2.5 (SD = 0.6) and 37% (SD = 3%) were correctly identified.

The interference tasks were performed twice by each participant: once in a single task condition (composed of a block of 50 trials in which no objects were presented haptically) and once concurrent with haptic encoding (in which the 50 objects for the primary recognition task were also presented for encoding, one per trial). The single interference task condition was performed before haptic encoding for around half of the participants in each group; it was done after the old/new recognition task for the remaining participants.

There were five interference groups. One did the *control (no-interference)* condition and another did a *motor interference* condition that required no response. The remaining three groups did *verbal, spatial,* and *haptic* interference tasks, all of which required people to make same/different judgments. Stimuli in the interference tasks were each presented for 3 s and verbal yes/no responses to the verbal, spatial, and haptic tasks were required immediately after each presentation. The participant's dominant hand rested on the inside of the box during encoding, except for the haptic and motor interference groups. The experimenter monitored compliance with the interference task instructions.

The *haptic interference task* involved deciding whether pairs of paper samples (e.g., white, recycled, magazine, newspaper, photo) were the same. Trials were presented pseudo-randomly from a set of 52 same pairs and 58 different pairs. The *motor interference* task required people to perform similar movements as for the haptic interference task and to rub the fingers of the dominant hand from the top to the bottom of the box. However, no paper samples were presented and they were not told to do a task. The *spatial interference* task simultaneously presented two of the 3D abstract shapes used by Shepard and Metzler (1971) on the computer screen. Participants decided whether each pair had the same shape. Same pairs were identical shapes but rotated by 40° in the plane from each other. Different pairs were mirror images of each other that were also rotated 40° in the plane from each other. The stimuli for the *verbal interference* task consisted of pairs of tri-syllabic pseudo-words that people had to read aloud. Participants decided whether each pair had the same syllables. In same pairs, both pseudo-words had the same syllables but in different order (e.g., TA-FA-LE/FA-LE-TA). In different pairs the consonants and vowels were switched across two syllables (e.g., NO-SI-NE/NI-NE-SO). At the beginning of the experiment, the no-interference group did six practice trials of the primary haptic task. The interference groups performed six practice trials of their interference task alone and then six practice trials of the interference task plus the haptic task.

Results

Analyses were conducted on corrected hit rates (hits minus false alarms) for the primary old/new haptic recognition task. An analysis of variance (ANOVA) was conducted with two between-subjects factors of object familiarity (high or low) and interference task at encoding (none, motor, haptic, spatial, or verbal). There was a significant effect of object familiarity, $F(1, 140) = 90.34$, $p < .001$, partial $\eta 2 = 0.39$. As expected recognition of highly familiar objects (0.73) was better than that of less familiar objects (0.46). There was also a significant effect of interference task, $F(4, 140) = 16.23$, $p < .001$, partial $\eta 2 = 0.32$. Finally, these two main effects were modulated by a significant interaction between familiarity and interference task, $F(4, 140) = 3.04$, $p = .02$, partial $\eta 2 = 0.08$) (see Figure 1), showing that the effects of the interference tasks depended on the type of objects presented. Post hoc Newman–Keuls analyses revealed that for the more familiar objects there was no significant difference among the no-interference (0.85), motor interference (0.86), verbal interference (0.72), and spatial interference (0.72) groups. However, all four groups were more accurate than the haptic interference group (0.50). In contrast, for the less familiar objects, the no-interference group (0.67) was more accurate than all four interference groups. The only significant difference among the motor interference (0.51), spatial interference (0.41), haptic interference (0.37), and verbal interference (0.32), groups was greater accuracy with motor interference than verbal interference.

Discussion

In Experiment 1 the disruptive effects of the different interference tasks on the encoding of haptic objects were modulated by object familiarity. Recognition memory for all objects was clearly disrupted by haptic interference at encoding, relative to a no-interference control task. In contrast, spatial, verbal, and motor interference all reduced accuracy relative to no interference for less familiar but not for highly familiar objects. This interaction may have been driven by task difficulty, with spatial, verbal, and motor interference having an influence only when recognition was particularly hard, whereas haptic interference was detrimental irrespective of task difficulty and despite different hands performing the interference and the primary tasks. This, in turn, suggests that haptic processes may have a more fundamental, unavoidable role during haptic encoding than spatial, verbal, and motor processes. In contrast, the disruptive effects of spatial, verbal, and motor interference under demanding dual-task conditions may result from competition for central resources under these circumstances.

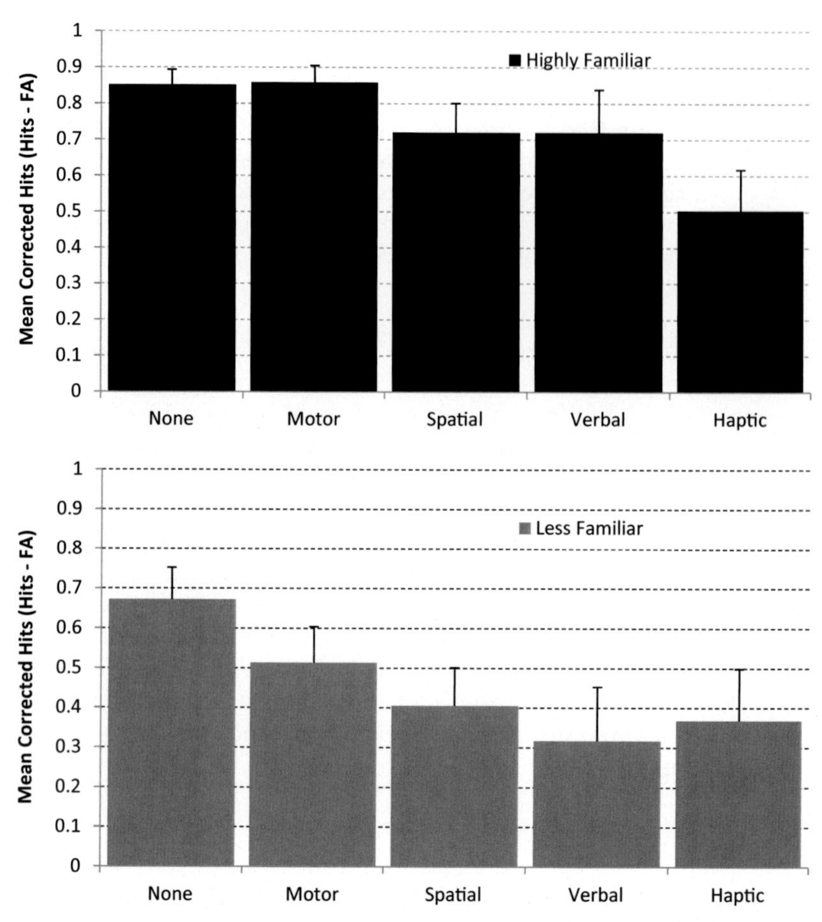

FIGURE 1 Corrected hits (hits minus false alarms) in Experiment 1 for the primary haptic recognition task for the no-interference condition and the four interference conditions (motor, haptic, spatial, and verbal), for the more (top) and less (bottom) familiar object sets separately. Error bars show 95% confidence intervals.

EXPERIMENT 2

In this study we compared the effect of three interference tasks (verbal, spatial, and motor) relative to a no-interference control condition using an object-matching task. Unusually for interference studies, these four conditions were manipulated within-subjects, removing inter-subject variability. The objects used were plastic scale models of familiar, nameable objects; the primary task of sequential matching has been used previously with these stimuli (Lawson, 2009). Unlike Experiment 1, the interference tasks were performed during the retention period between presenting the first and second object on each matching trial. Targeting interference at retention minimized the likelihood of peripheral problems in combining the primary, matching and secondary, interference tasks, which could disrupt encoding or retrieval processes. This study attempted to determine the nature of the perceptual representations involved during the rehearsal and maintenance of information required for haptic object recognition.

Method

Participants There were 32 participants who were students at the University of Liverpool. They participated for course credit or without reward.

Materials and Procedure Sixteen pairs of plastic 3D models of familiar objects such as tables and dogs were presented sequentially to participants. The objects were approximately hand-sized. Lawson (2009) provides further details about their production. Half of the trials were match trials in which the same object was presented twice (e.g., chair then chair). On the remaining mismatch trials the second object had a shape similar to the first but it could normally be distinguished haptically from it. The difficulty of shape discrimination was the medium context condition tested in Experiment 2 of Lawson (2009) and the two objects presented on a mismatch trial would usually be given a different basic-level name (see Experiment 1 in Lawson, 2009); examples included bath–sink, cup–jug and lizard–frog. Thus on mismatch trials, participants never had to distinguish between two exemplars of the same category (e.g., two different chairs) although the two objects always had related shapes. Participants made a rapid decision as to whether the two objects presented on a trial were the same or different. They responded with the left hand by making a speeded key press using a button box.

The 64 experimental trials were divided into four blocks of 16 trials with one block for each of the four interference conditions. The order of the four interference conditions was counterbalanced using a Latin square design across the participants. The order of the items was fixed. One order was used for half of the participants in each subgroup and the reverse order was used for the remaining participants. Each pair of objects was presented once in every block. Over the 64 trials, each pair was presented twice on match trials and twice on mismatch trials, with the two objects in a pair presented in a different order on each mismatch trial (so the four trials for the pig–dog pair were pig–pig, dog–dog, pig–dog. and dog–pig). The experimental trials were preceded by eight practice trials that had the same interference task as the first experimental trial. The experimenter explained each interference task immediately before each sub-block of trials that used that task.

On each trial the experimenter placed the first object behind an opaque glass screen so it was not visible to the participant then triggered the computer to say "go now." The participant then had 5 s to feel the first object with their right hand. A "stop now" cue was then played to them. This indicated that they had to stop feeling the first object and to start doing the interference task. This was done for 7 s in the retention interval between feeling the two objects. Another "go now" signal then indicated that they could stop the interference task and start to feel the second object which was presented in the same location as the first object until participants responded.

All participants completed four interference conditions: a control (*no-interference*) task and three interference tasks. In the *control task* participants waited quietly in the 7-s retention interval. In the *verbal interference task* participants were told a letter and were asked to name aloud as many different words as possible beginning with this letter during the retention period. A new letter was presented on each trial. This task should have disrupted any sub-vocal rehearsal strategy such as remembering the name of the first object. If participants simply retrieved the name of the first object from memory and rehearsed it, this interference task is a reasonable match for the underlying cognitive process because it required participants to retrieve single words from memory. In the *spatial interference task* participants were told a letter and had to visualize it as a giant capital letter and report aloud in which direction they would have to turn if they were walking clockwise around it, starting from the bottom left corner. For example, for T they should have responded "left," "right," "right," "left," "right." They were given further letters if they completed the first letter during the retention interval. Only straight-edged letters suitable for the task were used. In the *motor interference task* participants used the left hand to move jigsaw puzzle pieces as quickly as possible, one at a time, between two adjacent bowls. The spatial and motor interference tasks were designed to try to disrupt the maintenance of any spatial or motor representations of the first object. However, all three interference tasks drew on multiple processes; for example, the spatial task included verbal responses whereas the motor task had a spatial component.

Results

The analyses were conducted on mean correct response times (RTs) and percentage correct on the haptic matching task. Response times less than 1 s or greater than 10 s were removed as outliers (less than 1% of trials). An ANOVA was conducted with two within-subjects factors of matching (same or different) and interference task during the retention period (none, verbal, spatial, or motor). RTs were similar across the control task (4.3 s) and the three interference tasks (4.0 s for verbal, 4.0 s for spatial, and 4.1 s for motor) and there were no significant effects on the RT analysis. In the error analysis, there was a significant effect of matching, $F(1, 31) = 9.89$, $p = .004$, partial $\eta 2 = 0.24$, with greater accuracy on same trials (79.8%) than on different trials (70.6%). There was also a significant effect of interference task, $F(3, 93) = 10.89$, $p < .001$, partial $\eta 2 = 0.26$. There was no interaction between these two factors, $F(3, 93) = 1.83$, $p > .1$, partial $\eta 2 = 0.06$. Post hoc Newman–Keuls analyses revealed that the accuracy of haptic matching was reduced by spatial interference (66.2%) relative to the other three conditions, and that there was no difference between the no-interference (80.9%), verbal interference (79.1%), and motor interference (74.6%) conditions.

Discussion

Experiment 2 investigated the nature of the processes involved in rehearsing and maintaining haptically acquired object representations. The results were straightforward: spatial interference disrupted accuracy on object matching relative to a no-interference control condition whereas verbal and motor interference did not. In the secondary spatial interference task people had to imagine a capital letter and say which directions they would have to turn if they were to walk around it. This task involved both visual imagery and spatial transformations. In contrast, simply moving the hands during the motor interference task did not disrupt performance on the primary, matching task, whereas the spatial interference condition, which did disrupt matching, did not require movement. Although the objects used were generally familiar, nameable shapes (Lawson, 2009), verbal interference did not disrupt performance relative to the control no-interference task. This may have been because the primary matching task involved difficult shape discrimination (overall accuracy was 75%). The similarity of the two objects presented on different trials (e.g., a chair and then a bench; or a frog and then a lizard) meant that participants had to maintain detailed shape information to perform well. Merely remembering a name for the first object would often not have been sufficient to support accurate performance because these objects could not be named reliably and the same name would often have been given to different shaped objects (see Experiment 1 of Lawson, 2009).

Together, these results suggest that people did not retain their memory of the first object using motor or verbal strategies but that instead, spatial coding was important to the maintenance of accurate haptic shape representations; i.e., people retained and/or rehearsed spatial rather than verbal or motor representations. However, it is not possible to exclude an alternative hypothesis that the spatial interference task was simply harder than the other interference tasks and that this, in turn, resulted in less attention or processing resources being available for the primary matching task. This seems a reasonable critique of the no-interference and motor interference conditions. However, the word generation task used for verbal interference is demanding and is thought to tap executive functioning as well as verbal processes (Barry, Bates, & Labouvie, 2008).

The spatial tasks used in Experiments 1 and 2 required visual processing and/or visual imagery. The results from haptic span tasks reported by Paz et al. (2007) and Sebastián et al. (2008) and reviewed above showed that spatial interference was modality-specific. This suggests that even stronger spatial interference effects could have been obtained in the current studies if haptic rather than visual spatial tasks had been used. In future research it will be important to compare visual and haptic object recognition directly as primary tasks while using matched visual-spatial and haptic-spatial interference tasks to investigate whether modality-specific spatial interference also occurs for object recognition.

GENERAL DISCUSSION

There has been considerable progress in our understanding of haptic and tactile memory, as reviewed above. However, much of the research on memory for touch has presented simple, abstract stimuli and involved passive touch. These conditions are not optimal for this modality, so it is important to expand research on memory for touch to evaluate conditions more similar to those we experience in our everyday life. Tasks better suited to evaluating the strengths of touch perception allow unconstrained, active exploration and present 3D objects with rich information (for example, about material, weight, and size) in contrast to, for example, single finger exploration of 2D raised line drawings (e.g., Holtby & D'Angiulli, 2012; Lawson & Bracken, 2011).

We have focused on assessing studies of memory for touch that used interference or dual-task methodologies. Here, the findings are not wholly consistent. To illustrate this, consider four experiments which we have conducted, two of which were previously published (Lacey & Campbell, 2006) and two new studies that we report here. These experiments share similar methodologies. All four investigated the effects of interference on haptic memory in object processing tasks, with all four comparing a no-interference control condition with three to four different interference tasks. They all included verbal interference and at least two included haptic, visual, motor, and spatial interference conditions. Similar primary tasks (object matching and old/new recognition) and stimuli (complex, 3D objects varying in familiarity) were also used. Despite these commonalities across the four experiments, the results varied. Lacey and Campbell (2006) reported that familiar object recognition was not influenced by interference at encoding (although this may have been due to ceiling effects). In contrast, unfamiliar object recognition was disrupted by verbal interference and by visual interference at encoding, but not by haptic interference. Interference at retrieval had no main effect on accuracy. Experiment 1 here found that for the more familiar objects only haptic interference impaired performance, whereas for the less familiar objects all four interference tasks impaired performance. Finally, Experiment 2 described here found that only spatial interference disrupted performance, with no cost of motor or verbal interference.

These differences could result from variation in the point at which interference was applied, which ranged from encoding to retention to retrieval across the experiments. It could also be because Lacey and Campbell tested cross-modal haptic/visual recognition whereas Experiments 1 and 2 tested unimodal haptic recognition.

In addition, there was considerable variation in the tasks used to apply interference. For example, although articulatory suppression is generally used in memory research to produce verbal interference by preventing verbal rehearsal, none of these studies used it. Instead, Lacey and Campbell had participants passively listen to speech, Experiment 1 used a difficult non-word discrimination task, and Experiment 2 had participants generate words beginning with a specific letter. These three tasks varied in task difficulty and in the specific components of verbal processing that each tapped.

We raise these points to illustrate that if clear and wide-ranging conclusions are to be drawn from haptic interference studies, they need to use sets of standard secondary interference tasks and/or theoretically well-motivated tasks (see Chan & Newell, 2008) that precisely tap distinct cognitive processes. Furthermore, task difficulty should be matched across the different interference tasks and performance on the interference tasks should be monitored. None of our own experiments investigating haptic interference effects or those reported in the other 12 published studies investigating tactile and haptic interference studies that we reviewed above achieved all of these demanding conditions. The strongest design would include multiple primary tasks as well as multiple interference tasks, but this has rarely been done.

Notwithstanding these challenges to interpreting results from haptic interference tasks, we believe that a number of interesting conclusions can be made about the role of movement and verbal, haptic, and spatial processes in representing stimuli presented to touch. From the earliest experiments on touch, researchers have highlighted the role of movement in recognizing and identifying stimuli (Gibson, 1966; Kaas, Stoeckel, & Goebel, 2008; Nefs, Kappers, & Koenderink, 2001; Millar, 1999). For example, Millar (1999) suggested that a movement loop could be used as a rehearsal system for touch, similar to the phonological loop proposed for rehearsing verbal material (see Baddeley, 2000). This system could mentally rehearse executed movements to maintain a dynamic representation of haptically encoded stimuli and of exploration patterns. There is clear evidence that we encode certain stimulus properties such as the orientation in depth of an object relative to our own body (Craddock & Lawson, 2008; Ernst et al., 2007; Lawson, 2009, 2011; Newell et al., 2001). This indicates that we maintain some identity-irrelevant information about how object information was encoded (such as its position relative to ourselves), so there is no a priori reason to assume that motor processes are not also encoded. Nevertheless, the results from both Experiments 1 and 2 here and from the object manipulation condition of Lacey and Campbell (2006) did not support this proposal that we maintain motor representations. Motor interference tasks did not disrupt haptic object processing, with the exception of recognizing less familiar objects in Experiment 1. Further evidence consistent with this comes from Cecchetto and Lawson (2015). We recently found that producing an unseen sketch of a raised line drawing while it was being explored haptically did not reduce the accuracy of recognizing the drawing relative to a no-sketch control condition.

In contrast to this lack of evidence for the importance of motor processes during the acquisition and storage of information presented to the hands, haptic and spatial processes both appear to be important based on the deleterious effects of haptic and spatial interference (see Experiments 1 and 2 here; also Cohen et al., 2010; Gentaz & Hatwell, 1999; Ittyerah & Marks, 2007; Millar, 1974; Paz et al., 2007; Sebastián et al., 2008). Although Lacey and Campbell (2006) reported no effect of haptic interference, their interference task (manipulating an object in the non-encoding hand) is probably better described as a motor interference task because no response was required. The conclusion that haptic object memory representations have a spatial dimension consistent with the 3D structure of the stimuli converges with evidence from tactile working memory studies (e.g., Harris, Harris, & Diamond, 2001; Katus, Andersen, & Müller, 2012).

Finally, there is evidence that verbal and visual processes are also involved in acquiring and storing information perceived by touch, at least during the haptic encoding of less familiar objects (see Experiment 1 here; Experiment 1 of Lacey & Campbell, 2006; and Cohen et al., 2010; Gentaz & Hatwell, 1999; Holtby & D'Angiulli, 2012; Millar, 1974). However, as discussed above, many other studies have failed to detect disruptive effects of verbal or visual interference tasks, and so it remains unclear what factors determine whether these interference effects are observed.

The apparent diversity of effects of interference tasks on haptic object processing may be reconciled by assuming that alternative strategies may be used during the encoding and maintenance of haptic representations with the choice of strategies varying depending on the task used, the stimuli, and the individual. One strategy is to encode a haptic representation; alternative or additional strategies include the use of covert naming and verbal descriptions, as well as employing visuospatial and haptic-spatial imagery. These various processes may lead to distinct representations of the input being produced in parallel, but such representations would be short-lived. Their role may be to help to establish a single, more durable, modality-independent spatial representation that can then be maintained in short-term memory. This representation could subsequently be matched to both visual and haptic inputs and it could be transferred to a permanent memory store.

Acknowledgments

We would like to thank the Portuguese Foundation for Science and Technology for supporting the PhD research of the second author with the fellowship SFRH/BD/35918/2007, and Elizabeth Collier for helpful comments during preparation of this chapter.

References

Al-Yahya, E., Dawes, H., Smith, L., Dennis, A., Howells, K., & Cockburn, J. (2011). Cognitive motor interference while walking: a systematic review and meta-analysis. *Neuroscience and Biobehavioral Reviews*, *35*, 715–728.

Aleman, A., van Lee, L., Mantione, M. H. M., Verkoijen, I. G., & de Haan, E. H. F. (2001). Visual imagery without visual experience: evidence from congenitally totally blind people. *NeuroReport*, *12*, 2601–2604.

Anderson, D. E., Vogel, E. K., & Awh, E. (2011). Precision in visual working memory reaches a stable plateau when individual item limits are exceeded. *Journal of Neuroscience*, *31*, 1128–1138.

Andrade, J., Kemps, E., Werniers, Y., May, J., & Szmalec, A. (2002). Insensitivity of visual short-term memory to irrelevant visual information. *The Quarterly Journal of Experimental Psychology*, *55A*, 753–774.

Auvray, M., Gallace, A., & Spence, C. (2011). Tactile short-term memory for stimuli presented on the fingertips and across the rest of the body surface. *Attention, Perception, and Psychophysics*, *73*, 1227–1241.

Baddeley, A. D. (2000). The episodic buffer: a new component of working memory? *Trends in Cognitive Sciences*, *4*, 417–423.

Baddeley, A. D. (2007). *Working memory, thought and action*. Oxford: Oxford University Press.

Baddeley, A. D., Grant, S., Wight, E., & Thomson, M. (1975). Imagery and visual working memory. In P. M. A. Rabbitt, & S. Dornic (Eds.), *Attention and performance* (Vol. V). London: Academic Press.

Baddeley, A. D., & Hitch, G. J. (1974). Working memory. In G. Bower (Ed.), *The psychology of learning and motivation* (pp. 47–89). New York: Academic Press.

Ballesteros, S., & Reales, J. M. (2004). Intact haptic priming in normal aging and Alzheimer's disease: evidence for dissociable memory systems. *Neuropsychologia*, *42*, 1063–1070.

Ballesteros, S., Reales, J. M., & Manga, D. (1999). Implicit and explicit memory for familiar and novel objects presented to touch. *Psicothema*, *11*(4), 785–800.

Bancroft, T. D., Hockley, W. E., & Servos, P. (2012). Can vibrotactile working memory store multiple items? *Neuroscience Letters*, *514*, 31–34.

Bancroft, T. D., Hogeveen, J., Hockley, W. E., & Servos, P. (2014). TMS-induced neural noise in sensory cortex interferes with short-term memory storage in prefrontal cortex. *Frontiers in computational neuroscience*, *8*.

Bancroft, T., & Servos, P. (2011). Distractor frequency influences performance in vibrotactile working memory. *Experimental Brain Research*, *208*(4), 529–532. http://dx.doi.org/10.1007/s00221-010-2501-2.

Barry, D., Bates, M. E., & Labouvie, E. (2008). FAS and CFL forms of verbal fluency differ in difficulty: a meta-analytic study. *Applied Neuropsychology*, *15*, 97–106.

Bliss, J. C., Crane, H. D., Mansfield, P. K., & Townsend, J. T. (1966). Information available in brief tactile presentations. *Perception and Psychophysics*, *1*, 273–283.

Bliss, I., & Hämäläinen, H. (2005). Different working memory capacity in normal young adults for visual and tactile letter recognition task. *Scandinavian Journal of Psychology*, *46*, 247–251.

Bushnell, E. W., & Baxt, C. (1999). Children's haptic and cross-modal recognition with familiar and unfamiliar objects. *Journal of Experimental Psychology: Human Perception and Performance*, *25*, 1867–1881.

Cecchetto, S., & Lawson, R. (2015). Simultaneous sketching aids the haptic recognition of raised line drawings. *Perception*, *41*(12), 1514.

Chan, J. S., & Newell, F. N. (2008). Behavioral evidence for task-dependent "what" versus "where" processing within and across modalities. *Perception and Psychophysics*, *70*, 36–49.

Cohen, H., Voss, P., Lepore, F., & Scherzer, P. (2010). The nature of working memory for Braille. *PLoS ONE*, *5*(5), e10833.

Conrad, R., & Hull, A. J. (1968). Input modality and serial position curve in short-term memory. *Psychonomic Science*, *10*, 135–136.

Cornoldi, C., & Vecchi, T. (2003). *Visuo-Spatial working memory and individual differences*. Hove: Psychology Press.

Cowan, N. (1995). *Attention and memory: An integrated framework*. Oxford: Oxford University Press.

Cowan, N. (2001). The magical number 4 in short-term memory: a reconsideration of mental storage capacity. *Behavioral and Brain Sciences*, *24*, 87–185.

Cowan, N. (2008). What are the differences between long-term, short-term, and working memory? In W. S. Sossin, J.-C. Lacaille, V. F. Castellucci, & S. Belleville (Eds.), *Progress in brain research Essence of memory: Vol. 169*. (pp. 323–338). Amsterdam: Elsevier B.V.

Craddock, M., & Lawson, R. (2008). Repetition priming and the haptic recognition of familiar and unfamiliar objects. *Perception and Psychophysics*, *70*, 1350–1365.

Craddock, M., & Lawson, R. (2009a). Do left and right matter for haptic recognition of familiar objects? *Perception*, *38*, 1355–1376.

Craddock, M., & Lawson, R. (2009b). The effects of size changes on haptic object recognition. *Attention, Perception, and Psychophysics*, *71*, 910–923.

Craddock, M., Martinovic, J., & Lawson, R. (2011). An advantage for active versus passive aperture-viewing in visual object recognition. *Perception*, *40*, 1154–1163.

Dean, G. M., Dewhurst, S. A., Morris, P., & Whittaker, A. (2005). Selective interference with the use of visual images in the symbolic distance paradigm. *Journal of Experimental Psychology: Learning, Memory and Cognition*, *31*, 1043–1068.

Dean, G. M., Dewhurst, S. A., & Whittaker, A. (2008). Dynamic visual noise interferes with storage in visual working memory. *Experimental Psychology*, *55*, 283–289.

Deshpande, G., Hu, X., Lacey, S., Stilla, R., & Sathian, K. (2010). Object familiarity modulates effective connectivity during haptic shape perception. *NeuroImage*, *49*, 1991–2000.

Easton, R. D., Srinivas, K., & Greene, A. J. (1997). Do vision and haptics share common representations? Implicit and explicit memory within and between modalities. *Journal of Experimental Psychology: Learning, Memory, and Cognition*, *23*, 153–163.

Engle, R. W., Tuholski, S. W., Laughlin, J. E., & Conway, A. R. A. (1999). Working memory, short-term memory and general fluid intelligence: a latent-variable approach. *Journal of Experimental Psychology: General, 128*, 309–331.

Ernst, M. O., Lange, C., & Newell, F. N. (2007). Multisensory recognition of actively explored objects. *Canadian Journal of Experimental Psychology, 61*, 242–253.

Ester, E. F., Drew, T., Klee, D., Vogel, E. K., & Awh, E. (2012). Neural measures reveal a fixed item limit in subitizing. *Journal of Neuroscience, 32*, 7169–7177.

Gallace, A., & Spence, C. (2008). The cognitive and neural correlates of "tactile consciousness": a multisensory perspective. *Consciousness and Cognition, 17*, 370–407.

Gallace, A., & Spence, C. (2009). The cognitive and neural correlates of tactile memory. *Psychological Bulletin, 135*, 380–406.

Gallace, A., Tan, H. Z., Haggard, P., & Spence, C. (2008). Short term memory for tactile stimuli. *Brain Research, 1190*, 132–142.

Gallace, A., Tan, H. Z., & Spence, C. (2006). Numerosity judgments for tactile stimuli distributed over the body surface. *Perception, 35*, 247–266.

Garvill, J., & Molander, B. (1977). *A note on the relation between cross-modal transfer of learning and cross-modal matching*. Umea Psychological Reports, No 126. Sweden: University of Umea.

Gentaz, E., & Hatwell, Y. (1999). Role of memorization conditions in the haptic processing of orientations and the 'oblique effect'. *British Journal of Psychology, 90*, 373–388.

Gibson, J. J. (1966). *The senses considered as perceptual systems*. Boston: Houghton Mifflin.

Gilson, E. Q., & Baddeley, A. D. (1969). Tactile short term memory. *Quarterly Journal of Experimental Psychology, 21*(2), 180–184.

Green, A., & Vaid, J. (1983). Methodological issues in the use of the concurrent activities paradigm. *Brain and Cognition, 5*, 465–476.

Hannula, H., Neuvonen, T., Savolainen, P., Hiltunen, J., Ma, Y. Y., Antila, H., et al. (2010). Increasing top-down suppression from prefrontal cortex facilitates tactile working memory. *Neuroimage, 49*, 1091–1098.

Harris, J. A., Harris, I. M., & Diamond, M. E. (2001). The topography of tactile learning in humans. *Journal of Neuroscience, 21*, 1056–1061.

Harris, J. A., Miniussi, C., Harris, I. M., & Diamond, M. E. (2002). Transient storage of a tactile memory trace in primary somatosensory cortex. *Journal of Neuroscience, 22*, 8720–8725.

Heller, M. A. (1987). Improving the tactile digit span. *Bulletin of the Psychonomic Society, 25*, 257–258.

Hill, J. W., & Bliss, J. C. (1968). Modeling a tactile sensory register. *Perception and Psychophysics, 4*, 91–101.

Holtby, R. C., & D'Angiulli, A. (2012). The effects of interference on recognition of haptic pictures in blindfolded sighted participants: the modality of representation of haptic information. *Scandinavian Journal of Psychology, 53*, 112–118.

Hull, T., & Mason, H. (1995). Performance of blind children on digit-span tests. *Journal of Visual Impairment & Blindness, 89*, 166–169.

Ittyerah, M., & Marks, L. E. (2007). Memory for curvature of objects: haptic touch versus vision. *British Journal of Psychology, 98*, 589–610.

Kaas, A., Stoeckel, M., & Goebel, R. (2008). The neural bases of haptic working memory. In M. Grunwald (Ed.), *Human haptic perception: Basics and applications* (pp. 113–129). Basel: Birkhäuser.

Katus, T., Andersen, S. K., & Müller, M. M. (2012). Nonspatial cueing of tactile STM causes shift of spatial attention. *Journal of Cognitive Neuroscience, 24*, 1596–1609.

Kiphart, M. J., Hughes, J. L., Simmons, J. P., & Cross, H. A. (1992). Short-term haptic memory for complex objects. *Bulletin of the Psychonomic Society, 30*, 212–214.

Klatzky, R. L., & Lederman, S. J. (1987). The intelligent hand. In G. Bower (Ed.), *The psychology of learning and motivation* (21) (pp. 121–151). New York: Academic Press.

Klatzky, R. L., Lederman, S. J., & Metzger, V. (1985). Identifying objects by touch - an expert system. *Perception and Psychophysics, 37*, 299–302.

Kozhevnikov, M., & Blazhenkova, O. (2013). Individual differences in object versus spatial imagery: from neural correlates to real-world applications. In S. Lacey, & R. Lawson (Eds.), *Multisensory imagery* (pp. 299–318). New York: Springer.

Lacey, S., & Campbell, C. (2006). Mental representation in visual/haptic crossmodal memory: evidence from interference effects. *Quarterly Journal of Experimental Psychology, 59*, 361–376.

Lacey, S., Flueckiger, P., Stilla, R., Lava, M., & Sathian, K. (2010). Object familiarity modulates the relationship between visual object imagery and haptic shape perception. *NeuroImage, 49*, 1977–1990.

Lacey, S., Pappas, M., Kreps, A., Lee, K., & Sathian, K. (2009). Perceptual learning of view-independence in visuo-haptic object representations. *Experimental Brain Research, 198*(2–3), 329–337.

Lacey, S., & Sathian, K. (2013). Visual imagery in haptic shape perception. In S. Lacey, & R. Lawson (Eds.), *Multisensory imagery* (pp. 207–219). New York: Springer.

Lacey, S., & Sathian, K. (2014). Visuo-haptic multisensory object recognition, categorization and representation. *Frontiers in Perception Science, 5*, 730. http://dx.doi.org/10.3389/fpsyg.2014.00730.

Lacey, S., Stilla, R., Sreenivasan, K., Deshpande, G., & Sathian, K. (2014). Spatial imagery in haptic shape perception. *Neuropsychologia, 60*, 144–158.

Lawson, R. (2009). A comparison of the effects of depth rotation on visual and haptic three-dimensional object recognition. *Journal of Experimental Psychology: Human Perception and Performance, 35*, 911–930.

Lawson, R. (2011). An investigation into the cause of orientation-sensitivity in haptic object recognition. *Seeing and Perceiving, 24*, 293–314.

Lawson, R. (2014). Recognising familiar objects by hand and foot: haptic shape perception generalises to inputs from unusual locations and untrained body parts. *Attention, Perception and Psychophysics, 76*, 559–574.

Lawson, R., & Bracken, S. (2011). Haptic object recognition: how important are depth cues and plane orientation? *Perception, 40*, 576–597.

Lederman, S. J., & Klatzky, R. L. (1990). Haptic classification of common objects: knowledge-driven exploration. *Cognitive Psychology, 22*, 421–459.

Lederman, S. J., Klatzky, R. L., Chataway, C., & Summers, C. D. (1990). Visual mediation and the haptic recognition of two-dimensional pictures of common objects. *Perception and Psychophysics, 47*, 54–64.

Linden, D. E. J. (2007). The working memory networks of the human brain. *The Neuroscientist, 13*, 257–267.

Loomis, J. M., Klatzky, R. L., & Giudice, N. A. (2013). Representing 3D space in working memory: spatial images from vision, hearing, touch and language. In S. Lacey, & R. Lawson (Eds.), *Multisensory imagery*. New York: Springer.

Loomis, J. M., Klatzky, R. L., & Lederman, S. J. (1991). Similarity of tactual and visual picture recognition with limited field of view. *Perception, 20*, 167–177.

Mahrer, P., & Miles, C. (1999). Memorial and strategic determinants of tactile recency. *Journal of Experimental Psychology: Learning, memory, and Cognition*, *25*, 630–643.

Mahrer, P., & Miles, C. (2002). Recognition memory for tactile sequences. *Memory*, *10*, 7–20.

Manning, S. K. (1980). Tactual and visual alphanumeric suffix effects. *Quarterly Journal of Experimental Psychology*, *32*, 257–267.

Marmor, G. S., & Zaback, L. A. (1976). Mental rotation by the blind: does mental rotation depend on visual imagery? *Journal of Experimental Psychology: Human Perception and Performance*, *2*, 515–521.

Martinovic, J., Lawson, R., & Craddock, M. (2012). Time course of information processing in visual and haptic object classification. *Frontiers in Human Neuroscience*, *6*, 49.

McConnell, J., & Quinn, J. G. (2000). Interference in visual working memory. *The Quarterly Journal of Experimental Psychology*, *53A*, 53–67.

Miles, C., & Borthwick, H. (1996). Tactile short-term memory revisited. *Memory*, *4*, 655–668.

Millar, S. (1972). Effects of interpolated tasks on latency and accuracy of intermodal and cross-modal shape recognition by children. *Journal of Experimental Psychology*, *96*, 170–175.

Millar, S. (1974). Tactile short-term memory by blind and sighted children. *British Journal of Psychology*, *65*, 253–263.

Millar, S. (1999). Memory in touch. *Psicothema*, *11*, 747–767.

Monnier, C., & Picard, D. (2010). Constraints on haptic short-term memory. In *Eurohaptics 2010, part II Lecture notes in computer science* (6192) (pp. 94–98).

Nabeta, T., & Kusumi, T. (2008). Memory capacity for haptic common objects. *Perception*, *37*, 49.

Nairne, J. S., & McNabb, W. L. (1985). More modality effects in the absence of sound. *Journal of Experimental Psychology: Learning, Memory & Cognition*, *11*, 596–604.

Nefs, H. T., Kappers, A. M. L., & Koenderink, J. J. (2001). Amplitude and spatial-period discrimination in sinusoidal gratings by dynamic touch. *Perception*, *30*, 1263–1274.

Newell, F. N., Ernst, M. O., Tjan, B. S., & Bülthoff, H. H. (2001). View dependence in visual and haptic object recognition. *Psychological Science*, *12*, 37–42.

Newell, F. N., Woods, A. T., Mernagh, M., & Bülthoff, H. H. (2005). Visual, haptic and crossmodal recognition of scenes. *Experimental Brain Research*, *161*, 233–242.

Norman, J. F., Crabtree, C. E., Norman, H. F., Moncrief, B. K., Herrmann, M., & Kapley, N. (2006). Aging and the visual, haptic, and cross-modal perception of natural object shape. *Perception*, *35*, 1383–1395.

Norman, J. F., Kappers, A. M. L., Beers, A. M., Scott, A. K., Norman, H. F., & Koenderink, J. J. (2011). Aging and the haptic perception of 3D surface shape. *Attention, Perception, and Psychophysics*, *73*, 908–918.

Norman, J. F., Norman, H. F., Clayton, A. M., Lianekhammy, J., & Zielke, G. (2004). The visual and haptic perception of natural object shape. *Perception and Psychophysics*, *66*, 342–351.

Oberauer, K., & Göthe, K. (2006). Dual-task effects in working memory: interference between two processing tasks, between two memory demands, and between storage and processing. *European Journal of Cognitive Psychology*, *18*, 493–519.

Overvliet, K. E., Wagemans, J., & Krampe, R. T. (2013). The effects of aging on haptic 2D shape recognition. *Psychology and Aging*, *28*, 1057–1069.

Paivio, A. (1986). *Mental representations: A dual coding approach*. New York: Oxford University Press.

Paivio, A. (2007). *Mind and its evolution: A dual coding theoretical approach*. Mahwah, NJ: Lawrence Erlbaum Associates.

Pashler, H. (1994). Dual-task interference in simple tasks: data and theory. *Psychological Bulletin*, *116*, 220–224.

Paz, S., Mayas, J., & Ballesteros, S. (2007). Haptic and visual working memory in young adults, healthy older adults, and mild cognitive impairment adults. In *World haptics proceedings, Tsukuba, Japan* (pp. 553–554). Los Alamitos: IEEE Computer Society.

Pensky, A. E. C., Johnson, K. A., Haag, S., & Homa, D. (2008). Delayed memory for visual-haptic exploration of familiar objects. *Psychonomic Bulletin and Review*, *15*, 574–580.

Peterson, L. R., & Peterson, M. J. (1959). Short-term retention of individual verbal items. *Journal of Experimental Psychology*, *58*, 193–198.

Piazza, P., Fumarola, A., Chinello, A., & Melcher, D. (2011). Subitizing reflects visuo-spatial object individuation capacity. *Cognition*, *121*, 147–153.

Pietrini, P., Furey, M. L., Ricciardi, E., Gobbini, M. I., Wu, W. H. C., Cohen, L., et al. (2004). Beyond sensory images: object-based representation in the human ventral pathway. *Proceedings of the National Academy of Sciences USA*, *101*, 5658–5663.

Pring, L. (2008). Psychological characteristics of children with visual impairments: learning, memory and imagery. *British Journal of Visual Impairment*, *26*, 159–169.

Quinn, J. G., & McConnell, J. (1996). Irrelevant pictures in visual working memory. *The Quarterly Journal of Experimental Psychology*, *49A*, 200–215.

Quinn, J. G., & McConnell, J. (1999). Manipulation of interference in the passive visual store. *European Journal of Cognitive Psychology*, *11*, 373–389.

Quinn, J. G., & McConnell, J. (2006). The interval for interference in conscious visual imagery. *Memory*, *14*, 241–252.

Ricciardi, E., Bonino, D., Gentili, C., Sani, L., Pietrini, P., & Vecchi, T. (2006). Neural correlates of spatial working memory in humans: a functional magnetic resonance imaging study comparing visual and tactile processes. *Neuroscience*, *139*, 339–349.

Rokem, A., & Ahissar, M. (2009). Interactions of cognitive and auditory abilities in congenitally blind individuals. *Neuropsychologica*, *47*, 843–848.

Savolainena, P., Carlson, S., Boldt, R., Neuvonen, T., Hannula, H., Hiltunen, J., et al. (2011). Facilitation of tactile working memory by top-down suppression from prefrontal to primary somatosensory cortex during sensory interference. *Behavioural Brain Research*, *219*, 387–390.

Sebastián, M., Mayas, J., Manso, A. J., & Ballesteros, S. (2008). Working memory for visual and haptic targets: a study using the interference paradigm. In M. Ferre (Ed.), *Lecture notes in computer science* (5024) (pp. 395–399). Berlin, Heidelberg: Springer.

Seemüller, A., Fiehler, F., & Rösler, F. (2011). Unimodal and crossmodal working memory representations of visual and kinesthetic movement trajectories. *Acta Psychologia*, *136*, 52–59.

Seemüller, A., Müller, E. M., & Rösler, F. (2012). EEG-power and -coherence changes in a unimodal and a crossmodal working memory task with visual and kinesthetic stimuli. *International Journal of Psychophysiology*, *83*, 87–95.

Shepard, R. N., & Metzler, J. (1971). Mental rotation of three-dimensional objects. *Science*, *171*, 701–703.

Sinclair, R. J., & Burton, H. (1996). Discrimination of vibrotactile frequencies in a delayed pair comparison task. *Perception and Psychophysics*, *58*, 680–692.

Sperling, G. (1960). The information available in brief visual presentations. *Psychological Monographs*, *74*, 1–29.

Sullivan, E. V., & Turvey, M. T. (1972). Short-term retention of tactile stimulation. *Quarterly Journal of Experimental Psychology*, *24*, 253–261.

Swanson, H. L., & Luxenberg, D. (2009). Short-term memory and working memory in children with blindness: support for a domain general or domain specific system? *Child Neuropsychology*, *15*, 280–294.

Vecchi, T. (1998). Visuo-spatial imagery in congenitally totally blind people. *Memory*, *6*, 91–102.

Vecchi, T., Tinti, C., & Cornoldi, C. (2004). Spatial memory and integration processes in congenital blindness. *NeuroReport*, *15*, 2787–2790.

Wang, L., Bodner, M., & Zhou, Y. D. (2013). Distributed neural networks of tactile working memory. *Journal of Physiology - Paris*, *107*, 452–458.

Wang, L., Li, X., Hsiao, S. S., Bodner, M., Lenz, F., & Zhou, Y. D. (2012). Persistent neuronal firing in primary somatosensory cortex in the absence of working memory of trial-specific features of the sample stimuli in a haptic working memory task. *Journal of Cognitive Neuroscience*, *24*, 664–676.

Watkins, M. J., & Watkins, O. C. (1974). A tactile suffix effect. *Memory and Cognition*, *2*, 176–180.

Wijntjes, M. W. A., van Lienen, T., Verstijnen, I. M., & Kappers, A. M. L. (2008). Look what I have felt: unidentified haptic line drawings are identified after sketching. *Acta Psychologica*, *128*, 255–263.

Withagen, A., Kappers, A. M. L., Vervloed, M. P. J., Knoors, H., & Verhoeven, L. (2013). Short term memory and working memory in blind versus sighted children. *Research in Developmental Disabilities*, *34*, 2161–2172.

Woods, A. T., O'Modhrain, S., & Newell, F. N. (2004). The effect of temporal delay and spatial differences on cross-modal object recognition. *Cognitive, Affective, and Behavioral Neuroscience*, *4*, 260–269.

CHAPTER

19

Human Cortical Representation of Tactile Short-Term Memory for Stimulation Patterns on the Hand: Evidence From Magnetoencephalography

Ulysse Fortier-Gauthier[4], *Christine Lefebvre*[1], *Douglas Cheyne*[2], *Pierre Jolicoeur*[1,3]

[1]Centre de recherche de l'Institut universitaire de gériatrie de Montréal (CRIUGM), Montreal, QC, Canada; [2]Neuromagnetic Imaging Laboratory, Hospital for Sick Children, Toronto, ON, Canada; [3]Experimental Cognitive Science Laboratory, Département de Psychologie, Université de Montréal, Montreal, QC, Canada; [4]Centre de Recherche en Neuropsychologie et Cognition, Université de Montréal, Montreal, QC, Canada

INTRODUCTION

Tactile short-term memory (TSTM) is the capacity that allows us to maintain, for a few seconds, the properties of a mechanical stimulation delivered to the skin. Research findings on short-term memory (STM) suggest the involvement of modality-specific cortical networks (see Grimault et al., 2014; Nolden et al., 2013 for examples in the auditory modality; Grimault et al., 2009; Todd & Marois, 2004, for the visual modality; Hegner, Lutzenberger, Leiberg, & Braun, 2007; Hegner & Saur, 2007; Numminen et al., 2004; Preuschhof, Heekeren, Taskin, Schubert, & Villringer, 2006, for some in the tactile modality; as well as Klingberg, Kawashima, & Roland, 1996, for multimodal tactile, visual, and auditory; and Ricciardi et al., 2006, for visual and tactile). These networks typically consist of sensory-specific areas, cross-modality–associative areas, and frontal control areas, which are often part of the sensory processing pathway. For example, in the tactile modality, sensory information originating from the hand is thought to first reach the cortex at the primary somatosensory area (S1), on the postcentral gyrus contralateral to the stimulated hand. The information then proceeds, bilaterally and nearly simultaneously, to the secondary somatosensory areas (S2), located in the lateral opercula. The secondary somatosensory areas, in turn, project to motor and premotor areas, associative areas in the superior parietal cortex, and the frontal and parahippocampal areas through the insula (Burton, Fabri, & Alloway, 1995).

Early findings regarding the role of areas S1 and S2 has demonstrated that haptic exploration of rods with varying surface texture in a delayed matching to sample task produced sustained modulations in S1 of single neuron firing rates in nonhuman primates (Zhou & Fuster, 1996). However, Burton and Sinclair (2000) reported no sustained modulation of neuronal activity across a number of studies that could be attributed to memory in S1 or S2, beyond brief sensory frequency-selective modulation of single neuron activity for TSTM experiments using vibrotactile stimuli. At the most, frequency-selective activations in S2 faded 500 ms after the end of vibrotactile stimulation in nonhuman primates. Also, the retention of vibrotactile stimuli was not disrupted by a transcranial magnetic stimulation (TMS) pulse applied to contralateral S1, 900 ms after stimulation, or to ipsilateral S1 at any delay (Harris, Miniussi, Harris, & Diamond, 2002) during a task in which (human) participants had to compare the relative frequencies of stimuli sequentially presented to a single finger. Romo and Salinas (2003) reported that the neurons with the most consistent frequency-dependent

Mechanisms of Sensory Working Memory
http://dx.doi.org/10.1016/B978-0-12-811042-3.00019-0

sustained single-cell activity during a flutter discrimination task were located in prefrontal cortex, whereas activity in S1 and S2 were frequency-dependent only for a short period of time after the memory stimulation. Taken together, these results suggest that delay activity in the primary and secondary somatosensory cortex (S1 and S2) might be short-lived and represents transitory persistence of sensory information until a more stable representation can be established.

Neuronal activity supporting STM is expected to be sustained across the whole retention period to provide support for the information until retrieval. Despite the lack of apparent sustained activity past the early stages of retention in S2, reciprocal connectivity to parietal and frontal cortex makes it a plausible candidate for an involvement in STM by providing a bridge between sensory and control areas. Although its role in STM was suggested by linking S2 lesions to impaired tactile recognition of objects' texture and size (Horster & Ettlinger, 1987), more recent imaging studies have confirmed its implication in STM. Bonda, Petrides, and Evans (1996) found lower regional cerebral blood flow (rCBF) in S2 during a task where previously presented haptically explored three-dimensional nonsense patterns had to be recognized. However, the very low temporal resolution of the rCBF response may confound S2 transient activity after haptic palpation or during decision making with retention activity. Indeed, S2 seems more involved in the decision-making stage, comparing the memory and the test stimulations, than in the maintenance of the information in memory (Preuschhof et al., 2006; Romo, Hernández, Zainos, Lemus, & Brody, 2002).

Other regions involved in tactile STM have been revealed in imaging experiments of tactile recognition and information retention in the tactile modality. We first outline the type of study and imaging contrast and then summarize the results. Klingberg et al. (1996) contrasted rCBF levels between a task in which participants decided when a vibrating stylus frequency applied to the index finger was lower than the previous non-baseline stimulation and a control task in which descending frequency transitions had to be detected. Preuschhof et al. (2006) contrasted blood oxygenation level-dependent (BOLD) amplitude during the time interval between two vibrotactile stimuli applied to the index finger between a task in which participants indicated whether the second stimulus was lower or higher in frequency than the first, and another task requiring the subject to indicate the end of the second stimulus in a succession of two stimuli of identical frequency. Numminen et al. (2004) showed significant increases in the BOLD response by contrasting measures from an experimental task in which participants had to tell whether two stimulation triplets delivered in succession to the fingers were identical or different with a control task in which participants had to detect a stimulation on the opposite hand while receiving the same stimulation as in the experimental task. Reed, Klatzky, and Halgren (2005) found increased BOLD amplitudes in a task in which real-life object recognition by haptic palpation was compared with a task requiring them to localize an object with a similar palpation movement. The most recurrent regions in which a higher activation level was observed in the active memory condition were the left (Klingberg et al., 1996; Preuschhof et al., 2006) and right (Numminen et al., 2004) inferior frontal gyrus, the right (Reed et al., 2005) and left (Preuschhof et al., 2006; Reed et al., 2005) superior frontal gyrus, the right (Numminen et al., 2004; Reed et al., 2005) and left cingulate gyrus (Klingberg et al., 1996; Numminen et al., 2004; Reed et al., 2005), the right (Klingberg et al., 1996; Numminen et al., 2004; Reed et al., 2005) and left (Numminen et al., 2004; Preuschhof et al., 2006) inferior parietal lobule, the right (Numminen et al., 2004; Reed et al., 2005) and left premotor (Numminen et al., 2004) cortex, as well as several additional areas, depending on the specific nature of the task and stimuli. These regions appear to be part of higher-order connectivity networks of the brain, usually linked to functions used by multiple modalities.

The results of many studies comparing active memory against passive sensation suggest that the specialized tactile areas (S1 and S2) are coupled with areas that maintain spatial information, from the tactile as well as other modalities. These areas likely support a variety of processes, such as attention and multimodal information integration (Corbetta & Shulman, 2002; Klingberg et al., 1996; Ricciardi et al., 2006). Tactile sensory areas are not necessarily where the tactile information is maintained in TSTM. Tactile memory could also be mediated by stable neuronal representations that are maintained by the supra-modal areas revealed by the studies just reviewed. The recurrent presence of areas associated with spatial processing in short-term memory experiments has led some researchers to propose distinct memory stores for objects (features) and for spatial information (Smith et al., 1995). In fact, the frontal pole and motor and premotor areas have been linked to tactile object recognition, whereas the inferior parietal lobule and precuneus have been linked to tactile identification of objects' spatial location (Reed et al., 2005). Because of the nature of visual, auditory, and tactile stimuli in most STM experimental designs, which contain sensory object features associated to a distinct spatial location, the tasks allow for a strategic or automatic use of both object features and spatial location stores of information. The areas that have been reported for TSTM are possibly maintaining either spatial information or object features information that are of potential use not only for the retention, but also for the encoding as well as the retrieval of the stored tactile information. Therefore, it could be hypothesized that maintaining a tactile feature (texture, frequency, or intensity) that can be retrieved without spatial context requires a STM network different from maintaining only spatial information for which features of each individual stimulation are uninformative.

In this study, we explored the functional correlates of the maintenance of tactile stimulation. We sought to isolate regions that showed sustained activation during the retention interval of a simple same–different memory task.

We assumed that activity in the cortical regions directly involved in the maintenance of tactile representations would be proportionally modulated by increases in the quantity of information maintained. A similar approach has been successfully used for visual (Todd & Marois, 2004; Vogel & Machizawa, 2004; Grimault, et al., 2009) and auditory modalities (Lefebvre et al., 2013; Grimault, et al., 2014; Nolden et al., 2013). We used magnetoencephalography (MEG) to record brain activity during the retention interval of a tactile memory task in which we used three levels of memory load (0, 2, and 4). The increasing complexity of a pattern of pressure points delivered to the hand provided a manipulation of memory load while maintaining tactile features (frequency, intensity, and movement) constant. The use of stimulations with the same tactile features reduces the ambiguity provided by the information to be retained in TSTM, leaving the localization on the skin as the only useful information. We estimated effective memory load with Pashler's K value (Pashler, 1988; Rouder, Morey, Morey, & Cowan, 2011), which estimates the number of items successfully stored in STM (while correcting for guessing and response bias; see Pashler, 1988).

We used eight stimulators placed on one hand (left or right, counterbalanced across blocks). The memory load condition we call '0' was achieved by stimulating all eight locations because the resulting pattern could be considered a uniform, un-patterned stimulus. This led to a useful dissociation between the intensity of sensory stimulation and the effective memory load reflecting pattern complexity. The parametric manipulation of memory load allowed us to identify a small number of cortical regions that had sustained activity levels that changed with K, suggesting a likely involvement in the retention of tactile information. Because of the high temporal resolution of MEG, we could be confident that the activity we focused on reflected the retention of information, rather than encoding or retrieval, because we analyzed patterns of sustained activity long after the initial evoked responses, and well before the test stimuli were presented. The analysis of activity levels before the presentation of test stimuli ensures that the analyzed activity do not encompass retrieval, comparison, decision, and response-related activity. Consequently, this technique enabled us to explore regions involved in the retention of tactile information with a greater specificity than in previous research.

We also took advantage of the high temporal resolution of MEG to explore the activity during the early period of retention using the magnetic waveform components to guide us in selecting two early windows of interest to compare areas activated during these time windows with the ones that are activated later. The ephemeral participation of somatosensory areas to TSTM (Burton & Sinclair, 2000; Harris et al., 2002) seems to indicate a dynamic of tactile information flow from early somatosensory areas to other areas that provide longer-lasting retention for TSTM. This transformation from sensory input to sensory memory could help in understanding similar mechanisms in other modalities that appear to take place on a shorter time scale.

METHOD

Participants

We tested 14 participants, 7 of them women, aged 20–32 years (mean, 25.4 years), one of whom was left-handed. They were neurologically normal and were not taking neurologically active medication. All participants signed an informed consent form following the Université de Montréal Ethics Committee guidelines.

Material

We measured magnetic fields using a CTF275 whole-head magnetoencephalograph at a sampling rate of 1200 Hz, with synthetic third-order gradiometer noise correction (Vrba & Robinson, 2001) in a magnetically shielded testing room. In addition to MEG data, electroencephalogram data were collected from bipolar electrodes positioned at the outer canthi of the eyes, for the horizontal electro-oculogram (HEOG), and above and below the left eye for the vertical electro-oculogram (VEOG). Visual feedback was back-projected onto a translucent screen by a data projector located outside the shielded room. The tactile stimuli were delivered with a magnetically neutral pneumatic stimulator (Mertens & Lütkenhöner, 2000) with eight independent finger clips placed on one hand at a time during testing. The stimulated hand was changed after completing half the blocks. Clips were affixed to the first and second phalanges of each finger except the thumb. Stimulation was produced by the rise of a supple membrane displaced by a brief change in air pressure (100-ms stimulation). To keep from assuming a threshold of detection for the tactile stimulations, all timings and waveforms we report are time-locked to the moment the stimulation command was sent to the pneumatic stimulator, but because of the mechanical and pneumatic mechanisms, the actual stimulation was delivered with a delay of approximately 50 ms. Although the tactile stimulator was located outside the shielded room, the stimulator could be heard from inside. Because the stimulator sound did not provide pattern information for the memory task, no earplugs were used during testing.

Experimental Procedure

Participants were tested in a supine position in the MEG with the stimulated hand resting at the side. They provided responses with the non-stimulated hand (same with index finger/different with middle finger). Each trial began with a 500-ms delay period with a 100-ms jitter (400–600m). As illustrated in Figure 1, each trial had two tactile stimulation patterns, each presented for 100ms, consisting of 2, 4, or 8 simultaneously stimulated locations (stimulation load) selected at random from the eight possible locations. This pattern complexity accounted for memory load values of 2, 4, and 0, respectively (with only eight possible locations, memory and test were always identical and therefore no tactile pattern needed to be maintained). The first (memory) pattern was followed by an empty inter-stimulus interval (ISI) of 1800ms. After this retention interval, a test stimulation was presented. In half of the trials, this pattern was identical to the memory pattern; in the other half, it differed by one location (excluding the 8 location pattern, which was always identical). The task was to determine whether the two stimulation patterns were identical or different. The response had to be given in under 1900ms after the offset of the test stimulation. Finally, accuracy feedback was presented after the response for 1900ms. A minimal delay of 400ms after the second stimulation was imposed before the feedback to reduce blink-related artifacts. Each experimental block contained an equal number of the six trial types (memory match: same versus different; memory load: 2, 4, or 0) in pseudorandom order. A practice block of 18 trials preceded five blocks of experimental trials (84 per block) for each stimulated hand. The order of hand stimulation was alternated between participants. Thus, each participant performed a total of 840 experimental trials.

Analyses

Magnetic field data received a third-order gradiometer noise correction and was baseline corrected based on the mean during the 200ms before the presentation of the memory stimulation. Eye blinks and eye movements were excluded manually based on visual inspection of VEOG and HEOG channels. The magnetic field data were co-registered to the participant's anatomical magnetic resonance imaging (T1) brain scan using three fiducial coils placed at the nasion and the left and right ear, as well as points digitized from the participant scalp surface with a Polhemus Fastrak digitizer.

We identified four windows of interest for our source localization analyses. As stated previously, our main goal was to identify activity related to the maintenance of tactile stimulation. We therefore first chose the period going from 1000 to 1600ms after initial perceptual processing was finished and before processing of the second stimulation. We also identified three other windows of interest after careful examination of the MEG waveforms (Figure 2): the first large sensory peak, between 70 and 90ms after the memory pattern presentation, then the time window from 325 to 375ms, which is covering the first peak of a large component, and finally the window encompassing the largest activity before the sustained state, between 420 and 690ms. Source localization was performed using the event-related synthetic aperture magnetometry (erSAM) algorithm (Cheyne, Bakhtazad, & Gaetz, 2006). Maps of source activation

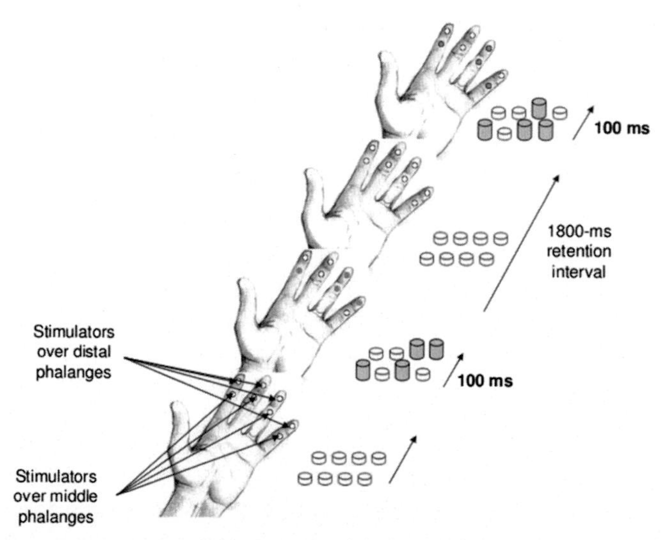

FIGURE 1 Illustration of the time course of a single trial during the experimentation. Participants were asked to memorize a memory pattern presented to a single hand for 100ms. After a retention interval (ISI) of 1800ms a test pattern was presented to the same hand. Participants had to tell whether both patterns were identical or different with the currently non-stimulated hand.

power in pseudo-Z units were produced for each memory load × stimulated hand × participant combination for the retention period between 1000 and 1600 ms using the full range of frequencies at a resolution of 2 mm and a 200-ms pre-stimulus baseline. We also produced maps for three other time windows preceding the sustained activity window, but this time using stimulation load as regressor instead of memory load. Frequency bands SAM for signals in the alpha (8–12 Hz), beta (18–24 Hz), and gamma (40–80 Hz) bands were also performed on the 1000–1600 ms retention interval. All localization maps were imported to the Talairach stereotaxic space. Individual maps of activity taking place during

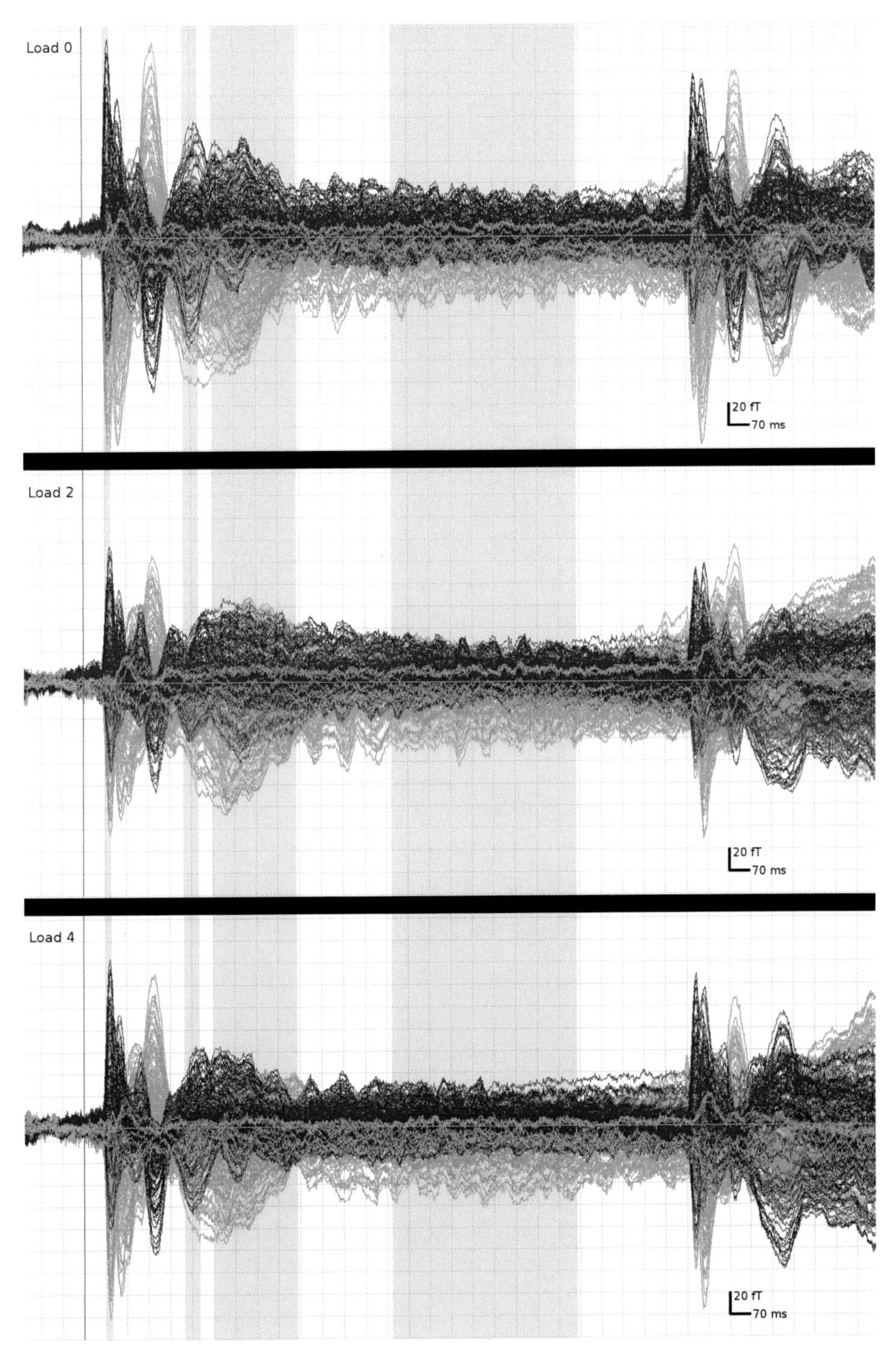

FIGURE 2 Average magnetic waveforms for the three memory load conditions. The time windows of analysis are displayed from 325 to 375 ms, 420 to 690 ms and 1000 to 1600 ms. We can see transient activity in the first two time windows that corresponds to increased activity with more stimulated locations. In the later part of the magnetic waveform before the test stimuli sensory input, we see sustained activity over baseline levels that corresponds to increased activity with memory load instead of stimulation load (i.e., number of stimulated locations).

the 1000–1600 ms retention interval were analyzed with a general linear model (GLM) and regressed against the individual K value of each participant for each memory load. Maps of early post-stimulation activity (first sensory peak, and first and second peak of the larger activation component) were analyzed with a GLM and regressed against the stimulation load. Participant, gender, and side of stimulated hand regressors were used to remove variability from the resulting localizations. We set the significance level at $p<.005$, uncorrected, to compensate for false positives resulting from the number of voxels tested while leaving some space for exploration. Figures are presented in neurological referential (right is right).

RESULTS AND DISCUSSION

Behavioral Results

We submitted the accuracy for each memory load level for each observer to a repeated-measures analysis of variance with memory load as the only factor. As expected, participants performed better for lower than for higher memory load with a load main effect on accuracy ($Acc_0=98\%$, $\sigma=2.01$; $Acc_2=86\%$, $\sigma=8.08$; $Acc_4=70\%$, $\sigma=6.95$; $F(2, 26)=144.69$, $p<.00001$), corroborating our assumption that stimulating all eight locations produced the easiest condition in the memory task. We retested the difference specifically between the load 2 and load 4 conditions (uncorrected $F(1, 13)=162.84$, $p<.00001$), which confirmed that accuracy was higher at load 2 than at load 4. We performed a similar analysis for response times. We found a main effect of load on the response times ($RT_0=685$ ms, $\sigma=133.96$; $RT_2=823$ ms. $\sigma=133.31$; $RT_4=870$ ms, $\sigma=154$; $F(2, 26)=61.72$, $p<.00001$). As expected, the fastest response times were found in the control condition (memory load 0), which did not require the memorization of information. The response times were also different between load 2 and load 4 ($F(1, 13)=26.47$, $p<.0002$).

The accuracy scores in the task were used to estimate individual Pashler K values, which were used in source-localization analyses. The mean of these values were: $K_0=0$, $\sigma=0$; $K_2=1.72$, $\sigma=0.24$; $K_4=2.65$, $\sigma=0.57$.

Localization of MEG Sources

Stimulation-Related Activity

Before looking into the retention period localization maps, we confirmed that our localizations were trustworthy by localizing the first tactile sensory peak from the magnetic waveforms between 70 and 90 ms after presentation of the memory pattern. In the source-localization map we looked for voxels in which activity increased linearly with the number of stimulated locations (2, 4, and 8), and which also interacted with the hand of stimulation. The expected pattern (given the coding scheme in the GLM analysis) was a positive correlation voxels for right-hand stimulations (shown in red) and a negative correlation for left-hand stimulations (shown in blue) in Figure 3. As expected, source localization revealed enhanced activity in the right postcentral gyrus for left-hand stimulations with increased stimulation load and enhanced activity in the left postcentral gyrus when the right hand was stimulated. These results confirm the general adequacy of our source-localization analysis approach.

Retention Interval Activity Source Localization

The source localization map of cortical areas likely involved in the maintenance of information in TSTM can be seen in Figure 4. In this analysis we looked for voxels with mean activity levels that increased with the estimated K values in a window of 1000–1600 ms, regardless of stimulated hand. The map in Figure 4 used a threshold of $p<.005$, uncorrected. However, the right premotor cortex and the frontal pole show sustained activity that fit individual K patterns during the retention period (at $p<.001$, threshold level, uncorrected). The premotor cortex has been linked to planning and execution of behaviorally relevant complex movements, such as reaching and grasping with the arm and hand (Matelli & Luppino, 2001; Filimon, Nelson, Hagler, & Sereno, 2007), as well as to finger tapping (Horenstein, Lowe, Koenig, & Phillips, 2009), and motor imagery (Hanakawa et al., 2002). Furthermore, there is good evidence that this general region receives input from primary sensory neurons, and it is thus plausible that the region could participate in retention of patterns of stimulation. Meanwhile, the frontal pole, as part of the default-mode network, has been tied to introspection, mind wandering, and monitoring of the environment (Buckner, Andrews-Hanna, & Schacter, 2008; Kelley et al., 2002; Mason et al., 2007). Interestingly, the frontal pole, which is sometimes deactivated for task-related activity, here was increasingly active with higher load. This could reflect that during retention, contrary to most event-related tasks in which external stimulus processing is important, participants wanted to reduce processing of external stimuli and wanted to focus attention on the internal representation of the memory template.

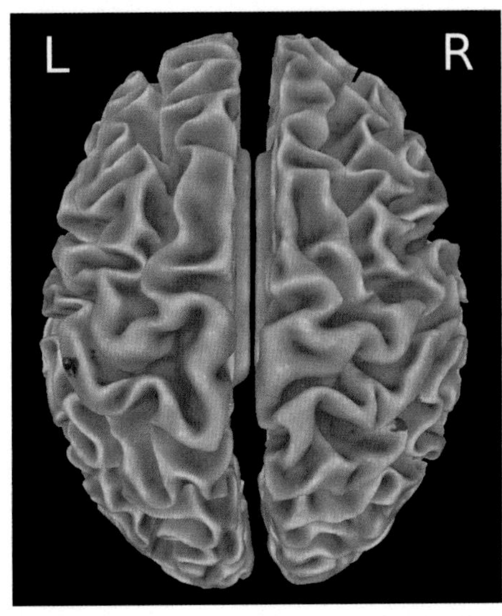

FIGURE 3 Map of areas that increase in activity with increase in stimulation load in the memory pattern during the period 70–90 ms after the memory pattern presentation. The results show the interaction of stimulation load by hand of stimulation, in which we would expect a positive correlation (red area) for a right-hand stimulation and a negative correlation (blue area) for a left-hand stimulation. This map represents areas that are activated by the initial sensory inputs. The map uses a $p < .001$ threshold.

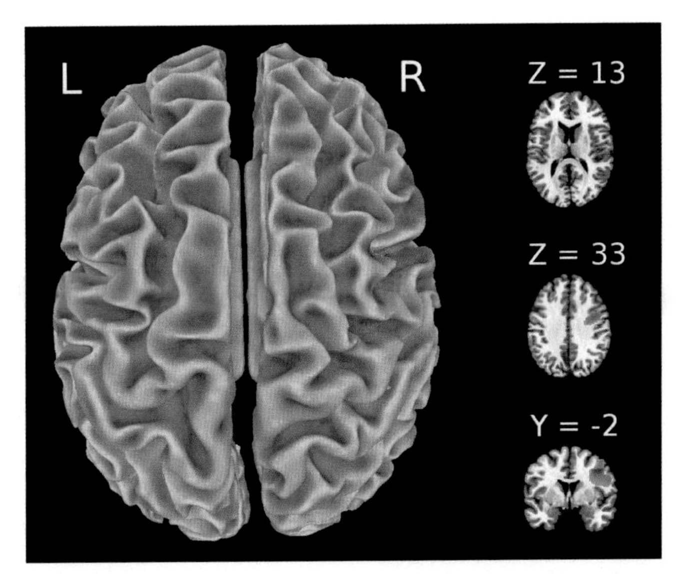

FIGURE 4 Map of areas that vary with an increase in the K value during the period between 1000 and 1600 ms after the memory pattern presentation. To the left, the three-dimensional map shows a top-down view of the brain with the front up. The frontal pole and right premotor activity increase with a larger value of K whereas the right and left middle frontal gyrus show reduced activity. The map uses a $p < .005$ threshold.

An alternative view would be that an egocentric spatial referential is sufficient for TSTM spatial representations and this enables participants to bias their cognitive resources toward that referential and away from external (visually based) referentials.

In addition to increases in the frontal pole and right premotor cortex, we found a reduction in activity with increasing K values at mirror areas on left and right middle frontal gyrus (see Figure 4). Involvement of these areas in the maintenance of tactile information, in combination with involvement of the frontal pole, could be a way to bias the attention toward an egocentric spatial and sensory referential system. These areas are close to the frontal eye field, which has a role in directing visuospatial attention and could be involved in visuospatial memory (Pierrot-Deseilligny, Műri, Rivaud-Pechoux, Gaymard, & Ploner, 2002). Also, disruption of the posterior dorsolateral prefrontal cortex (DLPFC) with TMS pulses has been reported to disrupt bimodal (visual-auditory) task performances (Johnson, Strafella, & Zatorre, 2007). Posterior DLPFC deactivation during TSTM maintenance could perhaps reduce interference from stimuli in the same or in other modalities. Because both areas also produce a significant effect with

TABLE 1 Clusters of Activity Related to Pashler's K Value during the Retention Period (1000–1600 ms) Ordered by Size

Location	Brodmann area	Talairach coordinates			Cluster size	
		X	Y	Z	*(at p<.005)*	*p* of peak
Right precentral gyrus	6	39	−2	37	1,328	<.0002
Frontal pole	10	1	64	13	367	<.0003
Left middle frontal gyrus	8	−27	36	45	110	<.002
Right middle frontal gyrus	8	19	26	33	88	<.002

stimulus load as the regressor, these two areas could instead be related to stimulus load, and index the number of stimulated locations during the retention period. These activations could provide, in combination with the right premotor area (which reflects K), a dual-memory representation of the tactile stimulation. Area details are listed in Table 1.

Localization of Oscillatory Activity in Different Frequency Bands

We also looked at activity in different frequency bands during the later retention interval for induced activity that would correlate with K values. We found no such significant activity for the alpha (8–12 Hz) or gamma band (40–80 Hz) but we did find areas that increased in activity with K value in the beta band (18–24 Hz) (see Figure 5 and Table 2). One of these areas extended from the right premotor cortex to the right anterior cingulate passing through the right cingulate gyrus. The other areas of interest were in the left precentral gyrus, left cerebellum, and right postcentral gyrus. These activations were localized using regular SAM and indicate activity for the beta band that was not phase-locked to the memory stimulation; hence, this analysis offers a different view of the ongoing activity during retention, which is otherwise lost in event-related averaging. The induced activity we see in Figure 5 seems to show that there is also activity in the left premotor cortex and cerebellum during retention, although this activity was not revealed as a sustained low-frequency equivalent in our erSAM maps, which highlight sustained event-related responses. This suggests these areas do not have the same role as the right premotor cortex during the retention interval. Induced beta band modulation has been argued to reflect tactile sensory gating, which has been linked to both stimulus-driven (Kisley & Cornwell, 2006) and attention-driven salience (Buchholz, Jensen, & Medendorp, 2014). The sensory gating effect (i.e., enhanced excitability) is accompanied by reduced beta band amplitude. However, sensory gating has a short duration and the beta reduction comes back to baseline before the localized retention interval. Instead, we observe in Figure 5 a sustained increase in beta band activity that correlates with K. Incidentally, parametric modulation of induced beta band activity was reported in the right frontal hemisphere during delayed match-to-sample task with vibrotactile stimulations (Spitzer, Wacker, & Blankenburg, 2010). Although the frequency code reported for the retention of vibrotactile information could not code for spatial patterns, somatotopic organization of the frontal areas could still retain the spatial information. Alternatively, the right anterior cingulate has been identified as part of a network that can extract and monitor sequences (Huettel, Mack, & McCarthy, 2002). The induced activity increasing with K value in the beta band of the right anterior cingulate, in combination with both right and left premotor areas, suggests that the memory pattern could be recoded and maintained as a sequence of motor plans identifying each location by the participants to be rehearsed and monitored.

Source Localization of Stimulation-Related Activity

Taking advantage of the temporal resolution afforded by MEG, we also explored two time windows preceding our retention interval target window. These time windows belong to the retention period, but the magnetic field activity we observed as well as previous research (Harris et al., 2002) suggests that something different is happening before about 1000 ms. We expected some form of recoding to take place from low-level sensory codes to representations of tactile patterns that require integration over several locations. The first time window, ranging from 325 to 375 ms after memory stimulation, corresponded to a large component in the magnetic waveforms after the initial sensory resolution and visible for each load. The second time window, from 420 to 690 ms, covered a period of increased activity before sustained activity of the retention period (from 700 ms onward). Areas active during both time windows tended to show negative regressions with K value, which suggests that during these periods activity was mainly driven by the absolute number of stimulated points, the stimulation load, which was respectively 8, 2, and 4 for memory loads 0, 2, and 4. For this reason, we present here (Figure 6 and Figure 7) the maps of both time windows' activity regressed against the stimulation load (2, 4, and 8). The maps show activity that is clearly different from the later retention interval. During the 325- to 375-ms time window, three regions regrouped most of the clusters (See Table 3 for details), all

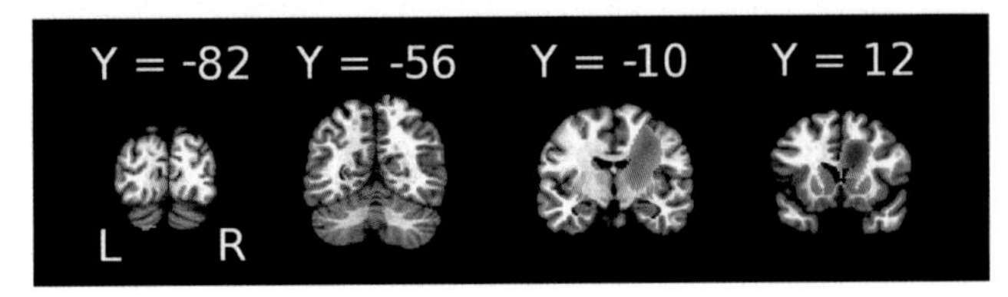

FIGURE 5 Beta band SAM map reveals some areas that have induced activity that increases with a larger K value. The main area extends from the right premotor and follow the right cingulate gyrus to the anterior cingulate. The map uses a $p<.005$ threshold.

TABLE 2 Clusters of Activity in the Beta Band Related to Pashler's K Value During the Retention Period (1000–1600 ms)

Location	Brodmann area	Talairach coordinates			Cluster size	*p* of peak
		X	Y	Z	(at $p<.005$)	
Right anterior cingulate	24	13	12	27	5,289	<.0002
Left precentral gyrus	6	−31	−10	59	123	<.0008
Left cerebellum	–	−15	−82	−39	41	<.003
Right postcentral gyrus	7	9	−56	67	25	<.004

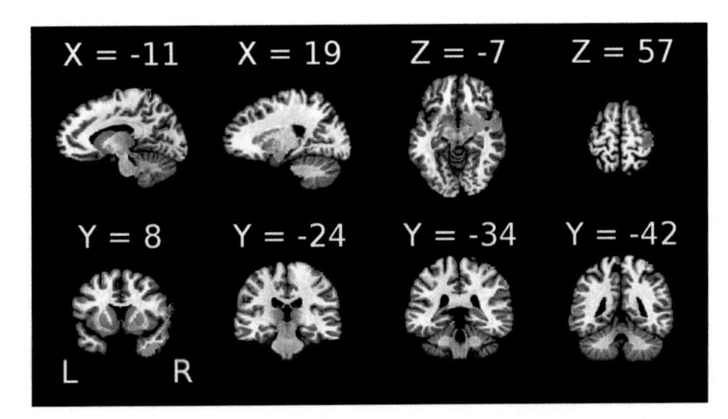

FIGURE 6 Map of areas increasing in activity with stimulation load in the memory pattern during the period between 325 and 375 ms after the memory pattern presentation. The map of early retention shows information contained in the control condition (8 stimulation points that are always identical) participating in brain activation whereas it holds no useful information for completing the task. The map uses a $p<.0005$ threshold.

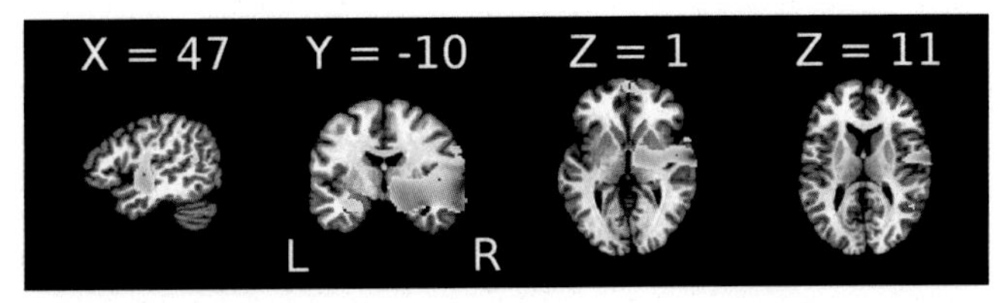

FIGURE 7 Map of areas increasing in activity with stimulation load in the memory pattern during the period between 420 and 690 ms after the memory pattern presentation. The map uses a $p<.005$ threshold.

increasing in activity with increasing number of location stimulated: the right and left S1, right and left parietal areas and a large region including the right lateral globus pallidum, the parahippocampal gyrus, and the superior temporal gyrus, and the inferior frontal gyrus. The time window is fairly early during retention and it is possible (if not likely) that ongoing activity in S1 continues to represent the strength of sensory stimulation delivered in the stimulus, although the presence of parietal activity suggests that tactile information has reached a higher level of integration. Parietal areas lesions have been reported to impair reaching movements as well as tactile discrimination (Ettlinger & Kalsbeck, 1962). The parietal area could provide a reference space for action relative to locations on the body that received tactile input, including cross-referencing with the visual and auditory systems (Jackson, Newport, Husain, Harvey, & Hindle, 2000). Likewise, the area spreading from the right superior temporal to the globus pallidus could be part of an egocentric spatial attention network (Karnath, Ferber, & Himmelbach, 2001; Karnath, Himmelbach, &

TABLE 3 Clusters Related to the Stimulation Load (2, 4, and 8) of Memory Stimulation During the 325- to 375-ms Period

Location	Brodmann area	Talairach coordinates X	Y	Z	Cluster size (at $p < .0005$)	p of peak
Right superior temporal gyrus[b]	38	61	8	−7	373[a]	<.000003
Right parahippocampal gyrus/ amygdala[b]	–	29	−4	−17	209[a]	<.00004
Right precentral gyrus[b]	44	61	10	9	83[a]	<.000009
Right superior temporal gyrus[b]	38	35	6	−41	59[a]	<.0001
Right lateral globus pallidus[b]	–	13	6	−1	21[a]	<.0002
Right precentral gyrus	4	37	−22	61	365	<.00002
Left postcentral gyrus[b]	7	−11	−50	67	180	<.000003
Left postcentral gyrus	3	−41	−24	−57	152	<.000005
Left superior temporal gyrus	22	−59	−42	7	74	<.00009
Left culmen[b]	–	−11	−28	−11	60	<.00007
Left cingulate gyrus[b]	31	−17	−22	45	56	<.0002
Right inferior parietal lobule[b]	40	61	−34	37	41	<.0002
Right superior parietal lobule[b]	7	19	−58	61	28	<.0002
Left superior frontal gyrus	6	−15	−10	67	23	<.0003

[a]*The first large cluster was broken into five subclusters at p < .0002.*
[b]*These clusters have similar but inverted polarity clusters when using K value for regression maps at p < .005.*

TABLE 4 Clusters Related to Stimulation Load (2, 4, and 8) in Memory Stimulation During the 420- to 690-ms Period

Location	Brodmann area	Talairach coordinates X	Y	Z	Cluster size (at $p < .005$)	p of peak
Right superior temporal gyrus[b]	21	65	−8	1	446[a]	<.0000004
Right parahippocampal gyrus[b]	–	31	−10	−13	504[a]	<.00002
Left parahippocampal gyrus[b]	–	−29	−10	−13	408	<.002
Frontal pole[b]	10	3	56	−1	367	<.002
Right superior temporal gyrus[b]	22	61	10	−3	53	<.0006
Left subcallosal gyrus	–	−11	8	−11	50	<.004
Right superior temporal gyrus[b]	39	49	−50	15	44	<.002
Left inferior frontal gyrus	46	−53	36	11	17	<.0009
Right fusiform gyrus[b]	19	47	−68	−13	12	<.0007

[a]*The first large cluster was broken into two subclusters at p < .0002.*
[b]*These clusters have similar but inverted polarity clusters when using K value for regression maps at p < .005.*

Rorden, 2002; Postle & D'Esposito, 2003) used to represent the encoded tactile pattern. However, because we omitted to plug the ears of the participants, auditory input from the pneumatic stimulation could contribute to early temporal activations (Nolden et al., 2013).

When looking at the 420- to 690-ms time window (Figure 7, Table 4), we saw a large cluster in the right temporal lobe that was present in the earlier time window (325–375 ms) but activation in S1 and parietal activations were no longer present. This is consistent with the sensor traces (Figure 2) in which activity in the later time window (420–690 ms) can be seen to begin in the earlier window (325–375 ms). Among the clusters of activity increasing with the stimulation load in the 420- to 690-ms window, the right fusiform gyrus is particularly interesting. The right fusiform gyrus, which is usually known as a specialized visual area involved in face processing, was also activated in late-blind participants during tactile discrimination tasks as well as visual imagery (Goyal, Hansen, & Blakemore, 2006). Activation of this area in the current work suggests that the region may be generally involved in encoding spatial patterns encoded in different sensory modalities. As reported for ventral areas, traditionally specialized areas linked to the visual system—processing of faces by the fusiform face area (Kanwisher, McDermott, & Chun, 1997; Gauthier, Tarr, Anderson, Skudlarski, and Gore, 1999; O'Craven & Kanwisher, 2000; Goyal et al., 2006), words by the visual word form area (Nobre, Allison & McCarthy, 1994; Cohen et al., 2000; Price & Devlin, 2003), and scenes by the parahippocampal place area (Epstein & Kanwisher, 1998; O'Craven & Kanwisher, 2000)—the area may serve a more general purpose than visual processing of faces and help recognize spatial patterns that are crucial to the identification of an object.

Another location of interest was the frontal pole, which was not activated in the earlier window (325–375 ms) but was now more active with increasing stimulation load. If we apply the same reasoning as for the 1000- to 1600-ms window, we may interpret this as a moment in the encoding process of the tactile information that no longer needs sensory input, so the frontal pole could work to reduce outside interference. The fact that the frontal pole follows the stimulation load (number of stimuli) instead of K (number of patterns retained) suggests that load 0 superfluous information is still represented strongly at this time. The frontal pole would have to be increasingly activated to filter out outside interference the more information is held in memory at this moment. Despite the overall tendency of increased activity for higher stimulation load, one cluster displayed reduced activity for increased stimulation load in the memory stimulation. Although the clusters in left inferior frontal gyri do not follow the stimulation load, it does not follow K, either (does not appear on the K regression map at $p < .005$). It appears to be an area that actually has a negative relation to the sensory load.

CONCLUSION

We explored brain areas that could be involved in the short-term retention of tactile information using MEG source localization. To isolate these areas from other unrelated areas, we assumed areas participating in the active retention of information would be more active when more information had to be maintained in memory. By regressing our localization maps to individual participant K values, we revealed areas that appear to carry tactile information during different phases of the retention interval. We underline the spatial nature of the stimuli because previous research report TSTM activity to be related to feature (e.g., frequency) retention (Klingberg et al., 1996; Preuschhof et al., 2006; Harris et al., 2002; Romo et al., 2002; Romo & Salinas, 2003) rather than spatial patterns, which could use an altogether different network.

One element that stood out from our results when we compared the two early retention time windows with the later one is that until 690 ms the level of activity generally increased with the number of stimulated points (stimulation load) on the hand from 2 to 8, reflecting total stimulation rather than the complexity of the pattern to be remembered, which should have been minimal at eight locations. This suggests that sensory information had not yet been recoded into a spatial pattern for retention and that the representation at that time still coded sensory intensity. This interpretation converges nicely with previous findings (Harris et al., 2002) that encoding is still being performed or at least that sensory areas are still being accessed to recall tactile information for about 700 ms after stimulation.

The initial two time windows revealed activity that could point to an egocentric spatial representation of the memory stimulation. The activity in bilateral S1 ceased to follow the stimulation load early whereas the right temporal lobe (and connected areas) were active until 690 ms, raising the possibility that the right temporal lobe may be acting as a memory buffer and a possible node in the circuit that recodes tactile sensations into a more abstract pattern code. Without hard evidence about the actual code used to represent the tactile information, we rely on the temporal evolution of activity. The initial activity in somatosensory areas ceases early to increase with stimulation load whereas superior temporal gyrus activity, reported as coding for egocentric spatial awareness (Karnath et al., 2001), carry over until 690 ms. Also, our stimuli applied pressure simultaneously over different locations in a similar way as a visual object for vision, possibly giving the stimulation a shape (a global spatial presence) rather than a sum of individual spatial locations. The hierarchical structures of sensory areas provide areas to represent objects without the detailed sensoritopic maps provided by primary sensory areas. Recoding information to higher hierarchical areas might come with a tradeoff between spatial resolution and sensory information complexity, but may also provide some protection against interference (Postle & Hamidi, 2007). Perhaps the activations in the right fusiform gyrus could be involved in the recoding of individual locations into a unified pattern code, leading to a memory code that can be integrated with spatial representations derived from other senses. Once so encoded, these representations can perhaps be maintained without relying on input from primary sensory regions, because S1 was no longer strongly active, thereby insulating tactile pattern memory from interference from ongoing inputs. In any case, some form of contextual binding by temporal structures would be in line with STM across modalities (Jonides et al., 2008).

The main retention window spanning the period from 1000 to 1600ms revealed two main areas positively related to tactile memory capacity (K). The right premotor area seemed to be the main area holding TSTM information during retention. The information was perhaps recoded in representations that could be used to enable motor actions involving the stimulated locations. These mental representations would then be easier to retain without interference (Klingberg & Roland, 1997) from incoming tactile stimulations because they would use different areas that are not shared with tactile sensory processes. The retention of tactile spatial information in the premotor cortex instead of other parietal areas, commonly seen in STM localization for other modalities, is possibly the result of the greater specialized use of the retained tactile information for motor planning above other cognitive uses. We rarely pay attention to tactile information for other everyday use than interacting with objects, whereas auditory and visual information is often recoded verbally for other mental manipulations. In counterpart, we can note that Braille readers use the parietal-occipital cortex to process and encode the tactile information meant for conscious lexical use (Burton et al., 2002). Nevertheless, the premotor cortex has been linked to the "what" stream (Reed et al., 2005), supporting the identification of tactile objects, which is consistent with the retention of the relative spatial locations of stimulations on the hand as a tactile object. Simultaneous pressure points on the ventral face of the fingers would typically be internally interpreted as defining the shape of an object held in hand. Although our task did not explicitly prevent verbal rehearsal of the information, the complexity and short time frame made it highly impractical as a strategy, and research with articulatory suppression confirmed that TSTM for stimuli similar to ours does not rely on verbal recoding (Auvray, Gallace, & Spence, 2011).

The frontal pole, which was also active in the 420- to 690-ms retention window, may have a different function during TSTM. During the early retention window, activity in the frontal pole increased with stimulation load whereas during the later retention period it instead increased with K. This suggests that the frontal pole takes on a flexible role that follows the evolving processing demands in different phases of the retention period. The role of the frontal pole reported in the literature (Buckner et al., 2008; Kelley et al., 2002; Mason et al., 2007) as part of introspection control could be compatible to the varying amount of effort required during the time course of the retention period to protect the internal tactile representations from external interference.

We observed (Figure 4) both lateral and bilateral activations later during retention (1000–1600ms), which could be explained by the different functions performed by the activated areas. The two mirror areas, negatively related to the K value, in right and left posterior DLPFC, could be related to an increased effort with increasing K value to reduce interferences from outside sources in other modalities. These regions would be inhibited to reduce information flow through them, thereby suppressing inputs from other sensory systems. This is particularly relevant because participants had to maintain fixation on a screen during the retention interval. When we contrast that activity with the one we see in the right premotor cortex, we wonder what led to the lateralization of the premotor activity, because both hands were stimulated (in different blocks). Although the memory pattern stimulation was applied to each hand in turn, the mental code that contributed to the retention was not bound to a particular side. The participant could recode the information into a pattern in the hemisphere ipsilateral to the dominant hand to protect the pattern representation from recurrent sensory and motor inputs. The fact that only one of our 14 participants was left-handed prevented us from exploring this conjecture. Another explanation could be that there is a hemispheric specialization for the nonverbal representations used during retention, whether tactile, motor, spatial (Serrien, Ivry, & Swinnen, 2006), or otherwise, mirroring the one for language and the selection of actions (Schluter, Krams, Rushworth, & Passingham, 2001).

As a side note, we found the capacity of participants to retain more information for a pattern of four locations compared with a pattern of two locations to be relatively weak. In visual and auditory modality the number of items retained before reaching a plateau is usually around three or four items (Vogel, Woodman, & Luck, 2001). We tend to believe that the reduced performance of the participants was most likely the result of the proximity of the stimulations on the fingers. The more locations one adds, the more likely that two of them will be on the same finger, which may result in stimuli that are more difficult to individuate. Because of the intensity of each individual stimulation and the size of the pneumatic stimulators, there may have been some overlap in the neurons' receptive fields stimulated by each location on a single finger (Johansson, 1978). This could reduce the performance to compare patterns with the same fingers stimulated during both memory and test stimulation, but with the test stimulation having the second stimulation on a finger shifted to another finger that was already stimulated.

The results obtained suggest that our approach of load modulation was successful in isolating likely areas implicated during TSTM. Active areas revealed with this approach during early and late retention periods have been linked in the literature to tactile memory tasks or tactile recognition tasks. The change of the relation between the cortical activity level and the K value after 1000ms supports the hypothesis of a recoding of the tactile information between 690 and 1000ms after the memory stimulation. Despite activity related to the number of locations stimulated in the early time window, primary somatosensory cortex do not appear to be actively holding TSTM information during retention when we look at activity varying with load at later time windows. The exact involvement and functional role of the areas active during memory retention remain speculative, but add to the growing evidence relating brain activity to TSTM.

References

Auvray, M., Gallace, A., & Spence, C. (2011). Tactile short-term memory for stimuli presented on the fingertips and across the rest of the body surface. *Attention, perception, & psychophysics*, *73*, 1227–1241.

Bonda, E., Petrides, M., & Evans, A. (1996). Neural systems for tactual memories. *Journal of Neurophysiology*, *75*, 1730–1737.

Buchholz, V. N., Jensen, O., & Medendorp, W. P. (2014). Different roles of alpha and beta band oscillations in anticipatory sensorimotor gating. *Frontiers in Human Neuroscience*, *8* (446). http://dx.doi.org/10.3389/fnhum.2014.00446.

Buckner, R. L., Andrews-Hanna, J. R., & Schacter, D. L. (2008). The brain's default network: anatomy, function, and relevance to disease. *Annals of the New York Academy of Sciences*, *1124*, 1–38.

Burton, H., Fabri, M., & Alloway, K. (1995). Cortical areas within the lateral sulcus connected to cutaneous representations in areas 3b and 1: a revised interpretation of the second somatosensory area in macaque monkeys. *The Journal of Comparative Neurology*, *355*, 539–562.

Burton, H., & Sinclair, R. J. (2000). Attending to and remembering tactile stimuli: a review of brain imaging data and single-neuron responses. *Journal of Clinical Neurophysiology*, *17*, 575–591.

Burton, H., Snyder, A. Z., Conturo, T. E., Akbudak, E., Ollinger, J. M., & Raichle, M. E. (2002). Adaptive changes in early and late blind: a fMRI study of Braille reading. *Journal of Neurophysiology*, *87*, 589–607.

Cheyne, D., Bakhtazad, L., & Gaetz, W. (2006). Spatiotemporal mapping of cortical activity accompanying voluntary movements using an event-related beamforming approach. *Human Brain Mapping*, *27*, 213–229.

Cohen, L., Dehaene, S., Naccache, L., Lehéricy, S., Dehaene-Lambertz, G., Hénaff, M. A., et al. (2000). The visual word form area Spatial and temporal characterization of an initial stage of reading in normal subjects and posterior split-brain patients. *Brain*, *123*, 291–307.

Corbetta, M., & Shulman, G. L. (2002). Controls of goal-directed and stimulus-driven attention in the brain. *Nature Neuroscience*, *3*, 201–215.

Epstein, R., & Kanwisher, N. (1998). A cortical representation of the local visual environment. *Nature*, *392*, 598–601.

Ettlinger, G., & Kalsbeck, J. E. (1962). Changes in tactile discrimination and in visual reaching after successive and simultaneous bilateral posterior parietal ablations in the monkey. *Journal of Neurology, Neurosurgery & Psychiatry*, *25*, 256–268.

Filimon, F., Nelson, J. D., Hagler, D. J., & Sereno, M. L. (2007). Human cortical representations for reaching: mirror neurons for execution, observation, and imagery. *NeuroImage*, *37*, 1315–1328.

Gauthier, I., Tarr, M. J., Anderson, A. W., Skudlarski, P., & Gore, J. C. (1999). Activation of the middle fusiform'face area'increases with expertise in recognizing novel objects. *Nature Neuroscience*, *2*, 568–573.

Goyal, M. S., Hansen, P. J., & Blakemore, C. B. (2006). Tactile perception recruits functionally related visual areas in the late-blind. *NeuroReport*, *17*, 1381–1384.

Grimault, S., Nolden, S., Lefebvre, C., Vachon, F., Hyde, K., Peretz, I., et al. (2014). Brain activity is related to individual differences in the number of items stored in auditory short-term memory for pitch: evidence from magnetoencephalography. *NeuroImage*, *94*, 96–106.

Grimault, S., Robitaille, N., Grova, C., Lina, J. M., Dubarry, A. S., & Jolicœur, P. (2009). Oscillatory activity in parietal and dorsolateral prefrontal cortex during retention in visual short-term memory: additive effects of spatial attention and memory load. *Human Brain Mapping*, *30*, 3378–3392.

Hanakawa, T., Immisch, I., Toma, K., Dimyan, M. A., Van Gelderen, P., & Hallett, M. (2002). Functional properties of brain areas associated with motor execution and imagery. *Journal of Neurophysiology*, *89*, 989–1002.

Harris, J., Miniussi, C., Harris, I., & Diamond, M. (2002). Transient storage of a tactile memory trace in primary somatosensory cortex. *Journal of Neuroscience*, *22*, 8720–8725.

Hegner, Y. L., Lutzenberger, W., Leiberg, S., & Braun, C. (2007). The involvement of ipsilateral temporoparietal cortex in tactile pattern working memory as reflected in beta event-related desynchronization. *NeuroImage*, *37*, 1362–1370.

Hegner, Y. L., Saur, R., Veit, R., Butts, R., Leiberg, S., Grodd, W., et al. (2007). BOLD adaptation in vibrotactile stimulation: neuronal networks involved in frequency discrimination. *Journal of Neurophysiology*, *97*, 264–271.

Horenstein, C., Lowe, M. J., Koenig, K. A., & Phillips, M. D. (2009). Comparison of unilateral and bilateral complex finger tapping-related activation in premotor and primary motor cortex. *Human Brain Mapping*, *30*, 1397–1412.

Horster, W., & Ettlinger, G. (1987). Unilateral removal of the posterior insula or of area SII: inconsistent effects on tactile, visual and auditory performance in the monkey. *Behavioral Brain Research*, *26*, 1–17.

Huettel, S. A., Mack, P. B., & McCarthy, G. (2002). Perceiving patterns in random series: dynamic processing of sequence in prefrontal cortex. *Nature Neuroscience*, *5*, 485–490.

Jackson, S. R., Newport, R., Husain, M., Harvey, M., & Hindle, J. V. (2000). Reaching movements may reveal the distorted topography of spatial representations after neglect. *Neuropsychologia*, *38*, 500–507.

Johansson, R. S. (1978). Tactile sensibility in the human hand: receptive field characteristics of mechanoreceptive units in the glabrous skin area. *The Journal of physiology*, *281*, 101–125.

Johnson, J. A., Strafella, A. P., & Zatorre, R. J. (2007). The role of the dorsolateral prefrontal cortex in bimodal divided attention: two transcranial magnetic stimulation studies. *Journal of Cognitive Neuroscience*, *19*, 907–920.

Jonides, J., Lewis, R. L., Nee, D. E., Lustig, C. A., Berman, M. G., & Moore, K. S. (2008). The mind and brain of short-term memory. *Annual Review of Psychology*, *59*, 193–224.

Kanwisher, N., McDermott, J., & Chun, M. M. (1997). The fusiform face area: a module in human extrastriate cortex specialized for face perception. *Journal of Neuroscience*, *17*, 4302–4311.

Karnath, H. O., Ferber, S., & Himmelbach, M. (2001). Spatial awareness is a function of the temporal not the posterior parietal lobe. *Nature*, *411*(6840), 950–953.

Karnath, H. O., Himmelbach, M., & Rorden, C. (2002). The subcortical anatomy of human spatial neglect: putamen, caudate nucleus and pulvinar. *Brain*, *125*, 350–360.

Kelley, W. M., Macrae, C. N., Wyland, C. L., Caglar, S., Inati, S., & Heatherton, T. F. (2002). Finding the self? an event-related fMRI study. *Journal of Cognitive Neuroscience*, *14*, 785–794.

Kisley, M. A., & Cornwell, Z. M. (2006). Gamma and beta neural activity evoked during a sensory gating paradigm: effects of auditory, somatosensory and cross-modal stimulation. *Clinical Neurophysiology*, *117*, 2549–2563.

Klingberg, T., Kawashima, R., & Roland, P. (1996). Activation of multi-modal cortical areas underlies short-term memory. *European Journal of Neuroscience*, *8*, 1965–1971.

Klingberg, T., & Roland, P. E. (1997). Interference between two concurrent tasks is associated with activation of overlapping fields in the cortex. *Cognitive Brain Research*, *6*, 1–8.

Lefebvre, C., Vachon, F., Grimault, S., Thibault, J., Guimond, S., Peretz, I., et al. (2013). Distinct electrophysiological indices of maintenance in auditory and visual short-term memory. *Neuropsychologia*, *51*, 2939–2952.

Mason, M. F., Norton, M. I., Van Horn, J. D., Wegner, D. M., Grafton, S. T., & Macrae, C. N. (2007). Wandering minds: the default network and stimulus-independent thought. *Science*, *315*, 393–395.

Matelli, M., & Luppino, G. (2001). Parietofrontal circuits for action and space perception in the macaque monkey. *NeuroImage*, *14*, S27–S32.

Mertens, M., & Lütkenhöner, B. (2000). Efficient neuromagnetic determination of landmarks in the somatosensory cortex. *Clinical Neurophysiology*, *111*, 1478–1487.

Nobre, A. C., Allison, T., & McCarthy, G. (1994). Word recognition in the human inferior temporal lobe. *Nature*, *372*, 260–263.

Nolden, S., Grimault, S., Guimond, S., Lefebvre, C., Bermudez, P., & Jolicœur, P. (2013). The retention of simultaneous tones in auditory short-term memory: a magnetoencephalography study. *NeuroImage*, *82*, 384–392.

Numminen, J., Schurmann, M., Hiltunen, J., Joensuu, R., Jousmaki, V., Koskinen, S. K., et al. (2004). Cortical activation during a spatiotemporal tactile comparison task. *NeuroImage*, *22*, 815–821.

O'Craven, K. M., & Kanwisher, N. (2000). Mental imagery of faces and places activates corresponding stimulus-specific brain regions. *Journal of Cognitive Neuroscience*, *12*, 1013–1023.

Pashler, H. (1988). Familiarity and visual change detection. *Perception & Psychophysics*, *44*, 369–378.

Pierrot-Deseilligny, C., Műri, R. M., Rivaud-Pechoux, S., Gaymard, B., & Ploner, C. J. (2002). Cortical control of spatial memory in humans: the visuooculomotor model. *Annals of Neurology*, *52*, 10–19.

Postle, B. R., & D'Esposito, M. (2003). Spatial working memory activity of the caudate nucleus is sensitive to frame of reference. *Cognitive, Affective, & Behavioral Neuroscience*, *3*, 133–144.

Postle, B. R., & Hamidi, M. (2007). Nonvisual codes and nonvisual brain areas support visual working memory. *Cerebral Cortex*, *17*, 2151–2162.

Preuschhof, C., Heekeren, H. R., Taskin, B., Schubert, T., & Villringer, A. (2006). Neural correlates of vibrotactile working memory in the human brain. *Journal of Neuroscience*, *26*, 13231–13239.

Price, C. J., & Devlin, J. T. (2003). The myth of the visual word form area. *Neuroimage*, *19*, 473–481.

Reed, C. L., Klatzky, R. L., & Halgren, E. (2005). What vs. where in touch: an fMRI study. *NeuroImage*, *25*, 718–726.

Ricciardi, E., Bonino, D., Gentili, C., Sani, L., Pietrini, P., & Vecchi, T. (2006). Neural correlates of spatial working memory in humans: a functional magnetic resonance imaging study comparing visual and tactile processes. *Neuroscience*, *139*, 339–349.

Romo, R., Hernández, A., Zainos, A., Lemus, L., & Brody, C. D. (2002). Neuronal correlates of decision-making in secondary somatosensory cortex. *Nature Neuroscience*, *5*, 1217–1225.

Romo, R., & Salinas, E. (2003). Flutter discrimination: neural codes, perception, memory and decision making. *Nature Reviews Neuroscience*, *4*, 203–218.

Rouder, J. N., Morey, R. D., Morey, C. C., & Cowan, N. (2011). How to measure working memory capacity in the change detection paradigm. *Psychonomic Bulletin & Review*, *18*, 324–330.

Schluter, N. D., Krams, M., Rushworth, M. F., & Passingham, R. E. (2001). Cerebral dominance for action in the human brain: the selection of actions. *Neuropsychologia*, *39*, 105–113.

Serrien, D. J., Ivry, R. B., & Swinnen, S. P. (2006). Dynamics of hemispheric specialization and integration in the context of motor control. *Nature Reviews Neuroscience*, *7*, 160–166.

Smith, E. E., Jonides, J., Koeppe, R. A., Awh, E., Schumacher, E. H., & Minoshima, S. (1995). Spatial vs. object working memory: PET investigations. *Journal of Cognitive Neuroscience*, *7*, 337–356.

Spitzer, B., Wacker, E., & Blankenburg, F. (2010). Oscillatory correlates of vibrotactile frequency processing in human working memory. *The Journal of Neuroscience*, *30*, 4496–4502.

Todd, J. J., & Marois, R. (2004). Capacity limit of visual short-term memory in human posterior parietal cortex. *Nature*, *428*, 751–754.

Vogel, E. K., & Machizawa, M. G. (2004). Neural activity predicts individual differences in visual working memory capacity. *Nature*, *428*, 748–751.

Vogel, E. K., Woodman, G. F., & Luck, S. J. (2001). Storage of features, conjunctions, and objects in visual working memory. *Journal of Experimental Psychology: Human Perception and Performance*, *27*, 92–114.

Vrba, J., & Robinson, S. E. (2001). Signal processing in magnetoencephalography. *Methods*, *25*, 249–271.

Zhou, Y. D., & Fuster, J. M. (1996). Mnemonic neuronal activity in somatosensory cortex. *Proceedings of the National Academy of Sciences*, *93*, 10533–10537.

CHAPTER

20

The Role of Spatial Attention in Tactile Short-Term Memory

Tobias Katus[1], *Søren K. Andersen*[2]

[1]Department of Psychology, Birkbeck College, London, United Kingdom; [2]School of Psychology, University of Aberdeen, Aberdeen, United Kingdom

This chapter advocates the view that executive control functions of short-term memory (STM) are equivalent to processes that mediate selective attention in perception. We will focus on evidence coming from our series of tactile STM experiments. However, because the investigation of tactile STM is an emergent field of research, most of the literature reviewed in this chapter will cover studies of tactile (perceptual) attention and visual STM.

The concept of STM encompasses processes involved in the storage, temporary maintenance, and mental manipulation of information that is no longer present in the sensory environment (Baddeley, 2003; Gazzaley, 2011; Gazzaley & Nobre, 2012; Jensen, Kaiser, & Lachaux, 2007; Zimmer, 2008). Selective attention, on the other hand, relates to functions that impose a goal-directed bias on the processing of events in the sensory environment (e.g., Carrasco, 2011; Desimone & Duncan, 1995). In brief, STM refers to internal representations whereas perception is associated with representations of external stimuli. Although these apparent differences seem to suggest that STM and perception may be distinct cognitive domains, the evidence reviewed in this chapter suggests otherwise.

There are two major perspectives on STM: structural and process-oriented. These theoretical streams are not mutually exclusive and coexist, because they focus on different aspects of sensory STM systems. The traditional view has emphasized structural aspects of STM[1] and produced a taxonomy of cognitive modules, for the more passive storage of information on the one hand and the more active processing of information on the other (Baddeley, 2000, 2003, 2012; Smith & Jonides, 1997; Smith & Jonides, 1999). The properties of storage components are to some degree modality-specific. Distinct buffers are thought to store visuospatial and phonological information (Baddeley & Andrade, 2000; Baddeley & Larsen, 2007; Repovs & Baddeley, 2006). A storage buffer for tactile information has not been integrated into Baddeley's framework. Nevertheless, structural accounts would characterize tactile STM as a modular system with a modality-specific component and a component of supramodal control, the so-called *central executive* (Repovs & Baddeley, 2006). The central executive is closely related to selective attention (Baddeley, 2003, 2012). To date, however, structural accounts of STM have not produced an elaborate taxonomy of central executive mechanisms.

The role of attention is emphasized by the more recent process-oriented view. Short-term memory is regarded as an emergent system that results from the attentional activation of long-term memory (LTM) representations that reside in posterior association areas (Cowan, 1995; Oberauer, 2002; reviewed by Jonides et al., 2008). Central to the emergent system view is that anatomical areas and neural control mechanisms otherwise subservient to perception and motor control are recruited by STM to allow for the storage and processing of memorized information (Awh & Jonides, 2001; Gazzaley & Nobre, 2012; Postle, 2006). For example, STM influences neural activity in the earliest sensory areas (Pasternak & Greenlee, 2005); furthermore, STM and perception rely on shared control mechanisms (Awh & Jonides, 2001; Awh, Vogel, & Oh, 2006; Gazzaley, 2011; Gazzaley & Nobre, 2012). This anatomical and functional overlap demonstrates that STM and perception are not distinct cognitive domains, a conclusion consistent with the emergent system view of STM (e.g., Postle, 2006).

[1] In the context of his structural framework, Baddeley prefers the label *working memory* over *STM*. For consistency, we prefer the (more neutral) label *STM* in this chapter.

Mechanisms of Sensory Working Memory
http://dx.doi.org/10.1016/B978-0-12-811042-3.00020-7

ATTENTION IN VISUAL STM

Selective attention acts at multiple stages in STM tasks both in the absence and presence of stimuli in the sensory environment (Awh et al., 2006; Gazzaley & Nobre, 2012; Griffin & Nobre, 2003). These stages include the phase of anticipation (Murray, Nobre, & Stokes, 2011) and encoding of memorized events (Zanto & Gazzaley, 2009), in the maintenance of mnemonic information during the retention delay (Awh, Anllo-Vento, & Hillyard, 2000; Jha, 2002), and during retrieval as well as comparison with information stored in STM (Kuo, Rao, Lepsien, & Nobre, 2009). Here we focus on the role of spatial attention during the retention delay: that is, when stimuli are absent.

Spatial effects have been reported in tasks that require the maintenance of locations (spatial STM) but also in tasks in which attributes from non-spatial feature dimensions had to be memorized (non-spatial STM). As for spatial STM, evidence comes from investigations of the attention-based rehearsal account, according to which spatial attention supports the online maintenance of locations in visual STM (Awh & Jonides, 2001; Awh et al., 2006; Theeuwes, Belopolsky, & Olivers, 2009). The attention-based rehearsal account states two key predictions (cf. Awh & Jonides, 2001; Awh et al., 2006). First, spatial attention is directed toward memorized locations throughout the retention delay of spatial STM tasks; and second, such spatial attention shifts are the consequence of an adaptive strategy and thus are beneficial for memory accuracy. As a consequence, memory accuracy should diminish if spatial attention is withdrawn from a memorized location.

The first prediction of attention-based rehearsal has been examined in studies using probe designs (Awh et al., 2000; Jha, 2002). Probes are task-irrelevant stimuli presented during the retention period to probe the allocation of attention. Probe stimuli that overlapped with memorized locations were reported to elicit enhanced P1 and N1 components relative to probes that overlapped with non-memorized locations (Awh et al., 2000). This spatial encoding bias in favor of stimuli near memorized locations suggests that focal shifts of attention accompany the maintenance of locations in STM. Investigations of the second prediction of attention-based rehearsal, however, produced mixed results (reviewed in Awh & Jonides, 2001; Awh et al., 2006; in opposition to Belopolsky & Theeuwes, 2009b). These studies employed dual-task designs to contrast STM accuracy when a memorized location was inside, rather than outside, the focus of (perceptual) attention. Selective impairment of spatial STM accuracy has been reported for secondary tasks that require attention shifts away from memorized locations (Awh, Jonides, & Reuter-Lorenz, 1998; Smyth & Scholey, 1994). For example, the study by Awh et al. (1998) combined a spatial memory task with a color discrimination task. Memory performance suffered when the stimulus relevant for color discrimination did not overlap with the memorized location compared with a condition in which memorized and perceptually attended stimuli did overlap. A functional significance of spatial attention to spatial STM, however, was questioned by other researchers (Belopolsky & Theeuwes, 2009b).

Even if the memorized attribute relates to a non-spatial feature dimension, selection between spatially segregated items in STM can cause focal shifts of spatial attention (Astle, Scerif, Kuo, & Nobre, 2009; Dell'Acqua, Sessa, Toffanin, Luria, & Jolicœur, 2009; Kuo et al., 2009). A series of electroencephalographic and functional magnetic resonance imaging studies employed the advantages offered by the retro-cue paradigm (reviewed by Lepsien & Nobre, 2006) to investigate selection in memory representations. Retro-cues are presented after a set of sample stimuli; some are relevant to the task (targets) whereas others are irrelevant (distracters). Because retro-cues indicate the target-defining attribute after the sample set has been encoded, retrospective selection occurs within mnemonic content. Kuo et al. (2009) presented a sample set consisting of stimuli with different shapes (Experiment 1) or colors (Experiment 2), followed by a centrally presented retro-cue[2] that indicated the target-defining attribute in a non-spatial dimension (color or shape). Participants were instructed to judge whether the cued attribute had been present in the sample set. The retro-cue elicited an N2pc component that reflected the location of the target, although the target's spatial attributes were irrelevant to the behavioral response. Thus, retrospective cueing of non-spatial (Astle et al., 2009; Dell'Acqua et al., 2009; Kuo et al., 2009) or spatial attributes (Griffin & Nobre, 2003) causes electrophysiological activity indicative of spatial coordinates of selected information in visual STM. It was argued that such attention effects reflect the spatial layout of neural top-down signals that bias features or objects in STM (Astle et al., 2009; Kuo et al., 2009). The obligatory representation of spatial attributes, regardless of task demands, appears to reflect a general principle for the organization of information in visual STM.

Above we outlined evidence that selective attention may act as a rehearsal mechanism to maintain information in visual STM. Before we address the question of whether this rehearsal strategy may also participate in the

[2] In the terminology used by Kuo et al. (2009), the stimulus informing about the target-defining attribute was labeled probe. To avoid confusion in this chapter, we refer to it as retro-cue, because this stimulus signaled the target *after* the sample set had been presented.

maintenance of tactile spatial STM, we will review the state of research on the neural mechanisms of selective attention—in particular, spatial attention—for tactile perception.

ATTENTION IN TACTILE PERCEPTION

The somatosensory system consists of the four sub-modalities: proprioception, pain, temperature, and touch. The sense of touch encodes mechanical pressure applied to the skin. Endogenous spatial attention in touch is a goal-driven cognitive function that enhances behavioral performance, thus increasing accuracy while reducing reaction times (Forster & Eimer, 2005; Sambo & Forster, 2011). As to the neural underpinnings of such behavioral effects, it was suggested that activity in sensory regions is governed by a gain control mechanism acting through neural amplification and/or suppression (Forster & Gillmeister, 2011; Johansen-Berg & Lloyd, 2000; Sambo & Forster, 2011). This modality-specific gain control mechanism is regulated via top-down biasing signals generated in frontoparietal neural networks associated with supramodal attention (Eimer, Forster, Fieger, & Harbich, 2004; Eimer, Forster, & van Velzen, 2003a; Sambo & Forster, 2011).

Selective attention alters hemodynamic responses in primary somatosensory cortex (SI) and secondary somatosensory cortex (SII) (Burton & Sinclair, 2000; Johansen-Berg & Lloyd, 2000; Schubert et al., 2008) and affects electroencephalographic activity originating in somatosensory cortical regions (Johansen-Berg & Lloyd, 2000; Sambo & Forster, 2011). Attention effects are generally larger in SII relative to SI (e.g., Johansen-Berg & Lloyd, 2000). Modulation of the event-related potential (ERP) resulting from spatial attention is commonly found at around 100–150ms after the onset of tactile stimuli. Components arising in this time range (P100 and N140) are at least partly generated by cell populations in bilateral SII (Allison, McCarthy, & Wood, 1992; Frot & Mauguière, 1999; Soto-Faraco & Azañón, 2013). The earlier P50 and N80 components are rarely affected by spatial attention (however, see Eimer & Forster, 2003a; Schubert et al., 2008).

The N140 is one of the most thoroughly examined components of the somatosensory ERP and manifests at around 140ms poststimulus over parietal regions (e.g., electrodes C5/6). Putative generators have been located in SII and the bilateral frontal lobes, including anterior cingulate gyrus and supplementary motor areas (Allison et al., 1992; Frot & Mauguière, 1999; García-Larrea, Lukaszewicz, & Mauguière, 1995; Waberski, Gobbelé, Darvas, Schmitz, & Buchner, 2002). Numerous studies found enhanced N140 amplitudes when a stimulated hand was in the focus of attention (Desmedt & Robertson, 1977; Eimer & Forster, 2003a; Eimer et al., 2004; Forster & Eimer, 2004, 2005; Josiassen, Shagass, Roemer, Ercegovac, & Straumanis, 1982; Michie, Bearpark, Crawford, & Glue, 1987; Zopf, Giabbiconi, Gruber, & Müller, 2004). Based on the observation that the N140 mirrors the pattern of modulation exhibited by the late directing attention positivity (LDAP) (see below) depending on variations in hand posture, it seems as if the N140 reflects the allocation of attention referenced to external space (Eimer et al., 2003a; Eimer et al., 2004). Although the N140 is a reliable marker of between-hands selection, it is less sensitive to spatial selection within a single hand (Desmedt & Robertson, 1977; Eimer & Forster, 2003b; Josiassen et al., 1982). Besides spatial attention, the N140 is also modulated by feature-based attention, such as the selection of stimulus intensity or frequency (Forster & Eimer, 2004), which suggests that processes of spatial and non-spatial attention operate in parallel in the sense of touch.

INVESTIGATING THE ROLE OF SPATIAL ATTENTION IN TACTILE STM

Research on tactile perception produced substantial insights as to how spatial attention shapes the neural processing of sensory events (reviewed in Johansen-Berg & Lloyd, 2000; Sambo & Forster, 2011). Research on STM in the visual modality demonstrated that spatial attention supports the maintenance of locations during the retention period of spatial memory tasks (Awh et al., 2000; Awh & Jonides, 2001; Awh et al., 2006; Jha, 2002; Theeuwes et al., 2009). In an attempt to bring these lines of research together, we evaluated the hypothesis that rehearsal of tactile locations may be mediated by a mechanism which is based on covert spatial attention, as in vision.

The attention-based rehearsal account (Awh & Jonides, 2001; Awh et al., 2006; Theeuwes et al., 2009) was evaluated in two tactile spatial memory experiments. The first study tested whether spatial STM triggers focal attention shifts toward memorized locations (Awh et al., 2000; Jha, 2002). Whether the disruption of such attention shifts impairs memory accuracy was assessed in a second experiment with a dual-task design (Awh et al., 1998; Smyth & Scholey, 1994). A third experiment tested whether tactile STM preserves spatial attributes of remembered information when memory for locations is not explicitly instructed (cf. Kuo et al., 2009). This potential spatial specificity of non-spatial

STM representations was examined through a task that required the feature-based selection between spatially segregated items in STM.

All studies used visual retro-cues (cf. Lepsien & Nobre, 2006) centered at gaze fixation. Retro-cues presented in the period after the presentation of tactile sample sets guide selection within internal (memory) representations. Central to our hypotheses were electrophysiological markers of spatial attention, such as lateralized components (evoked by visual cue stimuli) and the somatosensory N140 (to tactile stimuli). Lateralized components are marked by different amplitudes at electrodes located contralateral versus ipsilateral to the direction of spatial attention shifts (Harter, Miller, Price, LaLonde, & Keyes, 1989). Such amplitude differences are negative as for the N2pc (Eimer & Kiss, 2010; Luck & Hillyard, 1994a, 1994b) and anterior directing attention negativity (ADAN) (e.g., Eimer et al., 2003a) or positive as for the LDAP (e.g., Eimer et al., 2004). The N2pc, ADAN, LDAP, and somatosensory N140 are commonly interpreted as markers of perceptual and/or preparatory attention (e.g., Eimer et al., 2003b; Eimer et al., 2004, see also Murray et al., 2011). Modulation of these measures in memory tasks indicates that control processes for STM overlap with the mechanisms of (perceptual) selective attention.

STUDY 1: MEMORY FOR LOCATIONS CAUSES SHIFTS OF TACTILE ATTENTION

Study 1 (Katus, Andersen, & Müller, 2012a) tested whether the processing of sensory events is enhanced at memorized locations, as predicted by the attention-based rehearsal account. Participants performed the tactile spatial memory task described in Figure 1. Lateralized components evoked by the retro-cue were expected to mirror the direction of attentional shifts in STM. Task-irrelevant tactile probe stimuli were presented to the cued or uncued hand to probe the allocation of spatial attention during retention. Processing of probes was expected to reflect a spatial encoding bias in favor of stimuli presented near memorized locations (cued relative to uncued hand).

Two lines of evidence indicate that the focus of spatial attention shifted toward memorized locations in the retention period. First, visually presented retro-cues elicited the N2pc and LDAP components (Figure 2(a)). These lateralized effects reflected the cued hand and thus spatially specific access to tactile STM. In line with this observation, lateralized effects mirrored selected locations in visual STM in the period after spatially predictive retro-cues (Griffin & Nobre, 2003). Because retro-cues were symmetric in shape and presented after tactile samples had been presented to the left and right hands, effects time-locked to the onset of cues reveal shifts of endogenous, rather than exogenous, spatial attention.

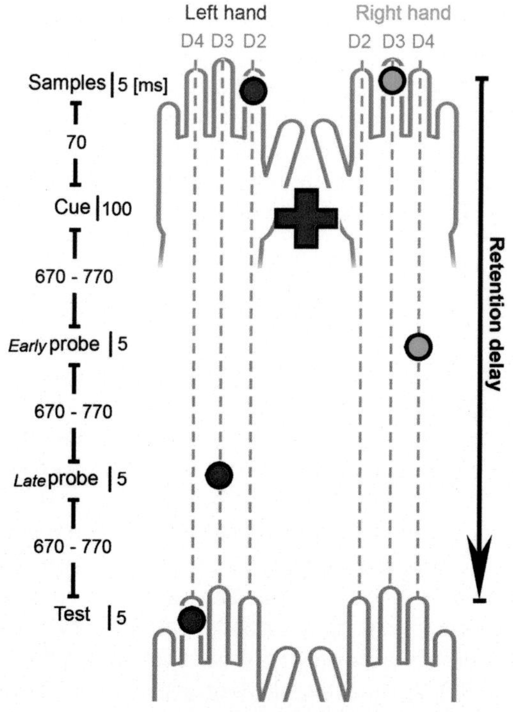

FIGURE 1 Spatial memory task design. Three stimulators were adhered to the index, middle, and ring fingers of the left and right hands. The stimulation sequence involved bilateral tactile sample stimuli (separately randomized between D2, D3, and D4 for each hand), a visual retro-cue (color change of the fixation cross), unilateral probe stimuli (presented early or late after the cue), and a unilateral test stimulus. Participants had to memorize the sample pulse's location at the currently task-relevant hand and compare this target with the location of the test. The color of the cue indicated the relevant hand (here: the left hand, owing to the red cue). Tactile stimuli at the task-relevant and irrelevant hand are symbolized by filled red and green dots, respectively. Probes were presented to the relevant and irrelevant hands with equal probability. The example illustrates a target-absent trial because the test did not match the left-hand sample's location.

Second, memory for locations guided somatosensory encoding in a way similar to the pattern of results produced by spatial attention in tactile perception. Attention-based rehearsal led to a spatial encoding bias that was reflected by the somatosensory N140 component (Figure 2(b)), which was larger for stimulation of the hand where a location was memorized relative to stimulation of the other (non-memorized) hand. Studies of perceptual attention showed that the N140 is larger when a stimulated hand is attended to rather than ignored (Desmedt & Robertson, 1977; Eimer & Forster, 2003a; Eimer et al., 2004; Forster & Eimer, 2004, 2005; Josiassen et al., 1982; Michie et al., 1987; Zopf et al., 2004). Our results are further in line with studies of attention-based rehearsal in visual STM. In those studies, the ERP to probe stimuli was modulated in the range of early visual components, such as the P1 and N1 (Awh et al., 2000; Jha, 2002). With regard to the anatomical overlap of memory and perception, it was suggested that sensory signals are stored in the same areas that encode such signals (Pasternak & Greenlee, 2005; Postle, 2006; Serences, Ester, Vogel, & Awh, 2009). The modulation of neural processing in sensory areas, recruited by STM, suggests a modality-specific biasing mechanism that alters the content of STM. Neural networks of endogenous selective attention in frontoparietal cortical regions are likely sources of top-down signals that bias memory traces in favor of relevant aspects and at the expense of aspects that are irrelevant to current behavioral goals (Eimer et al., 2003a; Eimer et al., 2004; Sambo & Forster, 2011).

Our finding that tactile spatial STM leads to focal shifts of tactile attention toward memorized locations confirms the first prediction of the attention-based rehearsal account. In line with studies of visuospatial STM (Awh et al., 2000; Jha, 2002), we found a sustained encoding bias during the retention period, here indexed by attentional N140 modulations for early probes, late probes, and test stimuli (Figure 2(b)). Comparison of early probes with late probes, however, revealed largest N140 modulations for early probe stimuli. In contrast, no time-variant encoding effects

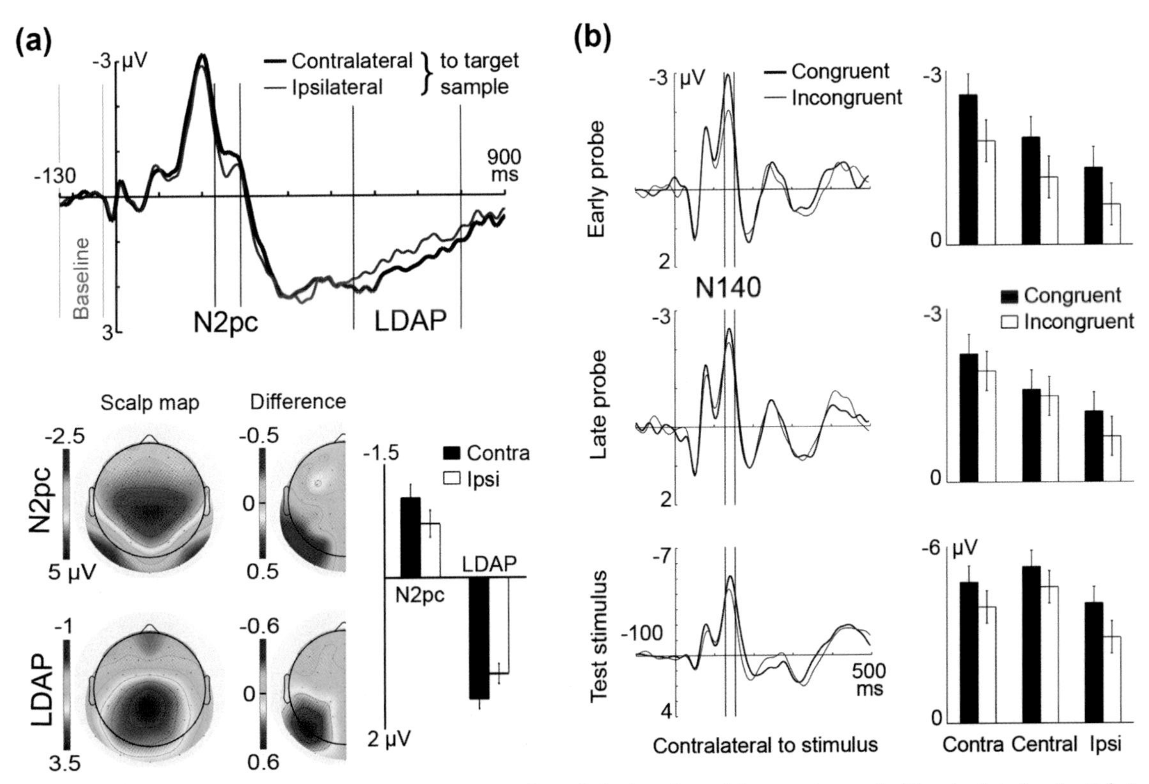

FIGURE 2 (a) The lateralized N2pc and LDAP components reflected the location of the target sample. Event-related potentials in response to visual retro-cues were recorded at lateral posterior electrodes contralateral (black line) and ipsilateral (gray line) to the cued sample stimulus. Isocontour voltage scalp maps show averaged amplitudes in the N2pc (top) and LDAP (bottom) time windows; the arrangement of data in these maps is equivalent to cueing of the right-hand sample. Difference maps were obtained by subtracting ipsilateral from contralateral recordings. Bar graphs show mean amplitudes of contralateral and ipsilateral ERPs in the N2pc and LDAP time windows. (b) Tactile stimuli presented during retention (probe stimuli) and after the retention period (test stimuli) reflected a spatial encoding bias in favor of the hand where the target sample was memorized. Event-related potentials elicited by early probes (top row), late probes (middle row), and test stimuli (bottom row) were recorded contralateral to stimulation. Spatially congruent stimuli (black line) were presented to the same hand as the target sample, whereas incongruent stimuli (gray line) were presented to the other hand. Bar graphs depict conditional mean amplitudes in the N140 time window for ERPs contralateral, central, and ipsilateral to stimulation. The encoding bias indexed by the N140 modulation was larger for early probes than for late probe stimuli. Note that scales differ on the *y*-axis. *Adapted with permission from Katus et al. (2012a).*

were reported in studies on visuospatial STM that used probe designs without retro-cues[3] (Awh et al., 2000; Jha, 2002). We hence speculated that retrospective selection between spatially segregated items may cause transient shifts of spatial attention (see Study 3). Such transient effects may contribute to the sustained impact of attention-based rehearsal in the early part of the retention period of spatial STM task.

STUDY 2: FUNCTIONAL SIGNIFICANCE OF SPATIAL ATTENTION FOR SPATIAL STM

Spatial attention is directed to memorized locations during the retention period of spatial STM tasks (e.g., Study 1). However, it is not clear whether such attention shifts are functionally significant for STM, or alternatively, are an epiphenomenon that is unrelated to memory accuracy. Study 2 (Katus, Andersen, & Müller, 2014) examined the hypothesis that covert attention shifts to memorized locations are mandatory for the optimal functioning of spatial STM.

The design of the dual-task experiment combined the spatial memory task used in Study 1 with a perceptual spatial attention task (Figure 3). Participants kept a location in memory while detecting transient changes (duration, 200 ms) in the frequency of ongoing vibrotactile stimulation (duration, 3500 ms). Test stimuli were presented 620 ms after the offset of the vibrotactile streams. Cue stimuli (duration, 600 ms) were presented between 290 and 390 ms after the sample set of the memory task and 690 ms before the stimuli of the perceptual task. The cues consisted of a retro-cue (for the memory task) and a pre-cue (for the perceptual task), and these cues could point in the same or different directions. If STM and perception tap the same resources of spatial attention, attention should be divided between the left and right hands with incompatible cues and be focused on a single hand with compatible cues. Performance is reduced under conditions of divided rather than focused attention (cf. Jans, Peters, & De Weerd, 2010; Störmer, Winther, Li, & Andersen, 2013). Thus, lower memory accuracy was predicted for the incompatible cue condition.

The disruption of attention-based rehearsal in the incompatible cue condition led to reduced memory performance relative to the compatible cue condition (Figure 4(a)). The incompatible cue interference effect suggests a functional significance of spatial attention for spatial STM in the sense of touch (compare studies in vision: Awh & Jonides, 2001; Awh et al., 1998; Smyth & Scholey, 1994; see also Belopolsky & Theeuwes, 2009b). Moreover, performance in the high-priority perceptual task depended on cue compatibility (Figure 4(b)). The incompatible cue interference

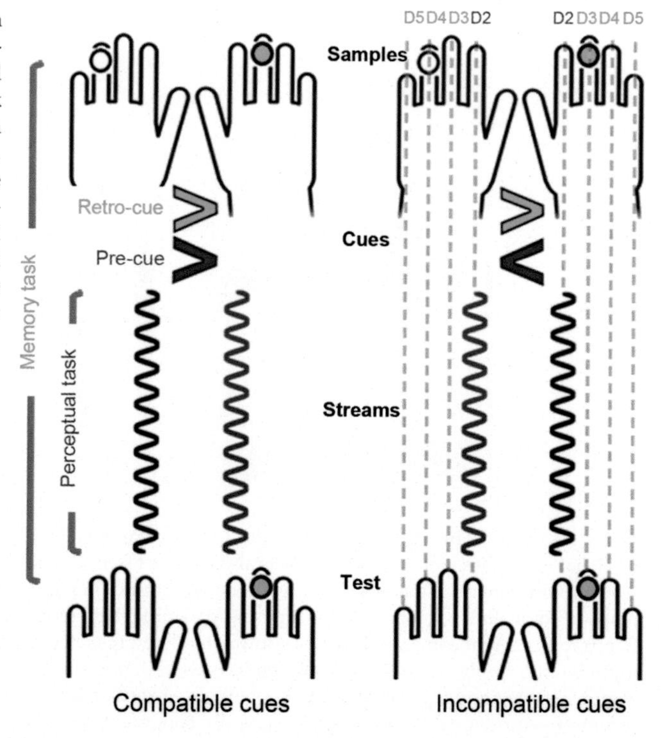

FIGURE 3 Dual-task experiment composed of a perceptual attention task, interleaved into the retention period of a spatial memory task. Simultaneously presented cue stimuli indicated whether the left or right hand was relevant in the STM task (retro-cue, green) and the perceptual task (pre-cue, red). Retro- and pre-cue could either point in the same direction (compatible cues, left) or into different directions (incompatible cues, right). The memory task (stimulators at D3 through D5) was identical to the one used in Study 1, except that test stimuli were always presented to the retrospectively cued hand (50% memorized finger and 50% non-memorized finger). The perceptual task (stimulator at D2) was prioritized and required speeded reactions to frequency changes in the vibrotactile stream at the prospectively cued hand. Verbal responses for the low-priority memory task were given after the retention period.

[3] In our study, the primary purpose of retro-cues was to rule out potential effects of exogenous attention that could occur if participants had to retain the location of a single unilateral sample stimulus.

effect on perceptual task performance was largest for early events and the effect diminished with increasing latency of event presentation after the cues. Incompatible cues were potentially more difficult to interpret than compatible cues, which might have delayed the preparation of spatial attention shifts. The physical complexity of cue stimuli was matched, however, because compatible and incompatible cues consisted of the same shape elements: namely, a red and a green arrow. Overall, behavioral data suggest that preparatory attention shifts (in direction of pre-cues) were hampered by spatial selection in STM (direction of retro-cues), and vice versa. This bidirectional interference indicates that common control mechanisms accommodate the spatial processing demands imposed by both tasks.

As shown in Figure 5, the side of the prioritized perceptual task (pre-cue direction) was mirrored by the N2pc, ADAN, and LDAP components of the ERP to compatible cue stimuli. Incompatible cues, however, did not elicit statistically significant N2pc and ADAN components, and only the LDAP still reflected the side of the pre-cue. Interpretation of the earliest lateralized effect (N2pc) may be difficult because physically different cue stimuli were used in the two compatibility conditions (van Velzen & Eimer, 2003). However, this confound does not concern subsequent lateralized components; ADAN and LDAP are insensitive to physical properties of attention-directing cues (van Velzen & Eimer, 2003). The reduced sizes of the ADAN and LDAP components, which are commonly understood as markers of anticipatory attention, suggest that prospective attention shifts were hampered by retro-cues that pointed in opposite direction of pre-cues. This entails the conclusion that spatial selection in mnemonic representations relies on the same control mechanisms that accomplish covert spatial orienting toward stimuli in the external world. In other words, we conclude that STM recruits tactile spatial attention to facilitate the maintenance of tactile locations during the retention period.

Why did visual studies on the functional significance of spatial attention to spatial STM produce mixed results? Evidence that spatial attention supports the maintenance of locations comes from a dual-task experiment in which participants performed a perceptual color discrimination task in the retention period of a spatial memory task (Awh et al., 1998). Performance was lower when the stimulus related to the perceptual task did not overlap with the memorized locations, compared with a condition in which perceptually attended and memorized locations did overlap. The authors suggested that reduced memory accuracy in the former condition was a consequence of the necessity to

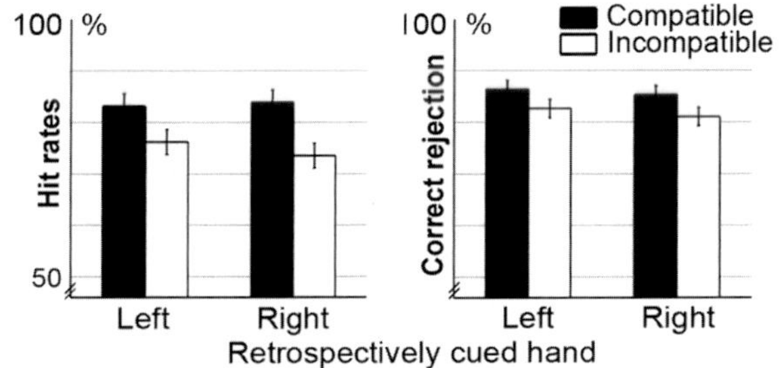

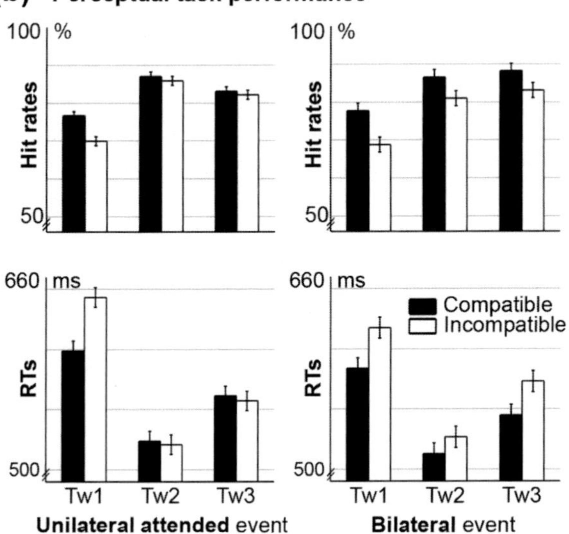

FIGURE 4 (a) Memory accuracy was reduced in incompatible cue trials (white bars) relative to compatible cue trials (black bars). Hit rates and correct rejection rates are shown for retro-cues indicating the left- or right-hand sample as relevant to the task. (b) Performance in the perceptual task was reduced with incompatible cues (white bars) relative to compatible cues (black bars). Separate rows show hit rates (upper row) and reaction times (lower row) to events in vibrotactile streams that required immediate responses. Such events were either inserted into the attended stream (unilateral attended, 33%) or simultaneously presented to both hands (bilateral, 33%). Events in the ignored stream (unilateral ignored, 33%) did not require behavioral responses. Events could occur during three time windows (Tw1-3: 500, 1500, and 2500 ms relative to stream onset; 33% each). *Adapted with permission from Katus et al. (2014).*

direct attention away from the memorized location. However, other authors questioned a functional significance of spatial attention for spatial STM (Belopolsky & Theeuwes, 2009b) and reported that spatial STM may lead to inhibition, rather than facilitation, of visual processing at memorized locations during the retention period. Moreover, it was concluded that even under circumstances in which facilitation occurs at memorized locations, such facilitation would not influence the accuracy of spatial STM (Belopolsky & Theeuwes, 2009b).

There are at least three reasons why we found a functional significance of attention-based rehearsal in our experiment despite the mixed findings reported in vision (Awh et al., 1998; as opposed to Belopolsky & Theeuwes, 2009b). First and most important, in our dual-task study, spatial stimulus attributes were explicitly relevant for both tasks. The cited visual experiments combined a spatial STM task with a non-spatial perceptual task based on the assumption that processing of non-spatial features (e.g., colors) can occur only within the focus of spatial attention (Tsal & Lavie, 1993). More recent evidence, however, indicates that the impact of feature-based attention on neural

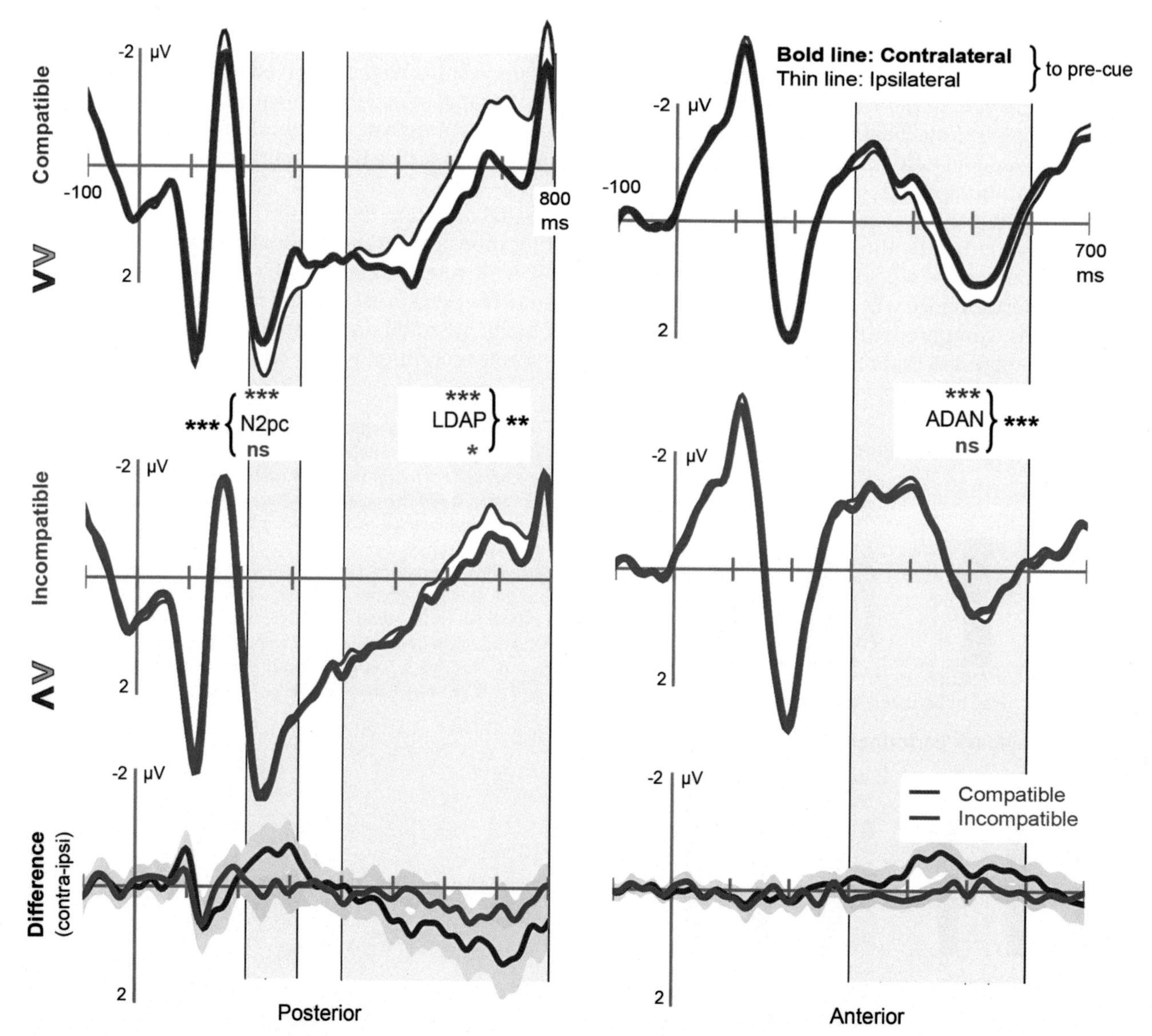

FIGURE 5 Event-related potentials elicited by visual cues. The side of the prioritized perceptual task (pre-cue direction) was reflected by the lateralized N2pc, ADAN, and LDAP components incompatible cue trials. These components were reduced in size with incompatible cues. Grand mean ERPs were measured contralateral (bold line) and ipsilateral (thin line) to the direction of the pre-cue at lateral posterior (left) and anterior (right) scalp regions. Separate graphs show the compatible (upper row) and incompatible cue condition (middle row). Difference waveforms (bottom row) were calculated by subtracting ipsilateral ERPs from contralateral ERPs. Red and blue shades reflect 95% confidence intervals for tests of difference values against zero (i.e., no lateralization), for the compatible and incompatible cue condition, respectively. Statistical significance of lateralized components in specific time windows is marked by asterisks in red (compatible cues) or blue (incompatible cues). Black asterisks indicate significant comparisons of component sizes between trials with compatible and incompatible cues (* for $p<.05$; ** for $p<.01$; *** for $p\leq.001$).

activity generated in early visual areas is not contingent on spatial attention (Andersen, Müller, & Hillyard, 2009) and affects all spatial locations equally even when this explicitly conflicts with task demands (Andersen, Hillyard, & Müller, 2013). Corresponding evidence from research on tactile perception suggests that the attentional selection of spatial and non-spatial attributes is mediated by parallel and independent processes (Forster & Eimer, 2004). Second, we sought to maximize the impact of prospective attention shifts on spatial STM by instructing participants to prioritize the perceptual task. This procedure should enlarge interference effects caused by a withdrawal of spatial attention from memorized locations, if shared mechanisms of spatial attention underlie the spatial processing demands imposed by both tasks. Third, simultaneously presented cues directed focal shifts of endogenous attention in STM and perception in the tactile experiment. In the current visual dual-task experiments, locations were not separately cued with regard to the two tasks.

The requirement to simultaneously memorize a location on one hand, while perceptually attending to a location on the other hand, caused a bidirectional interference between STM and perception in our tactile dual-task study. No interference was found in the visual study conducted by Belopolsky & Theeuwes (2009b). These discrepant results could, in principle, be explained by any of the aforementioned differences in task design. Our results confirm a functional significance of spatial attention for spatial STM in touch, and are thus in line with the second key prediction of the attention-based rehearsal account (Awh & Jonides, 2001; Awh et al., 2006; Theeuwes et al., 2009).

STUDY 3: NON-SPATIAL CUEING OF STM CAUSES SHIFTS OF SPATIAL ATTENTION

Spatial attention facilitates the maintenance of locations in STM—if locations are relevant to behavioral goals. However, studies of visual STM reported spatially specific effects in tasks in which memory for locations was neither explicitly instructed nor beneficial to behavioral performance (e.g., Astle et al., 2009; Kuo et al., 2009). In Study 3 (Katus, Andersen, & Müller, 2012b), we tested whether spatial attributes are represented in memory even when they are not directly task-relevant (Figure 6). To rule out contributions from attention-based rehearsal, participants were asked to memorize an attribute belonging to a non-spatial feature dimension, namely, intensity. Retro-cues were spatially uninformative and indicated whether the target sample stimulus had weak or strong intensity. Because the

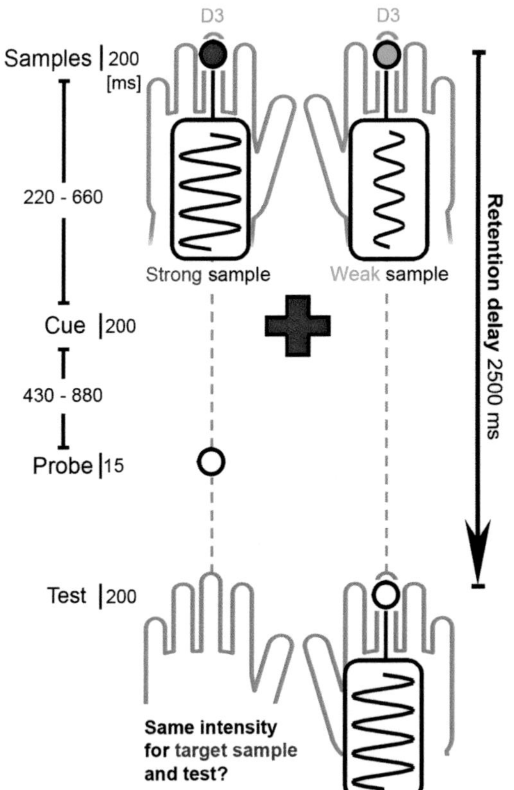

FIGURE 6 Non-spatial memory task. Two sample stimuli of different intensities were simultaneously presented to the middle fingers of the left and right hands. Participants judged whether the target sample and test stimulus had identical (50%) or different intensities (50%). The retro-cue indicated the target sample's intensity category, but not its location, by means of an arbitrary color mapping rule. In the example shown, the red cue indicates the relevance of the strong sample at the left hand. Unilateral probe stimuli had fixed intensity and were randomized to the left or right hand. Likewise, unilateral test stimuli could occur at the left or right hand with equal probability to rule out potential attention shifts in anticipation of test stimuli. Behavioral responses did not depend on whether target sample and test stimulus were located at the same hand (50%) or at different hands (50%). Intensities of weak and strong sample pulses, and also test stimuli, were not fixed, but varied from trial to trial. As a consequence of this intensity jitter within categories for weak and strong stimuli, the task could not be solved by merely memorizing a verbal label of the intensity category indicated by the retro-cue.

weak and strong sample stimuli had been randomized to the left and right hands in a trial-by-trial fashion, retro-cues guided feature (or object) based selection between spatially segregated items in tactile STM.

Selection between spatially segregated items in memory caused a spatial processing bias that reflects the configuration of the original sensory input. As shown in Figure 7(a), the N2pc component evoked by retro-cues mirrored the location of the retrospectively cued sample stimulus; similar results were found for the ADAN (not shown). Such evidence for a spatially selective memory access was flanked by results indicating the preferential processing of stimuli at the target sample's location. This spatial encoding bias was reflected by a modulation of the somatosensory N140 to probe stimuli (Figure 7(b)), which were presented at latencies overlapping with the early time bin in Study 1. No N140 modulation was found for test stimuli, which suggests that spatial attention was evenly distributed between the left and right hands at the end of the retention period. The transient nature of attention effects is of interest here because attention-based rehearsal is thought to have a sustained impact that prevails throughout the retention period (Awh et al., 2000; Jha, 2002). Because voluntary maintenance of locations was not required by the task, we conclude that selection of non-spatial feature attributes leads to spatially specific effects that reflect processes that are to some degree dissociable from those underlying attention-based rehearsal. Spatially specific modulation of the ERP as a result of selection of features such as shape and color has been previously reported in studies of visual STM. According to the explanation offered by Kuo et al. (2009), the activation of spatial coordinates in non-spatial memory tasks reveals feature- or object-based selection mechanisms that operate within the spatial layout of STM representations. Study 3 is consistent with this line of argument and points to the obligatory representation of spatial attributes in tactile STM in the context of a task with no apparent spatial processing or storage demands.

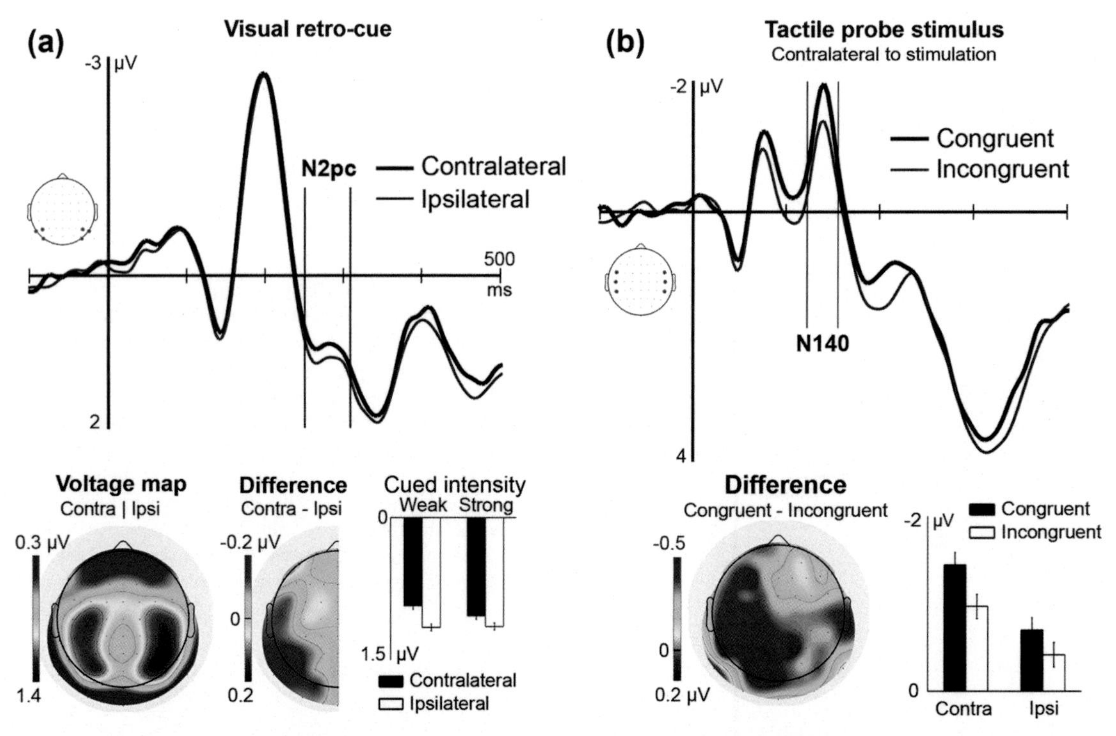

FIGURE 7 (a) The N2pc component elicited by the visual retro-cue indicated the location of the tactile sample with the cued intensity. Event-related potentials were measured at posterior scalp regions contralateral (black line) and ipsilateral (gray line) to the retrospectively cued sample stimulus. The voltage map depicts mean amplitudes in the N2pc time range for data arranged corresponding to cueing of the right-hand sample. The difference map shows amplitudes in the N2pc time range at contralateral minus ipsilateral electrodes. The bar graph shows mean amplitudes contralateral (black bars) and ipsilateral (white bars) to the target sample, for cueing of the weak versus strong stimulus of the sample set. (b) Probe stimuli reflected a spatial encoding bias indicating the target sample's location. Event-related potentials to task-irrelevant tactile probe stimuli were recorded over somatosensory scalp areas contralateral to stimulation. Spatially congruent probe stimuli were presented to the same hand as the target sample (black line); incongruent probes were delivered to the other, non-memorized, hand (gray line). The difference map for the N140 time range was obtained by subtracting voltage maps for incongruent probes from maps for congruent probes after arranging the data according to stimulation of the right hand. Bar graphs shows mean amplitudes in the N140 time range, recorded contralateral and ipsilateral to spatially congruent (black bars), and incongruent probe stimuli (white bars). *Adapted with permission from Katus et al. (2012b).*

OVERLAPPING FUNCTIONS OF STM AND PERCEPTION

Influential models of STM postulate that selective attention is implicated in memory operations (Baddeley, 2003; Cowan, 1995; Engle, Cantor, & Carullo, 1992; Oberauer, 2002). The possibility exists that information processing in internal (memory) representations is regulated by the same neural mechanisms that control information processing in external (perceived and/or anticipated) stimulus representations (Griffin & Nobre, 2003; Lepsien & Nobre, 2006; Postle, 2006). As a consequence, attentional biasing of information in tactile STM may produce a pattern of results that closely resembles the impact of selective attention on perception. Such a functional overlap is consistent with our result that selection in tactile STM modulates established markers of perceptual and/or preparatory attention (Katus et al., 2012a, 2012b, 2014).

Modulations of the ADAN and LDAP components have been reported in numerous studies of preparatory attention. Triggered by pre-cues, these components trace the direction of attention shifts toward anticipated sensory events (Eimer et al., 2003a; Eimer et al., 2004; Eimer, van Velzen, Forster, & Driver, 2003; Forster, Sambo, & Pavone, 2009; Harter et al., 1989; Hopf & Mangun, 2000; van Velzen & Eimer, 2003; van Velzen, Forster, & Eimer, 2002). The N2pc component, elicited by arrays of visual stimuli, indicated the location of relevant stimuli in studies of perceptual attention (Eimer & Kiss, 2010; Luck & Hillyard, 1994a, 1994b). Because these lateralized components are also sensitive to selection in visual STM (Dell'Acqua et al., 2009; Griffin & Nobre, 2003; Kuo, Stokes, & Nobre, 2012) and tactile STM, it seems that similar, if not equivalent, neural mechanisms control attentional selection in internal and external stimulus representations. Moreover, selection in STM guided sensory encoding, as reflected by modulations of the somatosensory N140 component. The influence of selection in STM on perception (Studies 1–3) and vice versa (Study 2) suggests that memory and perception are not separate cognitive domains that recruit distinct encapsulated processing modules. Rather, the close link between these domains indicates that memory and perception rely on shared control mechanisms.

THE ROLE OF SPATIAL ATTENTION IN STM FOR LOCATIONS: ATTENTION-BASED REHEARSAL

The series of studies reviewed in this chapter aimed to examine the role of spatial attention to STM in the sense of touch. Two studies with spatial memory tasks evaluated the attention-based rehearsal account. According to attention-based rehearsal, spatial attention mediates maintenance of spatial attributes in memory. Our results are generally in accordance with the attention-based rehearsal account. First, memory for locations led to focal attention shifts to memorized locations (Study 1). Attention shifts were time-locked to the presentation of cueing information and influenced sensory encoding during the retention period. Second, the disruption of attention-based rehearsal resulting from the requirement to direct attention away from memorized locations impairs memory performance (Study 2), entailing a functional significance of spatial attention to the maintenance of locations in STM. Thus spatial attention seems to bias information in tactile memory depending on the relevance of such information with regard to behavioral goals.

ARE EFFECTS OF ATTENTION-BASED REHEARSAL GENUINE MEMORY EFFECTS OR ARE THEY CAUSED BY ANTICIPATORY ATTENTION?

Investigations of attention-based rehearsal rely on tasks that involve a delayed comparison of spatial attributes associated with two stimuli: one presented before the retention period and one after retention. Attention shifts that occur during retention are interpreted as effects attributable to the memorization of the first stimulus (sample) rather than anticipation of the second one (test). In principle, anticipatory attention could contribute to effects of attention-based rehearsal. This confound is not specific to Studies 1 and 2, but a general limitation of spatial memory tasks that are required to evaluate the attention-based rehearsal account. Consider Study 1, in which test stimuli were presented with equal probability to the cued or uncued hand. This randomization cannot prevent the strategy to anticipate test stimuli at the cued hand because some of these test stimuli were distracters, whereas others were targets. It is obvious that test stimuli presented to the uncued hand were always distracters.

Anticipatory attention shifts to the cued hand are not an effective strategy in non-spatial memory tasks with designs such as that used in Study 3, in which test stimuli occurred with equal probability at the left or right hand, whereas their intensity—rather than location—was relevant to the task. Non-spatial attributes of test stimuli had to be processed in every trial, irrespective of the spatial relation between test stimulus and target sample. However, the fact that attention shifts can occur even when test stimuli cannot be anticipated is irrelevant to the evaluation of attention-based rehearsal because memory for locations was not instructed in Study 3. To conclude, spatially selective processes subservient to the maintenance of locations cannot be entirely isolated from preparatory attention shifts in spatial STM paradigms. Nevertheless, the time course of attention effects in the retention period in previous studies, including our own, suggests that the influence of memory for locations dominates over potential contributions from anticipatory attention.

Spatial attention shifts during the retention period are not time-locked to test stimuli. They are, however, time-locked to the onset of cues that allow for selection in STM. In Study 1, the duration of the retention period varied only slightly across trials, allowing for temporal prediction of test stimuli. A purely anticipatory account would expect strongest N140 modulations for late probes because these were presented just before test stimulus onset. Contrary to that, larger N140 modulations were found for early rather than late probes. In Study 2, test stimuli were always presented to the retrospectively cued hand, either to the memorized or a non-memorized finger (50% each). Again, largest effects of spatial attention were observed at short latencies after the cue stimuli and these effects diminished at longer latencies (Figure 4(b)). Furthermore, no evidence for spatial anticipation effects was reported in a visual STM experiment with a probe design. In the study conducted by Jha (2002), early probe stimuli were presented in the first half (400- to 800-ms latency) and late probe stimuli in the second half (2600- to 3000-ms latency) of the retention period (4800- to 5300-ms total duration). Early and late probes exhibited a sustained spatial encoding bias suggestive of attention-based rehearsal. This sustained effect of spatial STM can be reconciled with the variant time course of effects observed in our studies, because no retro-cues were presented in the study conducted by Jha (2002). In fact, we observed a spatial encoding bias throughout the retention period, although this encoding bias was largest early after the retro-cue. In summary, focal attention shifts during retention of spatial STM tasks do not occur time-locked to the end of the retention period. Theoretically, anticipatory attention may have a role in spatial STM tasks. In practice, however, potential contributions from anticipatory attention are outweighed by genuine memory effects owing to attention-based rehearsal.

SELECTION OF NON-SPATIAL ATTRIBUTES WITHIN THE SPATIAL LAYOUT OF STM

The preceding sections on attention-based rehearsal reviewed the role of spatial attention for spatial STM. Could spatial effects occur in the retention period of STM paradigms that avoid the known limitations of spatial memory tasks discussed in the previous section? Study 3 was a non-spatial memory task designed to investigate whether tactile STM preserves spatial attributes of memorized events, even if memory for spatial attributes is not required to solve the task. The observed spatially specific effects after spatially neutral retro-cues, which guided selection on the basis of intensity information, suggest that tactile STM is organized in a way that preserves the spatial layout of the original sensory input. Because similar results have been reported in vision (Dell'Acqua et al., 2009; Kuo et al., 2009), it appears that this spatially specific organization of memory content represents a general principle that applies to STM in both vision and touch. As discussed below, however, it is questionable whether spatial attention has a functional significance for the selection of non-spatial attributes in tasks where spatial attributes are not useful to achieve current behavioral goals.

The attention-based rehearsal account allows no prediction of spatially specific effects in the retention period of tasks with no explicit spatial processing and/or storage demands, as in our third experiment. Participants performed a delayed comparison of intensity information; yet, shifts of spatial attention accompanied selection between spatially segregated items in STM. The engagement of spatially specific processes in this case cannot be attributed to attention-based rehearsal. Furthermore, time courses of attentional N140 modulations differed between Studies 1 and 3 even though test stimuli were presented to the left and right hands with equal probability in both experiments. A sustained effect was found in the spatial STM task (Study 1), where the N140 was enhanced when the test stimulus presented after retention matched the target sample's location. On the contrary, a transient time course was observed in the STM task for stimulus intensity (Study 3) in which the N140 elicited by test stimuli did not differ between the cued and uncued hands. In conclusion, the role of spatial attention in non-spatial STM seems unrelated to attention-based rehearsal, in which spatial attention facilitates the maintenance of locations depending on current behavioral goals (Awh & Jonides, 2001; Awh et al., 2006; Theeuwes et al., 2009).

What functional role could spatial attention have in the context of a memory task for stimulus intensity? One possibility is that spatial attention mediates selection of non-spatial features. If selection of feature attributes cannot occur unless the respective stimulus is in the focus of spatial attention, spatial attention would have a functional significance for non-spatial STM. However, this is not supported by empirical evidence. Selection of features and locations is mediated by parallel and independent neural processes in the sense of touch (Forster & Eimer, 2004). A more plausible explanation is that spatial attention has a passive rather than an active role in non-spatial memory tasks. In this case, focal attention shifts would merely accompany but not cause selection of features or objects in STM. Spatial dependency of sustained maintenance processes that operate at the level of features and/or objects has been demonstrated in visual STM experiments by means of one particular lateralized component, the contralateral delay activity (CDA) (e.g., Jolicœur, Brisson, & Robitaille, 2008; Luck & Vogel, 1997; Vogel, McCollough, & Machizawa, 2005), also known as the sustained posterior contralateral negativity (e.g., Jolicœur, Sessa, Dell'Acqua, & Robitaille, 2006) or the contralateral negative slow wave (e.g., Klaver, Talsma, Wijers, Heinze, & Mulder, 1999). The most remarkable property of the CDA is its load dependence. The CDA increases with the increasing number of objects held in memory until an individual's limit in STM capacity is exceeded (Vogel et al., 2005). More interesting in the context of this chapter is that the CDA is a lateralized component and therefore spatially specific. Because features associated with one stimulus are stored as integrated objects in STM (Luck & Vogel, 1997), the CDA reflects the activation of spatial coordinates of currently memorized objects. In other words, the CDA is a spatially specific effect owing to neural processes that bias mnemonic representations of objects, consisting of bound spatial and non-spatial stimulus attributes (cf. Luck & Vogel, 1998; O'Craven, Downing, & Kanwisher, 1999; Schoenfeld et al., 2003).

Study 3 is consistent with evidence from visual STM (e.g., Dell'Acqua et al., 2009; Kuo et al., 2009) that tactile STM preserves spatial attributes in scenarios beyond attention-based rehearsal: that is, spatial STM. Spatial attention effects during retention are presumably caused by mechanisms of feature- or object-based selection that operate within the spatial layout of tactile STM. There are reasons to assume that such effects reflect a different aspect of spatial attention compared with attention-based rehearsal. The precise role of spatial attention in non-spatial STM, however, remains elusive. Spatially specific effects observed in Study 3 could be read as evidence for an organizational principle that produces chunks in STM by grouping information based on object membership (cf. Luck & Vogel, 1997).

MODALITY-SPECIFIC AND SUPRAMODAL MECHANISMS OF EXECUTIVE CONTROL IN STM

From a structural point of view, sensory STM systems are composed of at least two modules: a modality-invariant component of executive control and a modality-dependent storage component (Baddeley, 2003; Jonides, Lacey, & Nee, 2005; Smith & Jonides, 1997). Evidence from neuroimaging studies suggests that the latter component resides in sensory areas and is influenced by feedback from higher areas, resulting in a goal-driven modulation of mnemonic content (Gazzaley & Nobre, 2012; Jonides et al., 2005; Lepsien & Nobre, 2006; Postle, 2006). In the series of tactile experiments reviewed here, lateralized components evoked by cue stimuli consistently reflected selection in tactile STM. The late lateralized ADAN and LDAP components are associated with supramodal attention (Eimer & van Velzen, 2002; Jongen, Smulders, & van der Heiden, 2007; van Velzen et al., 2002); ADAN modulations accompanied selection in visual STM (Griffin & Nobre, 2003). Hence, modulation of ADAN and LDAP in our experiments indicates the involvement of supramodal control processes. However, it is unexpected that visual cues elicited N2pc components reflecting selection in tactile STM throughout all three studies outlined in this chapter. The N2pc was evoked with asymmetric arrow cues (Study 2) and symmetric color cues (Studies 1 and 3), with cues indicating spatial attributes of the target sample (Studies 1 and 2) and also non-spatial attributes such as intensity (Study 3). Although N2pc modulations to visual cues have been reported in a purely visual STM experiment (Kuo et al., 2009), here, the N2pc to visual cue stimuli reflected selection in tactile STM. What is the cause of this cross-modal effect, revealed by an early visual ERP component?

The N2pc has a posterior scalp distribution centered on lateral occipital cortex and is thought to reflect attentional filtering in visual cortex under control from higher cortical areas (Luck & Hillyard, 1994a, 1994b; Oostenveld, Praamstra, Stegeman, & van Oosterom, 2001). Source reconstruction performed in a magnetoencephalographic study (Hopf et al., 2000) identified two separate processes that contribute to the N2pc. Neural responses in parietal areas (180–200 ms) preceded activity in occipito-temporal areas (220–240 ms), which suggests that the N2pc is composed of two neural processes activated in cascade. The authors related the early parietal source to neural networks controlling the direction of attention shifts, whereas the later occipito-temporal response was interpreted as correlate of the implementation of filtering processes in visual areas. Feedback from the frontal eye fields, a further structure related to the control of selective attention, influences the generation of the macaque N2pc homologue in posterior cortical regions (Cohen, Heitz, Schall, & Woodman, 2009). This suggests that the neural generators of the N2pc in posterior areas receive top-down influences from modality-invariant control networks to regulate modality-specific processes such as spatial filtering of signals in visual cortex (Hopf et al., 2000; Luck & Hillyard, 1994b).

The N2pc is associated with the visual modality, in which spatial selection is based on an external frame of reference (Sambo & Forster, 2011). The use of an environmentally defined coordinate system is cognitively economical because vision encodes distal stimuli in the sensory environment. Touch, however, operates on two separable frames of reference (Eimer et al., 2003a; Heed & Röder, 2010; Sambo & Forster, 2011). Because touch encodes proximal stimuli that impinge on the skin's surface, locations are initially represented in a body-centered format and then translated to external space. Remapping of information coded in body coordinates takes the current body posture into account and is carried out between 70 ms and 360 ms post-stimulus by cell ensembles in posterior parietal cortex (Azañón, Longo, Soto-Faraco, & Haggard, 2010; Azañón & Soto-Faraco, 2008; Soto-Faraco & Azañón, 2013). Although body space is the initial representational format for locations in touch, the tactile representation of external space is central to many aspects of goal-directed behavior, because external space provides an adequate readout for motor systems (Soto-Faraco & Azañón, 2013). External space is furthermore a modality-independent code that allows integrating spatial representations across different modalities (Gallace, Soto-Faraco, Dalton, Kreukniet, & Spence, 2008; Ley, Bottari, Shenoy, Kekunnaya, & Röder, 2013) and is used by supramodal spatial attention (Driver & Spence, 1998; Eimer & van Velzen, 2002). The series of experiments described here demonstrates that the N2pc to visual retro-cues is sensitive to selection in tactile STM. This early posterior effect might point to an involvement of neural processes that are linked to the visual system. More specifically, we speculate that the N2pc reflects coordinates of memorized information in external space, and thus a spatial code that is common to vision and touch.

THEORETICAL PERSPECTIVES

In this section, the results of our studies are discussed in the context of theoretical models of attention and STM. Site-source models of selective attention (e.g., Macaluso, 2010) characterize control structures in supramodal cortical areas as the source of biasing signals that regulate modality-specific processes at the site where attention effects become evident: for example, in somatosensory areas. A similar hierarchical structure has been conceptualized in Baddeley's model of STM, in which the supramodal central executive controls information processing in subordinate modality-specific storage buffers (Baddeley, 2003, 2012). Modality-specific effects such as the spatial encoding bias observed in the retention period of spatial memory tasks indicate control mechanisms that operate within the storage component of STM. Modulation of the somatosensory N140 component of the ERP in our studies points toward a modality-specific gain control mechanism that is governed by biasing signals with source in frontoparietal neural networks of endogenous attention (Eimer et al., 2003a; Eimer et al., 2004; Sambo & Forster, 2011).

Sensory recruitment models postulate that sensory brain areas mediate the short-term storage of sensory signals (Jonides et al., 2008; Postle, 2006; Serences et al., 2009). Top-down influence on sensory areas may alter memory representations in a goal-directed fashion by biasing neural representations for items that are relevant to the task at hand (Courtney, Roth, & Sala, 2007). In experimental designs with retro-cues, the target-defining attribute is not known in the period between sample set and retro-cue. During this period, the competition between items in STM is not biased by top-down control. Competition becomes biased after the retro-cue, because this stimulus provides the critical information to distinguish between target and distracter items. The selective maintenance of relevant items could be achieved through amplification and/or suppression of neural processing within sensory areas (Gazzaley, 2011; Gazzaley & D'Esposito, 2007; Gazzaley & Nobre, 2012; Zanto & Gazzaley, 2009). The modulation of neural processing in brain regions responsible for storage functions indicates that the storage component of STM is a dynamic and adaptive system.

We provide evidence for a rehearsal mechanism that supports the maintenance of locations in tactile STM. This mechanism is not specific to STM, because it results from the recruitment of neural processes associated with spatial selective attention. As pointed out by Postle (2006), evidence for a reactivation mechanism based on attention and motor control in vision raises concerns as to whether it is necessary to maintain the theoretical construct of an inner scribe within the visuospatial sketchpad (Baddeley, 2003). The attention-based rehearsal mechanism in touch is similar to the one previously reported in vision (Awh et al., 2000; Awh & Jonides, 2001; Awh et al., 2006). Hence, it seems that spatial rehearsal is not a mechanism that is specific for a visual storage buffer. Spatial attention supports the maintenance of locations in the modalities of both vision and touch. However, because the neural underpinnings of rehearsal processes are at least to some degree modality-specific, it is likely that there are qualitative differences regarding how spatial filtering is implemented in visual versus tactile sensory areas.

Articulatory rehearsal was suggested as a mechanism that refreshes information contained in the auditory storage buffer, the phonological loop (Baddeley, 2003, 2012; Baddeley & Larsen, 2007). Articulatory rehearsal seems different from attention-based rehearsal. It has been characterized as a real-time process analogous to subvocal speech, with emphasis on temporal precision rather than spatial resolution (Baddeley, 2003; Postle, 2006; Smith & Jonides, 1999). Following the filter metaphor of selective attention, attention-based rehearsal can be described as a spatial filter applied to an internal stimulus representation. Spatial filtering of mnemonic information may involve motor codes and neural circuits of motor control (Belopolsky & Theeuwes, 2009a, 2011; Postle, 2006; Theeuwes et al., 2009). Cross-modal links observed in our studies may further point toward an engagement of control processes that operate on coordinates referenced to external space, the code for locations common to vision and touch.

CONCLUSIONS

In this chapter, selective attention was defined as the set of control functions found to modulate neural processing in perception: that is, during sensory encoding. Selective attention can accommodate many aspects of an STM control system commonly known as the central executive. For future research, we propose the strong null-hypothesis that the central executive is not significantly different from selective attention. However, some control functions at work during STM may have no counterpart in perception (Griffin & Nobre, 2003; Lepsien & Nobre, 2006). Such domain-specific processes indicate control functions of STM that may not overlap with (perceptual) selective attention. The first step to fractionate the central executive is therefore to substitute it by the well-defined construct of selective attention. Overall, our results can be reconciled with structural and also process-oriented models of STM. The latter emergent system view may be the more parsimonious theoretical account to explain our results. According to this view, tactile STM is a system for the storage and manipulation of information that emerges as a result of the engagement of frontoparietal neural networks that cause attentional activation of memory traces, which reside in somatosensory brain regions.

References

Allison, T., McCarthy, G., & Wood, C. C. (1992). The relationship between human long-latency somatosensory evoked potentials recorded from the cortical surface and from the scalp. *Electroencephalography and Clinical Neurophysiology*, *84*(4), 301–314.

Andersen, S. K., Hillyard, S. A., & Müller, M. M. (2013). Global selection of an attended feature is obligatory and restricts divided attention. *Journal of Neuroscience*, *33*(46), 18200–18207.

Andersen, S. K., Müller, M. M., & Hillyard, S. A. (2009). Color-selective attention need not be mediated by spatial attention. *Journal of Vision*, *9*(6), 1–7.

Astle, D. E., Scerif, G., Kuo, B.-C., & Nobre, A. C. (2009). Spatial selection of features within perceived and remembered objects. *Frontiers in Human Neuroscience*, *3*, 6.

Awh, E., Anllo-Vento, L., & Hillyard, S. A. (2000). The role of spatial selective attention in working memory for locations: evidence from event-related potentials. *Journal of Cognitive Neuroscience*, *12*(5), 840–847.

Awh, E., & Jonides, J. (2001). Overlapping mechanisms of attention and spatial working memory. *Trends in Cognitive Sciences*, *5*(3), 119–126.

Awh, E., Jonides, J., & Reuter-Lorenz, P. A. (1998). Rehearsal in spatial working memory. *Journal of Experimental Psychology: Human Perception and Performance*, *24*(3), 780–790.

Awh, E., Vogel, E. K., & Oh, S. H. (2006). Interactions between attention and working memory. *Neuroscience*, *139*(1), 201–208.

Azañón, E., Longo, M. R., Soto-Faraco, S., & Haggard, P. (2010). The posterior parietal cortex remaps touch into external space. *Current Biology*, *20*(14), 1304–1309.

Azañón, E., & Soto-Faraco, S. (2008). Changing reference frames during the encoding of tactile events. *Current Biology*, *18*(14), 1044–1049.

Baddeley. (2000). The episodic buffer: a new component of working memory? *Trends in Cognitive Sciences*, *4*(11), 417–423.

Baddeley, A. (2003). Working memory: looking back and looking forward. *Nature Reviews Neuroscience*, *4*(10), 829–839.

Baddeley, A. (2012). Working memory: theories, models, and controversies. *Annual Review of Psychology*, *63*, 1–29.

Baddeley, A. D., & Andrade, J. (2000). Working memory and the vividness of imagery. *Journal of Experimental Psychology: General*, *129*(1), 126–145.

Baddeley, A. D., & Larsen, J. D. (2007). The phonological loop unmasked? A comment on the evidence for a "perceptual-gestural" alternative. *Quarterly Journal of Experimental Psychology*, *60*(4), 497–504.

Belopolsky, A. V., & Theeuwes, J. (2009a). Inhibition of saccadic eye movements to locations in spatial working memory. *Attention, Perception and Psychophysics*, *71*(3), 620–631.

Belopolsky, A. V., & Theeuwes, J. (2009b). No functional role of attention-based rehearsal in maintenance of spatial working memory representations. *Acta Psychologica*, *132*(2), 124–135.

Belopolsky, A. V., & Theeuwes, J. (2011). Selection within visual memory representations activates the oculomotor system. *Neuropsychologia*, *49*(6), 1605–1610.

Burton, H., & Sinclair, R. J. (2000). Attending to and remembering tactile stimuli: a review of brain imaging data and single-neuron responses. *Journal of Clinical Neurophysiology*, *17*(6), 575–591.

Carrasco, M. (2011). Visual attention: the past 25 years. *Vision Research*, *51*(13), 1484–1525.

Cohen, J. Y., Heitz, R. P., Schall, J. D., & Woodman, G. F. (2009). On the origin of event-related potentials indexing covert attentional selection during visual search. *Journal of Neurophysiology*, *102*(4), 2375–2386.

Courtney, S. M., Roth, J. K., & Sala, J. B. (2007). A hierarchical biased-competition model of domain-dependent working memory maintenance and executive control. In N. Osaka, R. H. Logie, & M. D'Esposito (Eds.), *The cognitive neuroscience of working memory* (pp. 369–383). Oxford, New York: Oxford University Press.

Cowan, N. (1995). Attention and memory: An integrated framework. *Oxford psychology series* (Vol. 26). Oxford, New York: Oxford University Press.

Dell'Acqua, R., Sessa, P., Toffanin, P., Luria, R., & Jolicœur, P. (2009). Orienting attention to objects in visual short-term memory. *Neuropsychologia*, *48*(2), 419–428.

Desimone, R., & Duncan, J. (1995). Neural mechanisms of selective visual attention. *Annual Review of Neuroscience*, *18*, 193–222.

Desmedt, J. E., & Robertson, D. (1977). Differential enhancement of early and late components of the cerebral somatosensory evoked potentials during forced-paced cognitive tasks in man. *The Journal of Physiology*, *271*(3), 761–782.

Driver, J., & Spence, C. (1998). Attention and the crossmodal construction of space. *Trends in Cognitive Sciences*, *2*(7), 254–262.

Eimer, M., & Forster, B. (2003a). Modulations of early somatosensory ERP components by transient and sustained spatial attention. *Experimental Brain Research*, *151*(1), 24–31.

Eimer, M., & Forster, B. (2003b). The spatial distribution of attentional selectivity in touch: evidence from somatosensory ERP components. *Clinical Neurophysiology*, *114*(7), 1298–1306.

Eimer, M., Forster, B., Fieger, A., & Harbich, S. (2004). Effects of hand posture on preparatory control processes and sensory modulations in tactile-spatial attention. *Clinical Neurophysiology*, *115*(3), 596–608.

Eimer, M., Forster, B., & van Velzen, J. (2003a). Anterior and posterior attentional control systems use different spatial reference frames: ERP evidence from covert tactile-spatial orienting. *Psychophysiology*, *40*(6), 924–933.

Eimer, M., & Kiss, M. (2010). The top-down control of visual selection and how it is linked to the N2pc component. *Acta Psychologica*, *135*(2), 100–102.

Eimer, M., & van Velzen, J. (2002). Crossmodal links in spatial attention are mediated by supramodal control processes: evidence from event-related potentials. *Psychophysiology*, *39*(4), 437–449.

Eimer, M., van Velzen, J., Forster, B., & Driver, J. (2003b). Shifts of attention in light and in darkness: an ERP study of supramodal attentional control and crossmodal links in spatial attention. *Cognitive Brain Research*, *15*(3), 308–323.

Engle, R. W., Cantor, J., & Carullo, J. J. (1992). Individual differences in working memory and comprehension: a test of four hypotheses. *Journal of Experimental Psychology. Learning, Memory, and Cognition*, *18*(5), 972–992.

Forster, B., & Eimer, M. (2004). The attentional selection of spatial and non-spatial attributes in touch: ERP evidence for parallel and independent processes. *Biological Psychology*, *66*(1), 1–20.

Forster, B., & Eimer, M. (2005). Covert attention in touch: behavioral and ERP evidence for costs and benefits. *Psychophysiology*, *42*(2), 171–179.

Forster, B., & Gillmeister, H. (2011). ERP investigation of transient attentional selection of single and multiple locations within touch. *Psychophysiology*, *48*(6), 788–796.

Forster, B., Sambo, C. F., & Pavone, E. F. (2009). ERP correlates of tactile spatial attention differ under intra- and intermodal conditions. *Biological Psychology*, *82*(3), 227–233.

Frot, M., & Mauguière, F. (1999). Timing and spatial distribution of somatosensory responses recorded in the upper bank of the sylvian fissure (SII area) in humans. *Cerebral Cortex*, *9*(8), 854–863.

Gallace, A., Soto-Faraco, S., Dalton, P., Kreukniet, B., & Spence, C. (2008). Response requirements modulate tactile spatial congruency effects. *Experimental Brain Research*, *191*(2), 171–186.

García-Larrea, L., Lukaszewicz, A. C., & Mauguière, F. (1995). Somatosensory responses during selective spatial attention: the N120-to-N140 transition. *Psychophysiology*, *32*(6), 526–537.

Gazzaley, A. (2011). Influence of early attentional modulation on working memory. *Neuropsychologia*, *49*(6), 1410–1424.

Gazzaley, A., & D'Esposito, M. (2007). Top-down modulation in visual working memory. In N. Osaka, R. H. Logie, & M. D'Esposito (Eds.), *The cognitive neuroscience of working memory* (pp. 197–211). Oxford, New York: Oxford University Press.

Gazzaley, A., & Nobre, A. C. (2012). Top-down modulation: bridging selective attention and working memory. *Trends in Cognitive Sciences*, *16*(2), 129–135.

Griffin, I. C., & Nobre, A. C. (2003). Orienting attention to locations in internal representations. *Journal of Cognitive Neuroscience*, *15*(8), 1176–1194.

Harter, M. R., Miller, S. L., Price, N. J., LaLonde, M. E., & Keyes, A. L. (1989). Neural processes involved in directing attention. *Journal of Cognitive Neuroscience*, *1*(3), 223–237.

Heed, T., & Röder, B. (2010). Common anatomical and external coding for hands and feet in tactile attention: evidence from event-related potentials. *Journal of Cognitive Neuroscience*, *22*(1), 184–202.

Hopf, J. M., Luck, S. J., Girelli, M., Hagner, T., Mangun, G. R., Scheich, H., et al. (2000). Neural sources of focused attention in visual search. *Cerebral Cortex*, *10*(12), 1233–1241.

Hopf, J. M., & Mangun, G. R. (2000). Shifting visual attention in space: an electrophysiological analysis using high spatial resolution mapping. *Clinical Neurophysiology*, *111*(7), 1241–1257.

Jans, B., Peters, J. C., & De Weerd, P. (2010). Visual spatial attention to multiple locations at once: the jury is still out. *Psychological Review*, *117*(2), 637–684.

Jensen, O., Kaiser, J., & Lachaux, J.-P. (2007). Human gamma-frequency oscillations associated with attention and memory. *Trends in Neurosciences*, *30*(7), 317–324.

Jha, A. P. (2002). Tracking the time-course of attentional involvement in spatial working memory: an event-related potential investigation. *Cognitive Brain Research*, *15*(1), 61–69.

Johansen-Berg, H., & Lloyd, D. M. (2000). The physiology and psychology of selective attention to touch. *Frontiers in Bioscience*, *5*, 894–904.

Jolicœur, P., Brisson, B., & Robitaille, N. (2008). Dissociation of the N2pc and sustained posterior contralateral negativity in a choice response task. *Brain Research*, *1215*, 160–172.

Jolicœur, P., Sessa, P., Dell'Acqua, R., & Robitaille, N. (2006). On the control of visual spatial attention: evidence from human electrophysiology. *Psychological Research*, *70*(6), 414–424.

Jongen, E. M. M., Smulders, F. T. Y., & van der Heiden, J. S. H. (2007). Lateralized ERP components related to spatial orienting: discriminating the direction of attention from processing sensory aspects of the cue. *Psychophysiology*, *44*(6), 968–986.

Jonides, J., Lacey, S. C., & Nee, D. E. (2005). Processes of working memory in mind and brain. *Current Directions in Psychological Science*, *14*(1), 2–5.

Jonides, J., Lewis, R. L., Nee, D. E., Lustig, C. A., Berman, M. G., & Moore, K. S. (2008). The mind and brain of short-term memory. *Annual Review of Psychology*, *59*, 193–224.

Josiassen, R. C., Shagass, C., Roemer, R. A., Ercegovac, D. V., & Straumanis, J. J. (1982). Somatosensory evoked potential changes with a selective attention task. *Psychophysiology*, *19*(2), 146–159.

Katus, T., Andersen, S. K., & Müller, M. M. (2012a). Maintenance of tactile short-term memory for locations is mediated by spatial attention. *Biological Psychology*, *89*(1), 39–46.

Katus, T., Andersen, S. K., & Müller, M. M. (2012b). Nonspatial cueing of tactile STM causes shift of spatial attention. *Journal of Cognitive Neuroscience*, *24*(7), 1596–1609.

Katus, T., Andersen, S. K., & Müller, M. M. (2014). Common mechanisms of spatial attention in memory and perception: a tactile dual-task study. *Cerebral Cortex*, *24*(3), 707–718.

Klaver, P., Talsma, D., Wijers, A. A., Heinze, H. J., & Mulder, G. (1999). An event-related brain potential correlate of visual short-term memory. *Neuroreport*, *10*(10), 2001–2005.

Kuo, B. C., Rao, A., Lepsien, J., & Nobre, A. C. (2009). Searching for targets within the spatial layout of visual short-term memory. *Journal of Neuroscience*, *29*(25), 8032–8038.

Kuo, B.-C., Stokes, M. G., & Nobre, A. C. (2012). Attention modulates maintenance of representations in visual short-term memory. *Journal of Cognitive Neuroscience*, *24*(1), 51–60.

Lepsien, J., & Nobre, A. C. (2006). Cognitive control of attention in the human brain: insights from orienting attention to mental representations. *Brain Research*, *1105*(1), 20–31.

Ley, P., Bottari, D., Shenoy, B. H., Kekunnaya, R., & Röder, B. (2013). Partial recovery of visual-spatial remapping of touch after restoring vision in a congenitally blind man. *Neuropsychologia*, *51*(6), 1119–1123.

Luck, S. J., & Hillyard, S. A. (1994a). Electrophysiological correlates of feature analysis during visual search. *Psychophysiology*, *31*(3), 291–308.

Luck, S. J., & Hillyard, S. A. (1994b). Spatial filtering during visual search: evidence from human electrophysiology. *Journal of Experimental Psychology: Human Perception and Performance*, *20*(5), 1000–1014.

Luck, S. J., & Vogel, E. K. (1997). The capacity of visual working memory for features and conjunctions. *Nature*, *390*(6657), 279–281.

Luck, S. J., & Vogel, E. K. (1998). Response from luck and vogel. *Trends in Cognitive Sciences*, *2*(3), 78–79.

Macaluso, E. (2010). Orienting of spatial attention and the interplay between the senses. *Cortex*, *46*(3), 282–297.

Michie, P. T., Bearpark, H. M., Crawford, J. M., & Glue, L. C. (1987). The effects of spatial selective attention on the somatosensory event-related potential. *Psychophysiology*, *24*(4), 449–463.

Murray, A. M., Nobre, A. C., & Stokes, M. G. (2011). Markers of preparatory attention predict visual short-term memory performance. *Neuropsychologia*, *49*(6), 1458–1465.

O'Craven, K. M., Downing, P. E., & Kanwisher, N. (1999). fMRI evidence for objects as the units of attentional selection. *Nature*, *401*(6753), 584–587.

Oberauer, K. (2002). Access to information in working memory: exploring the focus of attention. *Journal of Experimental Psychology: Learning, Memory, and Cognition*, *28*(3), 411–421.

Oostenveld, R., Praamstra, P., Stegeman, D. F., & van Oosterom, A. (2001). Overlap of attention and movement-related activity in lateralized event-related brain potentials. *Clinical Neurophysiology*, *112*(3), 477–484.

Pasternak, T., & Greenlee, M. W. (2005). Working memory in primate sensory systems. *Nature Reviews Neuroscience*, *6*(2), 97–107.

Postle, B. R. (2006). Working memory as an emergent property of the mind and brain. *Neuroscience*, *139*(1), 23–38.

Repovs, G., & Baddeley, A. (2006). The multi-component model of working memory: explorations in experimental cognitive psychology. *Neuroscience*, *139*(1), 5–21.

Sambo, C. F., & Forster, B. (2011). Sustained spatial attention in touch: modality-specific and multimodal mechanisms. *The Scientific World Journal*, *11*, 199–213.

Schoenfeld, M. A., Tempelmann, C., Martinez, A., Hopf, J.-M., Sattler, C., Heinze, H.-J., et al. (2003). Dynamics of feature binding during object-selective attention. *Proceedings of the National Academy of Sciences of the United States of America*, *100*(20), 11806–11811.

Schubert, R., Ritter, P., Wustenberg, T., Preuschhof, C., Curio, G., Sommer, W., et al. (2008). Spatial attention related SEP amplitude modulations covary with BOLD signal in S1–a simultaneous EEG–fMRI study. *Cerebral Cortex*, *18*(11), 2686–2700.

Serences, J. T., Ester, E. F., Vogel, E. K., & Awh, E. (2009). Stimulus-specific delay activity in human primary visual cortex. *Psychological Science*, *20*(2), 207–214.

Smith, E. E., & Jonides, J. (1997). Working memory: a view from neuroimaging. *Cognitive Psychology*, *33*(1), 5–42.

Smith, E. E., & Jonides, J. (1999). Storage and executive processes in the frontal lobes. *Science*, *283*(5408), 1657–1661.

Smyth, M. M., & Scholey, K. A. (1994). Interference in immediate spatial memory. *Memory & Cognition*, *22*(1), 1–13.

Soto-Faraco, S., & Azañón, E. (2013). Electrophysiological correlates of tactile remapping. *Neuropsychologia*, *51*(8), 1584–1594.

Störmer, V. S., Winther, G. N., Li, S.-C., & Andersen, S. K. (2013). Sustained multifocal attentional enhancement of stimulus processing in early visual areas predicts tracking performance. *Journal of Neuroscience*, *33*(12), 5346–5351.

Theeuwes, J., Belopolsky, A., & Olivers, C. N. (2009). Interactions between working memory, attention and eye movements. *Acta Psychologica*, *132*(2), 106–114.

Tsal, Y., & Lavie, N. (1993). Location dominance in attending to color and shape. *Journal of Experimental Psychology: Human Perception and Performance*, *19*(1), 131–139.

van Velzen, J., & Eimer, M. (2003). Early posterior ERP components do not reflect the control of attentional shifts toward expected peripheral events. *Psychophysiology*, *40*(5), 827–831.

van Velzen, J., Forster, B., & Eimer, M. (2002). Temporal dynamics of lateralized ERP components elicited during endogenous attentional shifts to relevant tactile events. *Psychophysiology*, *39*(6), 874–878.

Vogel, E. K., McCollough, A. W., & Machizawa, M. G. (2005). Neural measures reveal individual differences in controlling access to working memory. *Nature*, *438*(7067), 500–503.

Waberski, T. D., Gobbelé, R., Darvas, F., Schmitz, S., & Buchner, H. (2002). Spatiotemporal imaging of electrical activity related to attention to somatosensory stimulation. *NeuroImage*, *17*(3), 1347–1357.

Zanto, T. P., & Gazzaley, A. (2009). Neural suppression of irrelevant information underlies optimal working memory performance. *Journal of Neuroscience*, *29*(10), 3059–3066.

Zimmer, H. D. (2008). Visual and spatial working memory: from boxes to networks. *Neuroscience and Biobehavioral Reviews*, *32*(8), 1373–1395.

Zopf, R., Giabbiconi, C. M., Gruber, T., & Müller, M. M. (2004). Attentional modulation of the human somatosensory evoked potential in a trial-by-trial spatial cueing and sustained spatial attention task measured with high density 128 channels EEG. *Cognitive Brain Research*, *20*(3), 491–509.

Index

'*Note*: Page numbers followed by "f" indicate figures and "t" indicate tables.'

I

L

M

N

O

P

R

S

T

U

V

W

Printed in the United States
By Bookmasters